Pulmonary Biology in Health and Disease

Springer Science+Business Media, LLC

E. Edward Bittar, MD

Department of Physiology, University of Wisconsin Medical School, Madison, Wisconsin

Editor

Pulmonary Biology in Health and Disease

With 162 Figures

Springer

E. Edward Bittar, MD
Department of Physiology
University of Wisconsin Medical School
Madison, WI 53706, USA
eebittar@facstaff.wisc.edu

Library of Congress Cataloging-in-Publication Data
Pulmonary biology in health and disease / edited by Edward Bittar.
p. ; cm.
Includes bibliographical references and index.
ISBN 978-1-4757-8133-5 ISBN 978-0-387-22435-0 (eBook)
DOI 10.1007/978-0-387-22435-0
1. Lungs—Physiology. 2. Lungs—Pathophysiology. I. Bittar, E. Edward.
[DNLM: 1. Lung—physiology. 2. Lung Diseases. 3. Muscle, Smooth—physiology. 4. Pulmonary Circulation. 5. Pulmonary Surfactants—physiology. WF 600 P98043 2001]
QP121 .P7725 2001
612.2—dc21 2001031418

Printed on acid-free paper.

Production coordinated by Chernow Editorial Services, Inc., and managed by Tim Taylor; manufacturing supervised by Erica Bresler.
Typeset by SNP Best-set Typesetter Ltd., Hong Kong.

9 8 7 6 5 4 3 2 1

ISBN 978-1-4757-8133-5 SPIN 10791572

Preface

Pulmonary Biology in Health and Disease was conceived as a companion to a handful of expensive, multivolume textbooks. This is part of the promising trend to publish shorter textbooks on the subjects of lung biology and remodeling. Whoever is familiar with human biology and the far-reaching consequences of the genome and postgenome revolutions is apt to concede that the centerpiece in remodeling lies in the field of molecular cardiobiology. The field of molecular cardiobiology includes the syndrome of chronic heart failure as well as ischemic cardioprotection. By analogy, the centerpiece in pulmonobiology is chronic asthma. Key topics in the present volume include signaling mechanisms regulating the endothelium and smooth muscle cells, inflammatory cells, mediators, airway surface liquid, and pharmacological therapy that focuses on how inflamed airways are altered.

Written primarily for predoctoral and postdoctoral graduates in the basic medical sciences, the medical student and postdoctoral physician, graduates in the allied sciences, nurses, pulmonologists, and physicians in critical care medicine, this book provides many of the fundamentals of contemporary pulmonology. It is divided into several parts devoted to the control of respiration, arterial chemoreceptors, muscles of ventilation, pulmonary physiology, and gas exchange in health, exercise, and disease. Special emphasis is placed on emphysema and its pathobiology, acute lung injury, asthma and inhaled toxicants. Because the field is always evolving, each chapter includes recommended readings that lead the reader to sources of additional information, such as the review on remodeling of the blood gas barrier by West and Mathieu-Costello.

The present state of affairs is, in many ways, crucial to the teaching of the biological sciences in our schools and universities. The information revolution is forcing us to prune, or remove, irrelevant details, to focus on fundamentals and principles, and, above all, to arouse interest in our students. With this attitude, the present book has been assembled. Although its scope is not exhaustive, the hope is that it will impel its readers to keep informed of progress in pulmonology.

Finally, I acknowledge a special debt of gratitude to the various contributors who have brought this text into being.

E. Edward Bittar
Madison, WI

Contents

Contributors

Laura Almaraz, MD
Departmento de Bioquímica y Biología Molecular y Fisiología, Facultad de Medicina, Universidad de Valladolid, Spain

Stephen L. Archer, MD
Professor, Department of Medicine, University of Minnesota, VA Medical Center, Minneapolis, MN 55417, USA

Peter J. Barnes, MA, DM, DSc, FRCP
Professor, Department of Thoracic Medicine, National Heart and Lung Institute, Imperial College, London SW3 6LY, UK

Paula Carvalho, MD
Assistant Professor of Medicine, Division of Pulmonary and Critical Care Medicine, Department of Medicine, University of Washington, Seattle, WA 98195, USA

Nirmal B. Charan, MD
Professor and Chief, Section of Pulmonary and Critical Care Medicine, VA Medical Center, Boise, ID 83702-4598, USA

Barbara A. Cockrill, MD
Instructor and Associate Physician, Department of Medicine, Massachusetts General Hospital, Boston, MA 02114, USA

H. Frederick Frasch, PhD
Postdoctoral Fellow in Anesthesiology, Center for Research in Anesthesiology, University of Pennsylvania Health System, Philadelphia, PA 19104, USA

Suman Goel, MB, BS
Research Associate, Department of Medicine, University of Illinois at Chicago, Chicago, IL 60612-7323, USA

Constancio González, MD, PhD
Professor, Departmento de Bioquímica y Biología Molecular y Fisiología, Facultad de Medicina, Universidad de Valladolid, Spain

TANYA GREATREX
Cardiff University, Cardiff School of Biosciences, Wales CF10 3XQ, UK

GABRIEL G. HADDAD, MD
Professor, Department of Pediatrics—Respiratory Medicine, Yale University School of Medicine, New Haven, CT 06520-8064, USA

JOHN L. HARWOOD, PHD
Professor, Cardiff School of Biosciences, Cardiff University, Wales CF10 3XQ, UK

HOMAYOUN KAZEMI, MD
Professor, Pulmonary and Critical Care Unit, Massachusetts General Hospital, Boston, MA 02114, USA

BRYAN E. MARSHALL, MD, FRCA, FRCP
Professor of Anesthesiology, Center for Research in Anesthesiology, University of Pennsylvania Health System, Philadelphia, PA 19104, USA

CAROL MARSHALL, PHD
Associate Research Professor of Anesthesiology, Center for Research in Anesthesia, University of Pennsylvania Health System, Philadelphia, PA 19104, USA

KEITH C. MEYER, MD
Associate Professor of Medicine, Section of Pulmonary and Critical Care Medicine, Department of Medicine, University of Wisconsin Medical School, Madison, WI 53792, USA

LLINOS W. MORGAN, PHD
Cardiff School of Biosciences, Cardiff University, Wales CF10 3XQ, UK

BERNHARD F. MULLER, MD
295 Judd Road, Milan, MI 48160, USA

EUGENE E. NATTIE, MD
Professor, Department of Physiology, Dartmouth Medical Center, Lebanon, NH 03756-0001, USA

ANA OBESO
Departmento de Bioquímica y Biología Molecular y Fisiología, Facultad de Medicina, Universidad de Valladolid, Spain

JOHN T. REEVES, MD
Professor, Department of Medicine and Pediatrics, University of Colorado Health Sciences Center, Denver, CO 80262, USA

RICARDO RIGUAL
Departmento de Bioquímica y Biología Molecular y Fisiología, Facultad de Medicina, Universidad de Valladolid, Spain

GIUSEPPE SANT'AMBROGIO, MD
Professor, Department of Physiology and Biophysics, University of Texas Medical Branch, Galveston, TX 77555-0641, USA

Dean E. Schraufnagel, MD
Professor, Section of Respiratory and Critical Care Medicine, Department of Medicine, University of Illinois at Chicago, Chicago, IL 60612-7323, USA

James R. Sheller, MD
Professor, Center for Lung Research, Pulmonary Medicine and Critical Care Unit, Vanderbilt University Medical Center, Nashville, TN 37232, USA

Gordon L. Snider, MD
Chief, Medical Service, VA Medical Center, Boston, and Professor and Vice-Chairman, Department of Medicine, Boston University School of Medicine, Boston, MA 02130, USA

Martin Tristani-Firouzi, MD
Associate Professor, Division of Pediatric Cardiology, School of Medicine, University of Utah, Salt Lake City, UT 84113, USA

André De Troyer, MD
Professor of Medicine, Chest Service, Erasme University Hospital, and Director, Laboratory of Cardiorespiratory Physiology, Brussels School of Medicine, B-1070 Brussels, Belgium

Nai-San Wang, MD, PhD
President, ChungTai Institute of Health Sciences and Technology, Taichung 406, Taiwan

E. Kenneth Weir, MD
Professor of Medicine, VA Medical Center, University of Minnesota, Minneapolis, MN 55417, USA

Brian J. Whipp, PhD, DSc
Professor, Centre for Exercise Science and Medicine, University of Glasgow, Glasgow G12 8QQ, UK

J.H. Widdicombe, DPhil
Professor, Cardiovascular Research Institute, University of California Medical Center, San Francisco, CA 94143-0130, USA

Deborah Yager
Division of Life Sciences, Lawrence Berkeley National Laboratory, Berkeley, CA 94720, USA

M. Younes, MD
Professor, Health Sciences Centre, Respiratory Hospital, Winnipeg, Manitoba, Canada R3A 1RB

1
Anatomy and Ultrastructure of the Lung

NAI-SAN WANG

The lung is unique among all internal organs by the way it exposes itself directly and continuously to the surrounding atmosphere in which hostile agents often abound. For its necessary inspiratory and expiratory movements, the lung has to be durable and efficient in gas exchange. Because gas exchange is essential for life, it has to be protected vigilantly. Although the lung is designed to maintain life with minimum effort, it also needs to respond to urgent needs, increasing its basic rates of function 10-fold or more within seconds in life-threatening situations.

These requirements are amply met in the structural layout of the human lung. Moreover, the lung performs many nonrespiratory metabolic functions, especially in the lining cells of the distal airway (Clara cells), alveoli (type II alveolar lining cells), and endothelium.

Because of its sturdiness and efficiency, vigilant defense, and enormous reserve, the lung can become irreversibly damaged before the person becomes symptomatic.

The Structural and Dynamic Analogy

The gross structural arrangement of the lung resembles that of a tree. The trachea is the trunk, the bronchi are its branches, and the alveoli its leaves. The trachea and bronchi are its rigid conducting system, and its major function of gas exchange occurs in the alveoli.

In terms of dynamics, the lung moves like a balloon in an expandable vacuum box, the thoracic cage. The lung behaves like a balloon because of its diffuse and integrated elastic fiber networks. Air is sucked in or blown out of the lung by the expanding or contracting volume of the box, which is similar to that of the bellows. By the flattening of the normally tented diaphragm and uplifting of the anteriorly slanted ribs at inspiration, the thorax increases its length, width, and depth, thereby increasing the volume.

Inspiration is an active process. Energy is required to contract diaphragmatic and other respiratory muscles to overcome the increasingly retracting force of elastic recoil and surface tension as alveoli expand. Expiration is a passive movement that mainly results from relaxation of the contracted respiratory muscles and recoiling of the stretched elastic fibers. The surface tension decreases as alveoli deflate to prevent a complete collapse of the air space.

The Gross Divisions of the Airway

The inspired air flows through the conducting zone. The exchange of oxygen and carbon dioxide takes place in the respiratory zone, and both air flow and gas exchange occur in the short transitional zone.

The Conducting Zone

The conducting zone consists of the trachea, bronchi, and bronchioles. The trachea is a straight airpipe that averages 2.5 cm in diameter and up to 25 cm in length. It is framed by 15–20 horseshoe-shaped cartilages.

The left and right main bronchi branch out from the lowest end of the trachea by a sharp cartilaginous septum or spur, the carina. Each main bronchus divides by an asymmetrical dichotomy into 3 lobar, 10 segmental, and up to 20 generations of smaller bronchi. At each bifurcation, the cross-sectional area of each daughter branch is decreased by a factor of 0.75, and the sum of that of the two increases by a factor of 1.5. The high airflow resistance in the central airway decreases rapidly toward the periphery of the lung due to this exponential increase in the cross-sectional area for airflow, especially in bronchi 2 mm or less in diameter.

A bronchiole starts when cartilage completely disappears from the bronchial wall. At about the same point bronchial glands also disappear. The terminal bronchiole is the smallest bronchiole still lined completely by bronchial epithelial cells. A terminal bronchiole and all its distal airspaces constitute a pulmonary acinus. The human lung has approximately 30,000 terminal bronchioles and acini.

The Transitional Zone

The bronchiole becomes the respiratory bronchiole when alveoli start to appear focally in the bronchiolar wall. Ciliated cells extend to the bronchioloalveolar junction. Cilia are short and decreased in number in these peripheral locations, but they maintain the continuity of the ciliary flow by a network arrangement.

The Respiratory Zone

The alveolar duct begins where alveoli completely replace the bronchiolar epithelial cells. After two to five divisions, the alveolar duct terminates in a semicircular blind end called the *alveolar sac*. Each alveolar sac is surrounded by four or more alveoli.

Divisions of the Lung: Lobes, Segments, and Lobules

The right lung divides into upper, middle, and lower lobes, and further into 10 segments. The lingular (middle) lobe of the left lung, which is secondary to compression by the heart, is small and often combined with the upper lobe. The apical and posterior segments of the upper lobe, and the anterior and medial basal segments of the lower lobe, in the left lung, are also often fused by the compression.

The Lobular Concept

The lung divides from lobes to segments, and from segments to smaller cone-shaped units successively. The lobule is the smallest lung unit surrounded by fibrous septi. A lobule is about 2cc in size and contains four to eight terminal bronchioles. The boundary of the lobule is best visualized in early pulmonary hemorrhage or pneumonia.

The cone-shaped unit has central bronchovascular bundle containing bronchus, pulmonary artery, nerves, and lymphatic channels. Adjoining cones share their septi, which contain pulmonary veins, nerves, and lymphatics. Each cone-shaped unit, large or small, regulates the amount of air and blood flowing through it semi-autonomously by coordinated neural and paracrine or neuroendocrine controls. This arrangement enables the lung to shut off the blood flow and collapse the air space to limit and segregate disease processes. The air space originally occupied by the diseased lung can then be utilized more efficiently by the healthy units.

The Course of Airway and Disease

The right main bronchus deviates only 20–30 degrees off the course of the trachea and then leads almost straight into the right lower lobe bronchus. The left main bronchus angles 40–60 degrees off the original course of the trachea, and extends longer than the right one to reach, around the heart, to the lung.

Aspiration is a common cause of pneumonia. When aspiration occurs in the supine position, the aspirated material follows the straightest and most dependent route into the right lower lobe bronchus and into the first vertically positioned orifice: the superior segment of the right lower lobe. In the lateral position, aspirated material follows the main bronchus into the posterior segment of the upper lobe of the dependent lung.

The Bronchial Wall

All major conducting airways have an innermost layer of bronchial mucosa lined mainly by ciliated cells, followed outwardly, in layers, longitudinal bundles of elastic fibers, circular bundles of smooth muscle cells, and groups of bronchial glands and rigid cartilage. All these structures, with nerves, vessels, and lymphatics, are embedded orderly in a loosely collagenated matrix. At inspiration, the large airways elongate and narrow to enhance collision of particles on the mucosa. At expiration, they shorten and widen to facilitate outflow of air. The loose matrix also allows pulmonary vessels to expand and accommodate rapid and manifold increases in blood flow without concomitant increases in resistance.

Approximately one opening of the duct of the bronchial gland can be found in a 1-mm square area in large human bronchi. The volume of the bronchial gland is estimated by the thickness of the bronchial gland divided by the distance between the basal lamina of the bronchial lining cells and the nearest perichondrium (Reid's index). In normal bronchi, Reid's index should be below 0.4; that is, the mucous gland should not occupy more than 40% of the submucosal area. The bronchial gland increases its size and production of mucus in chronic bronchitis and asthma.

Elastic fibers run longitudinally in the large bronchi, become helical in the small bronchi and bronchioles, and then branch extensively in the alveolar wall. The cartilage, which is the rigid component of the large air passage, becomes haphazard in shape, and continues to decrease rapidly in size with each bifurcation. As the cartilage mass decreases, the circular smooth muscle bundle becomes relatively prominent. In asthma, the effect of bronchoconstriction is most severe in the medium-sized bronchus. Smooth muscles becomes scarce in small airways. In chronic bronchitis, however, smooth muscle cells proliferate centrifugally and constrict small airways which causes breathing difficulties.

The Lining Cells of the Bronchi and Bronchioles

In the trachea and bronchi, at least half of the surface lining cells are ciliated cells (Breeze and Wheeldon, 1977). A scanning electron microscopic view of the surface of the large bronchus shows that cilia appear to cover the entire surface (Fig. 1.1). The mucous cell is the most common nonciliated cell. The ratio of ciliated to mucous cells is estimated to be between 7:1 and 25:1 in the large bronchus.

The Ciliated Cell

The ciliated cell is up to 20 μm in length and 10 μm in width, and contains about 100–200 cilia of 3–6 μm in length and 0.3 μm in width on the luminal surface. The nucleus is basal. The cytoplasm is rich in mitochondria, which accumulate apically to supply energy for ciliary movements.

A cilium has an intracytoplasmic basal body. A long striated rootlet extends from the basal body and anchors deep into the cytoplasm, as does a short basal foot that aligns itself sideward in the direction of the effective stroke of the cilium (Fig. 1.2). At the neck, or the portion where the cilium emerges from the cell, the cell membrane forms a rosarylike modification called *the necklace*. The cilium tends to break off, but it also regenerates itself from this neck portion. The main body of the cilium is formed by an axoneme that is surrounded by the extension of cell membrane. The axoneme consists of a central pair of single and nine peripheral doublets of microtubules (Fig. 1.3). Each doublet has an A subfiber, a complete tubule, and a B subfiber, an attached three-quarter circle. Two rows of side arms (the inner and outer dynein arms) protrude from the A subfiber toward the B subfiber of an adjacent doublet, and one radial spoke extends toward the central pair. The dynein arms are protein with high adenosine triphosphatase (ATPase) activity. At the distal end of a cilium are clawlike hooking devices or bristles (Kuhn, 1976).

FIGURE 1.1. A scanning electron microscopic view of the mucosal surface of a large bronchus shows that cilia cover most of the surface. Human bronchus ×1700.

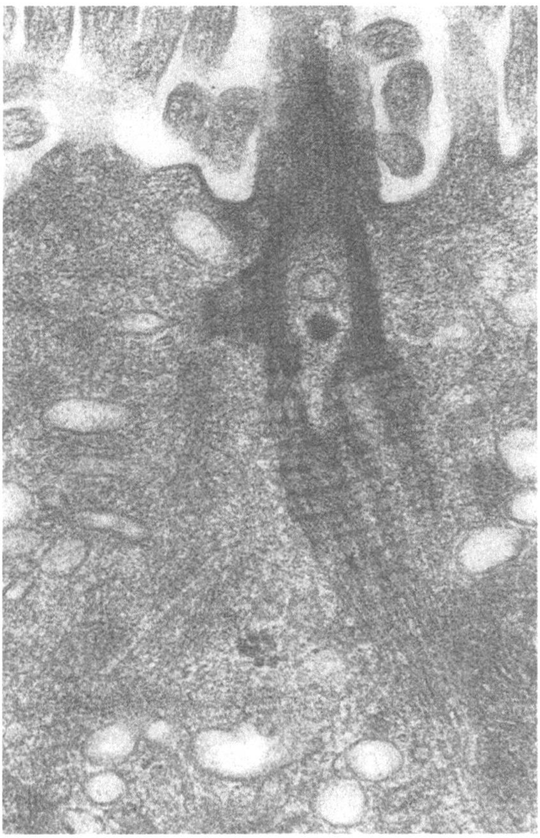

FIGURE 1.2. The intracytoplasmic basal body of a cilium shows a long striated rootlet that extends and anchors deep in the cytoplasm, and a basal foot that aligns itself sideward in the direction of the effective stroke of the cilium. Human bronchus. TEM ×110,000.

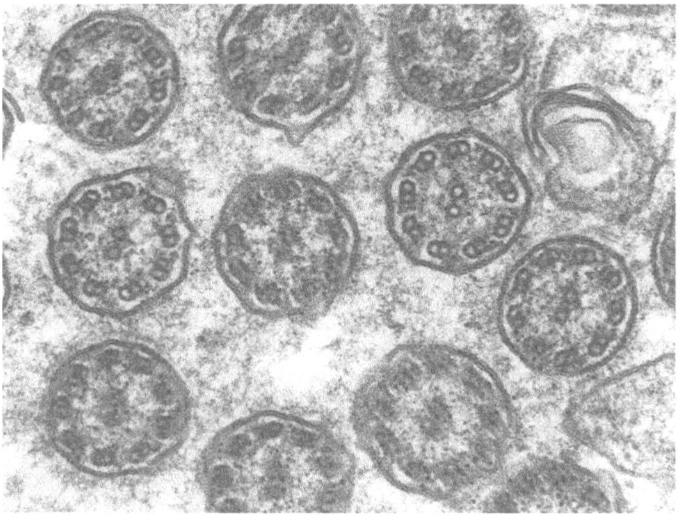

FIGURE 1.3. The cross-section of axonemes consists of a central pair of single, and nine peripheral doublets of microtubules. Each doublet has an A subfiber, a complete tubule, and a B subfiber, an attached three-quarter circle. From the A subfiber, two rows of side arms (the inner and outer dynein arms) protrude toward the B subfiber of an adjacent doublet. Human bronchus. TEM ×125,000.

The cilium bends and beats back and forth by sliding opposing groups of doublets in opposite directions. The sliding is accomplished by repeated attachments and detachments of dynein arms to adjacent doublets, similar to the motions used in jacking-up a car. The radial spoke functions similarly. The cilium at the recovery of preparatory stroke is curved and completely immersed in the watery (sol) layer of the mucus. The effective or propulsive stroke is done with cilia straight up, which allows the bristles to catch and propel the gel or viscous mucus on top of the sol layer forward.

Cilia and ciliated cells are constantly exposed to, and are vulnerable to many noxious agents, including cigarette smoke. Ultrastructural ciliary changes, therefore, are common and also varied, including compound or fused cilia, deranged axoneme with supernumerary or missing microtubules, and internalized (cytoplasmic) or shed cilia.

In Kartagener's syndrome (*situs inversus*, chronic sinusitis, and bronchiectasis), a congenital defect of dynein arms is responsible for the dysmotility of cilia which, in turn, causes stagnation of mucus and secondary infection. Moreover, congenital defects of radial spokes and other components of cilia have also been reported; however, some ciliary changes are reversible following treatment (Rossman et al., 1984). The differentiation between congenital and acquired ciliary changes may be difficult, and the diagnosis of dysmotile cilia syndrome or primary ciliary dyskinesia should be made cautiously. No consistent ciliary abnormality has been found in cystic fibrosis.

The Mucous Cell

The mucous cell has a basal nucleus, well-developed rough endoplasmic reticulum, Golgi apparatus, and abundant apical collections of mucous granules that form a "goblet" in appearance. The mucous granules are released into the bronchial lumen by the merocrine-type secretion. Mucus in the lung, presumably like that in the intestine,

is secreted for protection. Severe mucous cell hyperplasia may disrupt the ciliary flow, and an excessive amount of mucus may obliterate small scarred airways.

The mucous cell increases in number drastically in acute bronchial irritation, replacing almost all ciliated cells within days. The recovery of ciliated cells is probably fast, but more often chronic irritation persists and certain balances develop between the two processes. Because the rapid appearance of mucous cells following injury, and cells containing both mucous granules and cilia have been seen by electron microscopy (EM), Keenan et al. (1982) suggest that mucous cells are the precursor of ciliated cells; however, all ciliated and nonciliated cells in the bronchial mucosa can develop from the basal cell (Breeze and Wheeldon, 1977; Rennard et al., 1991).

The Dense Core (Neuroendocrine) Cell

The neuroendocrine cell (NEC), which is also named Feyrter's, or Kulchitsky cell, is primarily basal, but a small portion of the cell surface reaches the bronchial lumen. It has many dense core or owl-eyed secretory granules, 100–200 nm in size, which accumulate basally (Fig. 1.4). These granules may contain serotonin, bombesin, calcitonin, leuenkephalin, or other peptide hormones which are demonstrable by immunohistochemistry (IM). These peptides are released into the stroma to regulate the bronchial and vascular muscle tone.

Neuroepithelial Bodies (NEB)

These represent compact clusters of neuroendocrine cells and their supporting modified Clara cells that aggregate and bulge into the bronchial lumen like sentry posts at the bifurcation of airways (Fig. 1.4). NEB are innervated and may "sniff" the air and alter air and vascular flows by regulating the smooth muscle tone similar to but more efficiently than the individual dense core cell.

NEB are prominent in fetal or young lungs and increase their number and size following nicotine exposure and many types of chronic lung injuries. NEB and NEC, therefore, appear to participate in the growth and differentiation, and also repair and remodeling of the lung. The content of the neurosecretory granules differs in each situation.

Tumorlets (hyperplasia), carcinoid, and small cell carcinoma (dysplasia, neoplasia) constitute a series of gradual derangement of NEC. More often, however, they alter concomitantly with other cell types in all stages of injury and repair.

The Basal Cell

The basal cell of the bronchus is a pluripotential reserve cell that has LM and EM appearances similar to that of the epidermis of the skin. This cell is firmly affixed on the basement membrane by hemidesmosomes that enable the cell to survive mucosal injuries and to ensure a complete reconstruction of the bronchial mucosa. The basal cell is small and shows scanty organelles, scattered ribosome complexes, and rare tonofilaments.

Prolonged irritation stimulates the proliferation of the basal cell (hyperplasia). Dysplastic cells often show mixed organelles of different cell types by EM. This may explain why mixed adenosquamous and other cell types are common in a lung cancer and why immunoreactivity to keratin, carcinoembryonic antigen, and some neuropeptides can be co-expressed in many carcinomas of the lung.

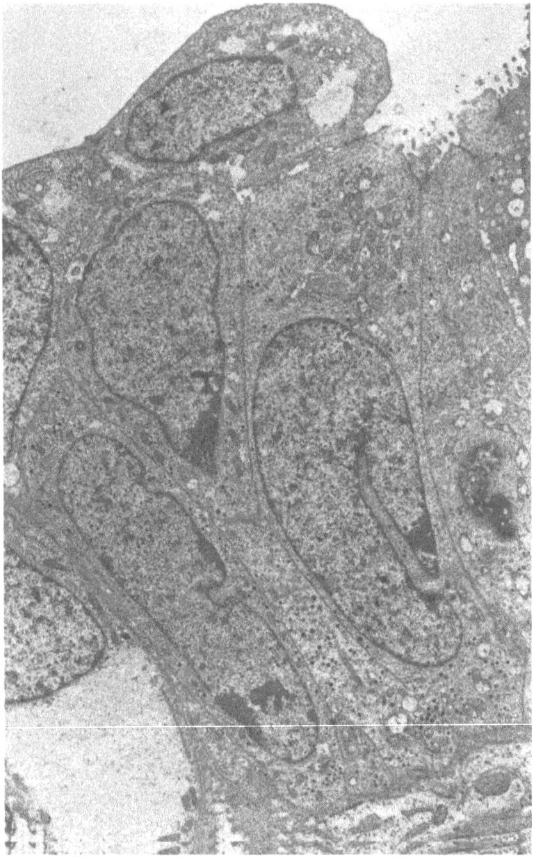

FIGURE 1.4. The neuroendocrine cell has many dense core or owl-eyed secretory granules, 100–200 nm in size, which accumulate basally. Neuroepithelial bodies (NEB) represent compact clusters of neuroendocrine cells and their supporting modified Clara cells. Human bronchus. TEM × 11,000.

Other Bronchial Lining Cells

Serous cells, which are rare in the bronchial mucosa, are mostly found in the bronchial gland. Brush cells are special type cells characterized by blunt microvilli and disklike or rodlike inclusions of unknown function. The term *intermediate cell* is often used in a broad sense to include cells with an unclassifiable phenoexpression.

The Clara Cell

The Clara cell is a nonciliated columnar cell found in terminal and respiratory bronchioles that replaces the disappearing mucous cell in small bronchioles. The Clara cell is taller than the ciliated cell in the terminal airway, is rich in endoplasmic reticulum and mitochondria, and contains a prominent Golgi apparatus with apically located secretory granules. The Clara cell secretes at least three different groups of proteins that range in size from 200 kDa to less than 10 kDa. The precise nature and function of these proteins is unclear, although some of them are similar to surfactant proteins secreted by the type II alveolar cells (*see* surfactant).

The Clara cell accumulates and detoxifies many inhaled or ingested noxious agents and are preferentially damaged when exposed to low concentrations of ozone, nitrogen dioxide (NO_2), or other chemicals. The Clara cell also serves as the reserve and reparatory cell in the small airway, which is a role similar to the basal cell proximally and the alveolar type II cell distally.

Cells of the Bronchial Gland

The lining cells of the bronchial gland duct are nonciliated and have mucous granules with a fibrillary appearance. Oncocytes with pink cytoplasm, which reflects an abundance of mitochondria, may form the distal segment of the bronchial duct. Onocytes are absent at birth but their number increases with age. They may regulate water and electrolytes of the mucus in the bronchial duct.

The acinus of the bronchial gland is formed distally by serous and proximally by mucous cells. The mucous cells in these glands are similar to those found on the bronchial mucosa. The bronchial gland normally produces more than 90% of mucus in the bronchus. The myoepithelial cell is a modified smooth muscle cell that forms a basketlike network around the acinus. These cells are innervated and contract to expel the acinar contents, mainly in response to the cholinergic stimulation.

The Epithelial Junctions

The epithelial cells that line all air passages and alveoli form a barrier to prevent leakage of water and solutes into the air space and prevent access of inhaled material to the interstitial space. The apical junctional complex includes the outermost tight junction, which forms a gasketlike seal, an intermediate junction with adhesion proteins in the 25–35 nm wide intercellular space and keratin fibril inserted buttonlike desmosome (Schneeberger, 1991). The gap junctions are transcellular channels that allow the exchange of small molecules and ions between adjacent cells.

The Vasculature of the Lung

The lung has a dual blood supply. One is the pulmonary circulation, which is a low-pressure system that accommodates total systemic venous return, and the other is the bronchial or nutritional system, which as a part of the systemic circulation has a high oxygen content.

Pulmonary Arteries and Veins

The pulmonary trunk arises from the right ventricle and almost immediately divides into left and right main pulmonary arteries. Pulmonary arteries follow and divide with bronchi and bronchioles, roughly in the same frequency and in similar diameters, out to the bronchioloalveolar junctions. From here, the blood flows into the alveolar capillary network.

The wall of a pulmonary artery is thinner than that of a systemic artery of corresponding size and has less smooth muscle cells but more elastic fibers. The normal systolic and diastolic pressures in the main pulmonary artery are approximately 20 and 10 mmHg, respectively (mean, 14 mmHg), as compared with 120 and 80 (mean, 90 mmHg) in the systemic circulation. Doubling of the resting flow volume only raises the pulmonary systolic pressure by 5 mmHg.

The pulmonary venules collect the capillary blood into the lobular septum. The small venules fuse and converge, forming larger pulmonary veins in the subsegmental septa. They group with the bronchi and pulmonary arteries at the segmental level, and proceed in their company to the hilum. The muscle layer of the pulmonary vein is scarce and irregularly arranged. The portion of the extrapulmonary vein near the left atrium is surrounded by cardiac muscle. The intrapulmonary arrangement of pulmonary and bronchial arteries and pulmonary veins resembles that of the portal system in the liver.

Bronchial Arteries and Veins

The origin of the bronchial artery varies considerably. The bronchial artery is classically the fused first pair of the intercostals arteries, which then descends and enters the wall of the lower trachea. The bronchial artery follows and nourishes the bronchial tree as far as the respiratory bronchiole. At the hilum, branches of the bronchial artery also radiate out and supply most portions of the mediastinal visceral pleura.

The bronchial veins mainly follow the course of the large bronchial arteries and drain into the azygos or hemiazygos. The dilated, prevenous capillaries in the bronchial wall are frequently fenestrated and leaky, as are those in the submucosal of the upper airway.

Vascular Shunts

The extent of shunting in the normal lung has been estimated to be less than 3% of the total cardiac output. Communications are most commonly found between bronchial and pulmonary small veins and capillaries. The thick-walled blockade or "sperr" artery between bronchial and pulmonary arteries is normally closed, but it will open during pulmonary arterial insufficiency, as there is with an acute pulmonary embolism, to perfuse the ischemic lung tissue (von Hayek, 1960).

With each injury and repair or neoplastic proliferation, the altered tissue is vascularized by the bronchial artery and more shunting channels are created. The left-to-right shunt in the lung, therefore, increases with age and can be substantial in patients with cancer, tuberculosis, or bronchiectasis.

The Alveolus

The smallest airway and pulmonary vessels mingle to form multifaceted, delicate honeycomblike alveoli (Fig. 1.5). The diameter of an alveolus ranges from 150 to 500 μm (average, 250 μm). A 70-kg man has about 300 million alveoli that constitute a gas-exchanging surface of approximately 143 m^2 and a lung volume of 4.3 L (Weibel, 1990).

The orifice of an alveolus is formed by thick elastic and collagen bundles that are the continuum of bronchial and bronchiolar elastic bundles (Fig. 1.6). From the orifice bundle, finer elastic fibers bulge out like a basket and are interwoven with capillaries (Fig. 1.7) (Wang and Ying, 1977). This network of elastic fibers and capillaries is plated (like wallpaper) on both sides by thin layers of Type I alveolar lining cell to form the alveolar wall. All elastic fibers are interconnected in all directions to form an integrated elastic network that is fundamental to the uniform expansion and elastic retraction of the lung in respiration.

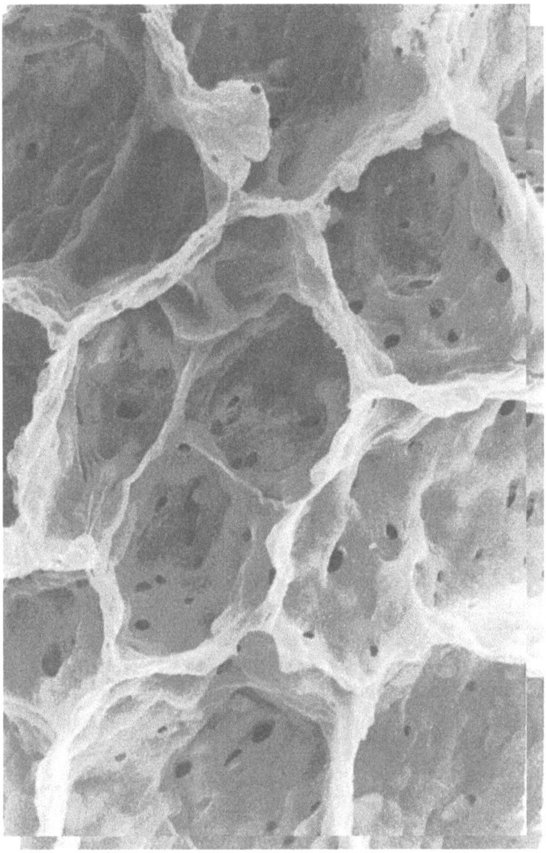

FIGURE 1.5. The smallest airway and pulmonary vessels mingle to form multifaceted, delicate honeycomblike alveoli with multiple pores of Kohn. An alveolus has a diameter between 150 and 500 μm (average, 250 μm). A 70-kg man has about 300 million alveoli. Mouse lung. SEM ×4000.

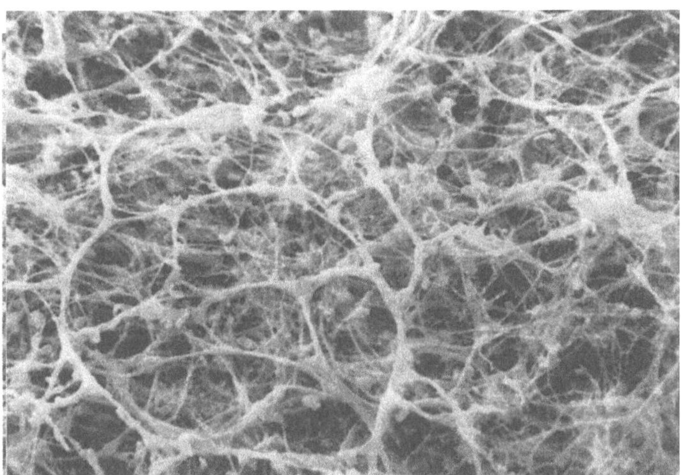

FIGURE 1.6. A lung digested with alkali to show the continuously tapering elastic and collagen network. Rabbit lung. SEM ×700.

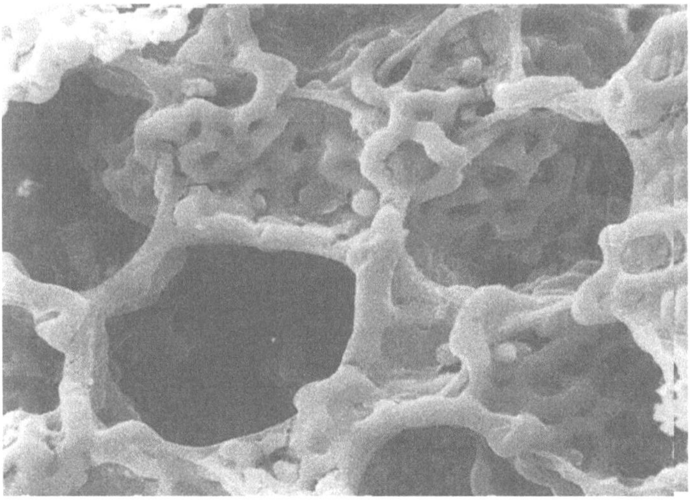

FIGURE 1.7. Latex particles are injected before digestion to show the interwoven capillary loops. Rabbit lung. SEM ×700.

The Alveolar Lining Cells

There are two types of alveolar lining cells. The Type-I alveolar lining cell, which resembles a fried egg, has a central flattened nucleus and a thin peripheral cytoplasm that reaches 50 μm in diameter. The Type-I cell constitutes 40% of the alveolar lining cells, but covers 90% of the alveolar surface (Meyrick and Reid, 1970). It has sparse surface microvilli and cytoplasmic organelles, but many pinocytic vesicles.

The Type-II alveolar lining cell, which is a cuboidal cell with diameter up to 15 μm, is characterized by many stubby surface microvilli, a large basal nucleus with a prominent nucleolus, abundant cytoplasm with mitochondria, well-developed endoplasmic reticulum and Golgi apparatus, and the most characteristic lamellar inclusion bodies, the precursor or surfactant (Fig. 1.8). It constitutes 60% of the alveolar surface cells, but covers only 5–10% of the alveolar surface (Myrick and Reid, 1970).

As in the bronchi, both Type I and II cells are joined by tight junctions that are underlined by a well-developed basal lamina, which prevents leakage of molecules larger than 1000 kDa.

In addition to secreting surfactant, the Type-II alveolar lining cell also serves as the reserve cell, which matures into the Type-I cell normally, and transports sodium from the apical to the basolateral surface to minimize alveolar fluid accumulation. The Type-II alveolar lining cell becomes hyperplastic in response to alveolar damage, participates actively in lung repair and remodeling, and may become dysplastic, or even neoplastic, with prolonged lung injury and fibrosis.

The Pulmonary Endothelium

The endothelium of the alveolar capillaries, which occupies a surface area of more than 140 m^2, is the largest and most dense vascular bed in the human body. The fine structure and permeability of the alveolar capillary endothelium is similar to that of the capillary endothelium elsewhere in the body.

The pulmonary endothelium selectively processes and modifies a wide range of substances, such as the conversion of angiotensin I to angiotensin II and the inactivation of bradykinin by the angiotensin-converting enzyme in caveolae (pinocytic vesicles) and on the microvilli of the luminal cytoplasmic membrane. The endothelium also clears serotonin, norepinephrine, prostaglandin (PG) E and F, adenine nucleotides, and some hormones and drugs. It also releases angiotensin II, PG I-2, endothelin, PDGF, and previously accumulated drugs and metabolites, and many more.

The endothelia of pulmonary arteries and veins differ from the capillary endothelium in that the pulmonary arteries and veins are subjected to much greater changes in the vascular intersurface area than do the capillaries. Both of them have more surface microvilli and cytoplasmic organelles than does the capillary endothelium, particularly on rod shaped membrane bound structures (Weibel-Palade bodies), which are the storage site of the coagulation factor VIII.

Alveolar Macrophages

All pulmonary macrophages originate from the monocytic series of the bone marrow. In acute irritation, they emerge directly from the circulation and traverse bronchial or alveolar walls to enter the airspace. Some of them, mainly in the chronic stage of injury, divide and emerge from the residential population of monocytic cells that arrived and settled earlier in the alveolar wall. Macrophages have elongated cellular processes called pseudopods, well-developed endoplasmic reticulum and Golgi apparatus, many membrane-bound structures that contain inflammatory mediators, and primary and secondary lysosomes.

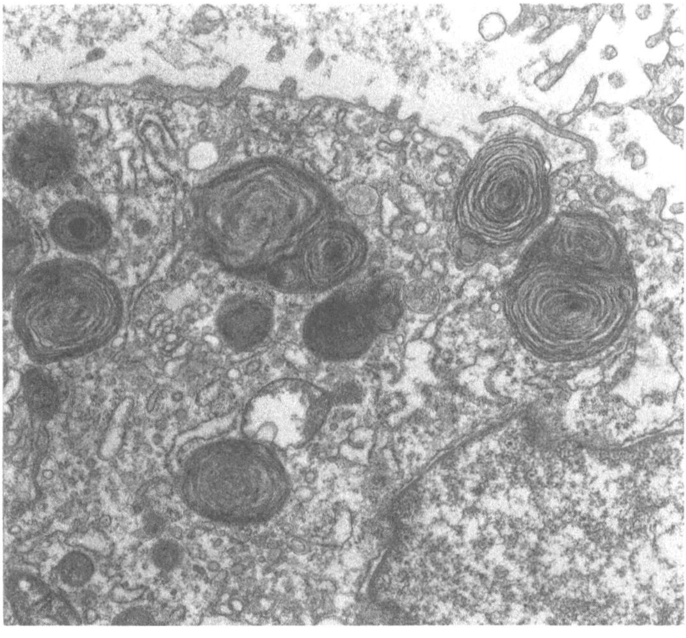

FIGURE 1.8. A portion of a Type-II alveolar lining cell showing stubby surface microvilli and the characteristic lamellar inclusion bodies, the precursor of surfactant. One of them is in the process of being excreted. Human lung. TEM ×20,000.

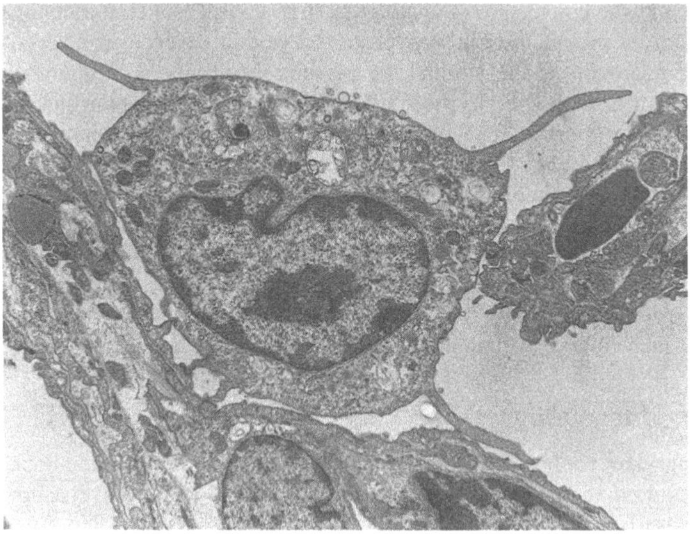

FIGURE 1.9. A macrophage, with elongated cellular processes called pseudopods is trespassing an alveolar pore. Mouse lung. TEM ×3500.

The macrophages move around on the alveolar and bronchial surfaces, they may trespass the alveolar pore (Fig. 1.9), and they engulf exogenous agents and damaged endogenous tissue debris. Once in the airspace, and especially after engulfment, macrophages cannot cross the intact epithelial layer back into the interstitium.

Other Cells and Structures in the Alveolar Wall

Mesenchymal cells, including fibroblasts, pericytes of capillaries, and myofibroblasts (contractile interstitial cells), occur in the alveolar septum. They are responsible for the maintenance and metabolism of the elastic and collagen fibers and proteoglycans in the alveolar wall. The rigid collagen fibers (as opposed to the soft elastic fibers) are present in the bronchovascular bundles, lobar and lobular septum, and pleura. Delicate collagen fibers are discernible only by EM in the alveolar wall.

Neutrophils, eosinophils, lymphocytes, plasma cells, basophils or mast cells, and fixed or migratory macrophages are normally rare in both the alveolar wall and bronchial interstitial space. Heavy sequestration and degranulation of neutrophils in the alveolar capillary wall may be responsible for insidious tissue lysis, such as elastolysis in pulmonary emphysema. Both mast cells and eosinophils are increased in bronchial asthma and other hypersensitivity diseases.

The Air–Blood Barrier

The alveolar arrangement provides nearly perfect and efficient exchange of oxygen and carbon dioxide between air spaces and red blood cells. It is estimated that 85–95% of the alveolar surface is covered with the pulmonary capillary network, giving an air–blood interface of about 126 m^2, which is a surface area about 70 times that of the skin. Only about one third of the capillary network is functional in the resting state, but it opens up rapidly with body exercise. A red blood cell passes through the capillary bed in 0.75 seconds.

The cytoplasm of endothelial and epithelial alveolar Type I cells is spread as thinly as possible, and their basal laminas are fused, forming an air–blood barrier with a mean thickness of 0.6 μm. When one considers that a red blood cell has an average diameter of 7 μm, one can appreciate the delicacy and fragility of the air–blood barrier.

Surfactant

The Type-II alveolar cell secretes its osmiophilic lamellar inclusion bodies into the alveolar space to form a partially crystallized hypophase of tubular myelin, which then spreads out into a thin surface layer. Surfactant is formed mainly by phospholipids, especially dipalmitoyl lecithin, with the addition of glycoproteins. At least three surfactant proteins (SP) have been purified and partially characterized. SP-A and SP-B are synthesized by Type-II alveolar and Clara cells, whereas SP-C proteins contribute to the biophysical functions of surfactant. SP-A also promotes the host defense mechanism of macrophages.

When the alveolus deflates, the phospholipids are compressed and aligned into a layer with hydrophilic and hydrophobic ends on each side at the air–liquid interface. This arrangement reduces the surface tension and prevents the collapse of the alveolus. At alveolar inflation, the orderly arrangement of the phospholipid molecules is disrupted, and the resulting increase in the surface tension assists the elastic recoil of the alveolus in expiration. The replenishment of surfactant about the alveolus and its presumed ascending flow toward the bronchiole is also helpful in alveolar clearance. Surfactant therefore plays several important roles in the stability and function of the alveolus.

Insufficient production of surfactant in prematurity results in a hyaline membrane disease with alveolar collapse and pulmonary edema. In the adult distress syndrome or a diffuse alveolar damage, an excessive leakage of fibrin and other capillary contents into the alveolar space interferes with the action of surfactant, despite a normal or even increased amount of surfactant in the alveolus.

Alveolar Regulation of Capillary Flow

Pulmonary blood flow is regulated mainly by contraction of small pulmonary arteries, especially in hypoxia. The regional alveolar blood flow, however, is regulated by other mechanisms. The contractile interstitial cells are attached to adjacent endothelial and epithelial cells. When stimulated, they contract to distort the capillary and disrupt the flow. The details of the control mechanisms, however, are not yet completely known.

The capillary flow is also disrupted at the extremes of inflation and deflation because of the interlocking or pinching of elastic fibers and capillaries. A widespread and prolonged collapse on hyperinflation of the lung is incompatible with life. Interruption of blood flow to focally collapsed or overinflated alveoli, however, is beneficial.

Alveolar Pores (of Kohn), Fenestrae, and Collateral Ventilation

Communications between adjacent alveoli are not present in a fetal lung. Pores (of Kohn) 2–15 μm in size start to appear in the alveolar wall soon after birth and appear to increase in number with age (see Fig. 1.5). Small alveolar pores are filled with surfactant, thus blocking the airflow between adjacent alveoli. In lobar pneumonia, edema fluid, fibrin, and bacteria spread rapidly between alveoli through the alveolar pores.

Alveolar pores larger than 15 μm are arbitrarily defined as *fenestrae*. Alveoli distal to an obstructed airway can receive air through fenestrae from alveoli ventilated by a nearby unobstructed bronchiole (collateral ventilation). Other channels of collateral ventilation include communications between adjacent alveolar ducts or small bronchioles, and between bronchioles and alveoli (Lambert's canals). These structures are most often found in the lungs of aged smokers or miners.

Lymphatics and Lymphoid Tissues

The lymphatics of the lung are divided into the pleural or superficial plexus in the visceral pleura and the deep or parenchymal plexus in the bronchovascular bundles and the lobar and lobular septa. The two systems drain separately toward hilar nodes. The alveolar walls do not have lymphatics.

Lymphatic capillaries are lined by large flattened endothelial cells with few organelles. Focally open or detachable junctions devoid of the basal lamina are characteristic of lymphatic capillaries. The collecting lymphatic channels resemble thin-walled veins with funnel-shaped valves.

Lymphoid cells accumulate in the bronchial mucosa to form bronchus-associated lymphoid tissue (BALT) that protrudes toward the lumen. Lymph nodes are found along the lymphatic pathways in the lung (intrapulmonary), hilum (hilar), carina (superior and inferior tracheobronchial), and along the trachea (peritracheal). Scalene lymph nodes in the supraclavicular space are frequently involved in metastatic lung cancer, either through lymphatic connections with the peritracheal lymph nodes or through direct adhesions between the visceral and the parietal pleura, often at the apex of the lung. The diffuse lymphangitic spread of cancer is not the result of massive nodal metastases involving obstructive retrograde spread, but rather represents multiple lymphangitic invasion following initial vascular dissemination of tumor cells.

Clearance of Inhaled Particles in the Lung

Inhaled particles larger than 10 μm impact on the mucous blanket of the upper airway. Particles of 2–10 μm settle on the peripheral airspaces. Particles less than 2 μm are not deposited readily and are mostly exhaled. In addition to the particle size, the chemical nature, concentration, and the duration of exposure are all important in the assessment of the retention of particles and their damage to the lung.

Macrophages in the alveolus move about by amoeboid motion and drift with the alveolar surfactant to reach the terminal bronchiole. All macrophages with their engulfed agents in the airway are swept upstream by ciliary movements, finally to be swallowed or expectorated.

Because one terminal bronchiole collects and concentrates the content of approximately 10,000 alveoli, the distal airway is often overburdened and inflamed. Examples include the *alveolar ductitis* of young cigarette or marijuana smokers, small-airways disease in the early stage of pneumoconiosis or cystic fibrosis, and *bronchiolitis obliterans* or incompletely resolved pneumonia from a variety of causes.

In these situations, bronchiolar and Type-I and -II alveolar lining cells engulf particles by a forced phagocytosis, which are then released into the interstitium. The epithelial barrier is often broken down. The particles that reach the interstitial space are engulfed by monocytic cells (intravascular macrophages) after they enter the capillaries. They are more commonly engulfed by interstitial macrophages and drain toward the hilar lymph nodes.

The centrifugal removeal of particles probably occurs only when the centripetal pathway is disturbed or overburdened. The clearance of particles and macrophages in this direction is less efficient; foreign particles are frequently found to accumulate in the pleura and lobular septum and to cause local fibrosis.

Nerves and Humoral Controls

The sympathetic and parasympathetic nervous systems form the pulmonary plexus at the hilum and then distribute around and along the airways and pulmonary vessels to reach the alveoli and pleura. The parasympathetic system has both sensory and motor nerves. Its afferent sensory nerve endings are present in the bronchial and alveolar walls and pleura (as stretch receptors) and the bronchial mucosa (as irritant receptors). The ganglia, which are located in the bronchial wall, reflexively integrate the local input and output. Its preganglionic afferent fibers in the vagus nerve, reach the sensory vagal nuclei in the medulla close to the respiratory center.

The preganglionic efferent fibers of the parasympathetic system arise from the motor vagal nuclei in the medulla. The postganglionic motor fibers, which are cholinergic, innervate and stimulate contraction of bronchial and bronchiolar muscle cells, cause secretion of the bronchial gland, and stimulate vasodilatation.

The sympathetic system is mainly efferent. Its preganglionic fibers from the central nervous system emerge from the second to sixth thoracic segments of the spinal cord to reach ganglions in the sympathetic trunk along each side of the vertebral column. The postganglionic fibers and endings are adrenergic and have two types of endings. The α-adrenergic stimulation induces constriction of bronchial and vascular smooth muscle cells, whereas β-adrenergic stimulation induces bronchodilatation and decreases secretion of the bronchial glands.

The distribution and function of the third nonadrenergic inhibitory or purinergic system are not completely known. Afferent and efferent nerve endings with dense core granule-containing cells have been identified in the paracrine system. Mast cells, macrophages, and other mononuclear cells in the bronchial and alveolar walls also release a variety of mediators, including serotonin and prostaglandins, which among their many effects also alter muscle tone. The reactivities of the bronchial muscle and pulmonary vasculature are influenced by many neural and humoral, and by both central and local factors. Their responses to stimuli in disease are frequently unpredictable.

The Pleural Cavity and Mesothelial Cells

The primitive body cavity or coelom, which are lined by mesothelial cells, appears early in the embryo. All constantly moving organs, such as the lung, heart, and intestines, subsequently develop into this cavity and are enveloped by the mesothelial cell layer as they do so. This arrangement renders these organs readily movable as well as pliable in size and shape during development and maturation.

The pleural cavity is a thin space between the visceral pleura, which covers the entire surface of the lung including the interlobar fissures, and the parietal pleura, which covers the inner surface of the thoracic cage, mediastinum, and diaphragm. The visceral pleura reflects at the hilum to continue as the parietal pleura. The opposing two layers of mesothelial cells are separated only by a layer of hyaluronic acid-rich fluid less than 20-μm thick.

The gross appearance of the pleura is smooth, glistening, and semitransparent. Based upon light microscopy, the pleura is typically divided into (1) a mesothelial layer, (2) a thin submesothelial connective tissue layer, (3) a superficial elastic layer, (4) a loose subpleural connective tissue layer, and (5) a deep dense fibroelastic layer. The presence and thickness of each layer varies regionally. The loose layer (fourth) is the plane of cleavage at decortication. The fifth layer frequently adheres tightly to blends into the parenchyma of the lung or the chest wall.

The mesothelial cells are stretchable and range in size from 16.4 × 6.8 to 41.0 × 9.5 μm. They may appear flat, cuboidal, or columnar. Mesothelial cells are characterized ultrastructurally by an abundance of elongated bushy microvilli 0.1 μm in diameter and up to 3 or more microns in length. The microvilli trap hyaluronic acid, which acts as a lubricant to lessen the friction between the moving lung and the chest wall. The cytoplasm is rich in pinocytic vesicles, mitochondria, and other organelles and prekeratin fibrils. The presence of dominant bushy microvilli (by EM) and of prekeratin fibrils (by IM), and absence of epithelial markers such as carcinoembryonic antigen, have proven useful in differentiating mesothelioma from metastatic adenocarcinoma in the pleural space.

The secretion and absorption of pleural fluid is governed by Starling's law. Large particles and cells such as fibrin molecules or macrophages are removed from preformed stomas directly connecting the pleural cavity with the lymphatics. The stomas are present only in specific areas of the parietal pleura, including mediastinal and infracostal regions, especially in the lower thorax. The entry of large particles into a dilated lymphatic lacuna through the stoma is facilitated by respiratory movements. The roof of the lacuna is formed by a network of thick collagen bundles that are covered by mesothelial cells on the pleural and endothelial cells on the lacuna side. These two layers of cells rupture readily in disease to increase the route of pleural clearance.

Summary

To live one must breathe continuously. The lung therefore exposes itself directly and constantly to the hostile surroundings.

To be efficient, the area of gas exchange at the surface is maximized to 143 m^2 by compartmentalization of a lung with a volume of 4.3 L into 300 million alveoli, and the air–blood barrier is minimized to 0.6 μm in thickness. Inspiration is executed by active contraction of the diaphragm, whereas expiration is associated with passive recoil of a uniformly stretched elastic fiber network and disrupted surfactant.

For the purpose of protection, the delicate alveoli lie inside a rigid thoracic cage, and are separated from the hostile environment by the nose, trachea, and up to 25 generations of branching airways. Agents present in the air are trapped on the mucus, engulfed by macrophages, and swept out by ubiquitous ciliated cells. Serious offenses are segregated and contained to the smallest possible lobules so that the remaining lung can function adequately.

The lung is also designed to maintain life with minimum effort, but can increase its rate of function 10-fold or more within seconds in life-threatening situations. Moreover, the lung performs many nonrespiratory metabolic functions, especially in Clara cells, Type-II alveolar lining cells, and endothelia. Because of sturdiness and efficiency, vigorous defense, and enormous reserve, the lung frequently becomes irreversibly damaged before the person becomes symptomatic.

References

Breeze, R.G., and Wheeldon, E.B. (1977) The cells of the pulmonary airways. Am. Rev. Respir. Dis. 116, 705–777.

Keenan, K.P., Combs, J.W., and McDowell, E.M. (1982) Regeneration of hamster tracheal epithelium after mechanical injury. Virchows Arch. Cell Pathol. 41, 193–214.

Kuhn, C. (1976) Ciliated and Clara cells. In: Lung cells in disease, Bouhuys, A. (ed.) pp. 91–108. North Holland, Amsterdam.

Meyrick, B., and Reid, L. (1970) The alveolar wall. Br. J. Dis. Chest. 64, 121–140.

Rennard, S.I., Beckmann, J.D., and Bobbins, R.A. (1991) Biology of airway epithelial cells. In: The lung: scientific foundations, Crystal, R.G., West, J.B., Barnes, P.F., Cherniack, N.S., and Weibel, E.R. (eds.) pp. 157–167. Raven Press, New York.

Rossman, C.M., Lee, R.M.K.W., Forrest, J.B., and Newhouse, M.T. (1984) Nasal ciliary ultrastructure and function in patients with primary ciliary dyskinesia compared with that in normal subjects and in subjects with various respiratory diseases. Am. Rev. Respir. Dis. 129, 161–167.

Schneeberger, E.E. (1991) Airway and alveolar epithelia cell junctions. In: The lung: scientific foundations, Crystal, R,G., West, J.B., Barnes, P.J., Cherniack, N.S., and Weibel, E.R. (eds.) pp. 205–214. Raven Press, New York.

Von Hayek, H. (1960) The human lung. (Krahl, V.E., trans.) p. 236. Hafner, New York.

Wang, N.-S., and Ying, W.L. (1977) A scanning electron microscopic study of alkali-digested human and rabbit alveoli. Am. Rev. Respir. Dis. 115, 449–460.

Weibel, E.R. (1990) Lung cell biology. In: handbook of physiology: the respiratory system. I. pp. 47–91. Washington DC.

Recommended Readings

Aicks, G.H. (2000) Cardiopulmonary anatomy and physiology. Mosby, St. Louis.

Gail, D.B., and Lenfant, D.J.M. (1983) Cells of the lung: biology and clinical implication. Am. Rev. Respir. Dis. 127, 366–367.

Gehr, P., Bachofen, M., and Weibel, E.R. (1978) The normal human lung: ultrastructure and morphometric estimation of diffusing capacity. Respir. Physiol. 32, 121–140.

Hlastala, M.P., and Robertson, H.T. (1998) Complexity in structure and function of the lung. Lung biology in health and disease series, vol. 121. Marcel Dekker, New York.

Ryan, U.S. (1982) Structural bases for metabolic activity. Ann. Rev. Physiol. 44, 222–239.

Wang, N.S. (1990) Scanning electron microscopy of the lung. In: Lung biology in health and disease, vol. 48. Lenfant, C., and Schraufnagel, D.E. (eds.) pp. 517–555. Marcel Dekker, New York.

2
Biology of Mammalian Airway Epithelium

J.H. Widdicombe

The entire surface of the airways is lined with epithelia of various forms (Harkema et al., 1991). Almost all are ciliated, and their role in mucociliary clearance has long been appreciated; however, the development of high-quality primary cultures and cell lines (Gruenert et al., 1995) has revealed many new functions. It is now realized, for instance, that airway epithelia are capable of affecting rapid changes in the depth and composition of the thin film of liquid that lines the airways. In addition, production of a wide range of macromolecules allows the epithelium to signal to underlying cells and tissues (e.g., smooth muscle, fibroblasts and leukocytes). It has been increasingly appreciated that alterations in the function of airway epithelium are central to the pathology of cystic fibrosis, asthma, and chronic bronchitis.

Surface Epithelium

Structure

There are stratified squamous, pseudostratified ciliated, nonciliated columnar, and olfactory epithelia in the nose (Harkema et al., 1991). In the trachea, bronchi and bronchioles, the height of the surface epithelium correlates with the airway diameter. In large species (e.g., humans) the epithelium is pseudostratified columnar in the trachea and large bronchi, with the cells being about 50 μm high and 5 μm across (see Fig. 2.1A). The pseudostratified appearance is created by the nuclei of basal cells that lie between the bases of the columnar cells, and have no contact with the lumen. As the height of the epithelium decreases distally, the basal cells disappear, and the epithelium becomes columnar. Progressive decreases in cell height lead to a cuboidal epithelium in the bronchioles (i.e., cell width and height are the same), and ultimately a squamous epithelium in the terminal bronchioles and alveoli. In small species (e.g., the rat and mouse) the tracheal epithelium is cuboidal (Fig. 2.1B) and resembles the bronchiolar

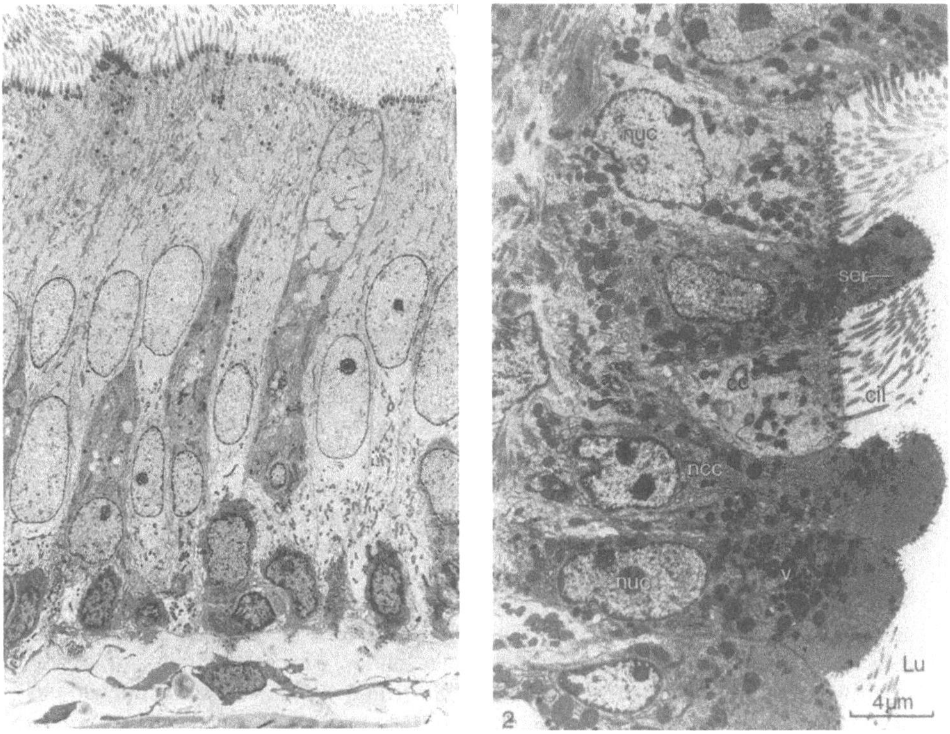

FIGURE 2.1. Tracheal epithelia. (A) Pseudostratified epithelium from dog. Most columnar cells are ciliated. The apical portion of a goblet cell with typical mucous granules is present, as are the bases of several other goblet cells. Small basal cells lie adjacent to the basement membrane. Scale bar = 20 μm. (*Source*: Reproduced with permission from Widdicombe et al., *Am. J. Physiol.* 241:C184–C192, 1981.) (B) Columnar epithelium from mouse. Clara cells comprise 50% of the total. Basal and goblet cells are absent. The histology resembles that of human bronchiole. Nuc = nucleus; cc = ciliated cell; ncc = nonciliated (Clara) cell; cil = cilia; v = secretory vesicle; Lu = lumen. (*Source*: Reproduced with permission from Pack et al., Cell Tissue Res. 208:65–84, 1980.)

epithelium of larger species. All cells that contact the lumen except those in the stratified squamous epithelium of the nose also contact the underlying basement membrane.

A combination of basal, ciliated, and one of the major secretory cell types (i.e., goblet, Clara, or serous cell) generally accounts for more than 90% of the cells of any given airway; however, at least eight other types of cell have been described (Breeze and Wheeldon, 1977; Harkema et al., 1991).

Basal cells are small, roughly spherical cells that are approximately 7 μm diameter and lie on the basement membrane between the ciliated and secretory cells (see Fig. 2.1A). They may comprise up to 50% of the cells in the tracheal and bronchial epithelia of the large mammalian species. Their cytoplasm is distinguished by bundles of tonofilaments, and they are connected to neighboring columnar cells by numerous desmosomes. They contain hemidesmosomes where they abut on the basement membrane. Extracellular anchoring filaments attach to the hemidesmosomes and form loops around subepithelial collagen fibers, thereby anchoring the basal cells to the lamina propia (Kawanami et al., 1979). The presence of desmosomes between basal and other cells and a correlation between numbers of basal cells and the height of airway epithelium suggests that one function of basal cells is to anchor the entire epithelium to the underlying tissues (Evans and Plopper, 1988).

Ciliated cells comprise roughly 30–60% of the total cells in all airways (see Fig. 2.1). They have a relatively electron-lucent cytoplasm that reflects a dearth of ribosomes. The nucleus, which is characterized by a smooth spherical shape, lies in the lower half of the cell. A well-developed Golgi apparatus lies just apical to the nucleus, and abundant mitochondria are present toward the top of the cell. The apical membrane of a ciliated cell contains about 200 membrane-bound cilia of characteristic structure (Sleigh et al., 1988), as well as abundant, long microvilli.

In larger species including humans, goblet cells are the typical secretory cells of the trachea and bronchi (see Fig. 2.1A). They contain large (300–1800 nm diameter) electron-lucent granules that store mucins. Immature or discharged goblet cells may contain only a few granules, but when mature, the entire upper two thirds to four fifths of the apical cytoplasm is packed with granules, which distend the apical portion of the cell and produce the characteristic goblet shape. In mature goblet cells, secretion is by "compound exocytosis" involving fusion of granules with both the apical membrane and other granules deeper in the cell.

In human bronchioles, and in all airways of small species (e.g., rabbit and mouse), goblet cells are replaced by Clara or serous cells (see Fig. 2.1B). The cytoplasm of Clara cells is characterized by abundant agranular endoplasmic reticulum and ovoid electron-dense granules. Serous cells contain electron-dense granules of about 600 nm diameter in the apical portion of the cell. In contrast to those of goblet cells, the granules of serous cells show no tendency to fuse with one another. Serous cells are common only in the tracheas of pathogen-free rats, human fetuses, and young hamsters (Jeffery, 1983), but they are found in small numbers in adult human bronchioles (Rogers et al., 1993). They are replaced by goblet cells in rats subjected to airway irritation (Huang et al., 1989).

Other epithelial cell types include oncocytes, brush, small granule, intermediate, transitional, and special-type cells. Brush cells are characterized by large microvilli. Their function is unknown and they have not been demonstrated convincingly in human airways. Neuroendocrine cells (Sorokin and Hoyt, 1989) contain small (i.e., 100–300 nm diameter) cytoplasmic granules in their bases. These granules typically have an opaque periphery surrounding an electron-dense core, and they contain serotonin, bombesin, and other peptides, as well as monoamines. Neuroendocrine cells are found at all levels of the airways, either individually or in clusters of up to 100. Aggregates of these cells are usually innervated. If so, they are known as neuroepithelial bodies. Special-type cells contain small numbers of electron-dense membrane-bound granules in the form of discs or rods (~300 × 50 nm) and have no contact with the lumen. They may be degenerating ciliated cells (Kawamata and Fujita, 1983). Oncocytes are large mitochondria-rich cells that are very rare in surface epithelium, but somewhat more common in glands. They probably develop into cysts and tumors. Cells that show no evidence of either ciliary or mucous differentiation have been referred to variously as intermediate, indifferent, or indeterminate cells, and are generally believed to be steps in the differentiation of basal into ciliated or secretory cells. So-called transitional cells possess both cilia and secretory granules. They presumably represent stages in the development of ciliated cells from either Clara or small mucous granule cells.

Several types of white blood cell are found in airway epithelium. Lymphocytes are the commonest, and their lack of organelles suggests that they are not producing

TABLE 2.1. Cell types of airway surface epithelium.

Cell type	Characteristic functions	Features
Ciliated	Cilia	Mucus propulsion ion transport
Goblet	Large electron opaque granules	Mucus secretion
Small mucous granule	Small mucous granules	Immature goblet cells stem cell
Clara	Abundant agranular ER small electron dense granules	Detoxification secretion of antiproteases and surfactant stem cell
Serous	Small electron-dense granules secretory canaliculi high levels of CFTR	Liquid secretion secretion of bactericidal compounds
Basal	No contact with lumen tonofilaments	Anchors epithelium stem cell
Brush	Large microvilli	Unknown
Neuroendocrine	Basal secretory granules containing peptides and monoamines	Chemoreception?
Special-type	Rod-shaped granules	Degenerating ciliated cells?
Oncocytes	Abundant mitochondria	Precancerous state?
Intermediate indifferent indeterminate	Lack of differentiation	Steps in differentiation of basal to other cell types
Transitional	Mucous granules and cilia	Stage in differentiation of small mucous granule cell to ciliated cell
Lymphocytes	Relatively undifferentiated	Immune response
Globular leukocyte	Electron dense granules	Combating nematodes
Dendritic cell	2–3 slender cytoplasmic processes	Presenting antigen

immunoglobulins. Globular leukocytes contain very electron-dense homogeneous granules about 600nm in mean diameter, with characteristic cytochemical reactions. They are derived from subepithelial mast cells, migrate into the lumen of inflamed airways, and may be involved in defense against nematode infections. Finally, dendritic cells, derived from bone-marrow precursor cells, present antigen to T lymphocytes (Sertl et al., 1986).

The different cell types of airway epithelium, their distinguishing features and probable functions are summarized in Table 2.1.

Development

Studies on the development of tracheal epithelia have been summarized (Plopper et al., 1986; St. George et al., 1994). In early fetal life, the only epithelial cell present in the airways is a nondifferentiated columnar cell, which shows little endoplasmic reticulum, small Golgi apparatus, no secretory granules, and an apical membrane that lacks microvilli or cilia. In large airways, the first differentiated cell to appear is the neuroendocrine cell, followed by ciliated cells, then secretory cells, with basal cells appearing last. It is believed that all the differentiated cell types are initially derived directly from the undifferentiated columnar cells. In all species, the major differentiated cells of adult epithelium are present at birth. The proportions of the different cell types change postnatally in some species. In others (e.g., the rhesus monkey) there is no change in the relative numbers of the different cell types after birth, but the cells undergo rapid cytodifferentiation and functional maturation. There is a progressive decline in the rate of cell division during fetal life. In the hamster, for instance, approximately 20% of cells are dividing on day 10 after conception, the earliest point at which sections can be obtained from the tracheal bud. By day 16, however, the day of birth, only about 1% of cells are dividing (McDowell et al., 1985). Adult airway epithelium has one of the lowest rates of division of any epithelium, and the cells have a correspondingly long average life span of 20–200 days (Ayers and Jeffery, 1988). By contrast, the life span is approximately 10 days in oral epithelium (Wright and Alison, 1984).

Function

Mucociliary Clearance

The airways are lined with a film of liquid from 5 to 100μm deep. This liquid is a complex mixture of secretions from cells of both the surface and glandular epithelia. Its components include plasma proteins, surfactant, DNA, antibacterial substances (such as lysozyme, lactoferrin and proteases), and mucins (Braga and Allegra, 1988). Visual inspection of the surface of monkey nasal mucosa led Lucas and Douglas (1934) to propose that the airway lining liquid consisted of two layers, with a mucous blanket lying over a watery sol layer bathing the cilia. They suggested that the periciliary liquid was of low viscosity and almost exactly the same depth as the length of the cilia. Thus, the cilia were able to beat in this layer, and during their effective strokes their tips contacted the underside of the mucus blanket and propelled it forward. Studies have essentially confirmed the two-layer hypothesis (see Fig. 2.2).

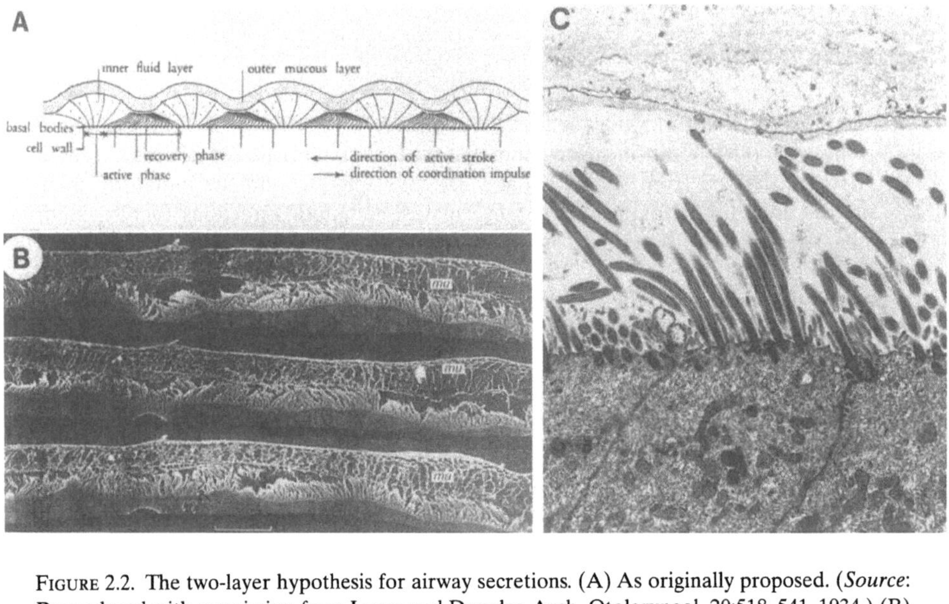

FIGURE 2.2. The two-layer hypothesis for airway secretions. (A) As originally proposed. (*Source*: Reproduced with permission from Lucas and Douglas, Arch. Otolaryngol. 20:518–541, 1934.) (B) Two layers seen in rapidly frozen specimens of rabbit tracheal cultures. Scale bar = 10μm. mu = mucous gel. (*Source*: Reproduced with permission from Sanderson and Sleigh, *J. Cell Sci.* 47:331–347, 1981.) (C) Two layers seen in an electron micrograph of rat trachea. (*Source*: Reproduced with permission from Yoneda, *Am. Rev. Resp. Dis.* 114:837–842, 1976.)

Mucins are responsible for the peculiar viscoelastic properties of the mucous blanket, allowing its propulsion by the cilia. These large (~500 kD) compounds are bound together in several ways, including hydrogen bonds, electrostatic and hydrophobic interactions, entwinement, disulphide bonds, and lectinlike bonding between carbohydrate and peptide portions of the molecules (Silberberg, 1988). The resulting gel has properties of both a solid (elasticity) and liquid (viscosity), and this combination is essential to its function (King and Rubin, 1994). Thus, the liquid properties of mucus allow it to be deformed and to entrap foreign particles. By contrast, elasticity is required for propulsion by the cilia (at ~5 mm/mm) toward the mouth (Wanner, 1986), where it is swallowed or expectorated. This process of mucociliary clearance thus keeps the surface of the airways clean. The mucous gel may be virtually absent in healthy, uninflamed airways (Bhaskar et al., 1985), but it becomes thicker in bronchitis, asthma and, cystic fibrosis, where mucous secretions may obstruct and plug the airways.

Regulation of Airway Liquid Content

Active liquid transport by epithelia is secondary to transepithelial active solute transport (Diamond, 1979). In healthy, uninflamed adult human airways the major solute transport process is active Na absorption (Yamaya et al., 1992). This makes the airway lumen electrically negative by approximately 30 mV, thereby inducing net passive movement of C1 toward the submucosa, both through (Uyekubo et al., 1998) and between the cells. The resulting transfer of salt creates an osmotic gradient locally across the epithelium that draws water from the lumen to the submucosa by osmosis. *In vitro*, active Na absorption brings about liquid absorption of 5–10 $\mu l \cdot cm^{-2} \cdot h^{-1}$, which would decrease the depth of the mucus layers *in vivo* at approximately 1 μm/min. *In vivo*, passive forces clearly must exist to counter active absorption of Na and liquid and prevent drying of the airway surface. Osmotic and hydrostatic pressures and surface tension are the most important such forces.

Interstitial hydrostatic pressures are generally around −2 mmHg, which suggests that the hydrostatic pressure gradient would also tend to remove liquid from the airway lumen. In the upper airways, evaporation will create osmotic gradients tending to draw liquid into the airway lumen. Research (Widdicombe, 1999) suggests, however, that in the absence of evaporation airway surface liquid may be markedly hypotonic with Na and Cl concentrations of around 50 mM. It has been suggested that when present as a thin film, airway surface liquid is prevented from moving down the transepithelial osmotic gradient by forces of surface tension generated by the cilia and other surface structures (e.g., mucous gel). In a cross-section through the ciliary bed, the total ciliary perimeter amounts to 600 m/cm^2 of epithelium. It has been calculated that this should generate a hydrostatic pressure of 2 atmospheres, which is equivalent to a transepithelial solute concentration gradient of 40 mOsm (Widdicombe, 1988). Thus, the presence of cilia ensures the presence of a periciliary sol.

Secretory forces come to predominate during inflammation (Fig. 2.3). Chloride secretion may be induced in the surface epithelium. Submucosal glands also secrete in response to inflammatory mediators and neurohumoral agents (Quinton, 1979; Finkbeiner and Widdicombe, 1992). Individual cat submucosal glands stimulated with cholinergic or α-adrenergic agents secrete at 10–20 nl/minute (Quinton, 1979). Assuming similar flow rates in human glands, then the gland density of 1/mm^2 (Tos, 1966) predicts a maximal gland secretion of 60–120 $\mu l \cdot cm^{-2} \cdot h^{-1}$ sufficient to increase the depth of airway surface liquid by 10–20 μm/minute (Wu et al., 1998). Extravasation and subepithelial edema, however, may be the dominant factors determining liquid flows across airway epithelium during inflammation. The presence of elevated subepithelial

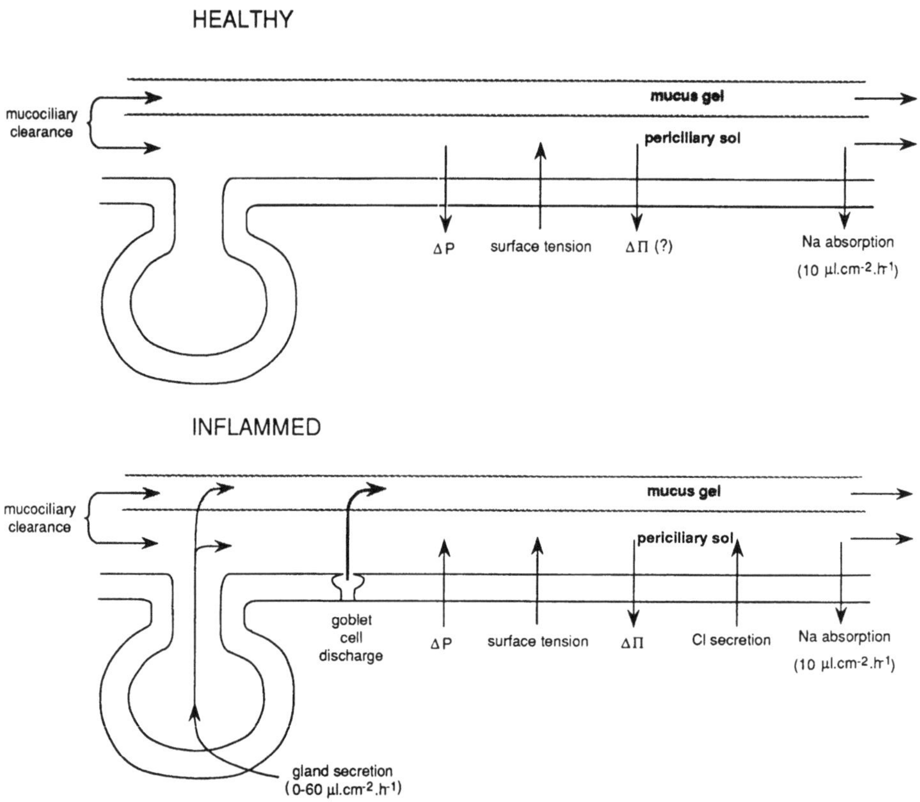

FIGURE 2.3. Routes of water movement across airway epithelium. Where possible, estimates of volume flows are given. [*Source*: Reproduced with permission from Widdicombe, in *Asthma and Rhinitis.* Busse, W.W. and Holgate, S.T. (eds), Blackwell, Cambridge, MA, 1993.]

hydrostatic pressure in inflamed airways is revealed by markedly dilated lateral intercellular spaces (LIS) (Laitinen et al., 1987); the same dilation is produced *in vitro* by small (5–20 cm H_2O) hydrostatic pressure gradients directed from the serosal to mucosal side of the epithelium (Kondo et al., 1992). In addition to dilating the LIS, such transepithelial pressure gradients disrupt tight junctions, thereby greatly increasing the permeability to water and macromolecules. The resulting bulk flows of liquid from serosa to mucosa are comparatively large and account for the presence of blood proteins in the lumen of inflamed airways.

Increased secretion of liquid in inflamed airways will tend to increase the depth of both the periciliary sol layer and the mucus blanket, thereby moving a thicker mucus blanket further from the cilia. Movement of mucus by ciliary beating will accordingly be less efficient, and mucus will increasingly move in response to gravity and airflow, resulting in its gradual accumulation.

Particles or noxious chemicals landing on the airway surface may cause reflexes or inflammatory mediator release in the immediate vicinity of the irritation. The resulting local secretion of liquid by glands and surface epithelium would "flush out" the source of irritation, and the excess liquid would be absorbed by Na absorption in the neighboring, unirritated regions.

Ion Transport

Active transport of ions and other solutes between the lumen and the subepithelial (serosal) space is one of the major functions of epithelia. Because of their potential role in regulating airway liquid content, the active ion transport processes of airway epithelia have been studied extensively (Welsh, 1987; Finkbeiner and Widdicombe, 1992; Widdicombe, 1996).

Ussing's short-circuit current technique is frequently the first approach to determining the types of active ion transport operating across a flat sheet of epithelium. The sheet is mounted between lucite half chambers so that the liquids bathing its mucosal (apical) and serosal (basolateral) surfaces are separated. Current flowing through an external circuit is adjusted to maintain the transepithelial potential difference at zero. This current is known as the short-circuit current (I_{sc}), and is equal to the sum of all the active transport processes operating across the tissue. To determine what active ion transport processes are responsible for the I_{sc}, one can employ specific transport blockers, ion substitution or measure transepithelial fluxes of radioactive tracers (Ussing and Zerahn, 1951).

Table 2.2 lists the predominant ion transport processes of airway epithelia, as determined from Ussing chamber studies on short-circuited tissues. Many epithelia show only active absorption of Na; Cl secretion cannot be induced by mediators. In others, active secretion of Cl is the major ion transport process. There is a tendency for larger airways to show Cl secretion rather than Na absorption, and a switch from Cl

TABLE 2.2. Major active ion transport processes of airway epithelia (under short-circuit conditions).

Only Cl Secretion
Fetal sheep trachea
Fetal dog trachea
Newborn dog trachea
Only Na Absorption
Pig nasal
Human nasal
Human bronchus
Dog bronchus
Pig bronchus
Human trachea
Sheep bronchus
Sheep trachea
Monkey trachea
Guinea-pig trachea
Cl Secretion and Na Absorption
Dog mainstem bronchus
Rabbit trachea
Ferret trachea
Cat trachea
Ox trachea
Dog trachea
Horse trachea

Source: For full references, see Finkbeiner and Widdicombe, 1992.

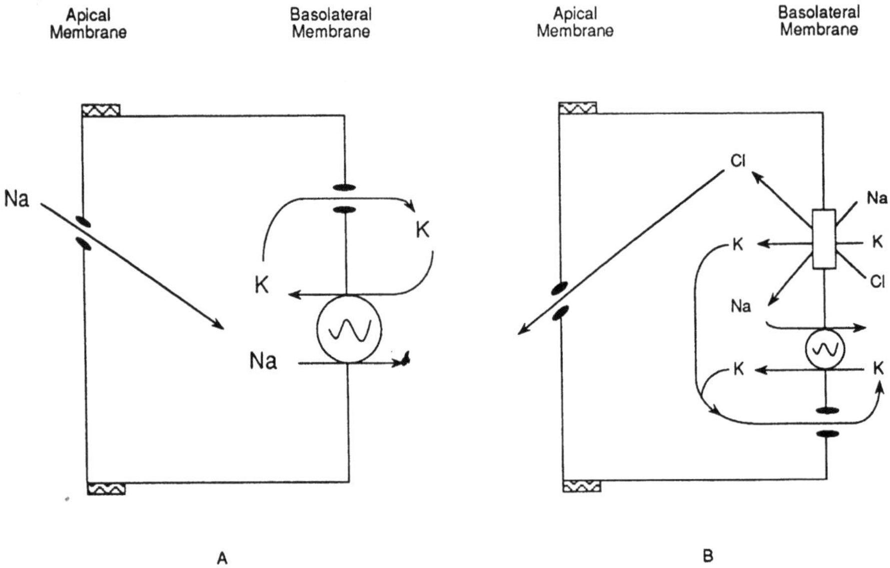

FIGURE 2.4. Major active ion transport processes of airway epithelia. (A) Cl secretion. (B) Na absorption. See text for details.

secretion to Na absorption during development. Chloride secretion may reflect airway inflammation as it is abolished by indomethacin, the inhibitor of prostaglandin synthesis. In healthy, uninflamed adult human airways, Na absorption may be the only major type of active ion transport present.

It is important to remember, however, that the active secretion of Cl seen under short-circuit conditions in Ussing chambers may be absent *in vivo*. Thus, the presence of a lumen negative transepithelial potential difference of approximately 30 mV depolarizes the apical membrane and may reverse the driving force for Cl movement across the apical membrane from outward (under short-circuit) to inward. Opening of Cl channels under these conditions can result in increased absorption of NaCl and water (Uyekubo et al., 1998).

Directional transport of ions by epithelia requires the presence of different transport proteins in their apical and basolateral membranes. The apical membrane of airway epithelia contains Na and Cl channels, but has negligible K conductance. The basolateral membrane, however, is K-selective containing cAMP- and Ca-activated K channels. The Na^+-K^+-ATPase and a Na^+-K^+-$2Cl^-$ cotransporter are also restricted to the basolateral membrane.

The way that this polarization of transport proteins brings about active secretion of Cl is illustrated in Figure 2.4A. Entry of Cl across the basolateral membrane is by cotransport with Na (and K). The energy in the transmembrane concentration gradient for Na allows Cl to be accumulated within the cells to a level greater than that predicted for passive distribution according to the apical membrane potential difference. There is therefore a net exit of Cl across this membrane, which is Cl-selective. The Na that enters by cotransport with Cl is removed from the cells by basolaterally located Na^+-K^+-ATPase. The ATPase also pumps in K, which, together with the K entering by Na^+-K^+-$2Cl^-$ cotransport, recycles through the basolateral membrane K channels.

The mechanism of active Na absorption by airway epithelia is illustrated in Figure 2.4B. Net entry of Na occurs via amiloride-sensitive channels in the apical membrane down both chemical and electrical gradients. The Na that enters is then extruded across the basolateral membrane by the Na^+-K^+-ATPase restricted to this membrane. As with Cl secretion, the K pumped in by the Na^+-K^+-ATPase recycles across the K-selective basolateral membrane.

Na absorption by airway epithelia is not under neurohumoral control. By contrast, chloride secretion is stimulated by a wide range of mediators (Widdicombe, 1991) acting predominantly through either Ca (e.g., bradykinin) or cAMP (e.g., isoproterenol). There is evidence that either second messenger can activate apical membrane Cl channels, basolateral K channels, and the Na^+-K^+-$2Cl^-$ cotransporter. When the level of active Cl secretion changes, it is important that the turnover rates of the different transport proteins alter in parallel so as to avoid marked changes in intracellular electrolyte (and water) content. Stimulation of Cl secretion by cAMP, for instance, will increase the turnover of the Na^+-K^+-$2Cl^-$ cotransporter, thereby increasing entry of Cl across the basolateral membrane. cAMP-dependent phosphorylation of apical membrane Cl channels occurs simultaneously and increases the amount of time they spend open, thereby matching the increased influx of Cl across the basolateral membrane with an increase in efflux across the apical membrane. The increased influx of Na on the Na^+-K^+-$2Cl^-$ cotransporter elevates $[Na]_i$ and stimulates the Na^+-K^+-ATPase so that pumped efflux of Na matches the increased influx on the Na^+-K^+-$2Cl^-$ cotransporter. (At normal levels of $[Na]_i$, the relation between Na^+-K^+-ATPase activity and $[Na]_i$ is very steep, and the necessary increase in Na^+-K^+-ATPase activity is achieved with little change in $[Na]_i$.) Increased turnover of the Na^+-K^+-ATPase and the Na^+-K^+-$2Cl^-$ cotransporter brings more K into the cells. Opening of basolateral K channels, however, causes a compensatory increase in K efflux. Thus, an increase in net Cl secretion is achieved with minimal changes in intracellular ion contents.

Although ion transport processes have been shown to mediate liquid transport across airway epithelia, it should be noted that there is as yet little firm evidence that changes in active ion transport alter mucociliary clearance.

Macromolecule Secretion

Three types of cells in the surface epithelium secrete macromolecules into the airway mucous layers (Basbaum and Finkbeiner, 1991). Goblet cells in the larger airways secrete mucins. Clara cells in the smaller airways contribute a variety of proteins including surfactant and antiproteases. The secretory product of serous cells is unknown, but by analogy with gland serous cells (Basbaum et al., 1990), they may secrete water and bactericidal proteins.

Airway goblet cell discharge has been studied by cytochemistry, and by measurement of radioactively-labeled high molecular weight glycoproteins released from explant cultures of airways that lack glands (e.g., hamster and guinea-pig tracheas) (Finkbeiner and Widdicombe, 1992). Most such studies have concluded that discharge of mucous granules is not under autonomic control, but can be induced by several irritants such as tobacco smoke, sulfur dioxide, ammonia vapor, and bacterial proteases (Kim, 1997). *In vivo* studies and the use of explant cultures, however, are complicated by the presence of multiple tissue types. For instance, irritants may act by stimulating sensory nerve terminals and releasing tachykinins by axon reflexes. A direct action of mediators on goblet cell secretion can be revealed only by using epithelial cell cultures. In goblet cells of cultured hamster trachea, cholinergic and adrenergic agents, prostaglandins, leukotrienes, histamine, serotonin, and permeant analogues of cAMP are without effect on mucus release (Kim, 1994). By contrast, cationic proteases, ATP, and changes in pH and osmolarity are all potent secretagogues (Kim, 1994).

Clara cells (Plopper et al., 1997) secrete pulmonary surfactant apoproteins, antileukoproteases, and a "Clara Cell Secretory Protein" (~10 kDa) of unknown function, which is homologous with uteroglobulin of rabbit uterus. Clara cell secretory granules are discharged by β-adrenergic agents and prostaglandins, but not by cholinergic agents. The agranular endoplasmic reticulum of Clara cells is rich in cytochrome P-450 monooxygenase that serves to metabolize harmful compounds.

Ciliary Motion

The structure of airway cilia and their mechanism of beating have been described in detail (Sleigh et al., 1988). In the airways, cilia are about 6 μm long and 250 nm in diameter, with a center-to-center spacing of approximately 250 nm. They beat at rates of up to 30 Hz in coordinated waves in fields of a few to several hundred cells. The ciliary beat cycle is illustrated in Figure 2.5. In the effective stroke, the cilia beat toward the mouth perpendicular to the airway surface. Their tips contain small claws (~30 nm) that contact the underside of the mucus blanket. At the end of the effective stroke there is a variable pause, the "rest phase," during which the cilia lie flat on the airway surface. During the recovery stroke the cilia move backward and to the right parallel to the airway surface till they return to the position corresponding to the start of the effective stroke. The duration of recovery stroke and the rest phase are approximately equal (Lansley et al., 1992), but the duration of the effective stroke is much shorter. In moving backward during their recovery strokes, the cilia hit the tips of resting cilia, stimulating them to begin their recovery strokes. Thus, the wave of ciliary activity passes

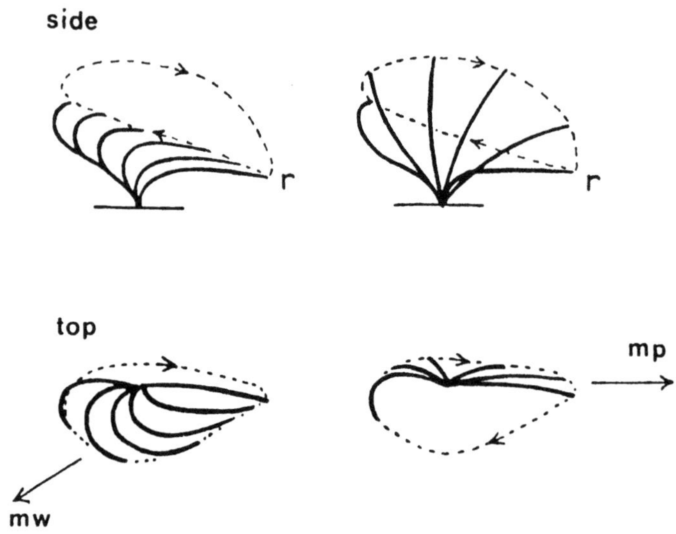

FIGURE 2.5. The beat cycle of an airway cilium seen from the side (top panels) and from above (bottom panels). In the recovery stroke (left panels), the cilium starts from the rest position (R) and moves backward in a clockwise direction. In the effective stroke (right panels), the cilium moves in a plane perpendicular to the airway surface, becoming maximally extended at midstroke. The metachronal wave of ciliary beating (mw) passes at an angle of 135 degrees to the direction of the effective stroke (see text for explanation). (*Source*: Reproduced with permission from Sleigh et al., *Am. Rev. Resp. Dis.* 137:726–741, 1988.)

backwards at an obtuse angle (135 degrees) to the right of the effective stroke (Sanderson and Sleigh, 1981).

Ciliary beat frequency is changed mainly by reducing the durations of the rest phase and recovery strokes (Lansley et al., 1992). At high rates of beating, the duration of the effective stroke may also be shortened. Neurohumoral mediators increase ciliary beat frequency by elevating either intracellular Ca or cAMP (Sanderson and Dirksen, 1989). The most potent stimulus, however, is mechanical deformation. Sanderson and colleagues (Sanderson et al., 1990) have shown that deformation of the apical membrane of a single cell raises Ca and increases the ciliary beat frequency in that cell, and then in the adjacent cells after a delay of around 0.5 seconds. A wave of increased $[Ca^{2+}]_i$ and ciliary beating eventually spreads for long distances over the epithelium. It is believed that these waves are generated by passage of IP_3 between cells via gap junctions (Boitano et al., 1992).

Mediator Release and Breakdown

In response to mediators such as bradykinin or platelet-activating factor, airway epithelium releases a range of arachidonic acid metabolites (Widdicombe, 1991). Of these, prostaglandin E_2 is quantitatively the most important, and it is released predominantly across the basolateral rather than the apical membrane. It may have a relaxing effect on underlying smooth muscle. Leukotriene B_4 is also released, and could be responsible for neutrophil migration in inflammation. A number of other arachidonic acid metabolites are released from dispersed isolated cells from airway epithelium, but the relevance of these results to intact tissue is highly questionable.

In several species, including humans, mechanical removal of the airway epithelium has been shown to increase the sensitivity of airway smooth muscle to a variety of contractile agents (Goldie and Fernandes, 1991). This has led to the suggestion that the epithelium produces an "epithelium-derived relaxing factor" (EpDRF). Reduced release of PGE_2 is at least partly responsible for the effects of epithelial removal in some species; however, there is evidence for relaxing factors in addition to this mediator. Their exact identity is unknown, though nitric oxide, the relaxant factor released by endothelium, does not seem to be one of them (Munakata et al., 1990).

Neuroendocrine cells (Scheuermann, 1997) contain a variety of mediators that are released across the basolateral membrane in response to hypoxia and other stimuli. Serotonin and gastrin-releasing peptide are quantitatively the most important of the released compounds, which may also include calcitonin, calcitonin gene-related peptide, endothelin, leu-enkephalin, and somatostatin.

Finally, within the last few years, airway epithelium has been shown to release a number of growth factors and cytokines, including interleukin-8 (a potent chemotactic agent for neutrophils), endothelin, epidermal growth factorlike growth factor, transforming growth factor beta 2, secretory component, granulocyte/macrophage colony stimulating factor, interleukin-6, and fibronectin (Robbins and Rennard, 1997).

An enzyme, enkephalinase, is found on the membranes of airway epithelial cells that degrades peptides released from sensory nerve terminals (Nadel, 1992). Inhibition of this enzyme with thiorphan *in vitro* increases the effects of substance P on gland secretion, smooth muscle contraction, and ciliary beat frequency. There is evidence that levels of epithelial neutral endopeptidase are reduced during inflammation. This would lead to elevated levels of peptides, which, will contribute to airway hyperreactivity by acting on sensory nerves and smooth muscle. Airway epithelial cell membranes also contain histamine N-methyltransferase, and may play a role in the breakdown of histamine (Ohrui et al., 1992).

Pathology

Noxious gases, viral infections, and inflammatory mediators all cause structural and functional changes in airway epithelium (Basbaum and Jany, 1990). The first sign of damage is often ciliary discoordination. In healthy airways, all ciliary fields beat at similar frequencies toward the mouth. In damaged epithelium, however, the beat frequency becomes increasingly variable, and some fields may even start to beat caudally (see Fig. 2.6). These changes may progress to complete cessation of ciliary motion, deciliation, and cell sloughing. Epithelia in which there has been a significant amount of cell death caused, for instance, by viral infections or mechanical abrasion, assume a deciliated, squamous histology. The other major type of ultrastructural change, which is typical of chronic bronchitis induced by smoking, is goblet cell hyperplasia in which the numbers of goblet cells increase at the expense of ciliated cells.

In the large airways, return to normal pseudostratified histology is brought about by division and differentiation of both basal (Inayama et al., 1988) and small mucous granule cells (Keenan et al., 1983). Clara cells are the progenitor cell responsible for renewal of bronchiolar epithelium (Evans et al., 1978).

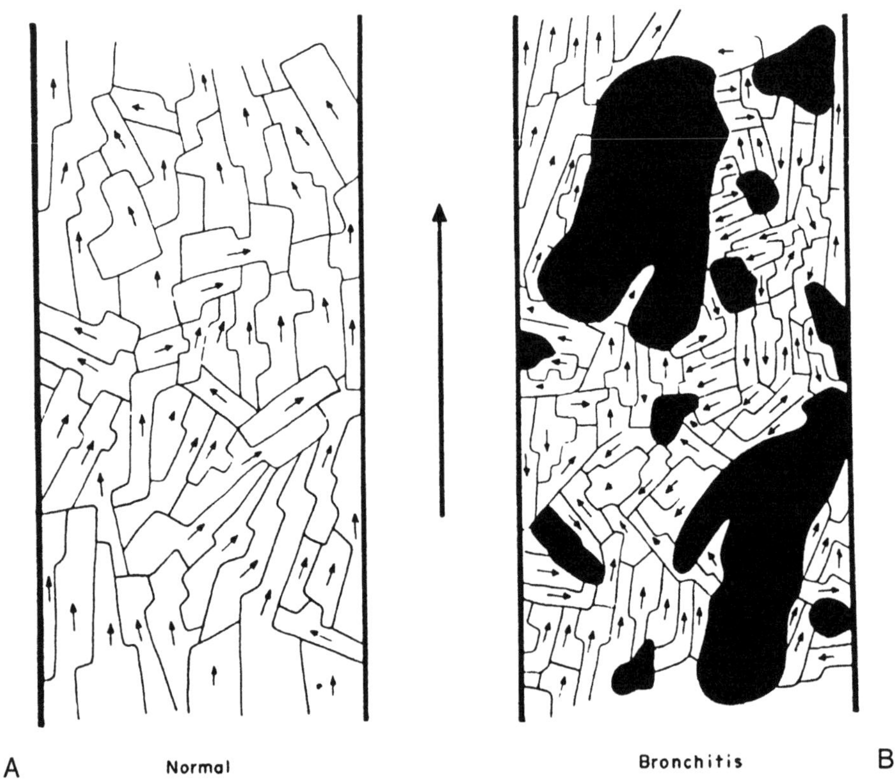

FIGURE 2.6. Alterations in mucociliary clearance in chronic bronchitis. (A) Ciliary fields in the trachea of a healthy rat. The small arrows show the direction of mucus transport in each field, and the large arrow points cranially. (B) Ciliary fields in an animal with bronchitis. Black areas indicate fields in which cilia are not beating. Note several fields now beat caudally. (*Source*: Reproduced with permission from Iravani and Melville, *Pharmacol. Ther.* 2:493–509, 1976.)

Sputum from asthmatics contains characteristic sheets of apparently normal airway epithelium known as *Creola bodies* (Naylor, 1962). Two mechanisms have been proposed to account for this shedding of epithelium. First, it may be caused by major basic protein of eosinophils, which is found in cytotoxic concentrations in the sputum of asthmatics (Frigas and Gleich, 1986). Second, elevated subepithelial hydrostatic pressures created by inflammatory exudates could break tight junctions and dislodge epithelial cells (Kondo et al., 1992). The degree of epithelial denudation in asthma has unfortunately never been quantified.

Cystic fibrosis (CF) is a genetic disease afflicting 1 in 2000 Caucasians. It also occurs at considerably lower frequencies in blacks and Asians. It is characterized by abnormal ion and water transport in many epithelia, although in 95% of cases death is due to occlusion of the airways with typically tenacious sticky mucous secretions (Boat et al., 1989). The gene responsible for CF has been cloned (Riordan et al., 1989), and it has been shown to encode for a 185 kDa protein (the cystic fibrosis transmembrane conductance regulator, or CFTR). The predicted structure of this protein contains 12 membrane-spanning segments, two nucleotide-binding folds, and an R-domain containing multiple sites for phosphorylation by protein kinases A and C. When expressed in a variety of cell types, CFTR induces a cAMP-activated C1 channel with biophysical properties identical to channels described earlier from human pancreatic, intestinal, and airway epithelia (Fuller and Benos, 1992).

In CF, the lack of functional CFTR in the apical membrane of airway epithelia results in reduced permeation of chloride (Widdicombe et al., 1985; Widdicombe, 1999). In addition, active absorption of Na is elevated (Boucher et al., 1986), although the relationship between this change and a defect in Cl channels is not understood. The exact mechanisms by which alterations in Cl and Na transport lead to airway pathology in CF are currently an area of much debate. One hypothesis states that failure of transcellular Cl movement results in higher than normal levels of Na and Cl in airway surface liquid. These elevated concentrations in turn inhibit the activity of endogenous microbials secreted by the epithelium, thereby promoting colonization by *Pseudomonas* (Widdicombe, 1999). A second hypothesis (Matsui et al., 1998) is that increased Na absorption in CF coupled with a predominantly paracellular movement of Cl leads to increased absorption of water from the airway lumen, a higher mucin concentration in airway surface liquid, greater viscosity of the mucous gel, inhibition of mucociliary clearance, and colonization of the accumulating mucus by *Pseudomonas* (see Fig. 2.7).

Submucosal Glands

Structure and Development

Although submucosal glands are absent from the airways of most small mammals, they provide the major source of the mucus needed for effective mucociliary clearance of inhaled particles in the larger airways of large mammals. In human trachea, for instance, the volume of secretory tissue in the glands has been estimated as 40 times that in the surface goblet cells (Reid, 1960).

The human lung starts as a ventral protrusion of the foregut about 4 weeks after fertilization. By 7 weeks the segmental bronchi are formed, and all the bronchi and bronchioles are formed at 16 weeks (Burri, 1985). Gland differentiation spreads centrifugally, with buds appearing at 10–12 weeks of fetal life, and secreting acini at 18 weeks (Bucher and Reid, 1961). By 25–28 weeks, the maximal number of between 3500

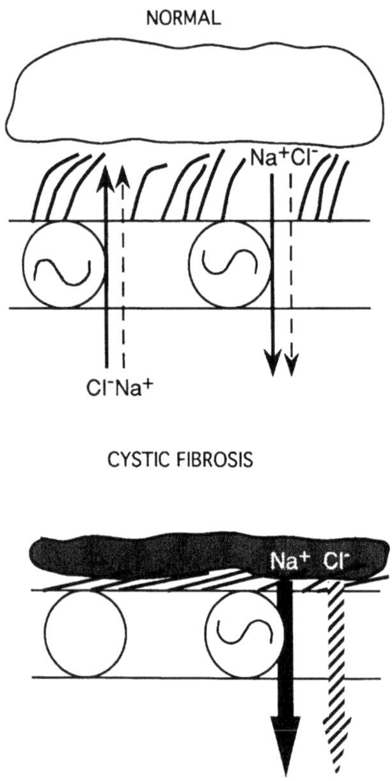

FIGURE 2.7. Schematic representation of airway epithelium during healthy conditions and during cystic fibrosis. Normally, the degree of hydration of the mucus gel is set by the balance between active CI secretion and active Na absorption. In cystic-fibrosis, CI secretion is absent and Na absorption is elevated. These changes will tend to remove the periciliary sol and condense the mucous blanket.

and 6000 glands is present in the trachea (Tos, 1966). At birth, this corresponds to a gland density of ground 10/mm², which declines with growth to an adult density of approximately 1/mm². Human airway glands are essentially completely differentiated at birth, and are evenly distributed along and around the trachea. The numbers of gland openings per unit surface area of human airways declines with increasing airway generation (Whimster, 1986), and they are absent from the bronchioles.

Figure 2.8 illustrates a human bronchial gland reconstructed from 4 μm serial paraffin sections (Meyrick et al., 1969). The gland opens onto the surface epithelium via a ciliated duct about 0.5 mm in length, which expands to form a "collecting duct," again about 0.5 mm long. The latter lacks ciliated cells, and receives about 15 major secretory tubules, which average 500 μm in length, and are lined with mucous cells. These tubules branch two or three times, with the branches ending in acini lined with serous cells. The external diameter of these secretory tubules is 50–75 μm, and the height of their cells averages 20–30 μm.

The apical pole of mature mucous cells is distended by electron-lucent granules, varying in diameter from 300–1800 nm. These granules show a tendency to fuse with one another and with the apical membrane. By contrast, serous cells contain smaller (300–1000 nm diameter) electron-dense granules, which do not distend the apical portion of the cell nor fuse with the apical membrane (Meyrick and Reid, 1970). Cross-sections through mucous tubules and serous acini of ferret airway glands are illustrated in Figure 2.9.

A prominent feature of the serous cells is their intercellular canaliculi (Meyrick and Reid, 1970). These tubes of about 1 μm diameter are extensions of the acinar lumen that pass between the lateral aspects of the cells. Apical membrane, which contains abundant microvilli, lines these canaliculi, and is separated from the cells' basolateral membranes by typical tight junctions. It is generally assumed that the canaliculi contribute to serous cell liquid secretion by amplifying the apical membrane. Canaliculi are accordingly sometimes referred to as "secretory capillaries."

The acini and tubules of the submucosal glands are invested with a covering of typical myoepithelial cells that lie in grooves in the basal aspect of the epithelial cells. In cat glands, contraction of these myoepithelial cells plays an important role in secretion (Shimura et al., 1986). Mitochondria-rich cells are found in the gland (Meyrick and Reid, 1970), and it has been suggested that they transport liquid. In humans, however, they average only about 6% of the duct cells, and their numbers increase with age (Matsuba et al., 1972). It seems likely, therefore, that they are oncocytes, which is a cell type found in small numbers in virtually all epithelia. Neuroendocrine cells and lymphocytes are also occasionally found in the glandular epithelium.

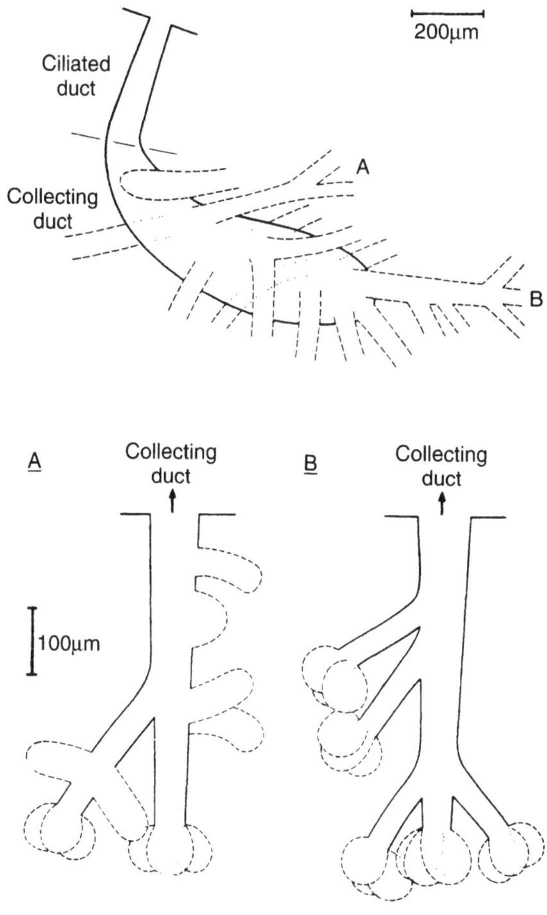

FIGURE 2.8. Reconstruction of human tracheal gland from serial cross-sections. (Top) The collecting duct and its major branches. (Bottom) Secretory tubules. Areas lined by mucous cells are shown by solid lines; hatched lines denote serous cells. The location of the secretory tubules in the gland as a whole are indicated by the labels A and B in the top panel. (*Source*: Reproduced with permission from Meyrick et al., *Thorax* 24:729–736, 1969.)

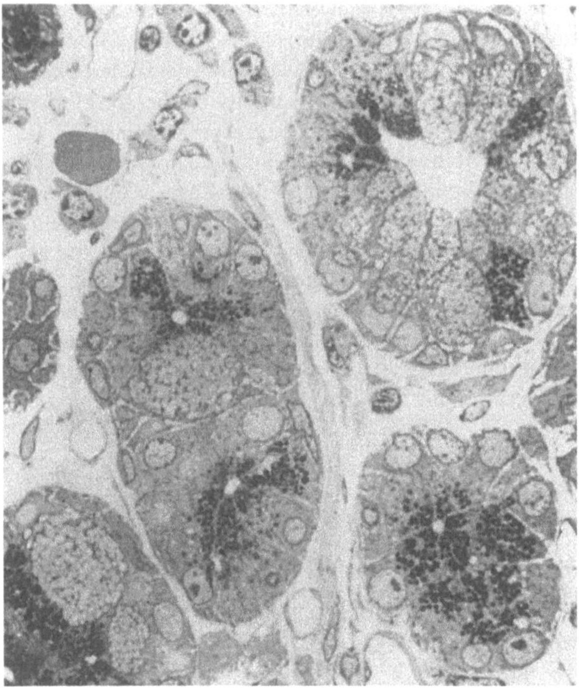

FIGURE 2.9. Section through ferret tracheal glands showing profiles of serous acini (s) and mucous tubules (m). Scale bar = 10 μm. (*Source*: Reproduced with permission from Basbaum, *Clin. Chest Med.* 7:231–237, 1986.)

Function

Ion and Liquid Transport

The distal location of serous relative to mucous cells suggests that the former secrete liquid that helps "flush out" mucus secreted by the latter. All acini possess a least a "demi-lune" of serous cells. There is other evidence for secretion of water by serous cells (Widdicombe et al., 1994). First, α-adrenergic agents selectively degranulate serous cells (Basbaum et al., 1981) and produce secretions that are more watery than are cholinergically induced secretions, which have lower viscosity and protein content (Leikauf et al., 1984). Second, serous cells contain much more CFTR than do mucous cells (Engelhardt et al., 1992), and they show higher levels of Cl secretion (Finkbeiner et al., 1994), and therefore presumably also water secretion.

Isolated swine airway gland cells accumulate Cl by a process inhibited by loop diuretics, inhibitors of Na^+-K^+-$2Cl^-$ cotransport (Yang et al., 1988). Thus, active transport of water by gland acini is presumably driven by the same Na-linked Cl secretion as found in surface epithelium. Cultured human gland cells in Ussing chambers secrete Cl in response to mediators in the potency sequence: methacholine > bradykinin > isoproterenol > phenylephrine (Yamaya et al., 1991).

Macromolecule Secretion

Cytochemical (Spicer et al., 1983) and immunocytochemical data (Perini et al., 1989) show that mucous cells contain mucins. By contrast, mucin core protein is scarce or undetectable in serous cells (Perini et al., 1989). Serous cells are instead a source of endogenous antibiotics, and additionally secrete a number of compounds that protect the airway surface from the harmful effects of bacterial and neutrophilic proteases (Basbaum et al., 1990). The basolateral membrane of serous cells contains receptors for secretory IgA, which is taken up by the cells and secreted into the gland lumen (Goodman et al., 1981). Finally, proteoglycans, which help package secretory granule contents, are released by serous cells and may play a role in mucus hydration and rheology (Basbaum et al., 1990). In short, most of the many nonmucin proteins in airway secretions probably originate from serous cells.

In vitro, release of high molecular weight $^{35}SO_4$-labeled material from pieces of airway wall is generally taken to represent gland secretion. Using this, and other, approaches several groups have found cholinergic agents to be more potent gland secretagogues than either α- or β-adrenergic agents (reviewed in Finkbeiner and Widdicombe, 1992). Other agents shown to stimulate mucus secretion from human airways include arachidonic acid, PGA_2, PGD_2, $PGF_{2\alpha}$, monoHETE, LTC_4, LTD_4, PGF-A, and histamine. Neutrophil elastase is a potent secretagogue of immortalized bovine serous gland cells (Sommerhoff et al., 1990). It also causes degranulation of mucous gland cells (Schuster et al., 1992).

Pathology

Increased airway mucus output is the hallmark of chronic bronchitis, and is associated with marked gland hypertrophy (Reid, 1960). There is also a conversion of gland serous cells into mucous cells.

Death from asthma is due more to occlusion of airways with mucous secretions than it is to contraction of smooth muscle (Dunnill et al., 1969). The increased mucin secretion is associated with gland hypertrophy as well as conversion of serous to mucous cells, although not to the same degree as in chronic bronchitis.

In advanced CF, submucosal glands show a degree of hypertrophy that matches that of chronic bronchitis (Sturgess and Imrie, 1982). In fact, the earliest detectable pulmonary manifestation of this disease is a dilation of gland lumina (Sturgess and Imrie, 1982). Further support for a role of submucosal glands in the airway pathology of CF is provided by the findings that secretion of Cl and water by cultures of human airway glands are greatly reduced in CF (Yamaya et al., 1991; Jiang et al., 1997). Thus, gland secretions are presumably comparatively dehydrated and viscous in this disease. They may therefore fail to detach from the gland openings or be poorly cleared by the cilia. Thus, the initial build up of airway mucus secretions in CF may be caused primarily by changes in function of the airway glands rather than the surface epithelium (see Fig. 2.10).

All three diseases discussed earlier share a common feature: The mucus-secreting apparatus of the glands increases relative to the liquid-secreting component. In CF this is due to a failure of CFTR in serous cells. In asthma and chronic bronchitis, it is caused by a conversion of serous to mucous cells. Nevertheless, one would predict relatively concentrated (i.e., dehydrated) gland secretions in all three diseases, which will presumably be of higher-than-normal viscosity, poorly cleared by the cilia, and therefore gradually accumulate.

Summary

The surface of the airways is bombarded by particles in the inspired-air; however, before these particles can contact and damage the epithelium they are trapped in a blanket of mucous that lines the airway surface. The beating of cilia in the apical mem-

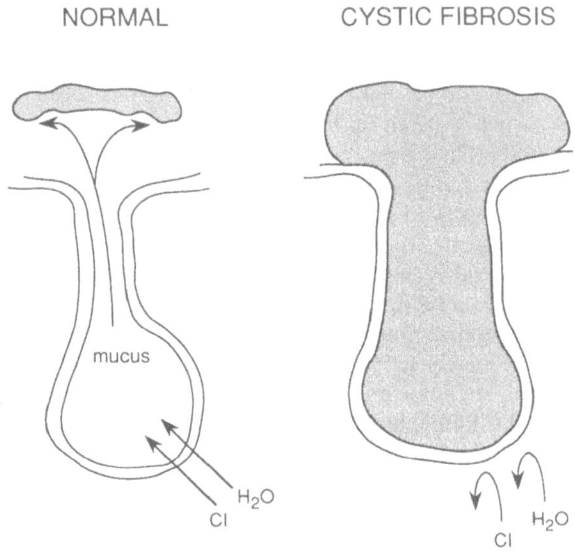

FIGURE 2.10. Role of glands in mucus accumulation in cystic fibrosis. In a normal gland, chloride secretion by serous cells in the acinus mediates the secretion of isotonic salt solution that flushes out mucins from the gland tubules. In cystic fibrosis, failure of acini to secrete Cl and water leads to a progressive build up of concentrated mucous secretions that distend the gland lumen. The dehydrated mucous secretions may fail to detach from the gland opening or may be inefficiently transported by the cilia, gradually accumulating on the airway surface.

branes of the airway epithelial cells propels the mucous blanket to the mouth, where it is swallowed or expectorated. The role of such "mucociliary clearance" in keeping the airways clean has been appreciated for many years. Since 1980, however, much new information has been obtained about the individual components of the system. Factors regulating ciliary beat frequency have been determined mainly by the work of Sanderson and colleagues (Sanderson and Dirksen, 1989). Methods for measuring mucus secretion have proliferated (Kim, 1994; Shimura and Takishima, 1994), providing increasingly sophisticated information on both gland and goblet cell function. Active ion transport by airway epithelium has been thoroughly characterized (Welsh, 1987; Widdicombe, 1996), and accurate measurements of transepithelial liquid flows have been made (Jiang et al., 1993; Jiang et al., 1997).

In addition to solid particles, the airway surface is exposed to many different gaseous toxins. The mucous gel doubtless provides some protection against these, but the finding of relatively high levels of cytochrome P-450 monooxygenase in Clara cells suggests that they may be the major defender against this type of attack (Plopper et al., 1997). It is also becoming clear that Clara cells secrete several protective compounds into the airway surface liquid.

Arachidonic acid metabolites, cytokines, and other mediators are all secreted by the airway epithelium. Some (e.g., ATP) are secreted into the airway lumen, where they may exert a paracrine influence on mucus secretion, ciliary motion and other functions (Grygorczyk and Hanrahan, 1997). Substances released across the basolateral membrane may have profound effects on the function of the underlying smooth muscle, fibroblasts, blood vessels and white blood cells. On the other hand, enzymes on the membranes of the epithelial cells play a role in terminating the action of mediators released by other cell types.

Cultures of airway epithelial cells have contributed greatly to our understanding of airway epithelial function, and have been used in studies of mucus secretion (Kim, 1994), ion transport (Widdicombe, 1990), ciliary motion (Sanderson and Dirksen, 1989) and mediator release (Robbins and Rennard, 1997). The best cultures of human surface epithelium are now virtually indistinguishable from native epithelium (Yamaya et al., 1992). Furthermore, cultures of specific cell types, such as Clara cells (Van Scott et al., 1989) or gland mucous cells (Finkbeiner et al., 1994), have been developed.

In conclusion, since 1980, the rapid development of new methods has greatly increased our understanding of airway epithelial function. The profound impact of this tissue on the function of surrounding cells and tissues has become appreciated. Finally, detailed information is now available on how malfunction of airway epithelium contributes to the pathology of CF, asthma, and chronic bronchitis.

References

Ayers, M.M., and Jeffery, P.K. (1988) Proliferation and differentiation in mammalian airway epithelium. Eur. Respir. J. 1, 58–80.

Basbaum, C., and Finkbeiner, W. (1991) Mucus production in the airways. In: Meetings on pathophysiology of pulmonary cells. Endothelial and mucus secreting cells, Junod, A., Olivieri, D., and Pozzi, E. (eds.) pp. 173–189. Masson, Milan.

Basbaum, C., and Jany, B. (1990) Plasticity in the airway epithelium. Am. J. Physiol. 259, L38–L46.

Basbaum, C.B., Jany, B., and Finkbeiner, W.E. (1990) The serous cell. Annu. Rev. Physiol. 52, 97–113.

Basbaum, C.B., Ueki, I., Brezina, L., and Nadel, J.A. (1981) Tracheal submucosal gland serous cells stimulated in vitro with adrenergic and cholinergic agonists: a morphometric study. Cell Tissue Res. 220, 481–498.

Bhaskar, K.R., O'Sullivan, D.D., Seltzer, J., Rossing, T.H., Drazen, J.M., and Reid, L.M. (1985) Density gradient study of bronchial mucus aspirates from healthy volunteers (smokers and nonsmokers) and from patients with tracheostomy. Exp. Lung Res. 9, 289–308.

Boat, T.F., Welsh, M.J., and Beaudet, A.L. (1989) Cystic fibrosis. In: The metabolic basis of inherited disease, Scriver, C.R., Beaudet, A.L., Sly, W.S., and Valle, D. (eds.) pp. 2649–2680. McGraw-Hill, New York.

Boitano, S., Dirksen, E.R., and Sanderson, M.J. (1992) Intercellular propagation of calcium waves mediated by inositol trisphosphate. Science 258, 292–295.

Boucher, R.C., Stutts, M.J., Knowles, M.R., Cantley, L., and Gatzy, J.T. (1986) Na transport in cystic fibrosis respiratory epithelia. Abnormal basal rate and response to adenylate cyclase activation. J. Clin. Invest. 78, 1245–1252.

Braga, P.C., and Allegra, L. (eds.) (1988) Methods in bronchial mucology. p. 407. Raven Press, New York.

Breeze, R.G., and Wheeldon, E.B. (1977) The cells of the pulmonary airways. Am. Rev. Resp. Dis. 116, 705–777.

Bucher, U., and Reid, L. (1961) Development of the mucous-secreting elements in human lung. Thorax 16, 219–225.

Burri, P.H. (1985) Development and growth of the human lung. In: Handbook of physiology. Section 3: the respiratory system, pp. 1–46. Fishman, A.P., Fisher, A.B., and Geiger, S.R. (eds.) American Physiological Society, Bethesda.

Diamond, J.M. (1979) Osmotic water flow in leaky epithelia. J. Membr. Biol. 51, 195–216.

Dunnill, M.S., Massarella, G.R., and Anderson, J.A. (1969) A comparison of the quantitative anatomy of the bronchi in normal subjects, in status asthmaticus, in chronic bronchitis, and in emphysema. Thorax 24, 176–179.

Engelhardt, J.F., Yankaskas, J.R., Ernst, S., Yang, Y., Marino, C.R., Boucher, R.C., et al. (1992) Submucosal glands are the predominant site of CFTR expression in the human bronchus. Nature Genetics 2, 240–247.
Evans, M.J., Cabral-Anderson, L.J., and Freeman, G. (1978) Role of the Clara cell in renewal of the bronchiolar epithelium. Lab. Invest. 38, 648–655.
Evans, M.J., and Plopper, C.G. (1988) The role of basal cells in adhesion of columnar epithelium to airway basement membrane. Am. Rev. Resp. Dis. 138, 481–482.
Finkbeiner, W.E., Shen, B.Q., and Widdicombe, J.H. (1994) Chloride secretion and function of serous and mucous cells of human airway glands. Am. J. Physiol. 267, L206–L210.
Finkbeiner, W.E., and Widdicombe, J.H. (1992) Control of nasal airway secretions, ion transport, and water movement. In: Treatise on pulmonary toxicology, vol. 1. Comparative biology of the normal lung, Parent, R.A. (ed.) pp. 633–657. CRC Press, Boca Raton.
Frigas, E., and Gleich, G.J. (1986) The eosinophil and the pathology of asthma. J. Allergy Clin. Immunol. 77, 527–537.
Fuller, C.M., and Benos, D.J. (1992) CFTR. Am. J. Physiol. 263, C267–C286.
Goldie, R.G., and Fernandes, L.B. (1991) Bioassay of airway epithelium-derived inhibitory factors. In: The airway epithelium, Farmer, S.G., and Hay, D.W.P. (eds.) pp. 505–526. Marcel Dekker, New York.
Goodman, M.R., Link, D.W., Brown, W.R., and Nakane, P.K. (1981) Ultrastructural evidence of trasport of secretory IgA across bronchial epithelium. Am. Rev. Resp. Dis. 123, 115–119.
Gruenert, D.C., Finkbeiner, W.E., and Widdicombe, J.H. (1995) Culture and transformation of human airway epithelial cells. Am. J. Physiol. 268, L347–L360.
Grygorczyk, R., and Hanrahan, J.W. (1997) CFTR-independent ATP release from epithelial cells triggered by mechanical stimuli. Am. J. Physiol. 272, C1058–C1066.
Harkema, J.R., Mariassy, A., St. George, J., Hyde, D.M., and Plopper, C.G. (1991) Epithelial cells of the conducting airways: a species comparison. In: The airway epithelium: physiology, pathology, and pharmacology, Farmer, S.G., and Hay, D.W.P. (eds.) pp. 3–39. Marcel Dekker, New York.
Huang, H.T., Haskell, A., and McDonald, D.M. (1989) Changes in epithelial secretory cells and potentiation of neurogenic inflammation in the trachea of rats with respiratory tract infections. Anat. Embryol. (Berl.) 180, 325–341.
Inayama, Y., Hook, G.E., Brody, A.R., Cameron, G.S., Jetten, A.M., Gilmore, L.B., et al. (1988). The differentiation potential of tracheal basal cells. Lab. Invest. 58, 706–717.
Jeffery, P.K. (1983) Morphologic features of airway surface epithelial cells and glands. Am. Rev. Resp. Dis. 128, S14–S20.
Jiang, C., Finkbeiner, W.E., Widdicombe, J.H., McCray, P.B., and Miller, S.S. (1993) Altered fluid transport across airway epithelium in cystic fibrosis. Science 262, 424–427.
Jiang, C., Finkbeiner, W.E., Widdicombe, J.H., and Miller, S.S. (1997) Fluid transport across cultures of human tracheal glands is altered in cystic fibrosis. J. Physiol. 501, 637–648.
Kawamata, S., and Fujita, H. (1983) Fine structural aspects of the development and aging of the tracheal epithelium of mice. Arch. Histol. Japon. 46, 355–372.
Kawanami, O., Ferrans, V.J., and Crystal, R.G. (1979) Anchoring fibrils in the normal canine respiratory system. Am. Rev. Resp. Dis. 120, 595–611.
Keenan, K.P., Wilson, T.S., and McDowell, E.M. (1983) Regeneration of hamster tracheal epithelium after mechanical injury. IV. Histochemical, immunocytochemical, and ultrastructural studies. Virchows Arch. 43, 213–240.
Kim, K., McCracken, K., Lee, B.C., Shin, C.Y., Jo, M.J., Lee, C.J., et al. (1997) Airway goblet cell mucin: its structure and regulation of secretion. Eur. Resp. J. 10, 2644–2649.
Kim, K.C. (1994) Epithelial cell goblet secretion. In: Airway secretion, Takishima, T., and Shimura, S. (eds.) pp. 433–450. Marcel Dekker, New York.
King, M., and Rubin, B.K. (1994) Rheology of airway mucus: relationship with clearance function. In: Airway secretion: physiological bases for the control of mucous hypersecretion, Takishima, T., and Shimura, S. (eds.) pp. 283–314. Marcel Dekker, New York.
Kondo, M., Finkbeiner, W.E., and Widdicombe, J.H. (1992) Changes in permeability of dog tracheal epithelium in response to hydrostatic pressure. Am. J. Physiol. 262, L176–L182.
Laitinen, L.A., Laitinen, A., and Widdicombe, J. (1987) Effects of inflammatory and other mediators on airway vascular beds. Am. Rev. Resp. Dis. 135, S67–S70.

Lansley, A.B., Sanderson, M.J., and Dirksen, E.R. (1992) Control of the beat cycle of respiratory tract cilia by Ca^{2+} and cAMP. Am. J. Physiol. 263, L232–L242.

Leikauf, G.D., Ueki, I.F., and Nadel, J.A. (1984) Autonomic regulation of viscoelasticity of cat tracheal gland secretions. J. Appl. Physiol. 56, 426–430.

Lucas, A.M., and Douglas, L.C. (1934) Principles underlying ciliary activity in the respiratory tract. II. A comparison of nasal clearance in man, monkey, and other mammals. Arch. Otolaryngol. 20, 518–541.

Matsuba, K., Takizawa, T., and Thurlbeck, W.M. (1972) Oncocytes in human bronchial mucous glands. Thorax 24, 181–185.

Matsui, H., Grubb, B.R., Tarran, R., Randell, S.H., Gatzy, J.T., Davis, C.W., et al. (1998) Evidence for periciliary liquid layer depletion, not abnormal ion composition, in the pathogenesis of cystic fibrosis airways disease. Cell 95, 1005–1015.

McDowell, E.M., Newkirk, C., and Coleman, B. (1985) Development of hamster tracheal epithelium. II. Cell proliferation in the fetus. Anat. Rec. 213, 448–456.

Meyrick, B., and Reid, L. (1970) Ultrastucture of cells in human bronchial submucosal glands. J. Anat. 107, 291–299.

Meyrick, B., Sturgess, J.M., and Reid, L. (1969) A reconstruction of the duct system and secretory tubules of the human bronchial submucosal gland. Thorax 24, 729–736.

Munakata, M., Masaki, Y., Sakuma, I., Ukita, H., Otsuka, Y., Homma, Y., and Kawakami, Y. (1990) Pharmacological differentiation of epithelium-derived relaxing factor from nitric oxide. J. Applied Physiol., 69, 665–670.

Nadel, J.A. (1992) Membrane-bound peptidases: endocrine, paracrine, and autocrine effects. Am. J. Resp. Cell Mol. Biol. 7, 469–470.

Naylor, B. (1962) The shedding of the mucosa of the bronchial tree in asthma. Thorax 17, 69–72.

Ohrui, T., Yamauchi, K., Sekizawa, K., Ohkawara, T., Maeyama, K., Sasaki, M., et al. (1992). Histamine N-methylltransferase controls the contractile response of guinea pig trachea to histamine. J. Pharm. Exp. Ther. 261, 1268–1272.

Perini, J.M., Marianne, T., Lafitte, J.J., Lamblin, G., Roussel, P., and Mazzuca, M. (1989) Use of an antiserum against deglycosylated human mucins for cellular localization of their peptide precursors: antigenic similarities between bronchial and intestinal mucins. J. Histochem. Cytochem. 37, 869–875.

Plopper, C.G., Alley, J.L., and Weir, A.J. (1986) Differentiation of tracheal epithelium during fetal lung maturation in the rhesus monkey macaca mulata. Am. J. Anat. 175, 59–71.

Plopper, C.G., Hyde, D.M., and Buckpitt, A.R. (1997) Clara cells. In: The lung: scientific foundations, Crystal, R.G., West, J.B., Weibel, E., and Barnes, P.J. (eds.) pp. 517–534. Lippincott-Raven, Philadelphia.

Quinton, P.M. (1979) Composition and control of secretions from tracheal bronchial submucosal glands. Nature 279, 551–552.

Reid, L. (1960) Measurement of the bronchial mucous gland layer: a diagnostic yardstick in chronic bronchitis. Thorax 15, 132–141.

Riordan, J.R., Rommens, J.M., Kerem, B.-S., Alon, N., Rozmahel, R., Grzelczak, Z., et al. (1989) Identification of the cystic fibrosis gene: cloning and characterization of complementary DNA. Science 245, 1066–1073.

Robbins, R.A., and Rennard, S.I. (1997) Biology of airway epithelial cells. In: The lung: scientific foundations, Crystal, R.G., West, J.B., Barnes, P.J., and Weibel, E.R. (eds.) pp. 445–457. Lippincott-Raven, Philadelphia.

Rogers, A.V., Dewar, A., Corrin, B., and Jeffery, P.K. (1993) Identification of serous-like cells in the surface epithelium of human bronchioles. Eur. Respir. J. 6, 498–504.

Sanderson, M.F., and Dirksen, E.R. (1989) Mechanosensitive and beta-adrenergic control of the ciliary beat frequency of mammalian respiratory tract cells in culture. Am. Rev. Resp. Dis. 139, 432–440.

Sanderson, M.J., Charles, A.C., and Dirksen, E.R. (1990) Mechanical stimulation and intercellular communication increases intracellular Ca^{2+} in epithelial cells. Cell Regulation 1, 585–596.

Sanderson, M.J., and Sleigh, M.A. (1981) Ciliary activity of cultured rabbit tracheal epithelium: beat pattern and metachrony. J. Cell Sci. 47, 331–347.

Scheuermann, D. (1997) Neuroendocrine cells. In: The lung: scientific foundations, Crystal, R.G., West, L.B., Weibel, E.R., and Barnes, P.J. (eds.) pp. 603–613. Lippincott-Raven, Philadelphia.
Schuster, A., Ueki, I., and Nadel, J.A. (1992) Neutrophil elastase stimulates tracheal submucosal gland secretion that is inhibited by ICI 200,355. Am. J. Physiol. 262, L86–L91.
Sertl, K., Takemura, T., Tschachler, E., Ferrans, V.J., Kaliner, M.A., and Shevach, E.M. (1986) Dendritic cells with antigen-presenting capability reside in airway epithelium, lung parenchyma, and visceral pleura. J. Exp. Med. 163, 436–451.
Shimura, S., Sasaki, T., Sasaki, H., and Takishima, T. (1986) Contractility of isolated single submucosal gland from trachea. J. Appl. Physiol. 60, 1237–1247.
Shimura, S., and Takishima, T. (1994) Airway submucosal gland secretion. In: Airway secretion. Physiological bases for the control of mucous hypersecretion, Shimura, S., and Takishima, T. (eds.) pp. 325–398. Marcel Dekker, New York.
Silberberg, A. (1988) Models of mucus structure. In: Methods in bronchial mucology, Braga, P.C., and Allegra, L. (eds.) pp. 51–61. Raven Press, New York.
Sleigh, M.A., Blake, J.R., and Liron, N. (1988) The propulsion of mucus by cilia. Am. Rev. Resp. Dis. 137, 726–741.
Sommerhoff, C.P., Nadel, J.A., Basbaum, C.B., and Caughey, G.H. (1990) Neutrophil elastase and cathepsin G stimulate secretion from cultured bovine airway gland serous cells. J. Clin. Invest. 85, 682–689.
Sorokin, S.P., and Hoyt, R.F. (1989) Neuropepithelial bodies and solitary small-granule cells. In: Lung cell biology, Massaro, D. (ed.) pp. 191–344. Marcel Dekker, New York.
Spicer, S.S., Schulte, B.A., and Chakrin, L.W. (1983) Ultrastructural and histochemical observations of respiratory epithelium and gland. Exp. Lung Res. 4, 137–156.
St. George, J.A., Wang, S., and Plopper, C.G. (1994) Development of the airway secretory apparatus: patterns of mucous cell differentiation. In: Airway secretion. Physiological bases for control of mucous hypersecretion, Takishima, T., and Shimura, S. (eds.) pp. 123–147. Marcel Dekker, New York.
Sturgess, J., and Imrie, J. (1982) Quantitative evaluation of the development of tracheal submucosal glands in infants with cystic fibrosis and control infants. Am. J. Pathol. 106, 303–311.
Tos, M. (1966) Development of the tracheal glands in man. Acta Pathol. Microbiol. Scand. 185 (suppl.), 1–130.
Ussing, H.H., and Zerahn, K. (1951) Active transport of sodium as the source of electric current in short-circuited isolated frog skin. Acta Physiol. Scand. 23, 110–127.
Uyekubo, S.N., Fischer, H., Maminishkis, A., Miller, S.S., and Widdicombe, J.H. (1998) Cyclic AMP-dependent absorption of chloride across airway epithelium. Am. J. Physiol. 275, L1219–L1227.
Van Scott, M.R., Davis, C.W., and Boucher, R.C. (1989) Na^+ and Cl^- transport across rabbit nonciliated bronchiolar epithelial (Clara) cells. Am. J. Physiol. 256, C893–C901.
Wanner, A. (1986) Mucociliary clearance in the trachea. Clin. Chest Med. 7, 247–258.
Welsh, M.J. (1987) Electrolyte transport by airway epithelia. Physiol. Rev. 67, 1143–1184.
Widdicombe, J.G. (1988) Force of capillarity tending to prevent drying of ciliary mucosa. In: The airways, Kaliner, M.A., and Barnes, P.J. (eds.) p. 597. Marcel Dekker, New York.
Widdicombe, J.H. (1999) Altered NaCl concentration of airway surface liquid in cystic fibrosis. News. Physiol. Sci. 14, 126–127.
Widdicombe, J.H. (1996) Ion transport by airway epithelia. In: The lung: scientific foundations, Crystal, R.G., West, J.B., Weibel, E., and Barnes, P.J. (eds.) pp. 39.31–39.12. Raven Press, New York.
Widdicombe, J.H. (1991) Physiology of airway epithelia. In: The airway epithelium, Farmer, S.G., and Hay, D.W.P. (eds.) pp. 41–64. Marcel Dekker, New York.
Widdicombe, J.H. (1990) Use of cultured airway epithelial cells in studies of ion transport. Am. J. Physiol. 258, L13–L18.
Widdicombe, J.H., Shen, B.-Q., and Finkbeiner, W.E. (1994) Structure and function of human airway mucous glands in health and disease. Adv. Struct. Biol. 3, 225–241.
Widdicombe, J.H., Welsh, M.J., and Finkbeiner, W.E. (1985) Cystic fibrosis decreases the apical membrane chloride permeability of monolayers cultured from cells of tracheal epithelium. Proc. Natl. Acad. Sci. U.S.A. 82, 6167–6171.

Wright, N., and Alison, M. (1984) The biology of epithelial cell populations. Clarendon Press, Oxford.

Wu, D.X.-Y., Lee, C.Y.C., Uyekubo, S.N., Choi, H.K., Bastacky, S.J., and Widdicombe, J.H. (1998) Regulation of the depth of the surface liquid in bovine trachea. Am. J. Physiol. 274, L388–L395.

Yamaya, M., Finkbeiner, W.E., Chun, S.Y., and Widdicombe, J.H. (1992) Differentiated structure and function of cultures from human tracheal epithelium. Am. J. Physiol. 262, L713–L724.

Yamaya, M., Finkbeiner, W.E., and Widdicombe, J.H. (1991) Altered ion transport by tracheal glands in cystic fibrosis. Am. J. Physiol. 261, L491–L494.

Yamaya, M., Finkbeiner, W.E., and Widdicombe, J.H. (1991) Ion transport by cultures of human tracheobronchial submucosal glands. Am. J. Physiol. 261, L485–L490.

Yang, C.M., Farley, J.M., and Dwyer, T.M. (1988) Acetylcholine-stimulated chloride flux in tracheal submucosal gland cells. J. Appl. Physiol. 65, 1891–1894.

Recommended Readings

Breeze, R.G., and Wheeldon, E.B. (1977) The cells of the pulmonary airways. Am. Rev. Respir. Dis. 116, 705–777.

Lundgren, C.E.G., and Miller, J.N. (eds.) (1999) Lung biology in health and disease. Vol. 132 in Lund biology in health and disease. Marcel Dekker, New York.

Sleigh, M.A., Blake, J.R., and Liron, N. (1988) The propulsion of mucus by cilia. Am. Rev. Respir. Dis. 137, 726–741.

Takishima, T., and Shimura, S. (eds.) (1994) Airway secretion. Physiological bases for the control of mucous hypersecretion. Lung biology in health and disease, vol. 72. Marcel Dekker, New York.

Widdicombe, J.H., and Widdicombe, J.G. (1995) Regulation of human airway surface liquid. Respir. Physiol. 99, 3–12.

3
Alveolar Surfactant

JOHN L. HARWOOD, LLINOS W. MORGAN, AND TANYA GREATREX

The Nature of Alveolar Surfactant

Alveolar surfactant is a remarkable and highly active surface material composed of lipids and proteins that are present in the fluid lining the alveolar surface of the lungs (Griese, 1999). It has several distinct features, and is carefully regulated under a range of normal physiological conditions throughout life. Indeed, the regulation of alveolar surfactant composition and its secretory rate into the alveoli is itself a superb example of metabolic regulation in practice.

Alveolar surfactant occurs in two major pools: extracellular and intracellular. By washing (lavaging) lungs it is possible to obtain preparations that contain the extracellular surfactant, which has been secreted into the alveoli and which, therefore, represents the functional material. Such material, however, will be contaminated by compounds from the upper airways, and it is difficult to obtain complete recovery of the total extracellular surfactant. Nevertheless, alveolar surfactant purified from lung lavages has a consistent composition from a large number of animal species. Moreover, the purified surfactant has the required physical properties expected of a material that produces a stable film of low surface tension that prevents alveolar collapse at end expiration *in vivo*. This is its biophysical function. Its nonbiophysical function is the protection of the lungs from injuries and infections caused by inhaled particles and microorganisms (Giese, 1999; Jobe and Ikegami, 2001).

Intracellular alveolar surfactant is present in the so-called lamellar bodies of the alveolar type-II epithelial cells. It has not been studied as much as the extracellular material, but, even so, material from quite a number of animal species has been exam-

TABLE 3.1. Basic composition of alveolar surfactant.

	wt. %	Types
Protein	10	Specific surfactant proteins SP-A, SP-B, SP-C, SP-D.
Lipid	90	About 90% phospholipid, of which DPPC makes up half. Phosphatidylglycerol noticeably present at about 10% of phospholipids.

ined. In general terms, it has a composition similar to the extracellular material that would be expected because it is the physiological precursor.

Alveolar surfactant is a lipid-rich lipoprotein mixture. It contains approximately 90% lipid, 10% protein, and small amounts of carbohydrate (as glycosyl groups on proteins) (Table 3.1). The lipid composition is dominated by phosphatidylcholine (PC), particularly its dipalmitoyl-molecular species (Fig. 3.1). The presence of significant amounts of phosphatidylglycerol is another interesting feature because it is not normally found in any more than trace amounts in mammalian tissues. The synthesis, secretion, and metabolism of dipalmitoylphosphatidylcholine (DPPC) and other surfactant lipids has been the subject of a number of comprehensive reviews (Harwood and Richards, 1985; Harwood, 1987; Post and van Golde, 1988; van Golde et al., 1988; Ballard, 1989; Batenberg, et al., 1990; Bourbon, 1991; Johansson and Curstedt, 1997). Although surfactant preparations purified from lung washings typically contain quite a number of proteins, most of these (e.g., albumin) appear to have derived from the plasma; however, several surfactant specific proteins (i.e., SP-A, SP-B, SP-C and SP-D) have been found. These are closely associated with surfactant lipids and appear to contribute to the overall surfactant properties. Indeed, studies suggest that the surfactant-associated proteins may play other important roles in surfactant biology. Several reviews can also be consulted for information about these proteins (Whitsett et al., 1985; BAL Coop. Group Steering Committee, 1990; Weaver and Whitsett, 1991; Wright, 1997; Haagsman and Diemel, 2001).

Alveolar Lipids

Most lipid extracts from mammalian tissues contain a range of compounds including neutral lipids, cholesterol, phospholipids, and sphingolipids. For alveolar surfactant the pattern is unusually simple (Table 3.2). Cholesterol is really the only important neutral

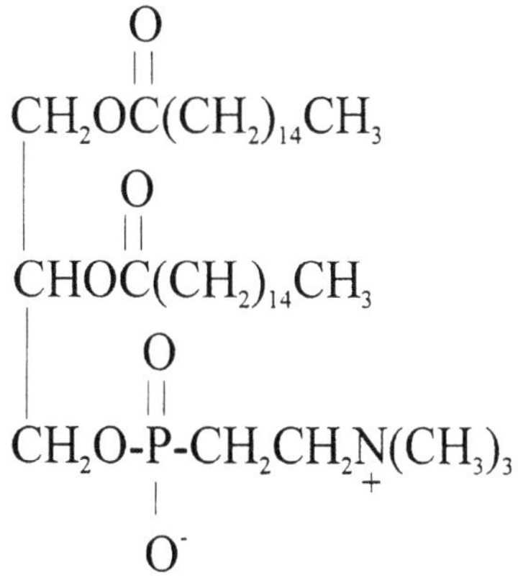

FIGURE 3.1. Dipalmitoylphosphatidylcholine: the major lipid component of alveolar surfactant.

TABLE 3.2. The lipid composition of some isolated alveolar surfactants.

	Dog	Human	Rat	Rabbit
Neutral lipid	13	10	11	14
Triacylglycerol	14	26	25	20
Cholesterol	86	71	57	70
Others	—	23	18	10
Phospholipids	87	90	89	86
Phosphatidylcholine	76	73	80	76
Phosphatidylethanolamine	6	3	4	5
Phosphatidylglycerol	7	12	9	8
Phosphatidylinositol	4	6	3	4
Others	7	6	4	7

The neutral lipid and phospholipid values are expressed as % of total lipid. The individual neutral lipids or phospholipids are shown as a % of these two classes.

lipid, but its level is rather insignificant compared with its normal amounts in, say, plasma membranes.

PC is by far the major phospholipid in surfactant. In most tissue extracts this phospholipid class would be expected to be present at up to 50% of the total, but it represents about three-quarters of all phospholipids in alveolar surfactant. The second-most abundant phospholipid is usually phosphatidylglycerol at about 10%. This is very unusual because phosphatidylglycerol is uncommon in animals. Because of its level in surfactant, phosphatidylglycerol has been suggested to be a "marker" component and to play some vital functional role. Both of these properties, however, are negated by its low levels in cat surfactants and various other air-breathing animals such as the chicken. In such cases, phosphatidylglycerol appears to be replaced by another negatively charged phospholipid, phosphatidylinositol. The latter is another significant component of surfactant (Table 3.2) and is formed by a biosynthetic route common to phosphatidylglycerol.

Other phospholipids are insignificant. Phosphatidylethanolamine, the second most prevalent membrane phospholipid in animals is only present in minor amounts, whereas sphingomyelin and phosphatidylserine are trace constituents. Diphosphatidylglycerol (cardiolipin), which can be regarded as a mitochondrial marker, is absent. One trace constituent that has been found in alveolar surfactant preparations is platelet-activating factor (PAF; 1-acyl, 2-acetylphosphatidylcholine) whose presence explains the platelet-activating properties of such isolates, especially from the fetal lung.

Each phospholipid class contains a variety of molecular species [i.e., different combinations of acyl (or alkyl) chains at the *sn*-1 and *sn*-2 positions of the glycerol backbone]. To a student it may seem tedious that such complications are mentioned. In terms of function, however, the nature of the exact molecular species is very important indeed (see later).

The fatty acid composition of phosphatidylcholines from different alveolar surfactants is shown in Table 3.3. It will be seen, clearly, that palmitate is the dominant fatty acid constituent. If you bear in mind that two acyl chains are present in each phosphatidylcholine molecule, more than half of the individual species are dipalmitoyl. In fact, together with the other saturated acids, myristate and stearate, disaturated phosphatidylcholines make up 60–80% of the total, compared with only 4% in the liver. There are good reasons for this when one considers the constant exposure of the surface layer of phospholipid in the alveoli to air and the tendency for unsaturated acyl

TABLE 3.3. Fatty acid composition of phosphatidylcholine fractions from some isolated alveolar surfactants.

Animal	Composition (% total fatty acids)						
	14:0	16:0	16:1	18:0	18:1	18:2	Others
Dog	5	67	4	7	10	1	6
Human	3	81	6	3	5	2	Traces
Pig	5	73	6	3	10	1	2
Rat	2	83	1	2	7	2	3
Rabbit	3	73	6	3	13	2	Traces

Fatty acids are abbreviated with the first figure indicating the number of carbons and the number after the colon showing the number of double bonds. 14:0 = myristate; 16:0 = palmitate; 16:1 = hexadecanoate (mainly palmitoleate); 18:0 = stearate; 18:1 = octadecenoate (mainly oleate) and 18:2 is mainly linoleate.

chains to oxidize. The other phospholipid classes are also quite saturated, although less so than phosphatidylcholine.

Other phospholipid variants (e.g., plasmalogens) have been reported, but they are quantitatively unimportant.

Why So Much Dipalmitoylphosphatidylcholine?

It is generally thought that the main functional agent of alveolar surfactant is DPPC. Although the biophysical basis of surfactant activity and the precise physiological mechanims of its action are still controversial (Bangham, 1991), a good correlation between the interfacial properties of surfactant and its physiological efficacy seems to exist. DPPC could be considered to be an ideal lipid to form a stable surface layer at the air–liquid interface in the alveolus. It is amphipathic, so it will naturally form a monolayer with the fatty acyl chains projecting into the air, whereas the polar phosphocholine moiety interacts with the aqueous phase. The negative charge on the phosphate group is balanced by a positive charge on the choline, thus rendering the molecule zwitterionic. Moreover, the dimensions of the polar part of the phospholipid are equal to those of the nonpolar (acyl) part that means that DPPC is cylindrical in shape and can easily pack into a monolayer. In addition, there is good hydrophobic bonding between the two palmitoyl residues.

The surface behavior of phosphatidylcholines, lung surfactant, and other derivatives can be measured by using a Wilhelmy balance and a hydrophobic Langmuir trough with cyclic compression–expansion (Gaines, 1966) or by the pulsating bubble technique (Enhorning, 1977). It has been suggested that it is important for phosphatidylcholines with a transition temperature (Tc; the temperature at which the acyl chains change from the gel to the liquid phase) greater than body temperature to be present for a stable, low surface tension film to be formed (Goerke and Gonzales, 1981). When mixtures of PC species were subjected to dynamic compression, the more fluid, unsaturated components were gradually squeezed out to leave a monolayer enriched in DPPC (Hawco et al., 1981). Indeed, the rate of "squeeze-out" and, hence, enrichment of the monolayer depends on the rate of compression–expansion. In fact, there is also a minimum proportion of DPPC that is needed to produce surface tensions low enough for lung stability. This is thought to be around 50% (Bourbon, 1991), which is a concentration easily met by natural alveolar surfactant. The unsaturated phospholipid con-

TABLE 3.4. Contribution of different surfactant components for lipid phase formation.

Component	Property	Function
DPPC-enriched monolayer	Rigid chains (below T_c)	Provides very low surface tension
		Stability at low surface tension
Unsaturated lipids	Acyl chains fluid	Monolayer spreading, tubular myelin formation
Acidic lipids + Ca^{++} + apoprotein	Bilayer inhomogeneities (nonbilayer states)	Monolayer spreading, tubular myelin formation
High DPPC content (other lipid mixtures)	Monolayer inhomogeneities Transient lipid clusters	Selective exclusion

Source: After Harwood and Richards, 1985, with permission.

stituents are thought to be needed in order to aid the initial adsorption and spreading of the surface film (King and Clements, 1972). Phosphatidylglycerol and unsaturated PCs would have this function *in vivo*. In most situations, phosphatidylinositol seems able to substitute for phosphatidylglycerol, although there are occasional situations where phosphatidylglycerol appears to be definitely superior. One such example is around birth, where newborns suffering from transient tachypnea (a mild and transient respiratory difficulty) were shown to have a specific deficiency in phosphatidylglycerol (Bourbon et al., 1990). Moreover, whatever the putative role of phosphatidylglycerol, its presence in surfactant (or amniotic fluid) is a valuable index of fetal lung maturity and is indicative that all the functional elements of surfactant are present. Thus, it is almost unknown to find respiratory distress of the newborn when surfactant phosphatidylglycerol is abundant.

The probable contributions of different lipids to surfactant function are summarized in Table 3.4.

The Biosynthesis of Surfactant Lipids

Phosphatidylcholine can be synthesized by several different routes, but the CDP-choline pathway is the main route in mammalian tissues (and the lung is no exception). The latter consists of three steps (Fig. 3.2). Of these, the final step is freely reversible and, therefore, unimportant in regulating the overall speed of phosphatidylcholine formation. The first step, which is catalyzed by choline kinase, usually has high rates and is not considered to be rate-limiting, whereas the cytidylyltransferase is. The evidence that cholinephosphate cytidylyltransferase is rate-limiting comes from several sources (Table 3.5).

Following results with the liver, where the regulation of cholinephosphate cytidylyltransferase through subcellular translocation has been well studied, the enzyme from the lung has been examined. When the enzyme was purified (Feldman et al., 1978) two forms of the cytidylyltransferase were found. The higher molecular weight form (5–50 $\times 10^6$) was more active than the other (190,000), and its activity was increased by preincubation with phosphatidylglycerol. Moreover, phosphatidylglycerol also shifted the apparent subcellular distribution so that a greater proportion was found in the microsomal fraction and less in the cytosol (Stern et al., 1976). This would all fit with the better-demonstrated liver situation; however, the role of translocation in the regulation of cholinephosphate cytidylyltransferase (CPCT) is still unresolved and some

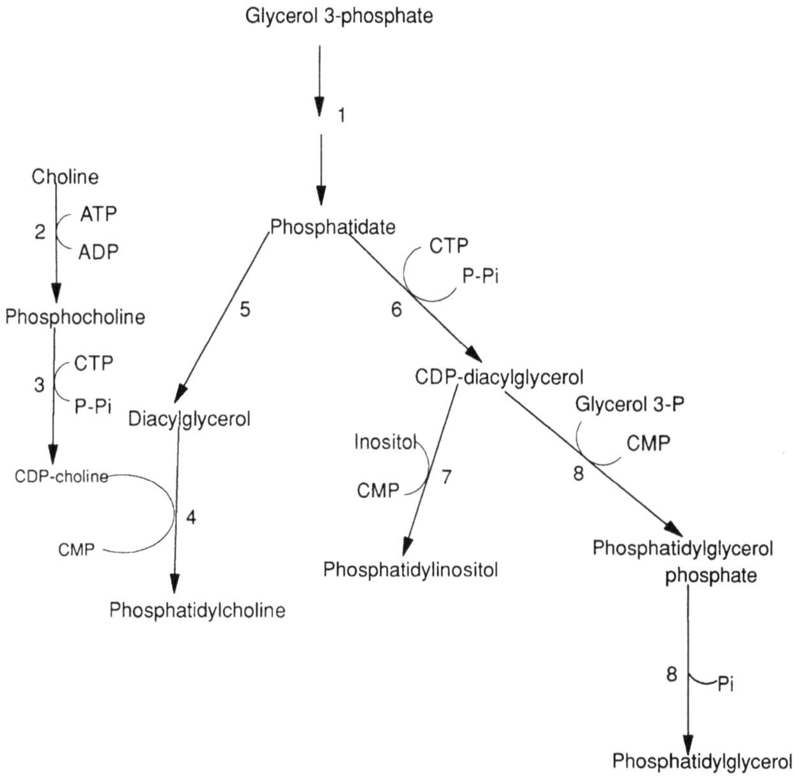

FIGURE 3.2. Biosynthesis of the main surfactant lipids.

workers believe that activation of pre-existing microsomal enzyme is more important, especially in connection with hormone-stimulated phosphatidylcholine synthesis (Post, 1987). Other studies have clarified the relationships between different forms of the enzyme and its subcellular location (Feldman et al., 1990). The microsomal form has the greatest activity. Moreover, the cytosolic form is believed to be translocated to the endoplasmic reticulum (Kalmar et al., 1994). Thus, CPCT is stimulated by some phospholipids modulated by phosphorylation, and its activity is increased by translocation from cytosol to endoplasmic reticulum.

Instead of phosphatidate being used as a source of diacylglycerol for zwitterionic phospholipids, it can itself be subject to a cytidylyltransferase reaction and converted

TABLE 3.5. Evidence that cholinephosphate cytidylyltransferase (CPCT) is rate-limiting for phosphatidylcholine formation.

- CPCT activity increases coincidently with phosphatidylcholine formation in developing lungs.
- CDP-choline is the smallest pool of the various choline intermediates.
- CPCT activity measured *in vitro* is usually lower than either choline kinase or phosphocholine transferase.
- Hormones that increase surfactant production also stimulate CPCT activity.
- The antagonistic actions of insulin and glucocorticoids on developing lung seem to be expressed at the level of CPCT.

to CDP-diacylglycerol (see Fig. 3.2). Because the intracellular concentration of this cytidylyltransferase is low, its activity may limit the overall rate of acidic phospholipid synthesis. Once CDP-diacylglycerol has been formed it is used as substrate by phosphatidylinositol synthase. Only *myo*-inositol (of the 16 possible isomers of inositol) is used for lipid formation. The activity of the lung enzyme rises gradually during the perinatal period to reach its highest level in the adult (Harwood, 1987). In contrast, the glycerolphosphate phosphatidyltransferase that catalyses the first of the two reactions to form phosphatidylglycerol showed a steeper rise in activity immediately before birth, just when this phospholipid became a significant component of surfactant.

Thus, the ratio of the acidic lipids, phosphatidylinositol and phosphatidylglycerol, changes markedly just before term. This alteration is produced mainly through a sudden increase in the production of phosphatidylglycerol. The cause of this change in metabolism has been investigated and two possible causative factors: CMP and inositol. Serum inositol levels decline just before birth, and such a change might reduce phosphatidylinositol synthesis (Bleasdale and Johnson, 1985). Doubts about this simple idea, however, come from the observation that inositol levels in lung tissue in fetal animals were the same as in adults (Bleasdale et al., 1982). As an alternative it has been shown that CMP stimulates the reverse reaction of phosphatidylinositol synthase to increase CDP-diacylglycerol levels. CMP is generated during PC synthesis (see Fig. 3.2), so it will be plentiful at a time close to parturition. These regulatory aspects of acidic phospholipid formation have been fully discussed (Bleasdale et al., 1985; Possmayer, 1987).

Remodeling of Newly Formed Phospholipids

Alveolar surfactant is remarkable for the unusual molecular species distribution in its phospholipid classes, especially PC. The operation of the CDP-choline pathway (Fig. 3.2), however, results in a mixture of PCs that mainly reflect the fatty acid composition of the diacylglycerols formed by phosphatidate phosphohydrolase. Thus, in order to generate large amounts of DPPC, the fatty acid composition must be changed. There are two obvious ways to do this (Fig. 3.3). Either a phospholipase A_2 removes the

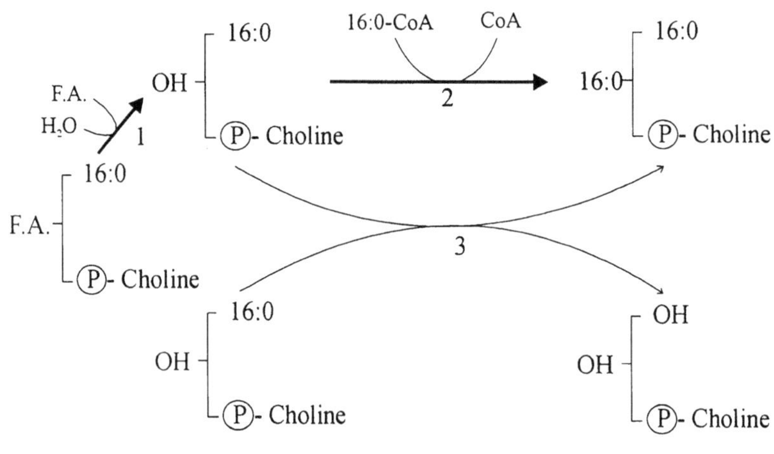

FIGURE 3.3. Possible pathways of remodeling to enrich phosphatidylcholine with palmitate (taken from Harwood (1987) with permission).

(mainly unsaturated) fatty acid from the *sn*-2 position of glycerol and this is replaced with palmitate through the action of a monoacyl(lyso)PC acyltransferase, or interlipid transfer takes place. In the latter reaction, two molecules of monoacylPC react together with one providing the source of the palmitate. Experiments measuring *in vitro* enzyme activities, with cultured Type-II cells from the lung and by double labeling techniques with lung slices, suggest that the phospholipase A_2/acyltransferase route is the main pathway for remodeling (see Batenburg, 1984).

In addition to PC, the alveolar surfactant phosphatidylglycerol is also enriched in palmitate, so it is likely that a similar type of remodeling also applies to this component.

Alveolar Surfactant Contains Special Proteins

The isolation of a lipid-rich surfactant pellet from lung lavage yields a preparation that contains many proteins, most of which have been identified as serum-derived. Extraction of lipid from the pellet leaves behind a major nonserum protein referred to as surfactant protein A (SP-A). In humans, SP-A consists of a major charge train of 9–13 proteins (molecular masses 34–36kDa) and a minor charge train of three proteins (molecular masses 28–30kDa). The charge and size heterogeneity of SP-A varies among species and primarily reflects the addition of one or two asparagine-linked oligosaccharide chains to a species-specific backbone of about 230 residues (Weaver and Whitsett, 1991). In airways, SP-A exists in thiol-dependent and non–thiol-dependent oligomers of $0.5–1.6 \times 10^6$kDa.

Unlike SP-A, two other surfactant proteins, SP-B and SP-C, are so hydrophobic that they remain in the lipid phase during standard extractions. SP-B is a protein of about 8kDa and forms thiol-dependent oligomers with dimers being the most common. SP-C is a mixture of three peptides of 33–35 amino acid residues, which differ only in their N-terminus (see e.g., Johansson et al., 1988). SP-C can aggregate *in vitro* to form non–thiol-dependent dimers.

A fourth surfactant-associated 43kDa hydrophilic protein, SP-D, has been found as a minor component in various alveolar surfactants; however, even though it shares some structural similarities to SP-A, it does not co-isolate with other surfactant components. It has been estimated that the concentration of SP-D in the rat lung is about 12% that of SP-A (Wright, 1997). No surface tension–lowering effect has been found, and SP-D is thought to have defense properties, as seen in other collectins (Johansson and Curstedt, 1997). Collectins are a family of collagenous carbohydrate binding proteins consisting of oligomers of trimeric subunits.

The Different Structures of Surfactant-Associated Proteins

The gene for SP-A has been sequenced from several species including humans (Weaver and Whitsett, 1991). It consists of five exons, with the coding sequence distributed among four. There may be more than one gene in humans and at least one pseudogene has been identified (Korfhagen et al., 1990). SP-A RNA encodes a protein of 248 amino acids (Fig. 3.4) in human, rat, and dog, and 247 amino acids in rabbit (see Weaver and Whitsett, 1991). The N-terminal domain contains a signal peptide of 20 amino acids followed by a short (7–10 residue) sequence. A cysteine in this latter part allows the interprotein disulphide linkage that gives rise to a trimeric structure that forms the basis of the oligomeric (six trimers) native structure. The N-terminal domain then joins a collagenlike part of the protein (Fig. 3.4) with 23–24 Gly-X-Y repeats. Finally, there is a 148-residue noncollagenlike region that includes the carbohydrate-recognition

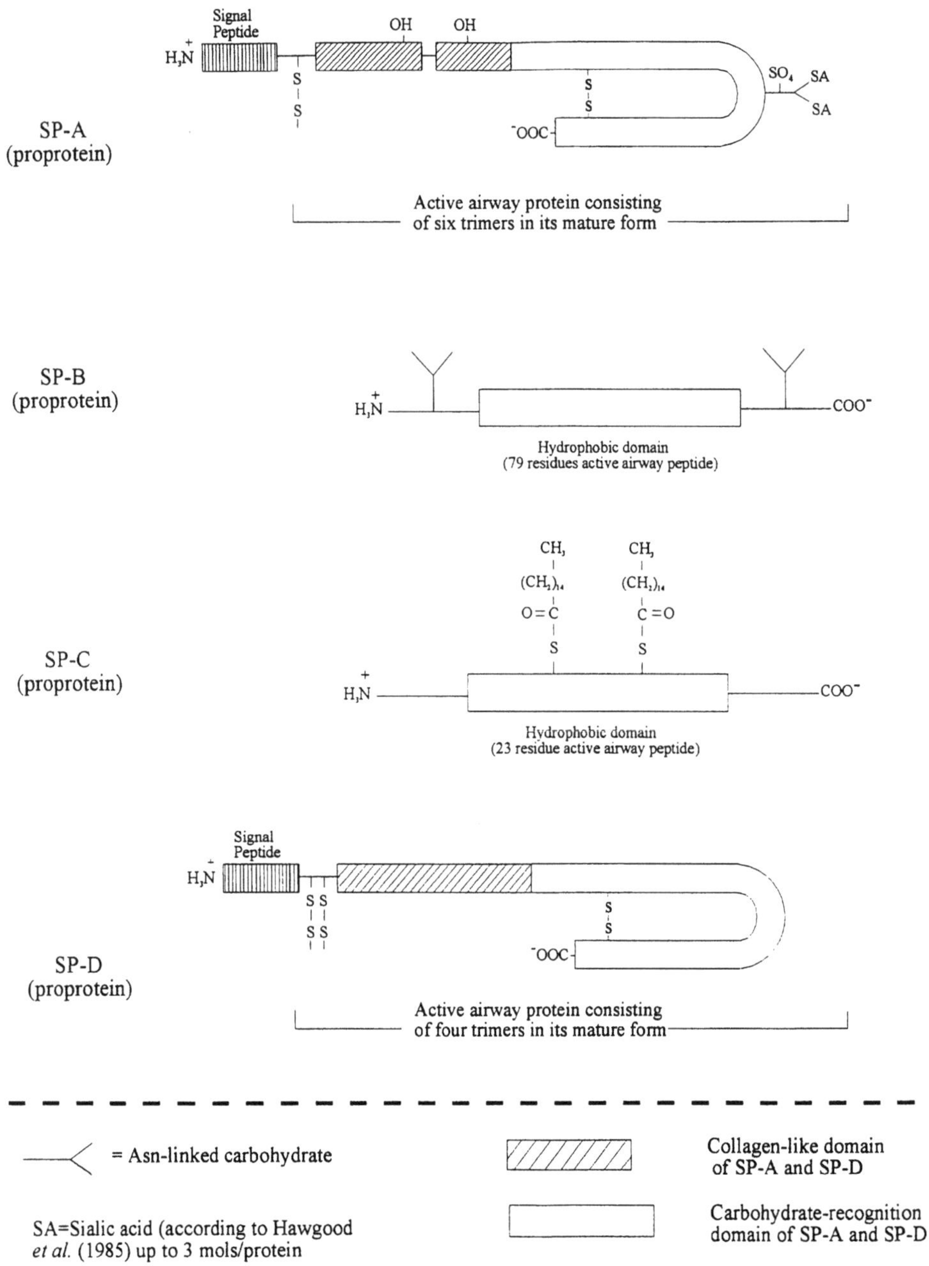

FIGURE 3.4. Simplified structures of the surfactant associated proteins (redrawn from information in Weaver and Whitsett (1991) with permission).

domain of 130 amino acids. There is a 24 amino acid "link" region located between the collagenlike and carbohydrate-recognition sequences that has been suggested to have a role in phospholipid binding (Ross et al., 1986). Other studies, however, have produced conflicting evidence for this proposed role (Whitsett and Weaver, 1991).

For SP-B, restriction mapping of human genomic DNA is consistent with a single gene (Pilot-Matias et al., 1989). Sequences for SP-B cDNA from other species show changes resulting in only one or two amino acid substitutions. The SP-B RNA encodes a protein of 381 amino acids in humans. The latter consists of a signal peptide of 20–23 residues, a propeptide of about 176 residues, the mature 79-residue SP-B peptide, and a C-terminal peptide of 102 amino acids. The mature peptide contains three regions: an amphipathic helix, an extremely hydrophobic region, and an 11-residue sequence similar to the active sites of two proteinase inhibitors (Emrie et al., 1989).

The human SP-C gene is organized into six exons that encompass approximately 2–7 kb (Glasser et al., 1988). The SP-C RNA encodes a protein of 197 amino acids in human which, unlike most proteins to be secreted, does not contain a signal peptide. A mature 35-residue peptide is eventually produced that has an extremely hydrophobic region of 23 amino acids (residues 11–33). This region is 43% valine and 35% leucine or isoleucine and is likely to form a rigid α-helix capable of spanning a membrane bilayer. Residues 5 and 6, which are cysteines, appear to be palmitoylated (Curstedt et al., 1990), thus, further increasing the hydrophobicity of SP-C.

Surfactant protein D is a 43 kDa, 355 amino acid glycoprotein. Like SP-A, it belongs to the collectin family of proteins and has collagenous and noncollagenous domains of approximately equal size. SP-D has a short N-terminal region, a collagenlike domain with 59 Gly-X-Y repeats, a short "link" region, and a much longer carbohydrate recognition domain (CRD) than SP-A. Unlike SP-A, there is no kink interrupting its collagenlike domain (Johansson and Curstedt, 1997).

The processing and secretion of surfactant proteins is discussed by Weaver and Whitsett (1991).

Possible Functions of Surfactant-Associated Proteins

It has been found that surfactant proteins can directly affect the physical properties of surfactant lipids both *in vitro* and *in vivo* (Table 3.6). Purified SP-B, SP-C, or mixtures of the two markedly enhance the rate of formation of surface films. SP-A addition could further enhance the activity. When measured in a Wilhelmy balance or a pulsating bubble surfactometer, SP-B was particularly effective at reducing surface tensions of phospholipid mixtures (Weaver and Whitsett, 1991). In confirmation of these *in vitro* observations, preparations of SP-B, SP-C, and surfactant lipids increase lung compliance and preserve the morphological integrity of distal airways in lungs from premature rabbits or surfactant-deficient rat lungs (Sarin et al., 1990). In addition, antibodies against SP-B caused respiratory failure when given to mice (Fujita et al., 1988). In view of the preceding, it is encouraging that preparations of surfactant lipids containing mixtures of SP-B and SP-C have been successfully used in clinical trials to treat

TABLE 3.6. Possible functions of surfactant proteins.

Function	SP-A	SP-B	SP-C
• Contribute to surfactant properties of phospholipids	+	+	+
• Formation of tubular myelin	+	+	–
• Facilitate turnover of surfactant phospholipids	+	+	+
• Inhibition of pulmonary surfactant secretion	+	–	–
• Facilitate phagocytosis of opsonized particles	+	–	–

Source: From Weaver and Whitsett, 1991, with permission.

premature infants suffering from respiratory distress syndrome and hyaline membrane disease. Treatment today is with exogenous surfactant replacement. This also is the case for a large number of pulmonary disturbances.

When surfactant is secreted it has been proposed that a tubular myelin structure is formed in the aqueous medium of the alveoli (see Fig. 3.5 later). This structure is thought to form spontaneously at the air–liquid interface from lamellar bodies when they are hydrated in the aqueous subphase (Johansson and Curstedt, 1997). Tubular myelin has also been proposed to be the precursor of the surface-active phospholipid monolayer, and both SP-A and SP-B are necessary to form tubular myelin from synthetic phospholipids (Suzuki et al., 1989). The precise role of tubular myelin in surfactant function, however, is still unclear (Weaver and Whitsett, 1991), even though it is a prominent structure in electron micrographs of lung tissue (Scarpelli and Mautone, 1984; Harwood and Richards, 1985).

After phospholipids have been secreted into the alveoli, they are gradually cleared, mainly by re-uptake into the Type-II epithelial cells. This process results in a half-life for DPPC of the order of about 12 hours for most animals. The re-uptake process appears to be stimulated by surfactant-associated proteins, especially SP-A. By contrast, this protein has also been shown to inhibit the secretion of surfactant phospholipids by Type-II cells *in vitro*. In addition, SP-A and SP-D have been suggested to have an important role in the immune defense system of the lung. Thus, SP-A increases phagocytosis of opsinized erythrocytes and bacteria by macrophages and monocytes (van Iwaarden et al., 1990). The receptor and SP-A domain involved in the receptor binding are probably different from those mediating the inhibition of surfactant secretion (see Weaver and Whitsett, 1991).

Secretion and Turnover of Surfactant

Various lines of evidence (e.g., microscopy or autoradiography) indicate that the various components of surfactant are packaged together in the lamellar bodies of the epithelial Type-II cells (Batenburg, 1984) before secretion by exocytosis into the alveoli. Confirmation of this came from experiments using radiolabeled precursors. It was found that the specific activities of PC, phosphatidylglycerol, phosphatidylinositol and cholesterol show identical changes in the alveolar lipids (Hass and Longmore, 1979) and that the appearance of DPPC and surfactant apoprotein parallel each other (King and Martin, 1980).

Lamellar bodies contain a very similar lipid composition to alveolar surfactant (see previous discussion), and isolated lamellar body fractions contain SP-A, SP-B, and SP-C. Such fractions are capable of forming tubular myelin (in the presence of calcium) and reducing surface tension to less than 12 mN/m, which suggests that the lamellar bodies contain the mature assembled phospholipid–protein complex (Farrell et al., 1990).

A general outline of secretion and turnover of surfactant, as well as the regulation of surfactant is shown in Fig. 3.5 and the regulation of surfactant secretion has been reviewed (Chander and Fisher, 1990). The control of surfactant secretion is not clearly understood and a large number of agents have been shown to have some activity in this regard. These include prostaglandins of the E series, cholinergic agents, and distension (stretch) of the alveolar epithelium (see Harwood and Richards, 1985). Most attention has been paid, however, to the role of β-adrenergic agents in the regulation of surfactant secretion (Walters, 1985). Epinephrine appears to be important at the time of birth in helping to establish normal breathing. It may also be involved in stress

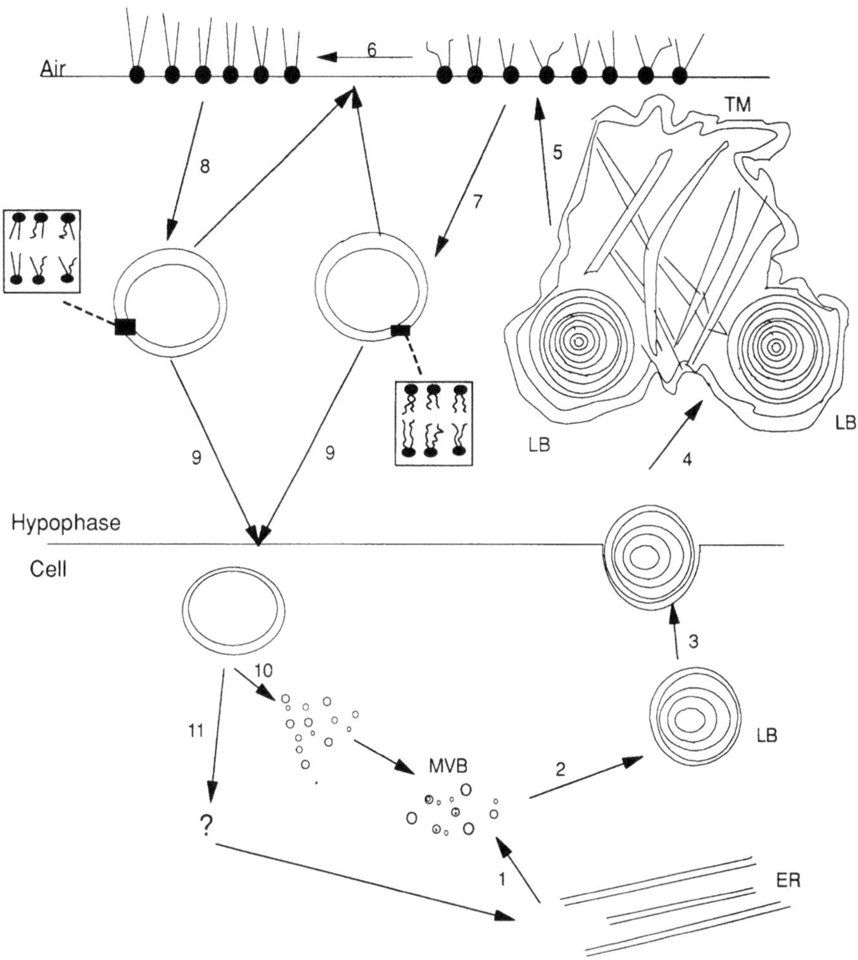

FIGURE 3.5. Diagrammatic illustration of the cycle of surfactant lipid secretion and re-utilization (taken from Harwood and Richards (1985) with permission).

reactions during which hyperventilation (and increased surfactant secretion) occur. It is unclear for the actual secretory process whether microtubules are involved or not (Harwood and Richards, 1985); however, experiments with cytochalasin B have shown that it will inhibit PC secretion in lung slices, presumably by disrupting the action of microfilaments. Such action has been shown to be concentrated around the lamellar bodies of Type-II cells and to become particularly prominent during surfactant secretion (see Walters, 1985).

Aspects of the metabolic and hormonal control of surfactant release have been covered by Bourbon (1991). In general, surfactant metabolism in the fetal lung differs from that in the adult lung. First, the kinetics of secretion are slower, second, the half-life is longer, and third, cycling is more efficient (Jobe and Igemani, 2001).

Although surfactant has long been known to have a relatively short half-life (King, 1974), there have been relatively few studies on its removal and/or catabolism. The possible ways in which surfactant can be removed or degraded are listed in Table 3.7. Although there is some evidence for each of these mechanisms, most results suggest that uptake and degradation/reutilization by Type-II cells is a major mechanism. With

TABLE 3.7. Suggested ways for the breakdown or removal of alveolar surfactant.

- Clearance up the airways.
- Uptake by alveolar macrophages (followed by catabolism).
- Breakdown within the alveoli.
- Uptake by Type-I or Type-II epithelial cells (followed by degradation).
- Movement across the alveolar epithelia and clearance through the lymphatics.
- Uptake and reutilization by Type-II epithelial cells.

Source: From Harwood and Richards, 1985, with permission.

regard to the lipid components the data have been discussed by Jobe and Jacobs (1984) and by Wright and Clements (1987). Internalization of SP-A by alveolar macrophages can occur by a mannose-dependent mechanism associated with enhanced phagocytosis of opsonized targets (Tenner et al., 1989). In contrast, uptake of SP-A by Type-II cells occurs by a mannose-independent receptor-mediated process (see Weaver and Whitsett, 1991). The picture is less clear for the other surfactant-associated proteins. Both SP-B and SP-C appear to enhance the uptake of phospholipids by Type-II cells in a dose-dependent manner. SP-C itself appears to have a faster turnover rate than PC, and both Type-II cells and alveolar macrophages appear to participate in its turnover (i.e., with the former for recycling and the second for degradation). It is noteworthy in the latter connection, however, that both SP-B and SP-C are completely resistant to proteinase degradation *in vitro* (see Johansson et al., 1988).

Respiratory Distress and Alveolar Surfactant

Because alveolar surfactant is synthesized only in the late gestation stages, cases of premature birth often result in acute respiratory distress syndrome (RDS) or hyaline membrane disease (see Raivo, 1983). The incidence of RDS is related to the immaturity of the fetus and will almost always occur with babies born before 31 weeks. Below this age the chances of survival are closely related to gestation age. The major problem experienced by the RDS baby is a difficulty in lung expansion. Although surfactant deficiency is a primary cause of RDS, a number of ancillary problems complicate successful treatment by simple replacement therapy (Table 3.8; see Morley, 1985). There is no doubt, however, that both prophylactic and therapeutic ("rescue") administration of surfactant are safe and effective. It remains controversial whether prophylaxis or rescue is superior, except in premature infants of 26 weeks or less, in whom multidose prophylaxis is superior to multidose rescue (see Corbet, 1992).

TABLE 3.8. Compounding problems in respiratory distress syndrome.

- Babies have difficulty in clearing fluid from lungs after birth.
- Few alveoli are present in the immature lungs.
- Lung edema develops with pulmonary hypertension.
- Poor muscle development makes expansion of the stiff lungs even more difficult. If a deep breath is made, the immature ribs are so soft that the chest wall is drawn in, thus reducing the lung volume.
- Although some surfactant is present, it has less saturated PC and less phosphatidylglycerol than normal.
- Substantial amounts of serum protein leak onto the alveolar surfaces.

Source: From Harwood and Richards, 1985, with permission.

A key fact for the obstetrician is to be able to predict, in advance, the risk of RDS. Because of biological variation, measurements of gestational age and fetal size, although useful, are unfortunately insufficient in themselves. Diagnosis of a high risk of RDS can influence the timing of elective preterm deliveries such that the incidence of RDS can be significantly reduced. However, amniocentesis (by which surfactant levels are assessed) is not without possible complications, and there are problems associated with some forms of treatment (e.g., glucocorticoids) that are used to accelerate lung maturation (Harwood and Richards, 1985). Some measurements of amniotic fluid which have been used to predict RDS are given in Table 3.9.

Two general methods have been used to treat RDS in infants. First, hormones have been utilized to hasten lung maturation and increase surfactant production. Second, replacement therapy with various artificial surfactants has been tried. The earlier work on hormones has been reviewed (Perelman et al., 1985; Harwood, 1987) and, because research has concentrated on replacement therapy with artificial surfactants, the latter topic will be emphasized here. There are useful reviews of this subject by Walti et al. (1991) and Giese (1999).

There have been more than 30 randomized controlled trials of the use of surfactant with more than 6000 babies having been enrolled in these studies. The results have been reviewed (Avery and Merritt, 1991; Morley, 1991) and the data subjected to meta-analysis (Hennes et al., 1991; Soll, 1991). There are considerable problems in comparing the results of individual trials mainly because of differences in the populations randomized, in exclusion criteria and in the outcomes assessed. Moreover, general differences relate to whether the surfactant used was natural or synthetic, and to whether the trial compares the prophylactic use or rescue use of surfactant (Colditz and Henderson-Smart, 1992). When outcomes were assessed, however, there appeared to be little significant difference for most criteria with the nature of the surfactant or its method of use. Mortality, pneumothorax, and intraventricular hemorrhage were all especially improved compared with untreated babies (see Colditz and Henderson-Smart, 1992). To take a specific example, where Infasurf (a bovine-derived surfactant) was administered at 90 mg/infant weighing less than 1 kg or 180 mg/infant greater than 1 kg, surfactant replacement therapy improved gas exchange in infants with RDS in addition to reducing pulmonary complications. Of 17 infants weighing 500–750 g who were given surfactant replacement therapy, one developed respiratory distress and 7

TABLE 3.9. Some measurements with amniotic fluid that have been used to predict respiratory distress syndrome.

Measurement	Comments
PC/Sph ratio	Significant number of false negatives. Ratio >2 assures lung maturity. Charring methods on TLC need to be modified for DPPC.
Relative % of PI and/or PG	Many babies lack both detectable PG and RDS.
Surfactant apoprotein (35 KDa)	Probably will gain increasing popularity. Sensitive if linked to ELISA assay.
Bubble stability (shake test)	Simple, low cost. Number of false negatives high.
Fluorescence polarization	High cost instrument but simple to use and reliable.
Enzyme assays	Phosphatidate phosphohydrolaze only useful constituent.

Source: From Harwood, 1987, with permission.
Abbreviations: Sph = sphingomyelin: PG = phosphatidylglycerol; PI = phosphatidylinositol; ELISA = enzyme-linked immunosorbent assay.

of the 17 died. In 10 unmatched controls in the same weight range, nine infants developed RDS and nine died (Ioli and Richardson, 1990).

Studies using prophylactic surfactant suggest that there may be benefit in administering more than one dose of surfactant (Kendig et al., 1989). It has been postulated that a good response after a subsequent dose may be seen because of effects of serum proteins leaking into the air spaces that could inhibit the activity of the first dose (Ikegami et al., 1986). The clearance rate of exogenous surfactant has been studied following multiple doses of an artificial surfactant, Exosurf (a mixture of DPPC with hexadecanol and tyloxapol). Turnover of the artificial surfactants appears to be similar for figures with other exogenous surfactants, including naturally derived materials (Ashton et al., 1992a). Hallman et al. (1986) had previously calculated a mean half-life of the phosphatidylglycerol component of exogenous human surfactant to be 30 hours. As a result of their studies, Ashton et al., (1992a) concluded that, after giving two doses of Exosurf, further doses might best be delayed by 2 days to get maximum benefit from the replacement therapy. Cost-benefit analyses show clearly that surfactant replacement saves both lives and money for the health services (Colditz and Henderson-Smart, 1992). Such considerations have been discussed fully by Maniscalco et al. (1989).

The general topic of replacement therapy in RDS was reviewed by Fujiwara (1984) and the connection between lack of surfactant lipids (especially DPPC) and respiratory distress clearly established. Nevertheless, further work to confirm original observations has continued, and the connection can now be made on a much firmer basis (Ashton et al., 1992b). Moreover, development of different mixtures for possible use as lung surfactants has continued with the use of phospholipid analogues (e.g., Turcotte et al., 1991). Because of the beneficial action of surfactant-associated proteins, such as SP-B (Yu and Possmayer, 1992) on surfactant dynamics, their inclusion may be beneficial in spite of the possibility of allergic responses (Strayer et al., 1986). Finally, the availability of data from large-scale trials, such as the 23 hospital multicenter trial on Exosurf (Stevenson et al., 1992), has given us a lot more data to demonstrate convincingly that surfactant replacement therapy is of real benefit to infants with RDS. A source of referencing work is Robertson and Taeusch (1995).

Other Diseases That May Involve Surfactant

Adult respiratory distress syndrome (ARDS) is a much more common complaint than RDS of the newborn. A conservative estimate counts about 150,000 cases per year in the United States alone, with an average mortality of 40% (Lachmann and Danzmann, 1984). There are many functional and pathological similarities between ARDS and RDS of the newborn, including a lack of alveolar surfactant. In ARDS, however, surfactant deficiency is a complication of lung injury rather than being a primary etiologic factor, as it is in neonatal RDS. In addition, RDS can be caused by protein inactivation of surfactant functions as a result of pulmonary edema. Nevertheless, the deficiency can be compensated for by the administration of exogenous surfactant (Lachmann and Danzmann, 1984). Such a course of action is increasingly recommended (e.g., Fuhrman, 1990; Giese, 1999).

The relative amounts of PC and PG have been shown to decrease in ARDS patients (Hallman et al., 1982, Pison et al., 1990). This decrease was more pronounced with an increasing severity of disease, which suggests that Type-II cell metabolic functions are altered in ARDS. It has also been demonstrated that SP-A and SP-B are decreased within ARDS patients compared to normal controls (Gregory et al., 1991). Evidence is increasing that this may be caused by cytokines such as TNF (Hallman et al., 2001).

TABLE 3.10. Changes in surfactant induced by pharmacological agents or diseases.

Agent/disease	Surfactant change
Respiratory distress syndrome	Levels severely depressed
Adult RDS	Severe reduction
Alveolar (lipo)proteinosis	Levels increased
Quartz dust	Levels increased up to 50 times
Asbestos dust	Levels increased up to 20 times
Glass powder	Levels unchanged or slightly decreased
Chlorphentermine	Accumulation*
Imipramine	Accumulation*
Amitryptamine	Accumulation*
Chlorpromazine	Accumulation*
Oxygen toxicity	Decreased surfactant
Ozone	Decreased phosphatidylcholine synthesis in Type-II cells
Nitrogen dioxide	Transient increase in surfactant
Phosgene	Initial decrease in surfactant
Paraquat	Large decrease in surfactant
3-Methylindole	Accumulation
Peritonitis	Decreased or altered surfactant
Acute pancreatitis	Decreased surfactant
Diabetes	Main implication for babies born to diabetic mothers
Bleomycin—used in cancer therapy	Increased surfactant
Radiation	Decreased or abnormal surfactant

Source: From Harwood and Richards, 1985, with permission.
* Accumulation of surfactant only marked for instilled drugs. Normal clinical doses of, for example, imipramine have little effect.

Sudden infant death syndrome (SIDS) is the most common cause of death in infancy: It affects about one baby in every 500 dying between 1 week and 2 years in the United Kingdom (Morley et al., 1982). It has been found that surfactant from SIDS babies contained significantly less phospholipid and DPPC and relatively more phosphatidylglycerol than did that of babies dying from other causes (Morley et al., 1982). These results have been confirmed, but it is not yet clear whether surfactant deficiency and abnormality is a primary or secondary phenomenon in SIDS (James et al., 1990).

Surfactant deficiency is also believed to have a role in respiratory syncytical virus–associated bronchiolitis (Dargaville et al., 1995), where patients display ARDS-like symptoms. The virus attacks the surfactant-producing epithelial cells of the lung, which could then affect synthesis of surfactant material (Panitch et al., 1993). Decreased SP-A, PC, and PG have been observed in these patients (Dargaville et al., 1996, Skelton et al., 1999).

In addition to the preceding, there are a wide range of pharmacological agents or diseases that have been shown to affect surfactant quality or quantity (see Harwood and Richards, 1985). These are listed in Table 3.10. Thus, there are many areas of interest in surfactant metabolism and function, and even though our information on this unique lipoprotein mixture has increased considerably over the last 20 years, it seems certain that the subject of alveolar surfactant biochemistry will continue to be of major importance in medicine in the future.

References

Ashton, M.R., Postle, A.D., Hall, M.A., Austin, N.C., Smith, D.E., and Normand, I.C.S. (1992a) Turnover of exogenous artificial surfactant. Arch. Dis. Child. 67, 383–387.

Ashton, M.R., Postle, A.D., Hall, M.A., Smith, S.L., Kelly, F.J., and Normand, I.C.S. (1992b) Phosphatidylcholine composition of endotracheal tube aspirates of neonates and subsequent respiratory disease. Arch. Dis. Child. 67, 378–382.

Avery, M.E., and Merrit, T.A. (1991) Surfactant replacement therapy. N. Engl. J. Med. 324, 910–912.

BAL Co-operative Group Steering Committee (1990) Am. Rev. Resp. Dis. 141, S169–S202.

Ballard, P.L. (1989) Hormonal-regulation of pulmonary surfactant. Endocrine Rev. 10, 165–181.

Bangham, A.D. (1991) Pattles's bubbles and Von Neergaard's lung. Med. Sci. Res. 19, 795–799.

Batenburg, J.J. (1984) Biosynthesis and secretion of pulmonary surfactant. In: Pulmonary surfactant, Robertson, B., van Golde, L.M.G., and Batenburg, J.J. (eds.) Elsevier, Amsterdam.

Batenburg, J.J., and Hallman, M. (1990) In: Pulmonary physiology: fetus, newborn, child, and adolescent, Scarpelli, E.M. (ed.) pp. 106–139. Lea and Febiger, Philadelphia.

Bleasdale, J.E., Maberry, M.C., and Quirk, J.G. (1982) Myo-inositol homeostasis in foetal rabbit lung. Biochem. J. 206, 43–52.

Bleasdale, J.E., and Johnston, J.M. (1985) In: Pulmonary development: transition from intrauterine to extraterrestrial life, Nelson, G.H. (ed.) pp. 47–73. Marcel Dekker, New York.

Bourbon, J.R., Francoual, J., Magny, J.F., Lindenbaum, A., Leluc, R., and Dehan, M. (1990) Changes in phospholipid composition of tracheal aspirates from newborns with hyaline membrane disease or transient tachypnoea. Clin. Chim. Acta 189, 87–91.

Bourbon, J.R. (1991) Nature, function, and biosynthesis of surfactant lipids. In: Pulmonary surfactant: biochemical, function, regulatory, and clinical concepts, Bourbon, J.R. (ed.) CRC Press, Boca Raton.

Chander, A., and Fisher, A.B. (1990) Regulation of lung surfactant secretion. Am. J. Physiol. 258, L241–253.

Colditz, P.B., and Henderson-Smart, D.J. (1992) Surfactant replacement therapy. J. Pediatr. Child Health 28, 210–216.

Corbet, A., and Long, W. (1992) Symposium on synthetic surfactant: introduction. J. Pediatr. 120, S1–S50.

Curstedt, T., Johansson, J., Persson, P., Eklund, A., Robertson, B., Lowenadler, B., and Jornvall, H. (1990) Hydrophobic surfactant-associated polypeptides: SP-C is a lipopeptide with two palmitoylated cysteine residues, whereas SP-B lacks covalently linked fatty acyl groups. Proc. Natl. Acad. Sci. U.S.A. 87, 2985–2989.

Dargaville, P.A., McDougall, P.N., and South, M. (1995) Surfactant abnormalities in severe viral bronchiolitis. Appl. Cardiopulm. Pathophysiol. 5 (suppl. 3), 20.

Dargaville, P.A., South, M., and McDougall, P.N. (1996) Surfactant abnormalities in infants with severe viral bronchiolitis. Arch. Dis. Childhood 72, 133–136.

Enhorning, G. (1977) Pulsatile bubble technique for evaluating pulmonary surfactant. J. Appl. Physiol. 43, 198–203.

Emrie, P.A., Shannon, J.M., Mason, R.J., and Fisher, J.H. (1989) cDNA and deduced amino acid sequence for the rat hydrophobic pulmonary surfactant-associated protein, SP-B. Biochim. Biophys. Acta 994, 215–221.

Farrel, P.M., Bourbon, J.R., Notter, R.H., Martin, L., Nogee, L.M., and Whitsett, J.A. (1990) Relationships among surfactant fraction lipids, proteins, and biophysical properties in the developing rat lung. Biochim. Biophys. Acta 1044, 84–90.

Feldman, D.A., Kovac, C.R., Dranginis, P.L., and Weinhold, P.A. (1978) The role of phosphatidylglycerol in the activation of CTP: phosphocholine cytidylyltransferase from rat lung. J. Biol. Chem. 253, 4980–4986.

Feldman, D.A., Rounsifer, M.A., Charles, L., and Weinhold, P.A. (1990) CTP: phosphocholine cytidylyltransferase in rat lung: relationship between cytosolic and membrane forms. Biochim. Biophys. Acta 1045, 49–57.

Fujita, Y., Kogishi, K., and Suzuki, Y. (1988) Pulmonary damage induced in mice by a monoclonal-antibody to proteins associated with pig pulmonary surfactant. Exp. Lung Res. 14, 247–260.

Fujiwara, T. (1984) In: Pulmonary surfactant, Robertson, B., van Golde, L.M.G., and Batenburg, J.J. (eds.) pp. 479–503. Elsevier, Amsterdam.

Fuhrman, B.P. (1990) Surfactant therapy. Crit. Care Med. 18, 682–683.

Gaines, G.L. (1966) Insoluble monolayers at liquid-gas interfaces. John Wiley and Sons New York.

Glasser, S.W., Korfhagen, T.R., Perme, C.M., Pilot-Matias, T.J., Kister, S.E., and Whitsett, J.A. (1988) Two SP-C genes encoding human pulmonary surfactant proteolipid. J. Biol. Chem. 263, 10326–10331.

Goerke, J., and Gonzales, J. (1981) Temperature dependence of dipalmitoyl phosphatidylcholine monolayer stability. J. Appl. Physiol. 51, 1108–1114.

Gregory, T.J., Longmore, W.J., Moxley, M.A., Whitsett, J.A., Reed, C.R., Fowler, A.A. et al. (1991) Surfactant chemical composition and biophysical activity in acute respiratory distress syndrome. J. Clin. Invest. 88, 1976–1981.

Griese, M. (1999) Pulmonary surfactant in health and human lung disease: state of the art. Eur. Resp. J. 13, 1455–1476.

Haagsman, H.P., and Diemel, R.V. (2001) Surfactant-associated proteins: functions and structural variation. Comp. Biochem. Physiol. Part A. 129, 91–108.

Hallman, M., and Epstein, B. (1982) Role of myo-inositol in the synthesis of phosphatidylglycerol and phosphatidylinositol in the lung. Biochem. Biophys. Res. Commun. 92, 1151–1159.

Hallman, M., Glumoff, V., and Ramet, M. (2001) Surfactant in respiratory distress syndrome and lung injury. Comp. Bio. Physiol. Part A. 129, 287–294.

Hallman, M., Slivka, S., Wozniak, P., and Sils, J. (1986) Perinatal development of myo-inositol uptake into lung cells: surfactant phosphatidylglycerol and phosphatidylinositol synthesis in the rabbit. Pediatr. Res. 20, 179–185.

Hallman, M., Spragg, R., Harrell, J.H., Moser, K.M., and Gluck, L. (1982) Evidence of lung surfactant abnormality in respiratory-failure—study of bronchoalveolar lavage phospholipids, surface-activity, phospholipase-activity, and plasma myoinositol. J. Clin. Invest. 70, 673–683.

Harwood, J.L., and Richards, R.J. (1985) Lung surfactant. Mol. Aspects Med. 8, 423–514.

Harwood, J.L. (1987) Lung surfactant. Prog. Lipid Res. 26, 211–256.

Hass, M.A., and Longmore, W.J. (1979) Surfactant cholesterol metabolism of the isolated perfused rat lung. Biochim. Biophys. Acta 573, 166–174.

Hawco, M.W., Davis, P.J., and Keough, K.M.W. (1981) Lipid fluidity in lung surfactant: monolayers of saturated and unsaturated lecithins. J. Appl. Physiol. 51, 509–515.

Hawgood, S., Efrati, H., Schilling, J., and Benson, B.J. (1985) Chemical characterisation of lung surfactant apoproteins: amino acid composition, N-terminal sequence, and enzymic digestion. Biochem. Soc. Trans. 13, 1092–1096.

Hennes, H.M., Lee, M.B., Rimm, A.A., and Shapiro, D.L. (1991) Surfactant replacement therapy in respiratory distress syndrome: meta-analysis of clinical trials of single-dose surfactant extracts. Am. J. Dis. Child. 145, 102–104.

Ikegami, M., Jobe, A., and Berry, D.A. (1986) A protein that inhibits surfactant in respiratory distress syndrome. Biol. Neonate 50, 121–129.

Ioli, J.G., and Richardson, M.J. (1990) Giving surfactant to premature infants. Am. J. Nurs. 90, 59–60.

James, D., Berry, P.J., Fleming, P., and Hathaway, M. (1990) Surfactant abnormality and the sudden infant death syndrome—a primary or secondary phenomenon. Arch. Dis. Child. 65, 774–778.

Jobe, A.H., and Ikegami, M. (2001) Biology of surfactant. Clin. Perinat. 28, 655–669.

Jobe, A.H., and Jacobs, H.C. (1984) Catabolism of pulmonary surfactant. In: Pulmonary surfactant, Robertson, B., van Golde, L.M.G., and Batenburg, J.J. (eds.) pp. 271–293. Elsevier, Amsterdam.

Johansson, J., Curstedt, T., Robertson, B., and Jornvall, H. (1988) Size and structure of the hydrophobic low molecular weight surfactant-associated polypeptide. Biochemistry 27, 3544–3547.

Johansson, J., and Curstedt, T. (1997) Molecular structures and interactions of pulmonary surfactant components. Eur. J. Biochem. 244, 675–693.
Johansson, J., Jornvall, H., Eklund, A., Christensen, N., Robertson, B., and Curstedt, T. (1988) Hydrophobic 3.7 KDa surfactant polypeptides structural characterisation of the human and bovine forms. FEBS Lett. 232, 61–64.
Kalmar, G.B., Kay, R.J., La Chance, A.C., and Cornell, R.B. (1994) Primary structure and expression of a human CTP-phosphocholine cytidylyltransferase. Biochim. Biophys. Acta 1219, 328–334.
Kendig, J.W., Notter, R.H., and Maniscalco, W.M. (1989) Clinical experience with calf lung surfactant. In: Surfactant therapy, Shapiro, D.L., Notter, R.H. (eds.) pp. 257–271. Alan Liss, New York.
King, R.J. (1974) The surfactant system of the lung. Fed. Proc. Fed. Am. Soc. Exp. Biol. 33, 2238–2247.
King, R.J., and Clements, J.A. (1972) Surface active material from dog lung. II. Composition and physiological correlations. Am. J. Physiol. 223, 715–726.
King R.J., and Martin, H. (1980) Intracellular metabolism of the apoproteins of pulmonary surfactant in rat lung. J. Appl. Physiol. 48, 812–820.
Korfhagen, T.R., Glasser, S.W., Bruno, M.D., McMahan, M.J., Clark, J.C., and Whitsett, J.A. (1990) Isolation and analysis of the murine surfactant protein-A genomic locus. Am. Rev. Respir. Dis. 142, 4, 2, A696.
Lachmann, B., and Danzmann, E. (1984) In: Pulmonary surfactant, Robertson, B., van Golde, L.M.G., and Batenburg, J.J. (eds.) pp. 505–548. Elsevier, Amsterdam.
Maniscalco, W.M., Kendig, J.W., and Shapiro, D.L. (1989) Surfactant replacement therapy: impact on hospital charges for premature infants with respiratory distress syndrome. Pediatrics 83, 1–6.
Morley, C.J. (1985) The role of β-adrenergic agents in the control of surfactant secretion. Biochem. Soc. Trans. 13, 1091–1092.
Morley, C.J. (1991) Surfactant treatment for premature babies: a review of clinical trials. Arch. Dis. Child. 66, 445–450.
Morley, C.J., Hill, C.M., Brown, B.D., Barson, A.J., and Davis, J.A. (1982) Surfactant abnormalities in babies dying from sudden infant death syndrome. Lancet, 1320–1322.
Perelman, R.H., Farrell, P.M., Engle, M.J., and Kemnitz, J.W. (1985) Developmental aspects of lung lipids. Ann. Rev. Physiol. 47, 803–822.
Pilot-Matias, T.J., Kister, S.E., Fox, J.L., Kroop, K., Glasser, S.W., and Whitsett, J.A. (1989) Structure and organisation of the gene encoding human pulmonary surfactant proteolipid SP-B. DNA 8, 75–86.
Pison, U., Obertacke, U., Brand, M., Seeger, W., Joka, T., Bruch, J., and Schmitneuerburg, K.P. (1990) Altered pulmonary surfactant in uncomplicated and septicemia-complicated courses of acute respiratory failure. J. Trauma Injury Infection Crit. Care 30, 191–226.
Possmayer, F. (1987) In: Phosphatidate phosphohydrolase, Brindley, D.N. (ed.) CRC Press, Boca Raton.
Post, M. (1987) Maternal administration of dexamethasone stimulates choline-phosphate cytidylyltransferase in fetal type II cells. Biochem. J. 241, 291–296.
Post, M., and van Golde, L.M.G. (1988) Metabolic and developmental aspects of the pulmonary surfactant system. Biochim. Biophys. Acta 947, 249–286.
Raivo, K. (ed.) (1983) Respiratory distress syndrome. Academic Press, New York.
Robertson, B., and Taeusch, H.W. (ed.) (1995) Surfactant therapy for lung disease. Marcel Dekker, New York.
Ross, G.F., Meuth, J., and Ohning, B. (1986) Purification of canine surfactant-associated glycoproteins A. Identification of a collagenase-resistant domain. Biochim. Biophys. Acta 870, 267–278.
Sarin, V.K., Gupta, S., Leung, T.K., Taylor, V.E., Ohning, B.L., Whitsett, J.A., and Fox, J.L. (1990) Biophysical and biological activity of a synthetic 8.7 KDa hydrophobic pulmonary surfactant protein SP-B. Proc. Natl. Acad. Sci. U.S.A. 87, 2633–2637.
Scarpelli, E.M., and Mautone, A.J. (1984) In: Pulmonary surfactant, Robertson, B., van Golde, L.M.G., and Batenburg, J.J. (eds.) pp. 119–170. Elsevier, Amsterdam.

Skelton, R., Holland, P., Darowski, M., Cheteuh, P.A., Morgan, L.W., and Harwood, J.L. (1999) Abnormal surfactant composition and activity in severe bronchiolitis. Acta Paediatrica (in press).

Soll, R.F. (1991) Surfactant treatment of RDS. In: Chalmers 1, ed., Oxford Database of Perinatal Trials, Version 1.2, Disk issue 5.

Stern, W., Kovac, C., and Weinhold, P.A. (1976) Activity and properties of CTP: choline phosphate cytidylyltransferase in adult and fetal rat lung. Biochim. Biophys. Acta 441, 280–293.

Stevenson, D., Walther, F., Long, W. et al. (1992) Controlled trial of a single dose of synthetic surfactant at birth in infants weighing 500–600 grams. J. Pediatr. 120, S3–S12.

Strayer, D.S., Merritt, T.A., Lwebuga-Mukassa, J., and Hallman, M. (1986) Surfactant-anti-surfactant immune complexes in infants with respiratory distress syndrome. Am. J. Pathol. 122, 353–362.

Suzuki, Y., Fujita, Y., and Kogishi, K. (1989) Reconstitution of tubular myelin from synthetic lipids and proteins associated with pig pulmonary surfactant. Am. Rev. Respir. Dis. 140, 75–81.

Tenner, A.J., Robinson, S.L., Borchelt, J., and Wright, J.R. (1989) Human pulmonary surfactant protein (SP-A), a protein structurally homologous to CIQ, can enhance FCR-mediated and CRI-mediated phagocytosis. J. Biol. Chem. 264, 13923–13928.

Turcotte, J.G., Lin, W.H., Pivarnik, P.E., Sacco, A.M., Shirali, S.S., Bermel, M.M., Lu, Z., and Notter, R.H. (1991) Chemical synthesis and surface activity of lung surfactant phospholipid analogs. II. Racemic N-substituted diether phosphonolipid. Biochim. Biophys. Acta 1084, 1–12.

van Golde, L.M.G., Batenburg, J.J., and Robertson, B. (1988) The pulmonary surfactant system—biochemical aspects and functional significances. Physiol. Rev. 68, 374–455.

van Iwaarden, F., Welmers, B., Verhoef, J., Haagsman, H.P., and van Golde, L.M.G. (1990) Pulmonary surfactant protein-A enhances the host-defence mechanism of rat alveolar macrophages. Am. J. Respir. Cell. Mol. Biol. 2, 91–98.

Walters, D.V. (1985) The role of β-adrenergic agents in the control of surfactant secretion. Biochem. Soc. Trans. 13, 1089–1090.

Walti, H., Couchard, M., and Relier, J-P. (1991) In: Pulmonary surfactant: biochemical function, regulatory, and clinical aspects, Bourbon, J.R. (ed.) pp. 385–429. CRC Press, Boca Raton.

Weaver, T.E., and Whitsett, J.A. (1991) Function and regulation of expression of pulmonary surfactant-associated proteins. Biochem. J. 273, 249–264.

Whitsett, J.A., Hull, W., Ross, G., and Weaver, T. (1985) Characteristics of human surfactant-associated glycoprotein-A. Pediatr. Res. 19, 501–508.

Wright, J.R. (1997) Immunomodulatory functions of surfactant. Physiol. Rev. 77, 931–962.

Wright, J.R., and Clements, J.A. (1987) Metabolism and turnover of lung surfactant. Am. Rev. Respir. Dis. 135, 426–444.

Yu, S-H., and Possmayer, F. (1992) Effect of pulmonary surfactant protein B(SP-B) and calcium on phospholipid adsorption and squeeze-out of phosphatidylglycerol from binary phospholipid monolayers containing dipalmitoylphosphatidylcholine. Biochim. Biophys. Acta 126, 26–34.

Recommended Readings

Griese, M. (1999) Pulmonary surfactant in health and human lung disease: state of the art. Eur. Res. J. 13, 1455–1476.

Hills, B.A. (1999) An alternative view of the role(s) of surfactant and the alveolar model. J. Appl. Physiol. 87, 1567–1583.

Notter, R.H. (2000) Lung surfactants: basic science and clinical applications. Marcel Dekker, New York.

Scarpelli, E.M. (1998) The alveolar surface network: a new anatomy and its physiological significance. Anat. Record 251, 491–527.

4
Regulation of the Airway Caliber

JAMES R. SHELLER

The ability of the airways to change their caliber has several important consequences that are seen most dramatically in diseases such as asthma or emphysema, where airways are extensively narrowed. Regulation of airway caliber is complex, and is accomplished by physical factors, neural reflexes, and local mediators.

Function of Airway Caliber Control

Why do the airways have a capacity to change their caliber? One likely function they have is to divert airflow away from areas of the lung that are not perfused with blood and therefore cannot participate in gas exchange, thereby improving ventilation/perfusion matching. This phenomenon has been demonstrated in dogs: When blood flow was interrupted to one part of the lung, airflow was altered as a consequence. The importance of this mechanism in humans is uncertain. Another conceivable advantage of the ability to change airway caliber is to allow a cough to be more effective in removing foreign objects from the airways. Narrowing of the airway in the region of a foreign object would increase the velocity of airflow past the object, thus enhancing the chances of blowing it out. Others have suggested that narrowing of the airways in response to inhalation of noxious chemicals such as chlorine gas might alter the pattern of distribution of the noxious agent, thereby ameliorating its damage. Airway tone is present in normal humans, but abolishing it with agents that relax airway smooth muscle does not dramatically alter short term lung function emphasizing the relative unimportance of airway tone in normal humans.

Airway Structure

Comprehending how airway caliber can be altered requires a knowledge of the structure of airways, and a passing acquaintance with the physics of gasflow. The airways in humans begin in the mouth and nose and extend through as many as 23 branches from the main intrathoracic airway, the trachea, to the gas exchange surfaces. Although the diameter of the airways decreases from the trachea (diameter: 15–20 mm) to as little as 0.5 mm in terminal bronchioles, the number of the smaller airways increases dramatically, so the total cross-sectional area of the smaller airways is much greater than that in the larger diameter airways (Fig. 4.1). This means that airflow is highest in the central larger airways and that most of the resistance to airflow normally lies in the central airways.

The airways are tubular structures with multiple branches to other tubes. They are lined with a pseudostratified ciliated columnar epithelium mixed with goblet cells that secrete mucus and are perforated by mucus gland duct openings, which can pour even more water and mucus onto the airway surface (Fig. 4.2). Occasional inflammatory cells, especially alveolar macrophages are found on the airway surface, and sensory nerve fibers lie just below or between epithelial cells. The nerve fibers are often abundant at airway branch points. Beneath the epithelial layer is an interstitium in which collagen, elastin, and inflammatory cells reside. A smooth muscle layer is next, where the fibers form a complex geodesic structure that surrounds the airway. Smooth muscle fibers that are oriented circumferentially thus will constrict the lumen when the muscle shortens. Fibers with a more longitudinal orientation both narrow the airway lumen and shorten the length of the airway. (In the trachea, a posterior membrane contains a transverse band of smooth muscle that can narrow the trachea). Cartilage provides stability in larger airways. The larger airways are contained in a bronchovascular sheath

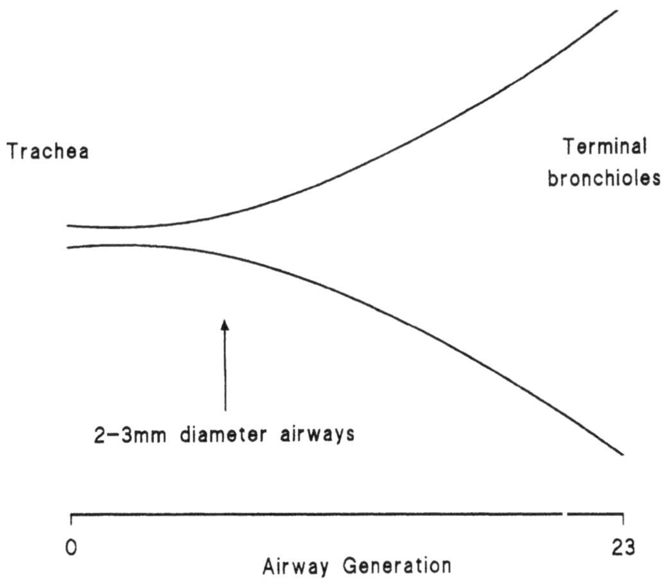

FIGURE 4.1. Schematic representation of the total cross-section of human airways of varying generation. Although the diameter of airways decreases as airway generation (branching) increases, the total cross-sectional area is greatest in the smallest airways.

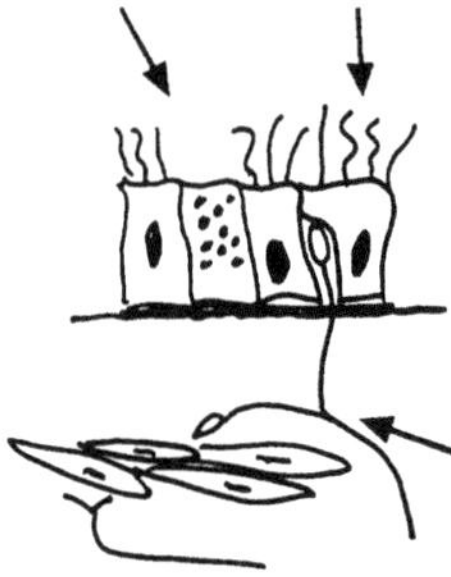

FIGURE 4.2. Cartoon of some of the cellular elements of an airway. Note the sensory nerve fiber between the two epithelial cells.

along with vessels, lymphatics, and nerves; these airways are exposed to pleural pressure. The larger airways have their own blood supply, viz., the bronchial circulation. Smaller airways, terminal bronchioles, are tethered or suspended by fibers that extend from the lung parenchyma. Thus, the composition and structural arrangement of the airways vary, depending on the size and location of the airway.

Neural Innervation

The neural innervation of the airways is complex and important for control of airway caliber. Fibers from the parasympathetic vagus nerve innervate the larger airways, and postganglionic vagal fibers are found near smooth muscle and mucus glands. Sympathetic fibers are also present, but their importance for airway caliber regulation in humans appears to be slight. Vagal innervation causes normal individuals to have smooth muscle tone; inhalation of atropine, which blocks acetylcholine (ACh) receptors, increases airway caliber in normal humans (the source of ACh is the vagal postganglionic efferent fiber). Sensory fibers abound in the airways and can be activated by mechanical or chemical stimuli; the sensory traffic results in reflex activation of vagal efferent fibers that in turn causes airway smooth muscle constriction and airway narrowing. Thus, it can be said, that the tone of bronchial smooth muscle under normal physiological conditions is largely governed by the autonomic nervous system, the predominant discharge being parasympathetic. During inspiration the bronchi dilate (sympathetic discharge), whereas during expiration they contract (parasympathetic). Ordinary examples of reflex bronchoconstriction include exposure to irritants and chemicals such as sulfur dioxide. The underlying mechanism is mediated by the cholinergic pathway. In striking contrast, the major determinants of tone in **isolated** human airways are cysteinyl-leukotriene (Cyst-LTs) and histamine which act as constrictors, and prostanoids which act as dilators (Schmidt and Rabe, 2000).

In addition to the classic adrenergic and cholinergic mediators released from autonomic nerves, a new class of mediators, grouped under the term *non—adrenergic, non—cholinergic* (NANC), has been discovered in the airways. Several of the NANC mediators have been identified as peptides, and they are similar to those found in the gut. They can be released from nerve terminals along with the classical mediators. For example, the smooth muscle relaxing peptide, vasoactive intestinal peptide (VIP), has been identified in cholinergic nerve terminals. It is presumably released together with ACh. In addition, neuropeptides can be released from sensory nerve terminals by an "axon reflex." Stimulation of one branch of the sensory nerve fiber results in activa-

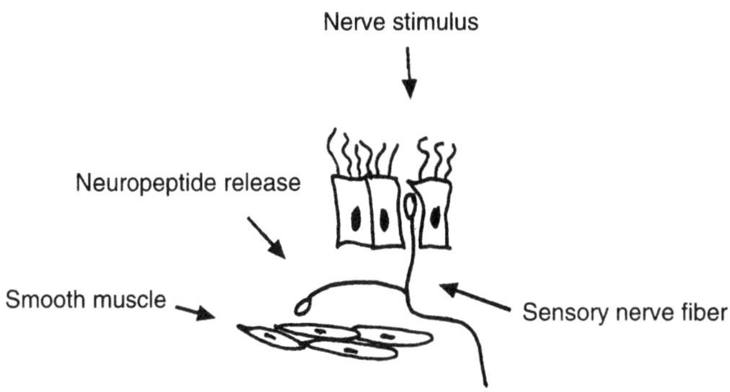

FIGURE 4.3. Cartoon of an axon reflex in an airway sensory nerve fiber. A stimulus to the epithelial branch leads to activation of the nerve fiber leading to smooth muscle, with consequent neuropeptide release and smooth muscle shortening.

tion of another branch, causing local release of neuropeptide transmitters (Fig. 4.3). The neuropeptides have potent effects on airway smooth muscle, both constrictor and relaxant. They can also increase vascular permeability (leading to edema) and can cause mucus secretion. The NANC system clearly influences airway caliber in experimental animals, and the role it plays in humans is now the subject of much research.

Although adenosine (A1 receptor), substance P, and numerous cytokines are known to act as bronchoconstrictors, their physiological role in the regulation of bronchial tone remains unclear. In addition, there is a circadian rhythm in bronchial tone: constriction is maximal at about 6 a.m. and dilation at about 6 p.m.

Inflammatory Cells

The airways contain numerous inflammatory cells that are capable of secreting a variety of peptide and lipid mediators that can have profound effects on smooth muscle tension, vascular permeablity, and mucus secretion. Cells such as mast cells are thought to be crucial for allergic mediated airway narrowing. Eosinophils, alveolar macrophages, and even airway epithelial cells can affect airway smooth muscle and therefore affect airway caliber.

Mechanisms of Airway Caliber Changes

Change in airway smooth muscle tone is an important means of regulating airway caliber, and airway smooth muscle itself is influenced by a variety of regulatory mechanisms, including mediators released from nerve terminals and local inflammatory cells, and from circulating hormones such as epinephrine. Because smooth muscle constriction is relatively rapid (seconds), it can modulate airway caliber on a minute-by-minute basis. It is assumed (probably not correctly) that rapid changes in airway caliber caused by smooth muscle agonists such as ACh, which contracts it, and albuterol, which relaxes it, are caused by smooth muscle alone. In normal humans only small changes in airway caliber can be detected after inhalation of albuterol, which suggests that smooth muscle tone is relatively slight.

Airway narrowing can also result from partial filling of the airway lumen by mucus or water (Fig. 4.4). Hypersecretion of mucus is important in the airway narrowing seen in diseases such as cystic fibrosis and chronic bronchitis from cigarette smoke. Mucus glands are under many of the same regulatory mechanisms as is smooth muscle. The airway can be narrowed by edema (swelling) and infiltration of the layer between the smooth muscle and airway wall. For example, several of the neuropeptides can cause

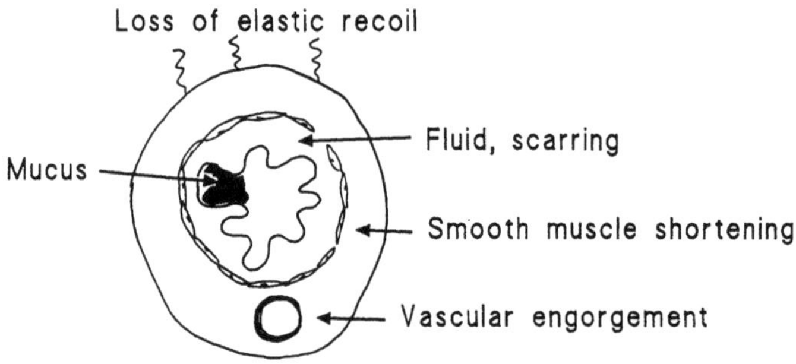

FIGURE 4.4. Cartoon of mechanisms of airway narrowing. Fluid accumulation or vascular engorgement in the bronchovascular sheath will compress the airway lumen. Narrowing of the lumen or partial filling with mucus or fluid will amplify the effect of smooth muscle shortening on resistance to airflow.

blood vessels to become leaky, which could fill the interstitium with fluid and compress the airway lumen. Engorgement of the bronchial vessels in the bronchovascular sheath may physically narrow airways; likewise, lymphatic enlargement could narrow airways. Loss of the tethering of airways in diseases such as emphysema can result in airway narrowing, especially during forced exhalation (see later).

Upper Airway Function

The upper airway serves as a passage for the inhalation of air (oxygen) in and exhalation of air (carbon dioxide) out of the lungs. In addition, the upper airway heats, humidifies, and filters the air and plays a key role in coughing, swallowing, and speech. In some people there may be partial collapse of the upper airway and increased upper airway resistance to airflow during sleep. The collapse can be severe enough, thus leading to sleep apnea (Pierce and Worsnop, 1999).

Measurement of Airway Caliber

Assessing airway caliber is an important physiologic and clinical task, but the most direct methods (e.g., measuring airway diameter under the microscope) are difficult to perform on living human beings! Anatomic methods have been used to investigate airway caliber experimentally and have been very useful, even crucial, in validating findings using techniques kinder to the individuals involved. Measurements can be made of airway resistance using a device known as a *body plethysmograph*. Physicians more commonly rely upon measures of forced expiratory flow that require a patient to inhale a deep breath, then forcibly exhale into or through some sort of device that measures volume or flow. (The reason why exhaled flow is measured will become apparent later.) How does this measurement reflect airway tone? The reason is actually a little involved, and has to do with how the airways are coupled to the pressuresthat cause airflow.

The resistance to laminar (nonturbulent) airflow in the lung (or in any tube) is given by Poiseuille's equation:

$$R = 8L\mu/\pi r^4$$

where L = the length, r = the radius of the tube
μ = the viscosity of the gas.

It is often pointed out that a change in the radius of the tube has a much more profound effect on resistance than does a change in length. This equation is rather complex to solve for the lung with its multiple branching airways of different dimensions, but resistance fortunately can be measured as the ratio of driving pressure divided by flow:

$$R = \Delta P/\text{Flow}$$

The driving pressure is the force or pressure required to push the air through the airways. The driving force for airflow in humans is the alveolar pressure minus atmospheric pressure (Fig. 4.5). The alveolar pressure is a function of pleural pressure, which is a function of muscular effort and the elastic recoil pressure of the lung. This needs some explaining. The lungs have a tendency to collapse from their fully inflated state, total lung capacity, to residual volume (the point at which no more air can be forced out of the lung). This recoil pressure diminishes as residual volume is approached (Fig. 4.6). During exhalation pleural pressure is positive and the recoil pressure of the lung sums with it to yield alveolar pressure. Thus, by knowing alveolar pressure and flow, the resistance of the airways can be calculated. This resistance has to be a reflection of airway caliber.

During a forced exhalation as flow occurs from the periphery of the lung to the mouth, some of the driving pressure is dissipated in direct proportion to the resistance of airways through which the gas has flowed. This results in a lower pressure within more central airways. At some point the pressure in the airway will equal that in the pleural space (Fig. 4.7). This point of equal pressure is called, remarkably, the *equal pressure point.* Because larger airways are exposed to pleural pressure, the airways downstream of the equal pressure point will be compressed, decreasing airway caliber. Once an equal pressure point has been generated further effort will not increase flow one bit because further increases in pleural pressure only serve to narrow the airway further. As lung volume diminishes, the recoil pressure of the lung diminishes, and the equal pressure point moves toward the periphery of the lung, allowing more of the central airways to be compressed. Flow therefore continues to decrease, leading to the characteristic shape of the expiratory flow volume curve, where flow is greatest at

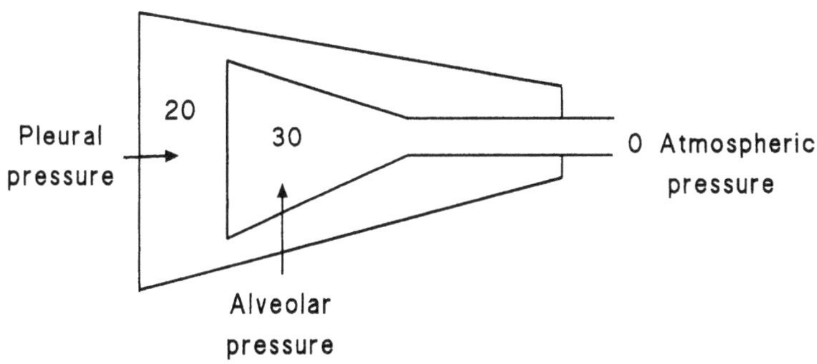

FIGURE 4.5. Schematic representation of the pleural space, alveoli, and conducting airways. Alveolar pressure during exhalation is the sum of the elastic recoil pressure of the lung and pleural pressure. The elastic recoil is 10 cm H_2O in this example at this lung volume. Thus, alveolar pressure is elastic recoil pressure, 10 cm H_2O, plus pleural pressure, 20 cm H_2O. It is 0 because pressures are referenced to atmospheric pressure.

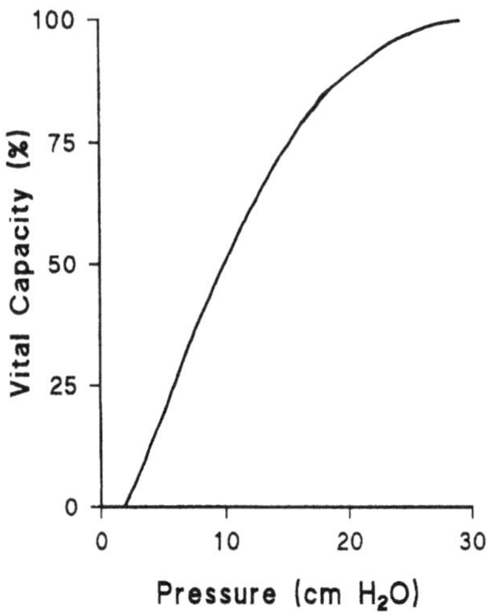

FIGURE 4.6. Volume pressure curve of the normal lung. The pressure required to hold the lung at a particular volume during deflation is known as the *recoil* pressure of the lung. This pressure decreases as lung volume shrinks toward residual volume.

high lung volumes and least near residual volume (Fig. 4.8). Thus, airway caliber control is a dynamic process, which can be influenced by lung volume and effort. The reason that airways are distensible in such a fashion may have to do with increasing the effectiveness of cough, and to allow expansion and lengthening of the airways as the lung expands.

The equal pressure point in normal humans is within the major bronchi. In disease where airway caliber is decreased and/or lung recoil diminished, the equal pressure point moves more peripherally, a greater length of airway is subjected to compressive forces, and airflow is reduced at all lung volumes in comparison with normal (Fig. 4.9). From measurements of forced exhaled flow, it is not possible to tell if increased airway

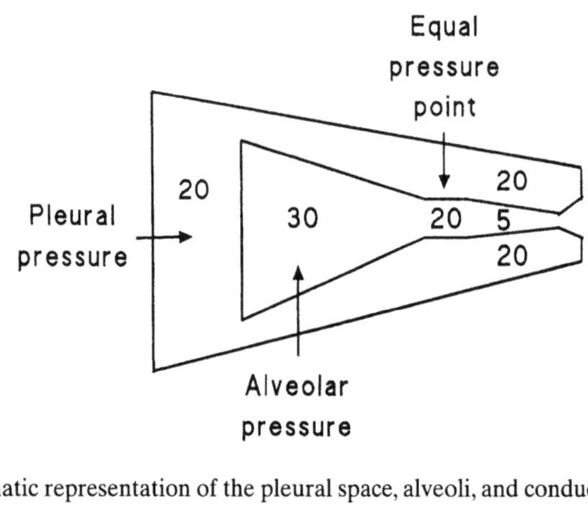

FIGURE 4.7. Schematic representation of the pleural space, alveoli, and conducting airways during a forced exhalation. As pressure decreases from alveoli to mouth (in order to overcome the resistance to flow in the airways), a point is reached when the pressure within the airway is equal to the pressure in the pleural space. From this equal pressure point onward to the mouth the airway will be subjected to increasing pressure tending to narrow it. Further increases in pleural pressure caused by muscular effort will not change the position of the equal pressure point or increase flow at this lung volume. (Lung volume is assumed to be measured in a body plethysmograph so as not to neglect compressive changes.)

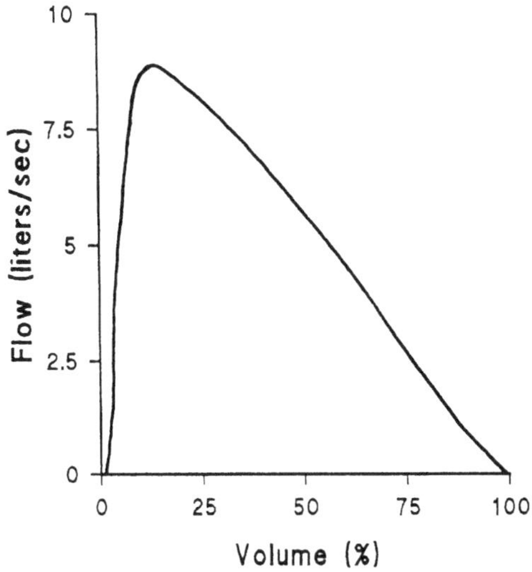

FIGURE 4.8. A plot of exhaled flow versus exhaled volume as a percentage of vital capacity. Exhaled flow is greatest near total lung capacity, but as more volume is exhaled and the lung shrinks in size, elastic recoil pressure is less, more compression of central airways occurs, and flow becomes progressively limited.

resistance or loss of elastic recoil or both is responsible for the decreased flow. Although an indirect and complex function of airway caliber, measurement of forced exhalation has proven to be very useful in clinical physiology as an estimate of airway diameter. One variable derived from forced exhalation is the volume exhaled in the first second, the FEV_1.

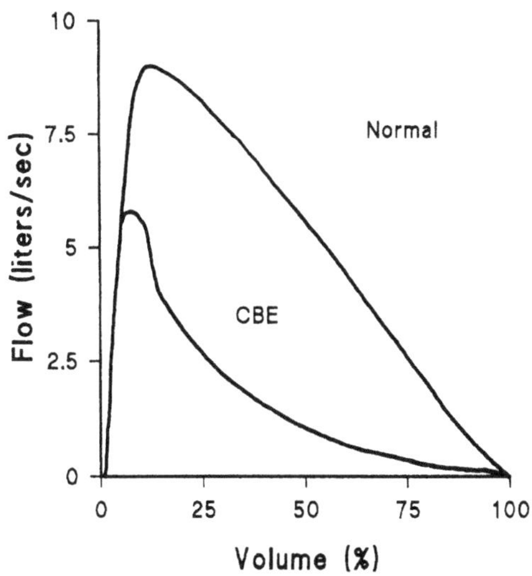

FIGURE 4.9. A plot of exhaled flow versus exhaled volume as a percentage of vital capacity from a normal individual and from a patient with obstructive lung disease, chronic bronchitis and emphysema (CBE). The flow at all lung volumes is reduced in the patient with CBE.

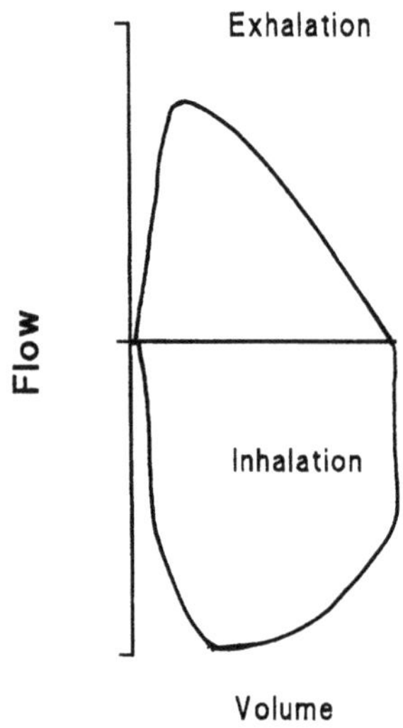

FIGURE 4.10. A plot of exhaled and inhaled flow versus volume as a percentage of vital capacity in a normal individual. Inhaled flow is not a function of lung volume; rather, it depends on inspiratory muscle effort because airways are not subject to compressive forces during inspiration when pleural pressures are negative.

It is instructive to contrast exhalation with inhalation: During inspiration pleural and alveolar pressures become negative, but pleural pressure is always more negative because of the elastic recoil pressure of the lung. Thus, there is always a positive distending pressure in the airway, no equal pressure point develops, and flow is limited by the ability of the inspiratory muscles to change pleural pressure, not by a compressed segment of airway (Fig. 4.10).

The tethering of the smaller airways to the surrounding lung parenchyma and the exposure of the larger airways to pleural pressure means that the airways are stretched open somewhat during a full inspiration. This increase in airway caliber caused by a deep breath is reflected by a fall in airway resistance measured at higher lung volumes. In addition, bronchoconstriction induced in a normal person can be reduced somewhat by a deep breath, presumably by causing temporary relaxation of the stretched smooth muscle (stress relaxation). Just the opposite may happen in asthma, when a deep breath causes further bronchoconstriction rather than airway dilatation. The reason for this is uncertain.

Nonspecific Airway Hyperreactivity

The regulation of airway caliber is most pronounced in persons with nonspecific airway hyperreactivity. Airway hyperreactivity is characterized by an extreme sensitivity of the airways to a variety of chemical or physical agents. If a patient with asthma, which is an inflammatory disease of the airways that is accompanied by increased airway reactivity, inhales increasing amounts of a smooth muscle constrictor, such as ACh, airway narrowing as measured by expired flow (or any other means) is caused (see Fig. 4.9). In contrast, normal persons have only small changes noted, even when large amounts of the agonist are inhaled.

Exercise and cool air can induce bronchoconstriction in individuals with increased nonspecific airway reactivity. The mechanism for exercise-induced bronchoconstriction appears to be heat loss from the airway. During normal breathing air is humidified and warmed by the nasal passages and larger airways so that by the time it reaches the first branching, the carina, the air is at body temperature (37°C) and is 100% saturated with water. This humidification and warming presumably protects the more delicate gas exchange surfaces from drying and cooling. With exercise the volume of air inhaled may increase 10–20-fold greater than resting values, and the increased airflow will overwhelm the ability of the upper airways to warm and humidify the air, especially when the air is cold and dry. It is currently thought that mediators released from inflammatory cells (e.g., as mast cells are responsible for the constriction). Exercise-induced bronchoconstriction occurs in proportion to the response to other agents (e.g., inhaled histamine or ACh). Thus, if a person reacts substantially to inhaled ACh, another nonspecific agent such as exercise will also provoke a substantial response.

Nonspecific airway reactivity is contrasted with "specific" airway reactivity in which persons develop airway narrowing to specific allergens (e.g., house dust). Although virtually all patients with asthma have nonspecific airway reactivity, those who have no skin test reaction to house dust, no IgE directed against house dust, will not respond to inhaled house dust (except perhaps as a nonspecific irritant) (see Fig. 4.11).

The mechanisms behind the development of nonspecific airway reactivity are unclear, although it presumably involves an exaggeration of the mechanisms of airway caliber regulation. Airways in patients with asthma often show an increase in the number and size of airway smooth muscle cells, a damaged epithelium, increased numbers of inflammatory cells (e.g., eosinophils and mast cells), increased amounts of mucus in the airway lumen, and a thickening of the subepithelial layer. Compare these findings with some of the mechanisms of bronchoconstriction given earlier and it will be noted that almost all are represented. The damaged epithelium may be the substrate for heightened sensory receptor sensitivity, which increases reflex bronchoconstriction. A degree of nonspecific airway hyperreactivity can be induced in normal people by viral infections of the airways or by transient exposure to ozone, which is an air pollutant, both of which can damage the airway epithelium and cause airway inflammation.

Attempts have been made to quantify short-term variation in airway caliber. For example, Macklem's group were able to measure respiratory impedance (Zrs) in both normal subjects and asthmatic patients before and after administering methacholine (a bronchoconstricting agent). The conclusions they reached were that the control of

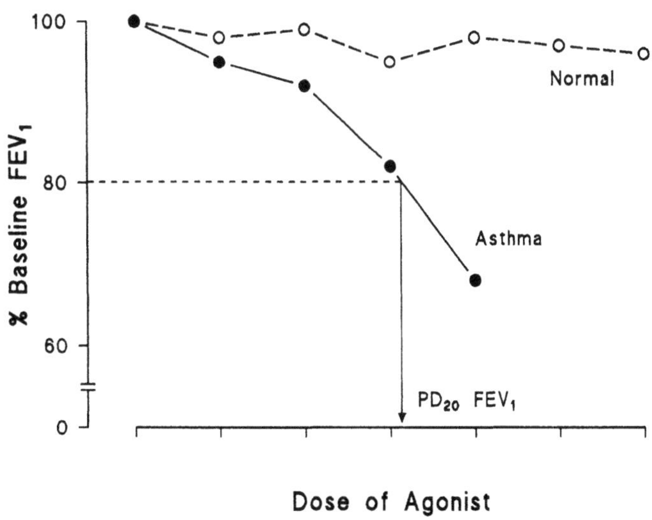

FIGURE 4.11. A plot of the FEV_1 versus dose of an inhaled bronchoconstrictor in a normal individual and in a person with nonspecific airway hyperreactivity. The normal person inhales increasing amounts of the agonist with little if any change in airway caliber. The person with airway hyperreactivity has a pronounced fall in FEV_1 with only small doses of agonist. The dose–response curve is often characterized by the dose of agonist, causing a 20% fall in FEV_1 as compared with baseline, which is shown on the graph by the arrow indicating the $PD_{20}FEV_1$.

airway caliber in normal lungs is homeokinetic in that variation is maintained within acceptable limits and that the configuration of the normal tracheobronchial tree is continuously changing. But ion the asthmatic patient this change is exaggerated. If, however, normal airway smooth muscle is activated and unloaded by increasing shortening velocity, its behavior resembles that of the asthmatic (Que et al., 2001).

Case Study 1: Asthma

A 13-year-old black female is brought by her parents to the physician with a complaint of episodic shortness of breath. She says that whenever she laughs or plays volleyball she has coughing fits and feels as if her chest is tight. After a bout of viral bronchitis that swept her school, she was sick with cough and noisy breathing for several weeks. Her teachers have commented on her noisy breathing. She has a history of hay fever and sneezes when she is around cats.

This classic history of asthma demonstrates several critical features of the disease, beginning with the episodic nature of the illness. Patients can be completely without symptoms at times, but crippled by airflow obstruction at others. The provoking features of bronchoconstriction—in her case, exercise, laughing (which requires a deep breath), and viral infections—are all typical. Note that she did not complain of difficulty getting air out; although airflow obstruction should result in this complaint, it is almost never prominent. Patients instead note chest "tightness," difficulty breathing, and noisy breathing. Cough is often prominent, troublesome, and may be the only complaint. The etiology of the cough is uncertain; sensory receptor sensitization and smooth muscle constriction have been evoked as mechanisms. It can be treated by agents that relax smooth muscle or by antiinflammatory drugs (e.g., steroids). Most children with asthma have atopy; IgE mediated skin and or nose and eye responses to allergens such as dust, grasses, trees, and animals. When questioned more directly, she and her parents may note that she also has symptoms of asthma when she comes in contact with cats, to which she is also allergic.

Her physical examination is unremarkable except for wheezing (rhonchi) on forced exhalation.

The physical examination of an asthmatic patient may reveal nothing abnormal; in fact, the physical exam is most helpful in making certain that no other illness (e.g., heart failure) is present. The girl did have wheezing when she blew out forcibly. This is not a specific feature of asthma, and any other disease that results in airway narrowing could also produce this finding (i.e., the wheezing noise being produced by turbulent airflow).

Forced expiratory flow was measured showing a decreased FEV_1 when corrected for her race, and inhalation of a β_2 agonist resulted in improvement in this measure.

Although the measures of forced expiratory flow can be normal in asthma, because the changes in airway caliber are often episodic, she did show some diminution in exhaled airflow. This was improved by a smooth muscle relaxing agent, which suggests that at this point in her disease the airway tone was caused by smooth muscle constriction as opposed to some other mechanism (e.g., obstruction of the airways with mucus). The reason for the increased muscle tone in asthma is usually unknown. During experimental allergen exposure there is evidence of mast cell mediator release; many of the mast cell mediators are potent constrictors of smooth muscle. Although normal individuals may improve a normal FEV_1 slightly with inhaled β_2 agonists, the substantial response to β_2 agonists in this patient implies considerable airway smooth muscle tone. In fact, this response to bronchodilators is analogous to the heightened airway narrowing that this patient would show should she inhale acetylcholine or histamine; a significant response to a bronchodilator implies increased nonspecific airway reactivity. It is usually not necessary to measure nonspecific airway reactivity in patients with asthma, although it is done for research purposes and in cases where the diagnosis is uncertain. Once again, inspect Fig. 4.11.

Case Study 2: Chronic Bronchitis and Emphysema

A 67-year-old white male visits his physician with complaints of a productive cough and shortness of breath (dyspnea) when he climbs a flight of stairs. He has no history of heart disease, but has smoked cigarettes for 50 years, usually consuming a pack per day. He dates the onset of his dyspnea to several weeks ago.

The pathophysiology of this common consequence of cigarette smoke pollution is quite different from asthma, despite the fact that airway narrowing is part of both illnesses. Chronic bronchitis and emphysema (CBE) are not associated with the dramatic airway narrowing of asthma, nor is the airway narrowing completely reversible. The airway and parenchymal damage caused by cigarettes in susceptible persons (perhaps 10% of smokers) is slow to develop. Although airway smooth muscle hypertrophy and hyperplasia is present in some patients with CBE, the airway narrowing is not fully reversible with inhaled β_2-agonists (or, for that matter, with any combination of medicines). Airway luminal narrowing in CBE results in part from mucus hypersecretion and fibrosis. In addition, there is loss of elastic recoil of the lung, so the tethering force tending to hold airways open is reduced, and the equal pressure point therefore tends to move more peripherally. The result is a decrease in measures of forced expiratory flow. Because the loss of airway caliber occurs slowly over decades, CBE is especially pernicious. Patients usually do not notice their loss of function; rather, they ascribe their dyspnea to "old age," and 50–75% of their lung function is lost, usually irretrievably, by the time they come to the notice of a physician. (The reason that 50% or more of lung function can be lost without symptoms reflects the enormous reserve capacity of the lung that is necessary for exertion, but not for resting activities.) There is some evidence that persons with nonspecific airway reactivity are at greater risk for loss of lung function with cigarettes. If the patient had had measurements of forced exhaled flow made at age 27, 37, and 47, then he would have shown an acceleration of the normal loss of FEV_1 which accompanies aging (perhaps 25 ml/year; elastic recoil is lost with age). His physician might then have been able to convince him to stop smoking before the irreversible loss of lung function became capable of causing symptoms (of course, no one should smoke, ever). The insidious onset of CBE can be detected by careful history; this patient actually stopped playing tennis at age 47 because of dyspnea, and golf at age 57 for the same reason.

Treatment with cigarette cessation, β_2-agonists, other smooth muscle relaxants, inhaled steroids, and antibiotics for infected sputum does help patients with CBE, but reversal of the disease is almost never achieved.

Summary

Airway caliber regulation is accomplished by a variety of neurohumoral and physical means. Airway smooth muscle causes rapid changes in airway caliber, and the smooth muscle is regulated by local release of mediators by inflammatory cells and nerve terminals. Airways can also be narrowed by scarring, engorgement of blood vessels and lymphatics sharing the same bronchovascular sheath, and by edema. The lumen of airways can be filled with mucus and inflammatory debris. Loss of lung elastic recoil results in decreased maximal expiratory flow. Airway caliber is affected by physical factors such as lung elastic recoil and pleural pressures. Exaggeration of airway caliber changes is characteristic of nonspecific airway hyperreactivity, which is a finding present in patients with asthma. Treatment of airway smooth muscle constriction is helpful in some diseases of the airways like asthma. A better understanding of the mechanisms of airflow regulation in obstructive lung diseases should lead to better and more specific therapy.

Recommended Readings

Black J.L., and Johnson, P.R. (1996) Airway smooth muscle in asthma. Respirology 1(3), 153–158.

Hicks, G.H. (2000) Cardiopulmonary anatomy and physiology. Mosby, St. Louis.

Macklem, P.T. (1998) The physiology of small airways. Am. J. Resp. Crit. Care Med. 157, S181–183.

Pare, P.D., Bai, T.R., and Roberts, C.R. (1997) The structural and functional consequences of chronic allergic inflammation of the airways. Ciba Fdn. Symp. 206, 71–86; discussion 86–89, 106–110.

Que, C.L., Kenyon, C.M., Olivenstein, R., and Macklem, P.T. (2001) Homeokinesis and short-term variability of human airway caliber. J. Appl. Physiol. 91(3), 1131–1141.

Schmidt, D., and Rabe, K.F. (2001) The role of leukotrienes in the regulation of tone and responsiveness in isolated human airways. Amer. J. Resp. Crit. Care Med. 161 (Pt 2), S62–67.

Sterk, P.J. (1998) The way forward in asthma: integrative physiology. Can. Resp. J. (suppl. 5A), 9A–13A.

Stulbarg, M.S., and Frank, J.A. (1998) Obstructive pulmonary disease. The clinician's perspective. Radiol. Clin. N. Am. 36, 1–13.

5
Receptors and Reflexes of the Respiratory Tract

GIUSEPPE SANT'AMBROGIO

Upper Airway Receptors and Reflexes

The upper airway comprises the extrathoracic portion of the airway and includes the nose, the nasopharynx, the oropharynx, the larynx, and the cranial half of the trachea. The upper airway has respiratory and nonrespiratory functions: the latter includes olfaction, air conditioning, chewing, swallowing, and vocalization. Some of these functions are sometimes in conflict with breathing and a well-developed control mechanism is often required to overcome these difficulties.

The larynx is an important component of upper airway reflexology, which is reflected in the abundant nervous supply and related reflex responses. The nose also has an important role, and these two components will constitute the main subject of this section.

The superior laryngeal nerve is the main site of laryngeal afferent activity. Clear respiratory modulation is found when recording from the peripheral cut end of the superior laryngeal nerve in most mammals. This activity occurs even when the larynx is functionally bypassed during tracheostomy breathing. It is greatly augmented, however, when breathing is diverted through the upper airway, especially when the animals perform inspiratory efforts against occluded upper airways. This demonstrates that the presence of airflow, and even more so changes in pressure, represent important stimuli for laryngeal afferents.

Four sensory modalities have been identified during recording from single units of the superior laryngeal nerve: pressure, flow (cold), "drive," and irritant. Drive receptors are those that are activated by the action, active and/or passive, of laryngeal or respiratory muscles. "Irritant receptors" lack a clear respiratory modulation and share some of their discharge pattern with the similarly denominated tracheobronchial endings. Flow/cold receptors are stimulated in most conditions during upper airway breathing, and remain silent during tracheal breathing, upper airway, and tracheal occlusion. The adequate stimulus for these endings is a decrease in temperature of the laryngeal mucosa: They remain silent at body temperature and start firing at lower temperature. Laryngeal pressure receptors are markedly stimulated during upper airway occlusion at end inspiration or expiration, depending on whether the adequate stimulus is positive (distending) or negative (collapsing) pressure. These receptors are frequently active during upper airway breathing, and most of them are responsive to negative pressure.

Laryngeal "drive" receptors are stimulated at each breath, even in the absence of airflow and pressure. By cold blocking the recurrent nerve or the external branch of the superior laryngeal nerve [SLN], reversible laryngeal paralysis has been used to show that it is the contraction of intrinsic laryngeal muscles that frequently stimulates these endings. Tracheal tug, which is motion due to the action of chest wall muscles transmitted to the larynx through the trachea, can also be identified as an excitatory influence for some of these endings. The term *drive* introduced to identify these endings describes the activating stimulus exerted on the respiratory center that drives both laryngeal and chest wall pump muscles. Recording from trigeminal afferents of the posterior nasal nerve, infraorbital nerve, and anterior ethmoidal nerve has disclosed the presence of nasal "drive" endings, which reflect the respiratory activity of nasal muscles. Other experiments have evaluated the effect of laryngeal cooling on breathing pattern and the maintenance of upper airway patency. Laryngeal cooling was found to have a depressive influence on the activity of the PCA muscle in anesthetized dogs (Sant'Ambrogio et al., 1995) and bronchoconstrictive effects in cats (Jammes et al., 1983). Moreover, laryngeal cooling produced a strong ventilatory depression in newborn puppies and adult guinea pigs (Orani et al., 1991). Whereas in the dog (adults and newborns) the prevailing or unique, afferent pathways for these reflexes was in the superior laryngeal nerve, in the guinea pigs there was a substantial contribution of nasal afferents (Orani et al., 1991). The reflex function of drive receptors could be estimated by comparing the response to tracheal occlusion before and after section or block of the SLN (Sant'Ambrogio et al., 1995). No significant differences in breathing pattern and activities of patency maintaining muscles could be detected between the two conditions, which suggests a scarce reflex contribution from drive receptors for other functions. However, we should consider the possible role of these receptors in other functions (e.g., vocalization). It became clear in some of the experiments designed to evaluate the role of pressure sensors and patency maintaining reflexes, that other areas of the upper airway were involved in addition to the larynx (i.e., the nose) (Horner et al., 1991).

In addition to receptors having a regular respiratory modulation, there are receptors that are either silent or randomly active in control conditions. They can, however, be promptly recruited when the laryngeal mucosa is exposed to mechanical or chemical stimuli (Anderson, 1990a,b). In fact, these receptors respond to several tussigenic stimuli and should be recognized as the triggering devises for the cough reflex from the larynx. Similar receptors have been described within the nasal cavity in cats and rats. A significant proportion of pressure and drive receptors can be stimulated by solutions of low osmolarity and inhibited by carbon dioxide (Anderson et al., 1990a,b). On the other hand, laryngeal irritant receptors are stimulated by solutions that lack permeant anions. Moreover, many of them could be stimulated by carbon dioxide. The experiments by Nolan et al. (1990) are of interest. They showed that introducing carbon dioxide into the isolated upper airway has strong excitatory effects on genioglossus activity while it depresses ventilation. These responses are similar to those induced by negative pressure in the upper airway. Nolan et al. (1990) suggested the possibility of an interaction of negative pressure and carbon dioxide in the case of an obstructive sleep apnea that "may act reflexly and synergistically to reverse the obstruction."

Innervation of the Tracheobronchial Tree and Lungs

The main nerve supply to the tracheobronchial tree and the lungs is through the vagus nerve with an additional contribution of sympathetic fibers. In a study on the pulmonary branches of the vagus nerve of the cat, based on observations with light and electron microscopy, Jammes et al. (1982) found a total of 5558 afferent fibers entering the lobes of the left lung. Of these only about 8.5% were myelinated (i.e., the unmyelinated fibers were found to be 10.8 times more numerous than their unmyelinated counterparts). In newborns the prevalence of unmyelinated fibers becomes even greater; for example, newborn kittens have only 10% of the number of myelinated fibers found in the adult cat (Marlot and Duron, 1979).

The extrathoracic trachea receives its afferent supply through the superior laryngeal nerve in its most cranial portion and the recurrent and pararecurrent laryngeal nerve in its caudal portion (Lee et al., 1992).

Taking into account different circumstances for their activation, conduction velocities, and blocking temperatures of their respective fibers, as well as their preferential circulatory accessibility, tracheobronchial and pulmonary receptors can be categorized as of four types: slowly adapting stretch receptors, rapidly adapting stretch receptors (or irritant), C-fiber bronchial receptors, and C-fibers pulmonary receptors.

Slowly Adapting Stretch Receptors

Discharge Pattern and Location

Slowly adapting stretch receptors (SARs) constitute the most conspicuous component of the afferent activity recorded from the peripheral cut-end of a cervical vagus nerve. In fact, the inspiratory augmenting pattern of activity recordable from the vagus nerve is essentially derived from these endings, which also accounts for most of the activity present at functional residual capacity (Fig. 5.1). The fibers of these receptors have conduction velocities characteristic of myelinated axons (Coleridge and Coleridge, 1986). Nervous conduction in these fibers is blocked at temperatures ranging from 5 to 10°C, which is another characteristic of myelinated axons (Pisarri et al., 1986). Although the vast majority of these receptors, which are located in intrathoracic airways, show a regular increase in discharge frequency during inspiration, some of them, with locations in the extrathoracic trachea, increase their firing rate during expiration

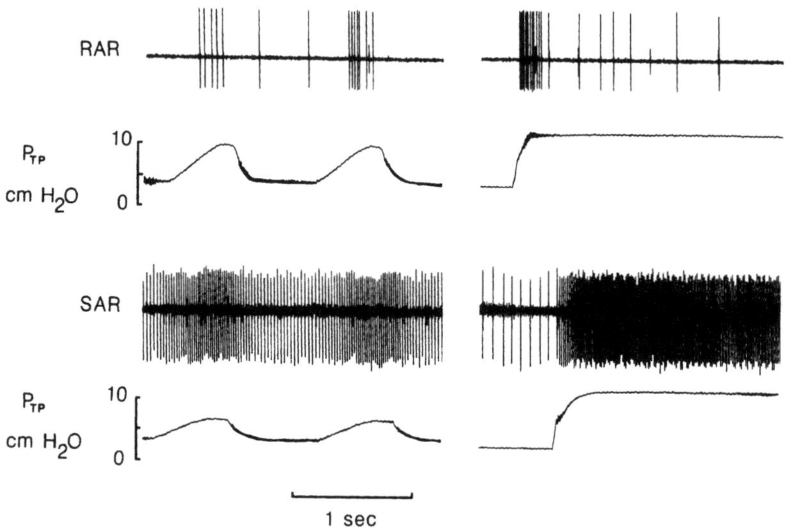

FIGURE 5.1. Action potentials recorded from a rapidly adapting receptor (RAR, top records) and a slowly adapting receptor (SAR, bottom records) of an adult opossum during passive ventilation with the chest open (records to the left) and a maintained, constant pressure, inflation (records to the right). P_{tp} = transpulmonary pressure. Note the irregular interspike interval of the RAR during passive ventilation and the rapid adaptation of the receptor discharge to a maintained pressure. Both types of receptor exhibit a respiratory modulation. (*Source*: Reprinted, with permission, from Farber et al., 1983, Am. J. Physiol., 245, R209).

(Sant'Ambrogio and Mortola, 1977). In either case, the increasing transmural pressure, which is higher in inspiration in the intrathoracic airways but higher in expiration in the extrathoracic trachea, dictates their activation. A sizable proportion of SARs (between 27 and 63% in the various species considered) (see Sant'Ambrogio, 1982 for refs.) maintain a discharge at functional residual capacity.

Slowly adapting stretch receptors have a long-lasting discharge in response to maintained lung inflation. These receptors show a rapid decline in activity immediately after the inflation that slows progressively into a sustained firing rate when challenged with a long-lasting distension (Fig. 5.2). This behavior identifies these endings as slowly adapting receptors and has been attributed to the viscoelastic properties of the microenvironment site of the endings.

Tracheal stretch receptors are uniquely localized within the membranous posterior wall; removal of the tunica fibrosa overlying the receptor field does not markedly affect receptor activation, which suggests a location of these endings within the trachealis muscle. Similar evidence was produced for SARs located in the main stem bronchus and lobar bronchi (Bartlett et al., 1976). Moreover, additional evidence based on reflex study and receptor recording strongly supports an association with the smooth muscle of these receptors (Bartlett et al., 1976).

Although SARs have been found throughout the tracheobronchial tree from the trachea to the terminal bronchioles, their concentration is greater within the larger airways (Fig. 5.3; see Sant'Ambrogio, 1982, for specific references). At variance with these results are the data of Ravi (1986) obtained in cats that support a prevalent location within the lung parenchyma. Keller et al. (1989) found in the guinea pig that the great majority (92%) of stretch receptors were located in small peripheral airways; the guinea pig would then "take a unique position among the species examined" so far.

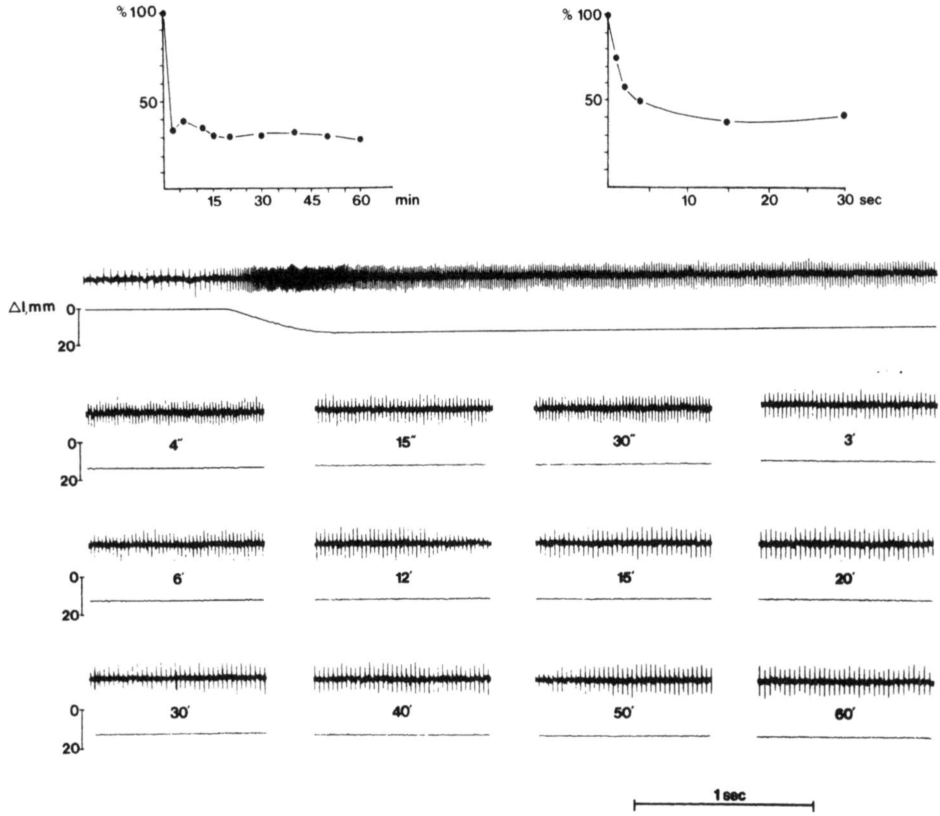

FIGURE 5.2. Action potentials recorded from a single vagal fiber originating from an extrathoracic tracheal stretch receptor while the posterior wall containing the ending is being stretched and maintained elongated for 1 hour. Successive recordings are shown for the indicated time intervals. The signal below the action potentials is the analog for the elongation. The two inserts show the time course of the firing rate of the receptor as a percentage of its value upon completion of elongation. The time scale of the inset at the right is expanded to show the decay of receptor's discharge during the initial 30 seconds. (*Source*: From Davenport et al., 1981, Respir. Physiol. 44, 334.)

Adequate Stimulus and Response to Transmural Pressure

Although the rate of firing of SARs is related to lung volume, a closer association exists between discharge rate and transpulmonary pressure, or even more so with circumferential tension; however, SARs differ in their response to transmural pressure. In fact, whereas SARs of large extrapulmonary airways (trachea, main stem bronchi, lobar bronchi) are active at functional residual capacity (FRC), most of the intrapulmonary SARs have a higher transpulmonary pressure threshold within the tidal volume range. In addition to responding to a maintained transmural pressure, the great majority of SARs also respond to its rate of change. They can respond to both the amount of distension (static response) and the rate at which the distension is introduced (dynamic response). SAR response characteristics are such that can be modeled by a force-transducing sensor in series with both a viscous element and an elastic element

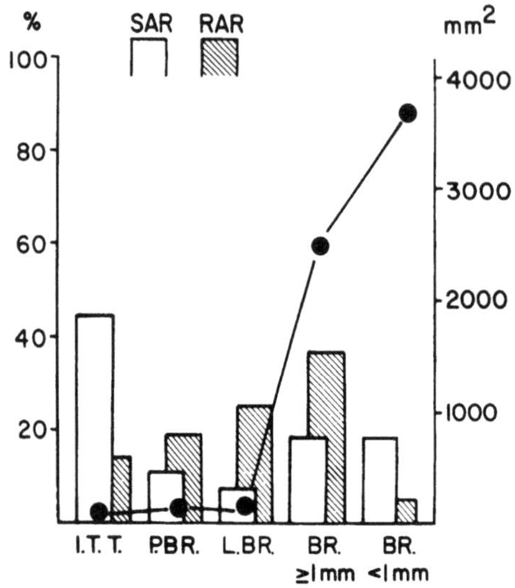

Figure 5.3. Location of SARs and RARs along the tracheobronchial tree of the dog (89, 90, 94). Distribution along various airways is shown as percentage of the total population studied for each type of receptor (left ordinate). Continuous line joining circles represents lateral surface area for various airways of one side calculated from data of Cumming and Semple (1973). ITT-intrathoracic trachea; PBR-principal bronchus; LBR-lobar bronchus; BR ≥ 1 mm, bronchi with diameter of 1 mm or larger (measured in collapsed lung); BR < 1 mm, bronchi with diameter less than 1 mm. (*Source*: From Sant'Ambrogio, 1982.)

arranged in parallel. The elastic element would account for the static response of the SARs, and the viscous element would account for their dynamic sensitivity.

Mechanical Factors Implicated in SAR Stimulation

The extrapulmonary portions of the tracheobronchial tree are made up by a series of U-shaped cartilaginous rings that support a membranous posterior wall. As mentioned earlier, SARs are found in the deeper layer of the posterior wall (i.e., the trachealis muscle).

The mechanical coupling between the cartilaginous rings and the posterior wall determines the response characteristics of SARs located in the extrapulmonary portions of the tracheobronchial tree. At zero transmural pressure, neither the cartilage nor posterior wall is at its resting position. In fact, the cartilage, which tends to expand, exerts a transverse stretch on the posterior wall that is thus elongated beyond its resting length. We may thus understand why most SARs are active at zero transmural pressure and increase their discharge rate either when a distending pressure or a collapsing pressure is applied across the airway wall (when the posterior wall is stretched, either outward or inward). This whole behavior results in an asymmetric response curve of extrapulmonary SARs to positive and negative pressures. The mechanical properties of the cartilaginous rings with their greater compliance at low negative pressure are the determinant reason for the asymmetric response curve (Fig. 5.4; Mortola and Sant'Ambrogio, 1979). The asymmetric response curve has significant physiological implications: Half of the trachea is outside the thorax and thus changes in transmural pressure during the breathing cycle differ from those in the intrathoracic trachea and bronchi. In fact, in these latter airways, transmural pressure at end expiration will be +4–+5 cm H_2O and will increase to a higher value at end inspiration leading to an increase of SAR activity. In the extrathoracic trachea the transmural pressure will be zero at both end expiration and end inspiration, and will become negative (collapsing) during inspiration and positive (distending) during expiration. Due to the asymmetry of the SAR response curve, there will be a decrease in receptor activation during inspiration, but an increase during expiration in the extrathoracic trachea (Fig. 5.4). It follows that activation of extrathoracic and intrathoracic airway SARs during a breathing cycle

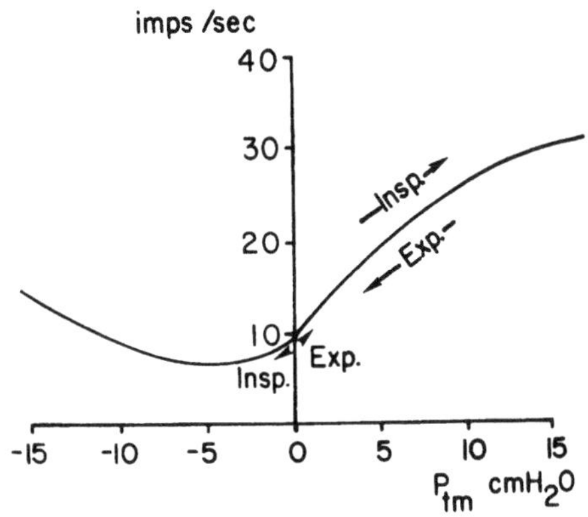

FIGURE 5.4. Relationship between transmural pressure and discharge rate of tracheal SARs. The arrows on the right side indicate the change in SAR discharge in the intrathoracic trachea during inspiration and expiration. The shorter arrows in the center indicate the change in activity of extrathoracic SARs during inspiration and expiration. In the extrathoracic trachea there are pressure changes, hence there are changes in SAR activity only when there is flow. (*Source*: From Sant'Ambrogio, 1982.)

is out of phase (i.e., activity of intrathoracic SARs increases during inspiration, whereas that of extrathoracic SARs decreases; the opposite will occur during expiration).

Response to Carbon Dioxide

CO_2 can either stimulate or inhibit airway stretch receptors. Changes in smooth muscle tone influence this response. In general, the inhibitory effect is especially strong when CO_2 is administered at hypocapnic levels (Bradley et al., 1976; Coleridge et al., 1978b). The effect of CO_2 is weaker at or above normocapnia.

Bronchial, but not tracheal, SARs are susceptible to inhibition by inhaled CO_2 (Bartlett and Sant'Ambrogio, 1976). Bronchial SARs are inhibited by CO_2 administered into the lumen, but not by CO_2 increases of the perfusing blood (Bartlett and Sant'Ambrogio, 1976). Using a preparation in which the pulmonary and systemic circulations were isolated with the possibility of controlling their respective CO_2 concentrations independently, Green et al. (1986) found that increasing P_{CO_2} in the pulmonary blood from 25 to 70 mmHg depressed SAR activity in an approximately linear fashion. This response was not modified by systemic P_{CO_2} and lung mechanics. It must be pointed out that in such a preparation, in which ventilation was kept constant, increases in CO_2 in the pulmonary blood are accompanied by concurrent increases in the alveolar and dead space air. The site of action of CO_2 on SARs (blood vs. air) has particular relevance for exercise hyperpnea, in which the pulmonary blood, but not the alveolar and dead space air, has an elevated P_{CO_2}.

The possibility that the inhibitory effect of CO_2 is mediated by changes in smooth muscle tone has been considered and generally excluded (Bradley et al., 1976; Coleridge et al., 1978b). When hypercapnia exerts a bronchoconstrictive effect through chemoreceptor stimulation, however, SARs increase their discharges.

Although SARs have been found to remain unaffected by changes in blood and air oxygen tensions, when hypoxia stimulates peripheral chemoreceptors, it causes a bronchoconstrictive response and SARs are readily stimulated (Fisher et al., 1983).

Properties of SARs in the Newborn

SARs in the newborn have a higher transpulmonary pressure threshold and lower discharge rate at any given pressure. This leads to a reduced activity during the breathing cycle; in fact, a major difference, as compared with the adult, is the scarcity or lack of activity of these endings at FRC. Factors that could explain these differences include the greater compliance of the extrapulmonary cartilaginous rings that lead to a lower tension at the trachealis muscle site of the receptors and the lower transpulmonary pressure at FRC (Fisher and Mortola, 1980).

Structure

Electron microscopy has been used to study presumptive afferent endings in smooth muscle of the dog's trachea (Krauhs, 1984). Serial thick sections demonstrated medullated fibers running between fascicles of the trachealis muscle, giving rise to unencapsulated endings with features that are considered typical of mechanoreceptors (i.e., large number of mitochondria, deposit of glycogen, and direct attachment of parts of the cell membrane to the basal lamina). The relationship of nerve to muscle appears complex; the endings cannot be described as "in series" or "in parallel."

Reflex Action and Functional Role

SARs activation leads to a shortening of inspiration and to a prolongation of expiration with a considerable effect on the breathing pattern (Euler, 1986), at least in anesthetized animals. These effects on breathing pattern are weaker in humans (Widdicombe, 1961).

Due to their activating influences on expiratory muscles, SARs may exert a stabilizing role on the shifts in FRC that occur with changes in posture (Davies et al., 1980). Activation of SARs also has a reflex inhibitory influence on tracheobronchial smooth muscle, which is a mechanism of possible importance in the regulation of airway resistance-dead space balance. SAR activity has been recorded and recognized from vagal fibers in humans (Guz and Trenchard, 1971).

Rapidly Adapting Receptors

Rapidly adapting receptors (RARs) have an irregular and scant activity in eupnea, but usually more during inspiration; their activity increases in hyperpnea (see Fig. 5.1).

Location Along the Tracheobronchial Tree and Within the Airway Wall

RARs are found at a greater concentration in the more proximal branches of the intrapulmonary tracheobronchial tree (see Fig. 5.3). Whereas SARs are uniquely located in the membranous posterior wall within the trachealis muscle, RARs are found throughout the circumference of the trachea and main stem bronchus (Sant'Ambrogio et al., 1978). RARs have fibers with conduction velocity between 12 and 40 m/second; nervous conduction is blocked in these fibers at temperatures between 7 and 3°C. These are characteristics of myelinated axons and are similar to those reported for SARs, although the mean velocity and blocking temperatures are somewhat lower (Sant'Ambrogio, 1982).

Local Probing, Changes in Transpulmonary Pressure, Irritant Agents

One characteristic of these receptors is their transient and irregular response to a maintained inflation (Fig. 5.1); in fact, the rapidly adapting response is the property often chosen to describe these endings. Another characteristic, that of "irritant" receptors, is the ability of these endings to respond to mechanical and chemical irritant stimuli. Mechanical irritant stimuli are those administered onto the airway mucosa with light probing instruments or by delivering inert substances (e.g., talc or carbon dusts) into the airway lumen. Chemical irritant stimuli that are more often used include: cigarette smoke, water, histamine, prostaglandins, 5-hydroxytryptamine, acetylcholine, and so on.

As indicated, RARs can be stimulated by gross deformations as those obtained with large distending and collapsing pressure as well as by discrete mechanical stimulation of the airway mucosa. The RAR activations caused with the two stimulations can often be separated by performing discrete and superficial lesions of the mucosa overlying the receptor field. The response to local probing is then abolished, whereas those to inflation and deflation are retained. Such results suggest a multibranched ending structure distributed to both superficial (epithelial) and deeper layers of the airways (Cumming and Semple, 1973, Sant'Ambrogio et al., 1978).

The response of RARs to transpulmonary pressure, as well as to its rate of change, is distinctly different from that of SARs. In fact, a step change in pressure results in a burst of action potentials with irregular intervals that subsides very rapidly. A mechanical model in which a force-transducing element is attached to a viscous element would account for this behavior.

Reduction in lung compliance has been documented to increase the activity of RARs (Sellick and Widdicombe, 1970); this is likely to be the reason for the greater activation of RARs in conditions of pulmonary congestion and edema. RARs show a greater response than other airway receptors to experimental lung congestion and edema; decreasing the concentration of plasma proteins can augment their activation (Kappagoda and Ravi, 1989).

A great variety of substances, either inhaled in the form of gas, aerosols, or fumes, or administered into the mixed venous or systemic blood, can activate RARs. In some cases the activation of these endings results as a direct effect of a particular agent; in other cases, it is the result of a secondary effect mediated by smooth muscle contraction. In fact, the activity of RARs (as that of SARs) is affected by variations in bronchomotor tone. Based on some evidence, the stimulatory effect of histamine on RARs, is mediated through a bronchomotor action (Coleridge et al., 1978a), whereas it is partially due to a direct effect according to another study, (Vidruk et al., 1977). Similar controversy exists in the case of prostaglandin $F_2\alpha$ (Sampson and Vidruk, 1977; Coleridge et al., 1978a).

An interesting characteristic of tracheobronchial RARs is their response to water and aqueous solutions. They can be activated by water instillation and, in fewer cases, also by isosmotic solutions of dextrose; the majority of these endings therefore behave as osmoreceptors rather than because of being activated lack of chloride (Pisarri et al., 1992). It is a situation entirely different from that encountered in the larynx, where "irritant" receptors, which are indistinguishable from those of the tracheobronchial tree for their response to mechanical stimuli, are uniquely behaving as lack of chloride receptors (Anderson et al., 1990a).

Another characteristic by virtue of which the two types of irritant receptors differ is their response to volatile anesthetics: laryngeal irritant receptors are vigorously stimulated by halothane (Nishino et al. 1993), whereas tracheobronchial receptors are inhibited. Reasons for these opposite effects are not yet clear.

Carbon dioxide has an inhibitory influence on these receptors, which is particularly potent at low hypocapnic levels (Coleridge et al., 1978b).

Properties of RARs in Newborns

Rapidly adapting receptors constitute a smaller proportion of the receptor population in newborns than they do in the adults of the same species (Fisher and Sant'Ambrogio, 1982). The decreased sensitivity of the newborn to tracheobronchial irritation, as seen through their ability to respond to tussigenic stimuli (Korpas and Tomori, 1979), is indeed consistent with sparse RAR activity. There is some difficulty, however, in justifying the higher frequency of sighing with the scant RAR activity during early development (Fleming et al., 1984). An explanation could be found in the low lung compliance of newborns, which derives from an end expiratory volume within the closing volume range (Fisher and Mortola, 1980).

Structure

As already mentioned, RARs are concentrated in the larger airways, particularly at the branching points. Electron microscopic observations of these areas in the cat revealed the presence of nonmyelinated fibers in the epithelium, both at its basal layer and close to the tight junctions near the lumen (Das et al., 1979). With light microscopy it is possible to see that several of these nerves are connected to myelinated fibers that reach into the vagus nerves. Das et al. (1979) concluded that these epithelial nerves are afferent because they degenerate after midcervical vagotomy, which would leave parasympathetic postganglionic fibers intact. Similar structures have been identified in most mammals, including humans. It is interesting that the mouse does not have epithelial nerves (Pack et al., 1984) and that ferrets have none in the trachea and only a few in the bronchi (Robinson et al., 1986); these two species lack a tracheobronchial cough reflex.

Reflex Action and Functional Role

The main reflex response elicited by RARs is cough; in fact, there is an excellent correlation between tussigenic areas and concentration of RARs, between types of stimuli capable of eliciting both RAR responses and cough. In addition, the observations that cough can be inhibited by blocking conduction (i.e., myelinated afferents), and that cough can be eliminated by levels of local anesthetics that do not interfere with the Hering-Breuer inflation apnea (mediated by the more deeply located SARa) are consistent with a tussigenic role of RARs (Hamilton et al., 1987).

Another reflex response mediated by these endings is sighing or augmented breaths. These deep sighs that occasionally occur in mammals are supposedly triggered by decreases in lung compliance and would play an important role in the homeostasis of lung mechanics (Glogowska et al., 1972). The deep sigh would be expected to re-open atelectatic areas responsible for the reduction in lung compliance.

C-Fiber Receptors

Definitions, Locations

C-fiber endings fall into two distinct groups on the basis of pharmacological and physiological criteria: pulmonary C-fiber endings (also called J-receptors; Paintal, 1969) and bronchial C-fiber endings. Pulmonary C-fiber receptors are designated as such when their corresponding fibers are nonmyelinated with a conduction velocity between 0.9 and 2.3 m/s, a blocking temperature around 0–3°C and possible activation within a short time interval by substances like capsaicin injected into the mixed venous blood. The delay in stimulation should be compatible for a location accessible through the pulmonary circulation (Coleridge and Coleridge, 1984). The term *J-receptor*, which is also used to identify these endings, stands for juxta-alveolar capillary to describe a location in the immediate vicinity of the alveolar–capillary wall (Paintal, 1969).

Bronchial C-fiber receptors correspond also to nonmyelinated fibers and are activated by substances like capsaicin injected into the mixed venous blood, but the delay in onset of stimulation is considerably longer, compatible with a location preferentially accessible through the systemic circulation (Coleridge and Coleridge, 1984). These receptors are located presumably within the wall of the tracheobronchial tree. For several of these endings a superficial location in the tracheal or main stem bronchial mucosa has been ascertained by light mucosal probing (Coleridge and Coleridge, 1984).

Mechanosensitivity of Pulmonary and Bronchial C-Fiber Receptors

Pulmonary C-fiber receptors of spontaneously breathing dogs have significant activity with a respiratory modulation; this is considerably decreased in the presence of artificial ventilation with the chest open. The activity of bronchial C-fiber receptors is scant and irregular with both spontaneous and artificial ventilation. Whereas lung inflation increases the activity of C-fiber pulmonary receptors and, to a lesser extent, that of bronchial C-fiber receptors, reducing lung volume below FRC is essentially without effect for both pulmonary and bronchial C-fiber receptors (Coleridge and Coleridge, 1984).

An increase in pressure in the lung interstitium, as occurs with pulmonary congestion or edema, is a strong activator of pulmonary C-fiber endings rather than bronchial C-fiber endings. The activation of pulmonary C-fiber endings by lung congestion and edema has raised the possibility that these endings may be involved in the limitation of exercise intensity through a reflex inhibitory influence on skeletal muscles (Paintal, 1969). It should be recognized, however, that lung congestion and edema are not normal constituents of exercise; In any case, the pattern of breathing induced by excitation of J-receptors, apnea followed by rapid, shallow breathing, does not resemble that observed in exercise.

Chemosensitivity of Pulmonary and Bronchial C-Fiber Receptors

In general, it can be said that whereas pulmonary C-fiber receptors have a greater mechanosensitivity, bronchial C-fiber endings have a higher chemosensitivity. The chemosensitivity of bronchial C-fiber receptors is well borne out by their responses to substances naturally present in, or released by, the lung as shown by Coleridge and Coleridge (1984). These workers drew a comparison between the responses of bronchial C-fiber receptors and cutaneous C-fiber endings following exposure (by injection into the systemic blood, or by inhalation) to humoral mediators of inflammation (e.g., histamine, prostaglandins, serotonin, and bradykinin). They found that pulmonary C-fibers were also stimulated by prostaglandins, especially by those of the E series, but not by histamine, serotonin, or bradykinin. In striking contrast, bronchial

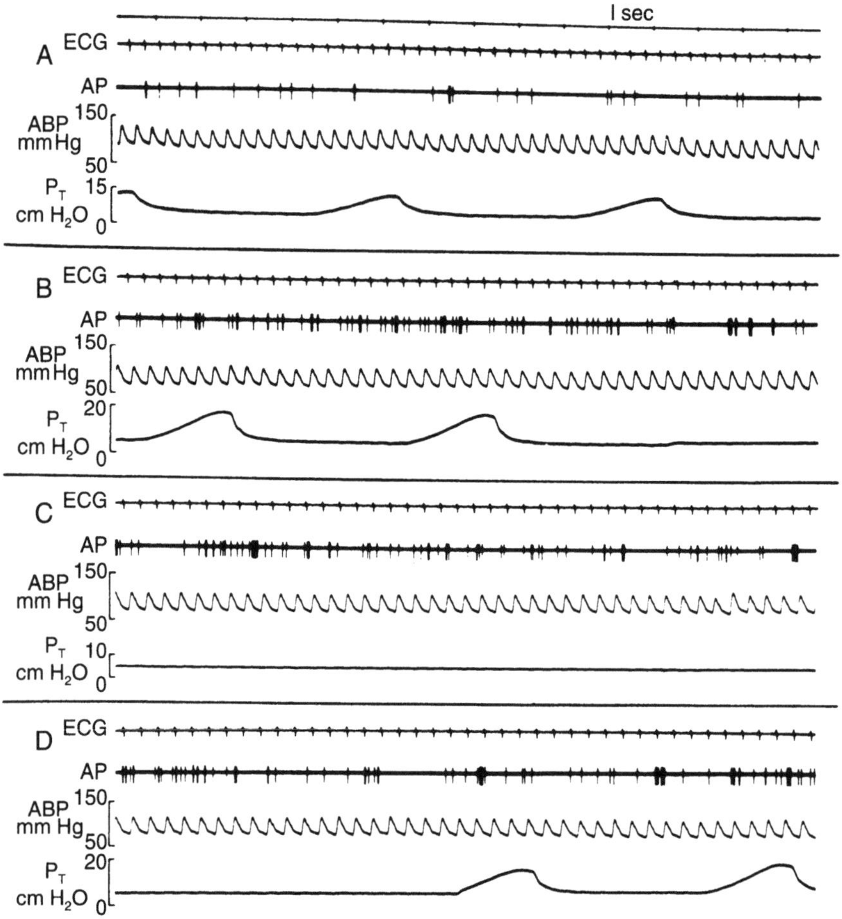

FIGURE 5.5. Stimulation of an airway C-fiber ending by inhalation of histamine aerosol (0.4% solution). ECG-electrocardiogram; AP-action potentials recorded from left vagal filament (the C-fiber ending was in the left lung); ABP-arterial blood pressure; P_T-tracheal pressure. (A) Control. (B) After 0.4% histamine aerosol had been administered for 30 seconds; the respiratory pump was turned off during the latter part of (B); the aerosol generator remained on until the end of (D). Interval of 5 seconds between (B) and (C) and also between (C) and (D). Note the lack of a respiratory modulated activity in (A) and the continued response of the receptor to histamine in the absence of any respiratory motion (C) and (D). (*Source*: Reprinted, with permission, from Coleridge et al., 1978a).

C-fiber endings were activated by administering 0.4% histamine aerosol, as illustrated in Figure 5.5.

In other studies, Coleridge et al. (1978b) found that both bronchial and pulmonary C-fiber receptors undergo a modest increase in activity when end tidal carbon dioxide is raised from 19 to 30 mmHg in a vascularly isolated lung. The possibility that these receptors represent CO_2 sensors for mixed venous blood and thus could be involved in exercise hyperpnea has been proposed by Trenchard (1986), but this remains a hypothesis to be validated.

Structure

Electron microscopic studies of the alveolar wall in human and rat lungs have shown the presence of a few nonmyelinated fibers, and even fewer structures, possibly identifiable as afferent terminals (Fox et al., 1980). On the other hand, the mouse lung has a good supply of nonmyelinated fibers and clearly identifiable afferent terminals within its alveolar walls (Hung et al., 1972). Nonmyelinated fibers with corresponding terminals have been found within the tracheal epithelium of humans (Rhodin, 1966) and in intrapulmonary airways of mice (Hung et al., 1973). These endings have structures identical to those of the alveolar wall.

Reflex Actions and Functional Role

Although the two sets of receptors (i.e., bronchial and pulmonary) to some extent have distinguishable properties, their reflex effects appear to be similar. In fact, both bronchial and pulmonary C-fibers endings, when activated by bolus injections of capsaicin into the mixed venous blood or the systemic circulation, lead to similar changes in the pattern of breathing: apnea followed by rapid, shallow breathing, bradycardia, hypotension, and bronchoconstriction (Coleridge and Coleridge, 1986). Additional responses include an increase in mucus secretion, a decrease in peripheral vascular resistance (not clearly elicited by bronchial C-fibers), and a profound depression of spinal reflexes. Controversy still exists as to the role of C-fiber afferents in the elicitation of cough (Karlsson et al., 1988). Indeed, some evidence suggests that pulmonary C-fiber receptor stimulation in cats inhibits cough (Tatar et al., 1988).

There is evidence showing that the spontaneous activity of C-fiber endings in dogs, even under normal conditions, plays a role in the control of the breathing pattern. Pisarri et al. (1986) found that, after a total block of myelinated afferents, a further block of nonmyelinated fibers could still prolong the duration of inspiration and expiration.

C-fiber receptors are primarily involved in the inflammatory changes mediated by axon reflexes (neurogenic inflammation). Manifestations of these processes include vasodilatation, edema, mucus secretion, and, in some species, smooth muscle contraction. Axon reflexes, in which branches of the same sensory endings act as both afferent and efferent terminals, are mediated by neuropeptides, such as substance P, neurokinin A, and calcitonin gene-related peptide. These peptides have been identified in afferent nerves of the tracheobronchial mucosa (Lundberg, 1990). It is not yet entirely clear whether they are only present in C-fiber terminals or also associated with RAR terminals. In other tissues, such as the skin and skeletal muscles, there are two types of nociceptors with myelinated and nonmyelinated fibers, which contribute to the responses elicited by pathological processes through axon and central reflexes (Widdicombe and Sant'Ambrogio, 1992).

Sympathetic Afferents

Action potentials associated with clear respiratory modulation have been recorded from higher *white rami communicantes* in dogs (Kostreva et al., 1975). The firing rate of these afferent fibers is linearly related to transpulmonary pressure and has a slow adaptation time course. Conduction velocity is characteristic of myelinated fibers. There is also evidence that some of the reflex actions evoked from pulmonary structures and not mediated through vagal afferents are contributed by sympathetic afferents (Kostreva et al., 1978).

Summary

From the oronasal openings to the terminal bronchioles, the airways vary considerably in structure and are subjected to mechanical and chemical influences that are considerably different. The extrathoracic airway undergoes pressure changes during the respiratory cycle, which differ from those of the intrathoracic airways. Both the larynx and pharynx are endowed with striated muscles capable of changing their caliber during the respiratory cycle. The tracheobronchial tree has smooth muscles for which a respiratory modulation is present, but just noticeable. The caliber of the nares is controlled through a neuromuscular mechanism, whereas that of the intranasal coaneis (from the Greek meaning funnel) is determined by neurovascular mechanisms.

The intrapulmonary airways (from the lobar bronchi to the terminal bronchioles) participate in the metabolism of lung parenchyma, and even when they are vascularized through the systemic circulation, they are subjected to the chemicohumoral environment of the pulmonary circulation.

The main sensory supply to the tracheobronchial tree and the lungs is through the vagus nerves. Most of the afferent fibers are nonmyelinated (C-fiber receptors), classified as bronchial or pulmonary on the basis of their circulatory accessibility. *Bronchial* indicates preferential accessibility through the systemic circulation; *pulmonary* through the pulmonary circulation. Whereas bronchial C-fiber receptors do not have any respiratory modulation, pulmonary C-fiber endings do in spontaneously breathing animals. Bronchial and pulmonary C-fiber receptors differ in their properties (mechano- and chemosensitivity), but their reflex effects are similar.

The myelinated, afferent, fast-conducting fibers are less represented within the vagus nerve; they account for most of the action potentials that arise from the tracheobronchial tree. Much of this activity originates from slowly adapting stretch receptors (SAR) and has a clear, regular respiratory modulation that is related to increases in transmural pressure which adapt slowly to a maintained distension. SARs are more concentrated within the smooth muscle of the larger airways. They play a major role at least in experimental animals in the regulation of the breathing pattern. Another insufficiently recognized role is the regulation of bronchomotor tone.

Rapidly adapting receptors (RAR) are another type of mechanoreceptor that show irregular respiratory modulation and adapt rapidly to a maintained increase in transmural pressure. They are more concentrated in larger intrapulmonary airways within the superficial layers of the mucosa. They can be activated by nociceptive mechanical stimuli and by various irritative substances and are often defined as "irritant" receptors. Cough is probably the main reflex mediated by these endings. They may also trigger periodic augmented breaths that occur spontaneously in response to decreases in lung compliance.

References

Anderson, J.W., Sant'Ambrogio, F.B., Mathew, O.P., and Sant'Ambrogio, G. (1990a) Water-responsive laryngeal receptors in the dog are not specialized endings. Respir. Physiol. 79, 33–44.

Anderson, J.W., Sant'Ambrogio, F.B., Orani, G.P., Sant'Ambrogio, G., and Mathew, O.P. (1990b) Carbon dioxide-responsive laryngeal receptors in the dog. Respir. Physiol. 82, 217–226.

Bartlett, D., Jr., Jeffery, P., Sant'Ambrogio, G., and Wise, J.C.M. (1976) Location of stretch receptors in the trachea and bronchi of the dog. J. Physiol. (Lond.) 258, 409–420.

Bartlett, D., Jr., and Sant'Ambrogio, G. (1976) Effect of local and systemic hypercapnia on the discharge of stretch receptors in the airways of the dog. Respir. Physiol. 26, 91–99.

Bradley, G.W., Noble, M.I.M., and Trenchard, D. (1976) The direct effect on pulmonary stretch receptors discharge produced by changing CO_2 concentration in dogs on cardiopulmonary bypass and its action on breathing. J. Physiol. (Lond.) 261, 359–373.

Coleridge, H.M., and Coleridge, J.C.G. (1986) Reflexes evoked from tracheobronchial tree and lungs. In: Handbook of physiology. Section 3: the respiratory system, vol. II. control of breathing, part 1. Cherniack, N.S., and Widdicombe, J.G. (eds.) pp. 395–429. American Physiological Society, Bethesda.

Coleridge, H.M., Coleridge, J.C.G., Baker, D.G., Ginzel, K.H., and Morrison, M.A. (1978a) Comparison of the effects of histamine and prostaglandin on afferent C-fiber endings and irritant receptors in the intrapulmonary airways. Adv. Exp. Med. Biol. 99, 291–305.

Coleridge, H.M., Coleridge, J.C.G., and Banzett, R.B. (1978b) Effect of CO_2 on afferent vagal endings in the canine lung. Respir. Physiol. 34, 135–141.

Coleridge, J.C.G., and Coleridge, H.M. (1984) Afferent vagal C fibres innervation of the lungs and airways and its functional significance. Rev. Physiol. Biochem. Pharmacol. 99, 1–110.

Cumming, G., and Semple, S.J. (1973) Disorders of the respiratory system. Chapter one. Blackwell Scientific Publications, Oxford.

Das, R.M., Jeffery, P.K., and Widdicombe, J.G. (1979) Experimental degeneration of intraepithelial nerve fibers in cat airways. J. Anat. 128, 259–267.

Davies, A., Sant'Ambrogio, F.B., and Sant'Ambrogio, G. (1980) Control of postural changes of end expiratory volume (FRC) by airways slowly adapting mechanoreceptors. Respir. Physiol. 41, 211–216.

Euler von, C. (1986) Brain stem mechanisms for generation and control of breathing pattern. In: Handbook of physiology. Section 3: the respiratory system, vol. II, control of breathing Cherniack, N.S., and Widdicombe, J.G. (eds.) pp. 1–67. American Physiological Society, Bethesda.

Fisher, J.T., and Mortola, J.P. (1980) Statics of the respiratory system in newborn mammals. Respir. Physiol. 41, 155–172.

Fisher, J.T., Sant'Ambrogio, F.B., and Sant'Ambrogio, G. (1983) Stimulation of tracheal slowly adapting stretch receptors by hypercapnia and hypoxia. Respir. Physiol. 53, 525–339.

Fisher, J.T., and Sant'Ambrogio, G. (1982) Location and discharge properties of respiratory vagal afferents in the newborn dog. Respir. Physiol. 50, 209–220.

Fleming, P.J., Goncalves, A.L., Levine, M.R., and Woollard, S. (1984) The development of stability of respiration in human infants: changes in ventilatory responses to spontaneous sighs. J. Physiol. (Lond.) 347, 1–16.

Fox, B., Bull, T.B., and Guz, A. (1980) Innervation of alveolar walls in the human lung: an electron microscopic study. J. Anat. 131, 683–692.

Glogowska, M., Richardson, P.S., Widdicombe, J.G., and Winning, A.J. (1972) The role of the vagus nerves, peripheral chemoreceptors, and other afferent pathways on the genesis of augmented breaths in cats and rabbits. Respir. Physiol. 16, 179–196.

Green, J.F., Schertel, E.R., Coleridge, H.M., and Coleridge, J.C.D. (1986) Effect of pulmonary arterial Pco_2 on slowly adapting pulmonary stretch receptors. J. Appl. Physiol. 60, 2048–2055.

Guz, A., and Trenchard, D. (1971) Pulmonary stretch receptor activity in man: a comparison with dog and cat. J. Physiol. (Lond.) 213, 329–343.

Hamilton, R.D., Winning, A.J., and Guz, A. (1987) Blockade of "alveolar" and airway reflexes by local anesthetic aerosols in dogs. Respir. Physiol. 67, 159–170.

Horner, R.L., Innes, J.A., Murphy, K., and Guz, A. (1991) Evidence for reflex upper airway dilator muscle activation by sudden negative pressure in man. J. Physiol. (Lond.) 436, 15–29.

Hung, K.S., Hertweck, M.S., Hardy, J.D., and Loosli, C.G. (1972) Innervation of pulmonary alveoli of the mouse lung: an electron microscopic study. Am. J. Anat. 135, 477–496.

Hung, K.S., Hertweck, M.S., Hardy, J.D., and Loosli, C.G. (1973) Ultrastructure of nerves and associated cells in bronchiolar epithelium of the mouse lung. J. Ultrastr. Res. 43, 426–437.

Jammes, Y., Fornaris, E., Mei, N., and Barrat, E. (1982) Afferent and efferent components of the bronchial vagal branches in cats. J. Auton. Nerv. Syst. 5, 165–176.

Jammes, Y., Barthelemy, P., and Delpierre, S. (1983) Respiratory effects of cold air breathing in anesthetized cats. Respir. Physiol. 54, 41–54.

Kappagoda, C.T., and Ravi, K. (1989) Plasmapheresis affects responses of slowly and rapidly adapting receptors to pulmonary venous congestion in dogs. J. Physiol. 416, 79–91.
Karlsson, J.-A., Sant'Ambrogio, G., and Widdicombe, J. (1988) Afferent neural pathways in cough and reflex bronchoconstriction. J. Appl. Physiol. 65, 1007–1023.
Keller, E., Kohl, J., and Koller, E.A. (1989) Location of pulmonary stretch receptors in the guinea pig. Respir. Physiol. 76, 149–158.
Korpás, J., and Tomori, Z. (1979) Cough and other respiratory reflexes, Karger, S. Basel.
Kostreva, D.R., Hopp, F.A., Zuperku, E.J., Igler, F.O., Coon, R.L., and Kampine, J.P. (1978) Respiratory inhibition with sympathetic afferent stimulation in canine and primate. J. Appl. Physiol. 44, 718–724.
Kostreva, D.R., Zuperku, E.J., Hess, G.L., Coon, R.L., and Kampine, J.P. (1975) Pulmonary afferent activity recorded from sympathetic nerves. J. Appl. Physiol. 39, 37–40.
Krauhs, J.M. (1984) Morphology of presumptive slowly adapting receptors in dog trachea. Anat. Rec. 210, 73–85.
Lee, B.-P., Sant'Ambrogio, G., and Sant'Ambrogio, F.B. (1992) Afferent innervation and receptors of the canine extrathoracic trachea. Respir. Physiol. 90, 55–65.
Lundberg, J.M. (1990) Peptide and classical transmitter mechanisms in the autonomic nervous system. Arch. Int. Pharmacodyn. 303, 9–19.
Marlot, D., and Duron, B. (1979) Postnatal development of vagal control of breathing in the kitten. J. Physiol. (Paris) 75, 891–900.
Mortola, J.P., and Sant'Ambrogio, G. (1979) Mechanics of the trachea and behaviour of its slowly adapting stretch receptors. J. Physiol. (Lond.) 286, 577–590.
Nishino, T., Anderson, J.W., and Sant'Ambrogio, G. (1993) Effects of halothane, enflurane, and isoflurane on laryngeal receptors in dogs. Respir. Physiol. 91, 247–260.
Nolan, P., Bradford, A., O'Regan, R.G., and McKeogh, D. (1990) The effects of changes in laryngeal airway CO_2 concentration on genioglossus muscle activity in the anaesthetized cat. Exp. Physiol. 75, 271–274.
Orani, G.P., Anderson, J.W., Sant'Ambrogio, G., and Sant'Ambrogio, F.B. (1991) Upper airway cooling and *l*-menthol reduce ventilation in the guinea pig. J. Appl. Physiol. 70, 2080–2086.
Pack, R.J., Al-Ugaily, L.H., and Widdicombe, J.G. (1984) The innervation of the trachea and extrapulmonary bronchi of the mouse. Cell Tissue Res. 238, 61–68.
Paintal, A.S. (1969) Mechanism of stimulation of type J pulmonary receptors. J. Physiol. (Lond.) 203, 511–532.
Pisarri, T.E., Jonzon, A., Coleridge, H.M., and Coleridge, J.C.G. (1992) Vagal afferent and reflex responses to changes in surface osmolarity in lower airways of dogs. J. Appl. Physiol. 73, 2305–2313.
Pisarri, T.E., Yu, J., Coleridge, H.M., and Coleridge, J.C.G. (1986) Background activity in pulmonary vagal C-fibers and its effects on breathing. Respir. Physiol. 64, 29–43.
Ravi, K. (1986) Distribution and location of slowly adapting pulmonary stretch receptors in the airways of cats. J. Auton. Nerv. Syst. 15, 205–216.
Robinson, N.P., Venning, L., Kyle, H., and Widdicombe, J.G. (1986) Quantitation of the secretory cells of the ferret tracheobronchial tree. J. Anat. 145, 173–188.
Rhodin, J.A.G. (1966) Ultrastructure and function of the human tracheal mucosa. Am. Rev. Respir. Dis. 93, 1–15.
Sampson, S.R., and Vidruk, E.H. (1977) Chemical stimulation of rapidly adapting receptors in the airways. In: The regulation of respiration during sleep and anesthesia, Fitzgerald, R.S., Gautier, H., and Lahiri, S. (eds.) pp. 281–290. Plenum, New York.
Sant'Ambrogio, G. (1982) Information arising from the tracheobronchial tree of mammals. Physiol. Rev. 62, 531–569.
Sant'Ambrogio, G., and Mortola, J.P. (1977) Behavior of slowly adapting stretch receptors in the extrathoracic trachea of the dog. Respir. Physiol. 31, 377–385.
Sant'Ambrogio, G., Remmers, J.E., De Groot, W.J., Callas, G., and Mortola, J.P. (1978) Localization of rapidly adapting receptors in the trachea and main stem bronchus of the dog. Respir. Physiol. 33, 359–366.
Sant'Ambrogio, G., Tsubone, H., and Sant'Ambrogio, F.B. (1995) Sensory information from the upper airway: role in the control of breathing. Respir. Physiol. 102, 1–16.

Sellick, H., and Widdicombe, J.G. (1970) Vagal deflation and inflation reflexes mediated by lung irritant receptors. Q. J. Exp. Physiol. 55, 153–163.
Tatar, M., Webber, S.E., and Widdicombe, J.G. (1988) Lung C-fibre activation and defensive reflexes in anaesthetized cats. J. Physiol. (Lond.) 402, 411–420.
Trenchard, D. (1986) CO_2/H^+ receptors in the lungs of anesthetized rabbits. Respir. Physiol. 63, 227–240.
Vidruk, E.H., Hahn, H.L., Nadel, J.A., and Sampson, S.R. (1977) Mechanisms by which histamine stimulates rapidly adapting receptors in dog lung. J. Appl. Physiol. 43, 397–402.
Widdicombe, J.G. (1961) Respiratory reflexes in man and other mammalian species. Clin. Sci. 21, 163–170.
Widdicombe, J.G., and Sant'Ambrogio, G. (1992) Mechanoreceptors in respiratory systems. In: Adv. compar. and environ. physiol., vol. 10. Ito, F. (ed.) pp. 111–135. Springer-Verlag, Berlin.

Recommended Readings

Bartlett, D., Jr., Jeffery, P., Sant'Ambrogio, G., and Wise, J.C.M. (1976) Location of stretch receptors in the trachea and bronchi of the dog. J. Physiol. (Lond.) 258, 409–420.
Coleridge, H.M., and Coleridge, J.C.G. (1986) Reflexes evoked from tracheobronchial tree and lungs. In: Handbook of physiology. Section 3: the respiratory system, vol. II, control of breathing, part 1, Cherniack, N.S., and Widdicombe, J.G. (eds.) pp. 395–429. American Physiological Society, Bethesda.
Goldie, R.G., Rigby, P.J., Fernandes, L.B., and Henry, P.J. (2001) The impact of inflammation on bronchial neuronal networks. Pulm. Pharmacol. Ther. 14, 177–182.
Karlsson, J.-A., Sant'Ambrogio, G., and Widdicombe, J. (1988) Afferent neural pathways in cough and reflex bronchoconstriction. J. Appl. Physiol. 65, 1007–1023.
Mortola, J.P., and Sant'Ambrogio, G. (1979) Mechanics of the trachea and behaviour of its slowly adapting stretch receptors. J. Physiol. (Lond.) 286, 577–590.
Pierce, R.J., and Worsnop, C.J. (1999) Upper airway function and dysfunction in respiration. Clin. Exp. Pharmacol. Physiol. 26, 1–10.
Sant'Ambrogio, G. (1982) Information arising from the tracheobronchial tree of mammals. Physiol. Rev. 62, 531–569.
Sant'Ambrogio, G., Tsubone, H., and Sant'Ambrogio, F.B. (1995) Sensory information from the upper airway: role in the control of breathing. Respir. Physiol. 102, 1–16.

6
The Control of Respiration

Gabriel G. Haddad

Regulation of respiration is a subject that has fascinated investigators for centuries. How the "act of breathing" takes place, what are the mechanisms involved, what are the neural substrates in the central neurons system (CNS) that are basic to fetal breathing, the successful first breaths at birth, and the maintenance of respiration thereafter are only a few questions that many investigators have attempted to address for many years. The earliest experiments that have been recorded go back to Galen (A.D. 130–200), a physician who made observations on breathing after spinal cord injury on gladiators. Much later, investigators such as Lorry and LeGallois made important observations on experimental animals. Lorry found no changes in breathing after removal of the cerebellum and LeGallois, in his "experiments on the principle of life," methodically dissected out the region that is crucial for respiration (LeGallois, 1813). He also made interesting observations on how animals breathe when exposed to lack of O_2 at various ages; in fact, he was the first to note that newborns are much more resistant to anoxia than adults (LeGallois, 1813). Other investigators in the field included Budge, Langendorff, and Lumsden and Pitts at the beginning of the twentieth century. Of course, Ramon y Cajal also contributed to our understanding of respiratory neurobiology when he noted that the region that corresponded to that of the ventrolateral *nucleus tractus solitarius* appeared first in evolution with air-breathers.

We have learned enormously in the past century, and this is due to a large number of experiments performed on animals as well as to "experiments in nature" and observations made at the bedside. In order to present some of this knowledge in limited space but also comprehensively, I will not categorically divide up this chapter by organ or in a traditional way (i.e., detailing our current information regarding the brain, the carotids, the lung receptors etc.), especially because this book has other chapters where such topics will be reviewed in detail. I have elected instead to divide this chapter into three sections. The first section will present our current understanding of respiratory control based upon six concepts that I have distilled from our current knowledge. The second section will detail some ideas related to respiratory control in early life and with development. The third provides examples of aberrant or abnormal respiratory control conditions, and some of the consequences in terms of tissue hypoxia.

Overall Concepts of Respiratory Control

Physicians and scientists will increasingly need to be knowledgeable in respiratory control and regulation for a number of reasons. For the pediatrician and internist, there are now well-known clinical situations and conditions in which one or more elements of the overall control of breathing are stressed, defective, or immature. This is especially significant either in the critically ill or at either end of the age spectrum. Examples of this abound. Consider a few examples (e.g., apnea, including that of prematurity, upper airway obstruction during sleep, severe asthma, hypoventilation and heart failure, and hypoxemia resulting from various reasons). The pediatrician would also be interested in this area because the transition from fetal to neonatal life is very complex and involves various ideas and concepts of respiratory control.

In addition to these clinically oriented driving needs, the investigator and scientist would be interested in this area because of at least two reasons: first, the respiratory feedback system and neural network is both challenging to understand and can be a prototype or a model system for understanding other oscillating behaviors. Second, because of major advances in methodologic and technical approaches, our ability to answer questions that we have so far not been able to is markedly enhanced. Hence, our understanding of the neural substrates involved in respiration and in the neurobiology of breathing will no doubt increase in measurable ways.

Instead of describing each of the elements of the respiratory control system, I shall describe the function of the system succinctly in terms of concepts that have been elaborated over the past century, ever since Legallois and the "Noeud Vital." These are considered overall concepts that are based on data obtained from experiments on respiration or from other oscillating behaviors such as locomotion or swimming in vertebrates or invertebrates.

CONCEPT 1

Respiration is controlled via a negative feedback system with a controller in the central nervous system.

The overall aim of the respiratory feedback system is to maintain blood gas homeostasis in a normal range in the most economical way, from both energy consumption and mechanical standpoints. To accomplish this, the feedback system utilizes both an afferent and an efferent limb (Fig. 6.1). The afferent limb (e.g., the carotid bodies, the airways) is made up of tissues that have receptor endings and can send information to the central controller about functional parameters (e.g., the magnitude of stretch of the airways, the stretch of muscle fibers, the temperature of the environment, or of the extracellular space). The carotid bodies that inform the central controller of the O_2 level represent another important part of the afferent system. Both airway and carotid body sensors compare on-line and on a breath-by-breath basis actual and baseline or programmed "set" signals to generate error signals. These organs then feed this information to the central nervous system. For example, the carotid sinus nerve fibers, which

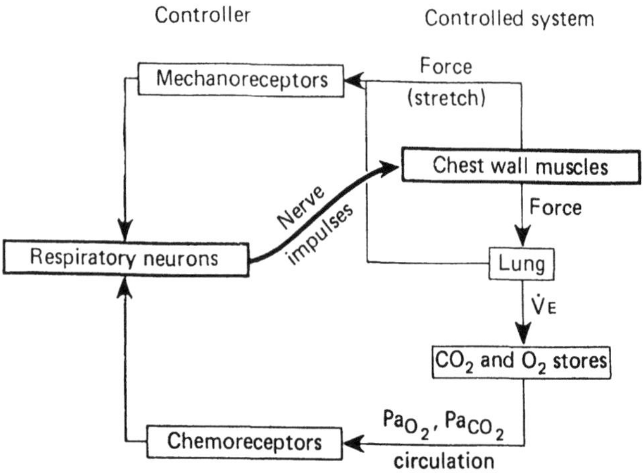

FIGURE 6.1. Flow diagram of the respiratory control system (*Source*: Pulmonary Diseases and Disorders, Alfred Fishman, ed. McGraw Hill, New York, 1980.)

carry impulses generated by the carotid glomus cells, synapse in the medulla oblongata. The efferent loop (i.e., the respiratory muscles and their innervation) is the part of the feedback system that is responsible for the execution of the decision made centrally. There are many muscles of respiration, and the external intercostal muscles and the diaphragm are only two of them. Activity and timing of the airway muscles are very critical in determining airway resistance and patency, thus regulating the magnitude of ventilation.

The term *negative* feedback refers to the fact that the controller attempts to rectify deviations from normality. If CO_2 increases (e.g., because of lack of ability to remove CO_2 adequately, such as from airway obstruction), the output of the controller is increased in an attempt to increase ventilation and hence decrease CO_2. If O_2 drops in the blood and the observed Po_2 is very different from a "programmed" Po_2 that carotid *glomus* cells recognize as abnormal, then these cells will increasingly fire in order to stimulate central neurons and phrenic nerve activity.

CONCEPT 2

Central neuronal processing and integration in the brainstem is hierarchical in nature.

Respiratory muscles are recruited to perform different tasks at different times. For example, the diaphragm and some abdominal muscles are activated during tidal breathing as well as during expulsive maneuvers (e.g., coughing and straining). Such muscle actions take place when an individual jumps and splints the thorax and abdomen. In other conditions, respiratory muscles can be totally inhibited. For example, CO_2 responsiveness is decreased substantially while delivering a speech because respiratory muscles are recruited mostly for speech, sometimes at the expense of respiration. Bottle- or breast-feeding in the young is sometimes associated with a reduction in ventilation and a drop in arterial Po_2 because of inhibition of respiratory muscles and breathing efforts. If presented with a number of neurophysiologic signals (representing options about various needs), the central controller can *enhance* or *reduce* the response to certain stimuli at the expense of or to support others. There is therefore a hierarchy that is used by brainstem networks in determining the response of the respiratory system at any one time. This hierarchy is clearly based on the neurophysiology of controlling and integrating neurons in the brainstem, which are receiving afferent volleys from numerous others. Temporal and spacial summation of excitatory and inhibitory postsynaptic potentials will be critical in deciding the behavior of these neurons. In addition, their cellular and membrane properties will indeed affect their output, as will be described later. Along the same vein, changes in the state of consciousness modulate the ability of the brainstem to respond to afferent stimuli. For example, trigeminal afferent impulses are less inhibited by cortical influences during quiet sleep than they are during REM sleep or wakefulness. Thus, the effect of trigeminal stimulation on respiration is more pronounced in quiet sleep. Age is similarly very important. The response of the brainstem to stimuli varies with maturation and thus with cortical input to brainstem structures.

CONCEPT 3

Although brainstem neurons are heavily modulated by central and peripheral inputs, some brainstem neurons are endowed with cellular and membrane properties that allow them to beat spontaneously or have oscillatory potentials. These properties probably play an important role in generating rhythmic respiratory neuronal behavior.

The central controller has two basic functions that are inherently linked, namely, integration of afferent information, and generation and maintenance of respiration. These functions are thought to occur in anatomically different locations (i.e., different neuronal groups?), but it is not yet known in precise mechanistic terms how each takes place. For example, it is not known how the central generation of respiration takes place, where it is located, or how incoming information is integrated. The respiratory controller may be an **ensemble** of neurons that either form an emergent network or are endogenous or conditional bursters. In the former case, respiratory neurons would not be expected to have special inherent membrane **properties** (e.g., bursting properties) that make their membrane potential spontaneously oscillate. Rather, the output of the network would oscillate because of the special **interconnections** and synaptic interactions among these respiratory neurons (Dekin and Haddad, 1990). In the latter case, respiratory neurons, similar to those forming the sinus node of the heart, would have properties that make them individually "burst" or oscillate, even if they are disconnected from all other neurons (endogenous burster or pacemaker) (Fig. 6.2). A conditional burster, by definition, is a neuron that oscillates only when exposed to certain chemicals (e.g., neurotransmitters). The properties of these neurons are also very critical in shaping the output of the network itself, irrespective of the properties of the respiratory network as a whole. The study of cellular, synaptic, and network properties of neurons in the brainstem, therefore, is likely to yield information needed to understand how the respiratory vital center paces the diaphragm and other muscles (Fig. 6.3) (Dekin and Haddad, 1990).

Although, as described earlier, the exact nature of these respiratory neurons and how they operate is not known, additional data suggest that the respiratory rhythm is generated by an oscillating network in the ventrolateral formation of the medulla oblongata. The region that seems to be essential for the rhythm is the pre-Botzinger Complex, as all cranial nerve activity ceases totally after this region is separated from lower brainstem levels (Feldman et al., 1990; Feldman et al., 1991; Smith et al., 1991;

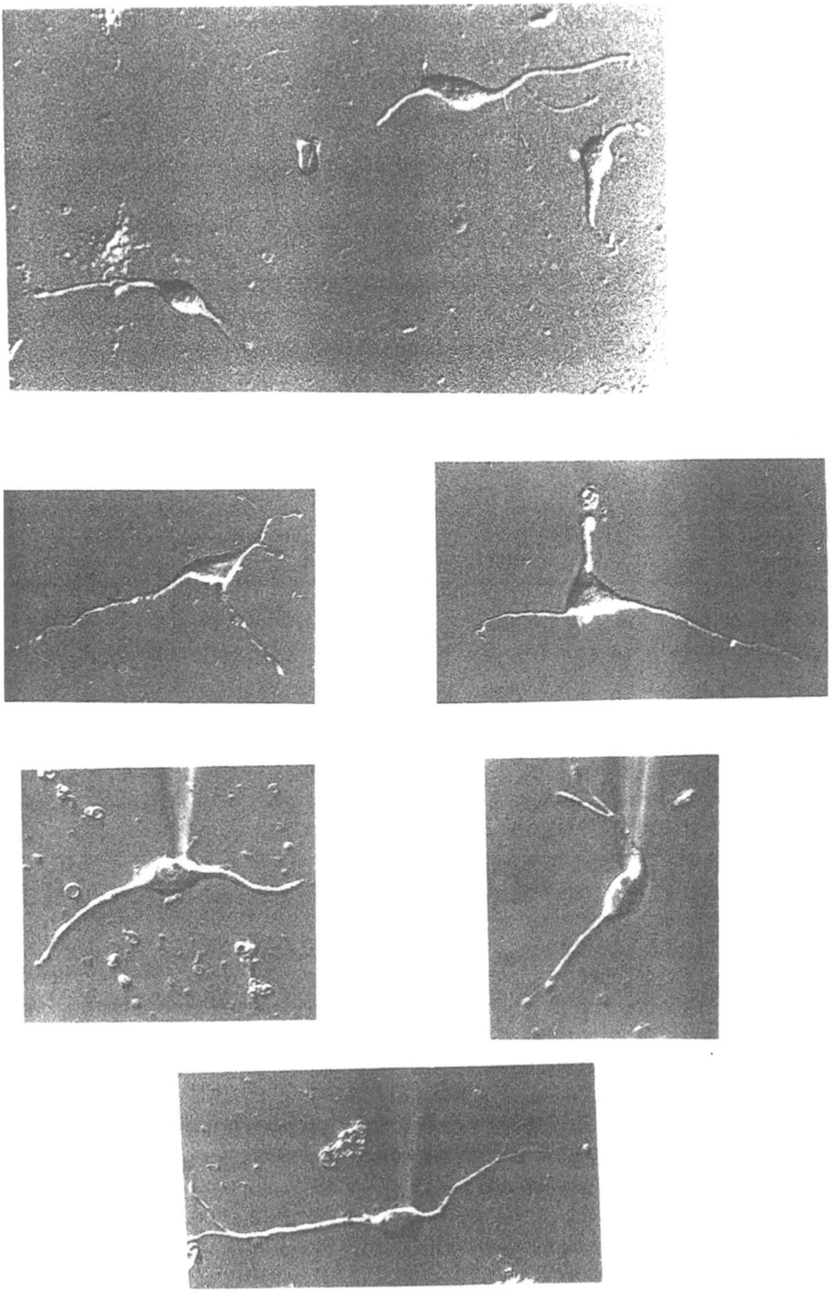

FIGURE 6.2. Freshly dissociated cells from the brainstem. Such cells could then be patch-clamped and their membrane properties studied *in vitro*, separated from synaptic interactions as *in vivo* or culture.

Richter et al., 1992). A number of questions clearly remain to be answered, such as, What are the properties of individual neurons in this area, how interconnected are these with others, and What is the nature of their synapses with neurons in the brainstem and other more rostral regions? Feldman, Smith, and colleagues have attempted to answer a number of these questions. For example, we know now that glutamatergic receptors (AMPA) and glutamate as a ligand play an important role in inducing the respiratory rhythm (Funk et al., 1995; Ge et al., 1998; Reckling et al., 1998). Whether

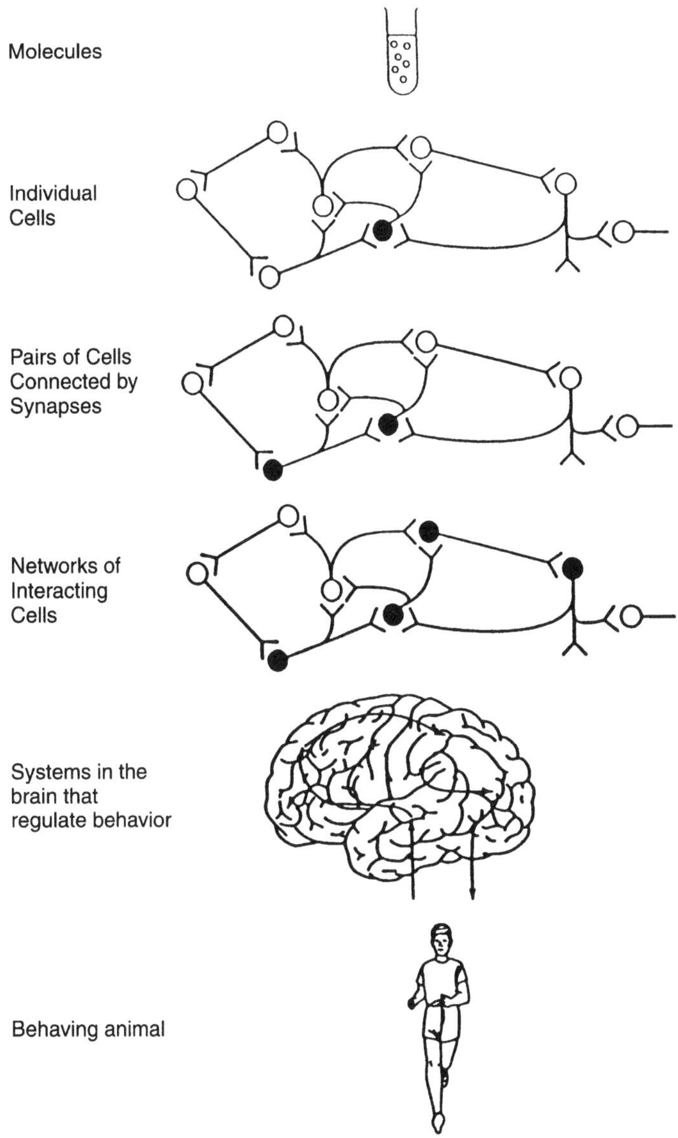

FIGURE 6.3. Ways of studying respiration or other brain functions. Approach can span from studying molecules in test tube to behaving human with single cells or network of cells in between. (*Source*: The neuron, Levitan and Kaczmarek, eds., Oxford, 1991.)

other ligand-receptor systems play an important role (e.g., glutamate) is not known at present.

CONCEPT 4

Respiratory rhythm generation in central neurons is most likely the result of an integration between network, synaptic, cellular, and molecular characteristics, as well as properties of brainstem and other involved neurons.

It is clear from studies in vertebrates and invertebrates that rhythmic movements, whether in locomotion, jump, pyloric rhythmicity, or heart beat, are generally based on either the action of endogenous burster neurons (or conditional bursters) or on a network of neurons that, by virtue of their connections, oscillate (Selverston et al., 1976; Calabrese, 1979; Getting, 1981; Getting, 1983a, 1983b; Grillner, 1985; Robertson et al., 1985). In the case of networks, there is evidence that strongly suggests that central neurons in vertebrates or invertebrates can no longer be considered as just "followers" (Dekin and Haddad, 1990). Such neurons have properties that will change or alter the characteristics of the input. Hence, the input–output relation will depend on the characteristics of the neurons involved.

A striking example of such an oscillating network is a set of neurons that have been well studied in *Tritonia diomedia* (Getting, 1981; Getting, 1983a, 1983b; Dekin and Haddad, 1990). *Tritonia* is a marine slug, which tries to escape once a starfish touches it. Its central pattern generator (CPG), which is responsible for escape swimming, has been extensively studied by Getting, who observed that the functional interaction of one set of synapses, between C_2 and VSI neurons is related to the fact that VSI-B neurons possess membrane properties that delay the onset of C_2 excitation. The delay is an important feature of the time course of the overall function of the CPG and motor program (Fig. 6.4). Indeed, VSI-B neurons possess an intrinsic mechanism, whereby an A-current, first described in molluscan neurons (Dekin and Haddad, 1990), appears with membrane depolarization and then quickly inactivates. Hyperpolarization of the membrane potential interrupts this inactivation. The resting membrane potential of the VSI-B neuron is sufficiently negative to abolish A-current inactivation, but not to turn it on. When however the barrage of (EPSPs) coming from C_2 neurons arrives and depolarizes the membrane of VSI-B neurons, the A-current is activated. This causes VSI-B initially to display a period of reduced excitability during which no action potentials fire. Reduced excitability, which may last for seconds, effectively delays the transfer of excitation from one part of the network to the other.

For dorsal and ventral medullary respiratory neurons, there is convincing evidence that membrane currents can shape their repetitive firing activity (Dekin and Haddad, 1990). These include both the classic sodium and potassium currents responsible for the action potential as well as an A-current, two types of calcium currents, calcium-activated potassium currents, inward rectifier currents, ATP-sensitive K^+ currents, and other currents (Dekin and Haddad, 1990; Jiang and Haddad, 1991). There is general agreement about the presence of these channels in respiratory neurons because many of these channels were studied in identified respiratory neurons *in vivo* after their initial demonstration in brain slices (Dekin and Haddad, 1990). So far, however, no assignment of a role for these currents in shaping the activity of the dorsal respiratory group (DRG) neurons has been made. Further studies require the performance of both *in vivo* and *in vitro* experiments.

The possibility has been considered that excitation may be responsible for the firing activity of "late" inspiratory neurons in the DRG (Dekin and Haddad, 1990). Were this

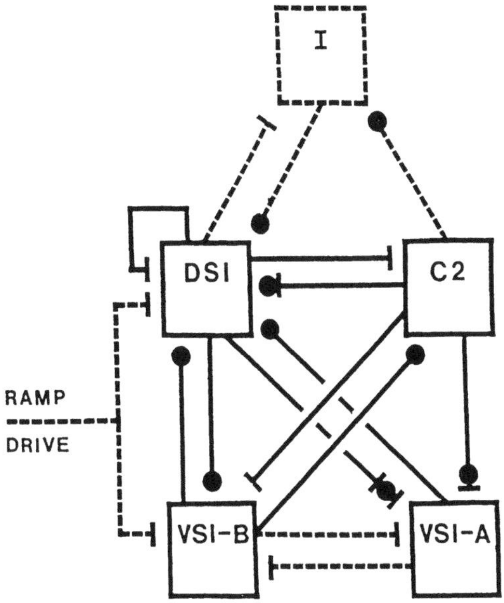

FIGURE 6.4. Network diagram showing the interconnections between various premotor neuronal types in the Tritonia diomedia. T-bars = excitatory synapses; filled dot = inhibitory synapse; mixed symbols = multiaction synapse. (*Source*: Modified from *Tritonia* swimming, a model system for integration within rhythmic motor systems, Getting and Dekin, In Model neural networks and behavior, Selverston, A., ed. Plenum, New York, 1985.)

true, it would then be possible that the A-current in these neurons works in conjunction with other processes (e.g., synaptic facilitation) to shape a ramplike excitatory synaptic drive to phrenic motoneurons. In the case of bulbospinal neurons in the guinea pig or rat DRG, the magnitude of delayed excitation can be modulated over a wide range of the membrane potential. During expiration, DRG neurons receive synaptic inhibition that hyperpolarizes their membrane potential level to between –60 to –80 mV (Fig. 6.5) (Haddad and Getting, 1989; Dekin and Haddad, 1990). The magnitude of the delay would vary between 25 and 300 ms, respectively, over this membrane potential range, that is to say, within the limits of the expiratory phase, the magnitude of delayed excitation can vary by at least an order of magnitude. Assuming, then, that individual neurons or members of the pool possess slightly different membrane potential levels before being depolarized, then they would be expected to show differences in delayed excitation. This would lead to gradual recruitment of the ensemble spike activity within a pool during the course of depolarization, which, in turn, would be translated into a ramplike excitatory synaptic drive to phrenic motoneurons.

The data relating to the cellular and membrane properties and synaptic efficiency of newborn neurons helps to account for the differences in the integrated output between newborn and adult neural networks. For example, both active and passive cellular properties in newborn DRG cells are different from those in adults (Haddad and Getting, 1989; Haddad et al., 1990; Haddad, 1991). Newborn brainstem neurons show less spike frequency adaptation, less inward rectifier currents, different after-hyperpolarization, a wider action potential waveform, and no delayed excitation (Haddad and Getting, 1989; Dekin and Haddad, 1990; Haddad, 1991). Such maturational changes are associated with those seen in several variables of cytosolic and membrane structure and function. For example, consider the distribution of ion channels and receptors on cell

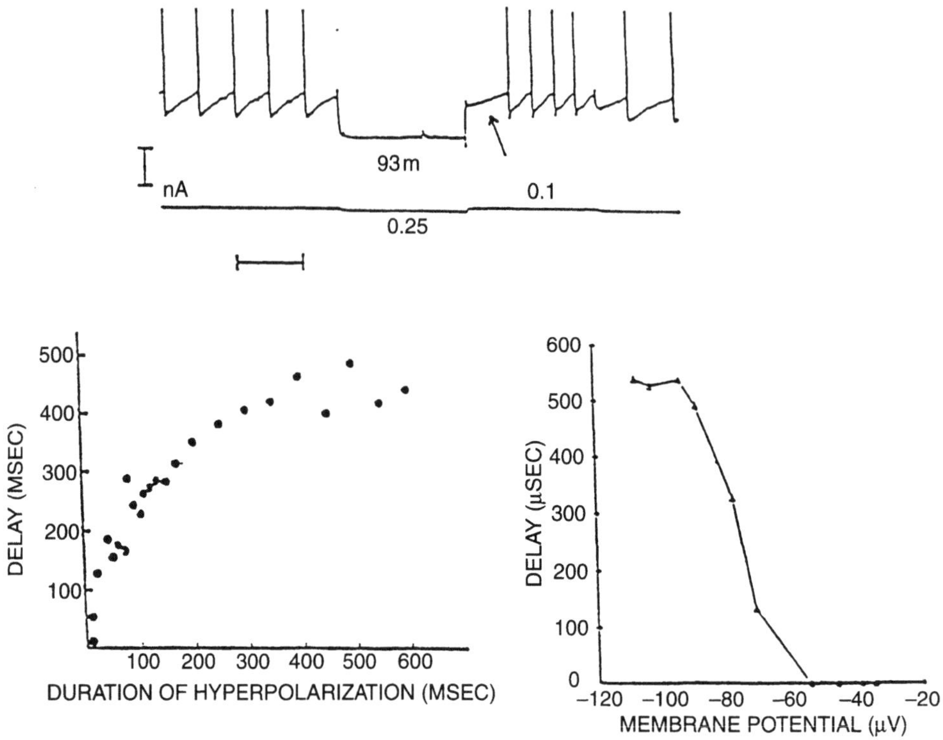

FIGURE 6.5. Delayed excitation in neurons of the nucleus *tractus solitarius* (arrow). This delayed excitation depends on duration of hyperpolarization and membrane potential (lower panels). (*Source*: J. Neurophysiol. 62, 1213–1224, 1989.)

membranes. The structural and functional nature and regulation of ion channels change in early life with age. Because of the absence of the ramp neural drive from phrenic discharge in the young rat, one attractive hypothesis is that the phrenic discharge pattern is due to the lack of delayed excitation in the newborn (Dekin and Haddad, 1990). It could also be that the phrenic discharge pattern in the newborn is a reflection of immature and inefficient central synapses. This is evidenced by a reduction of the current required to induce postsynaptic potentials and that these postsynaptic potentials cannot be maintained with prolonged stimulation as they are in the adult. For a detailed discussion of respiration in infancy, see Stocks (1999).

CONCEPT 5

Afferent impulses are not essential for the generation of breathing activity, but are important for the modulation of respiration.

Many afferent messages converge on the brainstem at any one time. Chemoreceptors and mechanoreceptors in the larynx and upper airways sense stretch, air temperature, and chemical changes over the mucosa, and relay this information to the brainstem. Afferent impulses from these areas travel through the superior laryngeal nerve and the tenth cranial nerve (vagus). Changes in O_2 or CO_2 tensions are sensed at the carotid and aortic bodies, and afferent impulses travel through the carotid and aortic sinus nerves. Thermal or metabolic changes are sensed by skin or mucosal receptors or by hypothalamic neurons, and are carried through spinal tracts to the brainstem for integration. Furthermore, afferent information to the brainstem needs to be both formulated and sensed by the peripheral nervous system. As an example, the major sensors of CO_2 in the body lie on the ventral surface of the medulla oblongata; therefore, the main feedback regarding CO_2 levels and homeostasis come from the brainstem itself.

It is well established that afferent information is not a prerequisite for the generation and maintenance of respiration. Rhythmic phrenic activity can be detected for hours when the brainstem and spinal cord are removed from the body and maintained *in vitro*. Other experiments on chronically instrumented dogs *in vivo* in which several sensory systems are simultaneously blocked (vagal afferent block, 100% O_2 to eliminate carotid discharges, sleep to eliminate wakeful stimuli, and diuretics to alkalinize the blood) indicate that afferent information is not necessary to stimulate the inherent respiratory rhythm in brainstem respiratory networks; however, both *in vitro* and *in vivo* studies demonstrate that, in the absence of afferent information, the inherent rhythm of the central generator (respiratory frequency) is slowed down. Chemoreceptor afferents can therefore play an important role in modulating respiration and rhythmic behavior.

CONCEPT 6

The efferent respiratory control system, respiratory musculature, is a potential site of both neuromuscular failure and respiratory regulation, especially during stress in early life.

Effective ventilation requires the coordinated interaction between the respiratory muscles of the chest and those of the upper airways and neck. For example, it is well known that the activation of upper airway muscles occurs prior to and during the initial part of inspiration. Indeed, the genioglossus contracts to move the tongue forward, thereby increasing the patency of the airways; the vocal cords abduct to reduce laryngeal resistance; and laryngeal muscles modulate expiratory flow to influence lung volume and function.

A multitude of neuromuscular changes take place early in life, most notably in the muscle cells, the neuromuscular junction, the nerve terminals and synapses, and the chest wall. Because muscle and chest wall properties change with age, it is therefore likely that neural responses can be influenced by pump properties, especially that these muscles execute neural commands. One of the important maturational aspects of respiratory muscles is their pattern of innervation. In the adult, one muscle fiber is innervated by one motoneuron. In the newborn, however, each fiber is innervated by two or more motoneurons, and the axons of different motoneurons can synapse on the same muscle fiber, whence the term *polyneuronal innervation*. Synapse elimination takes place postnatally. In the case of the diaphragm, the adult type of innervation is reached by several weeks of age, depending on the animal species. The time course of polyneuronal innervation of the diaphragm in the human newborn is not known (Sieck and Fournier, 1991).

The neuromuscular junctional folds, postsynaptic membranes, and acetylcholine receptors and metabolism undergo major postnatal maturational changes. The acetylcholine quantal content per end-plate potential is lower in the newborn than it is in the adult rat diaphragm. The newborn diaphragm is also more susceptible to neuromuscular transmission failure than the adult, especially at higher frequencies of stimulation. The reason for this is not yet clear.

The chest wall in newborn infants is highly compliant (Bryan and Gaulthier, 1985; Watchko et al., 1991). Because of this, and because young infants spend a large proportion of time in REM sleep during which the intercostal muscles are inhibited, there is little splinting of the chest wall for diaphragmatic action. With every breath in supine infants (especially in REM sleep), therefore, the chest wall is sucked in paradoxically at a time when the abdomen expands. This creates an additional load on the respiratory system, resulting in higher work of breathing per minute ventilation in the infant than in the adult. Some investigators consider this to be an important reason for the newborn's susceptibility to muscle fatigue and respiratory failure.

Control of Respiration: Developmental Aspects

In the time since 1980, there has been a substantial increase in developmental studies of the regulation of respiration. The main reason for this is that pre- or postnatal development is rapid; consequently, changes must be rapid.

The developmental studies that have been performed span a whole range of questions and relate to a number of subsystems underlying ventilatory control. For example, questions on the development of brainstem neurons, respiratory muscles, afferent receptors and systems (including the carotid bodies), and the end-organs (lungs) have been addressed. It is now clear that the subsystems are immature at birth and that profound changes occur postnatally (Hughson et al., 1998). The following is a brief account of some of the important age-dependent alterations.

Central Neurophysiologic Aspects

Although *in vitro* studies of the neonatal rat (whole brainstem preparation) were not directed at understanding the neonate in particular, these studies have shed some light on fundamental issues pertaining to the control of respiration in the newborn (Smith et al., 1991). In fact, it is now recognized that the brainstem *in vitro* of the young rat, which is very immature in the first week of life, does not need any external or peripheral drive for the oscillator to discharge. The intrinsic respiratory rate (as judged by cranial nerve output), however, is markedly reduced. From these studies it emerges that peripheral or central (rostral to the medulla and pons) input is needed to maintain the respiratory output at a much higher frequency.

It is perhaps more important that it has been hypothesized that even though the basis of the respiratory rhythm generator in the adult is a network of neurons, the system could be a hybrid of a network and pacemaker-type of oscillator in the neonate (Feldman, 1990; Feldman, 1991; Smith et al., 1991). This idea stems from experiments showing that the respiratory rhythm is not abolished in the neonatal animal pre-exposed to blockers of inhibitory mediators such as GABA or bathed in Cl^--free solutions (Feldman, 1990). This means that respiratory rhythm generation in the newborn is not based on inhibitory connections, but rather on pacemaker cells. This is only one hypothesis, and it is presently far from proven.

Another interesting observation is that the discharge pattern of each neuronal unit in the neonate, as seen in extracellular recordings, differ from that in the adult in two major ways. First, the inspiratory discharge is not ramp in shape, but increases very fast and decreases within the same breath. The second is that it is extremely brief, sometimes limited to even a few action potentials (Farber, 1988). The reasons for this inspiratory discharge are somewhat obscure, but they could be related to the properties of the cells, their network, or their microenvironment. In addition to differences in inspiratory discharge, expiratory units discharge weakly and often appear only after imposing an expiratory load (Farber, 1985; Farber, 1981). In sharp contrast, adult expiratory counterparts are much more active, even at rest.

Because the discharge pattern of central neurons in the adult or neonate is affected by peripheral input, including input from the vagus nerve, and because the respiratory feedback system can operate on a breath-by-breath basis, one question that arises is whether the lack of myelination affects function in neonatal nerve fibers that subserve the feedback system. It does, both because of lack of myelination and potential delays in signaling, as well as because inspiratory and expiratory discharges are so fast that they exclude the effect of peripheral information on the CNS *within* the same breath. It is clear, then that peripheral information may affect subsequent breaths and hence an important question is whether breath-by-breath feedback in the young and the adult are equipotent. Next to nothing is known about this.

Differences between neonates and adults are also observed in response to endogenous stimuli such as responses to neurotransmitters or modulators. Immature and young animal respond differently to neurotransmitter, as revealed by Farber (1990) in focused mainly work on the opossum (Farber, 1990). Glutamate injected in various locations in the brainstem, even in large doses, induces respiratory pauses whereas the response is stimulatory in the older mature animal. Inhibitory neurotransmitters such as GABA exert age-dependent effects in the opossum. Cl^--mediated inhibition in the isolated brainstem is completely suppressed in the adult, but not in the neonate (Richter et al., 1991), although the pattern of each burst of activity changes. These differences in response are not yet understood because there are too many variables in these experiments (e.g., the size of the extracellular space, receptor development and sensitization).

Peripheral Sensory Aspects

In this section, I shall review data on the primary O_2 sensor in the body—the carotids. I will show that there are major differences between the newborn and the adult vis-à-vis the response of the carotids to low O_2 as well as with respect to the importance of this organ in overall respiratory function, and even survival of the organism in early life. A more detailed treatment of the subject is available in Chapter 7.

Recordings from both single fiber afferents and sinus nerves have shown major differences between the fetus and newborn and between the newborn and the adult. It has been demonstrated that fetal chemoreceptor activity occurs in the normal fetus and that a large increase in activity can be evoked by decreasing Po_2 in the ewe (Blanco et al., 1984a, 1984b; Blanco et al., 1988). The estimated response curve was considerably left-shifted compared with the adult such that Pao_2 values below 20 Torr were required to initiate an increase in discharge. As would be predicted, the large increase in Pao_2 at the time of birth virtually shuts off chemoreceptor activity in the newborn. This decreased sensitivity, however, does not last long, and a normal adultlike sensitivity takes place by 1–2 weeks after birth (Blanco et al., 1984a, 1984b; Hanson et al., 1987; Blanco et al., 1988). The mechanism(s) for the maturation of these peripheral sensors have yet to be elucidated. There are a number of factors, external or endogenous, that probably play a role in this process. For example, arterial chemoreceptors are subject to external and hormonal influences, which may affect the sensor or alter tissue Po_2 within the organ. Tissue Po_2 during the neonatal period shows a decrease over the first week after birth, which might be an important factor in the increase in hypoxia sensitivity. This change in tissue Po_2, however, is attributable to a change in vessel size because little or no change occurs in the characteristics of the carotid body vasculature during this period (Clarke et al., 1990). Neurochemicals may also play a major role as modulators of chemosensitivity; for example, endorphins are known to decline in the newborn period, and the effect of exogenous endorphin is inhibition of chemoreceptor hypoxia sensitivity (Pokorski et al., 1981).

Evidence is available, however, that chemosensitivity of the newborn chemoreceptor is less than that in the adult even in the absence of external modulatory factors (hormonal or neural). Nerve activity of rat carotid bodies *in vitro*, following transition from normoxia to hypoxia is about fourfold greater in carotid bodies harvested from 20-day-old rats as compared with 1–2-day-old rats (Kholwadwala and Donnelly, 1992). This corresponds well with the maturational pattern of the respiratory response to hypoxia in the intact animal (Eden and Hanson, 1987) and suggests that major maturational changes occur *within* the carotid body itself. Histologic, biophysical, and neurochemical changes occur with development, but the significance of these changes are very dependent on assumptions of how the organ senses Po_2. For example, the maturational increase in chemosensitivity may be attributed to a maturational change in the biophysical properties of glomus cells. In one model, hypoxia directly inhibits a membrane-localized K^+ channel that is active at rest, and the resulting depolarization leads to calcium influx, enhanced secretion, and increased neural activity in adult carotid cells (Gonzalez et al., 1992). Evidence supporting this mechanism has been obtained from cultured, adult rabbit glomus cells whose outward (or K^+ current) is apparently inhibited by hypoxia. By comparison, glomus cells harvested from immature animals (e.g., rats) show a decrease in whole-cell K^+ current during hypoxia, but the decrease in K^+ current is attributed to decreased activation of a Ca^{2+}-dependent K^+ current, and not to a specialized K^+ channel sensitive to Po_2 (Fig. 6.6) (Peers, 1990; Ganfornina and Lopez-Barneo, 1991).

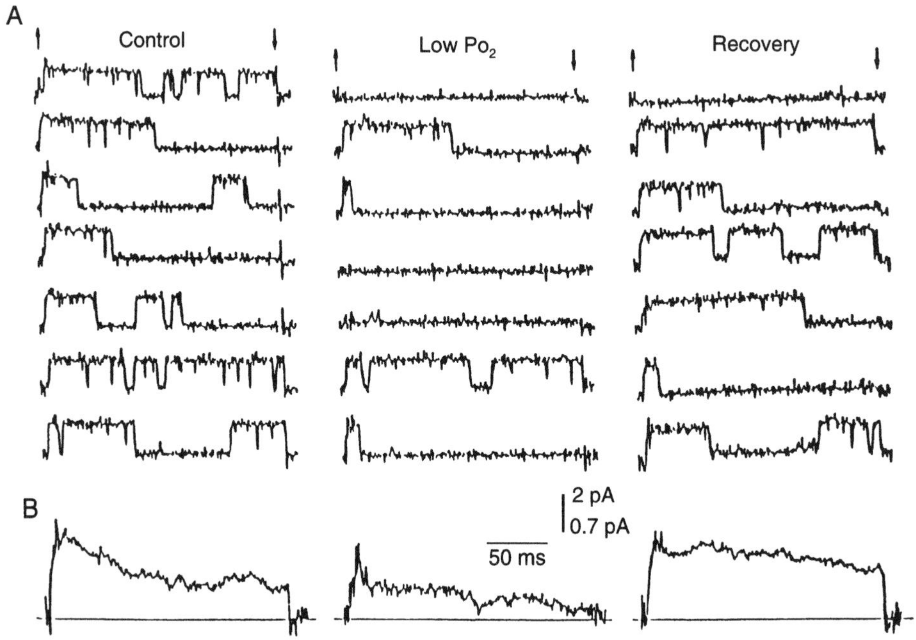

FIGURE 6.6. K^+ channels in excised membrane patches of carotid glomus cells responding to low Po_2. (*Source*: Proc. Natl. Acad. Sci. USA 88, 2927–2930, 1991.)

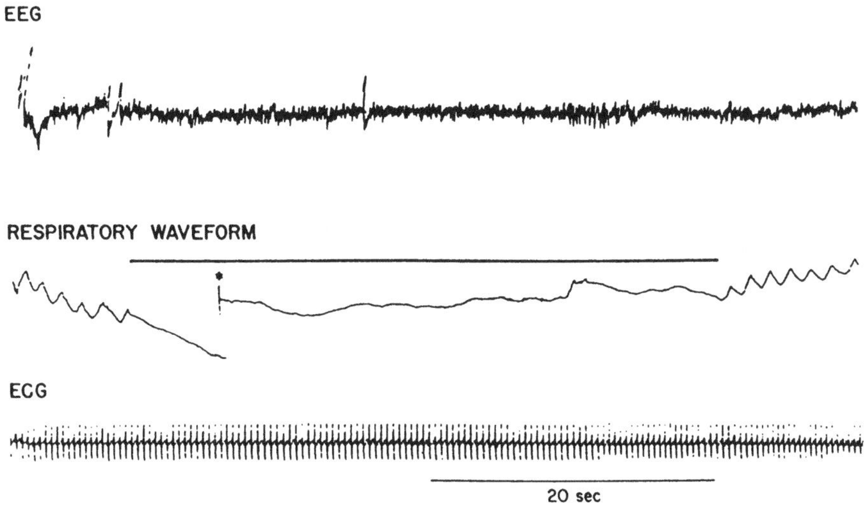

FIGURE 6.7. Prolonged apneas seen in piglets with carotid sinus denervation. (*Source*: J. Appl. Physiol. 68, 1048–1052, 1990.)

In comparison with the adult, peripheral chemoreceptors assume a greater role in the newborn period. Although not essential for initiation of fetal respiratory movements (Blanco et al., 1988), peripheral chemoreceptor denervation in the newborn period results in severe respiratory impairments and a high probability of sudden death. After denervation lambs fail to develop a mature respiratory pattern (Bureau and Begin, 1982; Bureau et al., 1985a, 1985b, 1986), and, more importantly, suffer a 30% mortality rate, weeks or months following surgery. In other species, denervation also leads to lethal respiratory disturbances (Hofer, 1984, 1985; Donnelly and Haddad, 1986; Haddad and Donnelly, 1988; Donnelly and Haddad, 1990). For instance, denervated rats suffer from severe desaturation during REM sleep (Hofer, 1984), and piglets suffer periodic breathing with profound apneas during quiet sleep (Fig. 6.7) (Donnelly and Haddad, 1990). Of particular interest is the observation that these lethal impairments only occur during a fairly narrow developmental window, and denervation before or after this window period only results in relatively minor alterations in respiratory function (Haddad and Donnelly, 1988; Donnelly and Haddad, 1990). This window of vulnerability, which is unmasked by denervation in the newborn period, gives support to the speculation that the sudden infant death syndrome (SIDS) may be due to an immaturity or malfunction of peripheral chemoreceptors (Naeye et al., 1976; Cole et al., 1979; Perrin et al., 1984).

Pathophysiology and Disease States

Respiratory Pauses and Apneas

Although there are numerous studies on apnea in the newborn and adult human, there are still major controversies. The length of the respiratory pause that has been defined as apnea has varied and has been subject to debate. *Apnea* can be defined statistically as a respiratory pause that exceeds 3 standard deviations of the mean breath time for an infant, a child at any particular age, or an adult. This definition requires data from a population of subjects, lacks physiologic significance, and does not differentiate between relatively shorter or longer respiratory pauses. On the other hand, the definition of apnea may be based on the sequelae of pauses, such as associated cardiovascular or neurophysiologic changes. Such a definition relies on the functional assessment of pauses and is, therefore, more relevant clinically. Because infants have higher O_2 consumption (per unit weight) than the adult and relatively smaller lung volume and O_2 stores, it is possible that relatively short (e.g., seconds) respiratory pauses, which may not be clinically important in the adult, can be serious in the very young or premature infant.

Independent of age, respiratory pauses are more prevalent during sleep than during wakefulness. The frequency and duration of respiratory pauses, however, depend on the sleep state. Respiratory pauses are more frequent and shorter in REM than in quiet sleep. They are more frequent in young infants than in older children or adults.

Although the pathogenesis of respiratory pauses remains debatable, there is consensus about certain observations. Normal full-term infants, children, and adults exhibit respiratory pauses during sleep. Some think that respiratory pauses and breathing irregularity are a "healthy" sign and that the complete absence of such pauses may be indicative of abnormalities. This is akin to the concept of heart rate variability and that a lack of short- or long-term variability in heart rate can be a sign of disease or immaturity. The reality is that prolonged apneas can be life threatening. The pathogenesis of these apneas may be related to the clinical condition of the patient at the time of their onset, the associated cardiovascular (systemic or pulmonary) changes, and the chronicity of the clinical condition. Whether the etiology is central or peripheral is also important. Prolonged apneic spells require therapy and, optimally, treatment should involve the underlying pathophysiology.

Whether apneas or respiratory pauses are short or life threatening, or clinically significant or irrelevant, the pathogenesis of apneas can vary. The etiology maybe in the CNS, in the periphery (e.g., in the airways) or in the coordination between peripheral and central events. Upper airway obstruction (UAO), for example, is an entity characterized by a partial or complete lack of airflow, not because of the absence of phrenic output but because of obstruction in the airways per se. This is very different from abnormal (or lack of) airflow resulting for the failure of phrenic impulses reaching the diaphragm. It is of the utmost importance to distinguish between these two conditions if optimal therapy is to be provided.

Upper airway obstruction (UAO) during sleep is seen with increasing frequency in children and adults. In contrast to adults, in whom the etiology of UAO often remains obscure, many children have anatomic abnormalities, the most common cause in children being tonsillar and adenoidal hypertrophy. Other associated abnormalities include craniofacial malformations, micrognathia and muscular hypotonia. The usual site of UAO in both infants and adults is the oropharynx, between the posterior pharyngeal wall, the soft palate, and the genioglossus. During sleep (especially REM sleep), upper airway muscles, including those of the oropharynx, lose tone and trigger an episode of UAO.

The Sudden Infant Death Syndrome (SIDS)

As its name indicates, this is a condition that occurs unexpectedly in infants with a peak incidence between 2 and 3 months of age. It is the most frequent cause of death between 1 and 2 months of age in the industrialized countries. It occurs mostly in the cold months of the year, presumably during sleep, with a slight preponderance in male infants. It is an interesting observation that the frequency of SIDS depends on race, being low in Asians and much higher in Hispanics and Blacks. Maternal risk factors include poor pre-natal care and cigarette smoking during pregnancy. Almost one third of infants dying of SIDS are low-birth-weight infants. On postmortem, 70–100% of infants dying of SIDS have petechial hemorrhages limited to the surfaces of intrathoracic organs, including the pericardium, lungs and thymus. Age-matched controls either have many fewer petechiae, or none at all. Because of the location of these petechiae and because of experimental data showing that increased negative intrathoracic pressures are required to induce such petechiae, some investigators have argued that airway obstruction must have preceded death, and some believe that such as event might have even led to demise. The fact that there have been abnormalities on detailed postmortem examination in the brain of these infants is of major importance. For example, about 80% of their brains are heavier than 95% of age-matched controls. SIDS infants (at least some of them) have increased gliosis, especially in the brainstem. In addition, they have delayed myelination in a number of brain areas. Finally, SIDS brains show an overabundance of dendritic spines in neurons, which again suggests that there may be delayed maturation (or defective development) of their CNS. Whether these alterations and abnormalities are causative to an abnormal regulation of cardiorespiratory function remains unclear. It is interesting to note, however, that siblings of SIDS have many fewer spontaneous arousal responses during a 24-hour period. Infants who had an aborted SIDS episode (arrested but resuscitated in time) have a higher heart rate, a smaller heart rate variability, and less "noise" in the pattern of their heart rate, indicating that they may have an abnormality in the autonomic regulation of their vital functions.

O_2 Deprivation and Cell Injury

A number of pathophysiologic conditions are known to lead to respiratory failure with hypercapnia and tissue O_2 deprivation. Almost all cardiorespiratory diseases potentially can produce failure of this system. This outcome may be deleterious to other organs because of the ensuing acidosis and hypoxia; however, it is the hypoxia that should be avoided at all cost because human tissues, especially the CNS, have a relatively low tolerance in a microenvironment devoid of O_2 (Haddad and Jiang, 1993; Haddad and Jiang, 1994; Jiang et al., 1994; Jiang and Haddad, 1994; Xia and Haddad, 1994; O'Reilly et al., 1995; Chidekel, 1997; Fung and Haddad, 1997; Haddad, et al., 1997; Haddad and Jiang, 1997; Jiang and Haddad, 1997; Ma and Haddad, 1997; O'Reilly, 1997; Banasiak and Haddad, 1998; Chow and Haddad, 1998; Friedman et al., 1998; Mironov and Richter, 1998; Ma et al., 1999; Mironov and Richter, 1999).

Since 1990 we have learned a great deal about the effect of lack of oxygenation on various mammalian and nonmammalian (vertebrate and nonvertebrate) tissues, as well as at various ages, including fetal, postnatal, and adult. There is a vast array of cellular and molecular responses to lack of O_2. From an organism point of view, the carotid bodies discharge and affect ventilation when the Pao_2 falls less than 50 Torr. It is probably the case that other tissues in the body generally do not respond or react to a Pao_2 greater than 50 Torr. Indeed, most tissues start "sensing" a decrease in Pao_2 only less than 35–40 Torr. For example, the brain, which is one of the most sensitive tissues to lack of O_2, has a resting (no hypoxia induced) interstitial O_2 tension, probably in the range of 20–35 Torr, depending on age, area (white vs. grey matter), neuronal metabolism, temperature, proximity to blood vessels, and so on.

Despite advances in our understanding of the effects of lack of oxygenation on tissue metabolism, excitability, and function, major questions still remain unanswered. These include mechanisms that lead to injury or those that protect tissues from it. This area of research is very complex. For example, consider the nervous system, in which several mechanisms are activated during O_2 deprivation. Membrane biophysical events such as those involving Na^+ and K^+ channels, and others such as increased anaerobic metabolism, increased intracellular levels of H^+ and Ca^{2+}, increased concentrations in extracellular neurotransmitters (e.g., glutamate and aspartate), free radical production, activation of kinases, proteases, and lipases, injury and destruction of important cytoskeletal proteins, and fetal gene regulation of a number of proteins (e.g., c-fos, NGF, HSP-70, -actin) are just some events that take place during lack of O_2 (Haddad and Jiang, 1993; Haddad and Jiang, 1994; Jiang and Haddad, 1994; Jiang, et al., 1994; Xia and Haddad, 1994; O'Reilly et al., 1995; Chidekel, 1997; Fung and Haddad, 1997; Haddad and Jiang, 1997; Haddad, et al., 1997; Jiang and Haddad, 1997; O'Reilly, 1997; Ma and Haddad, 1997; Banasiak and Haddad, 1998; Chow and Haddad, 1998; Friedman et al., 1998; Mironov and Richter, 1998; Ma et al., 1999; Mironov and Richter, 1999). These topics are discussed at length by Dahan (1998), and in a volume edited by Altose and Kawakami (1999).

Summary

A number of concepts about respiratory control can now be formulated based on experiments performed by a large number of respiratory physiologists in the past century. Although the respiratory controller in the CNS does not need any afferent feedback for initiation of rhythmic activity, it does need a number of afferent mechanisms to achieve the observed activity at rest and during increased activity. We also now do know that the carotids are important O_2 sensors in the body, that they mediate a number of functions beside the level of Pao_2 (e.g., temperature and blood pressure regulation), and that neuromuscular failure can contribute to the overall failure of respiratory function when the respiratory muscles are under substantial work load. Furthermore, there is evidence that the "noeud vital" may be in the rostral medulla, but that this is greatly influenced by rostral brain regions such as the hypothalamus, the amygdala, and the neocortex. How this respiratory control system develops and changes with age, from fetal life to old age, is not fully understood, but studies are in progress and headway is being made. The introduction of genomics as a strategy to link differential ventilatory function to molecular structure has been dramatically changing the direction of research in lung biology with special regard to respiratory control (Tankersley, 1999). Inbred murine models with mutations and polymorphism are found to have altered breathing.

References

Banasiak, K.J., and Haddad, G.G. (1998) Hypoxia-induced apoptosis: effect of hypoxic severity and role of p53 in neuronal cell death. Brain Res. 797, 295–304.

Blanco, C.E., Dawes, G.S., Hanson, M.A., and McCooke, H.B. (1984a) The response to hypoxia of arterial chemoreceptors in fetal sheep and newborn lambs. J. Physiol. (Lond.) 351, 25–37.

Blanco, C.E., Hanson, M.A., Johnson, P., and Rigatto, H. (1984b) Breathing pattern of kittens during hypoxia. J. Appl. Physiol. 56, 12–17.

Blanco C.E., Hanson, M.A., and McCooke, H.B. (1988) Studies of chemoreceptor resetting after hyperoxic ventilation of the fetus in utero. In Chemoreceptors in respiratory control, Riberio, J.A., and Pallot, D.J. (eds.) pp. 221–227. Croom Helm, London.

Bryan, A.C., and Gaulthier, C. (1985) Chest wall mechanics in the newborn. In: Lung biology in health and disease, Roussos, C., and Macklem, P.T. (eds.) pp. 871–888. Marcel Dekker, New York.

Bureau, M.A., and Begin, R. (1982) Postnatal maturation of the respiratory response to O_2 in awake newborn lambs. J. Appl. Physiol. 52, 428–433.

Bureau, M.A., Lamarche, J., Foulon, P., and Dalle, D. (1985a) Postnatal maturation of respiration in intact and carotid body chemodenervated lambs. J. Appl. Physiol. 59, 869–874.

Bureau, M.A., Lamarche, J., Foulon, P., and Dalle, D. (1985b) The ventilatory response to hypoxia in the newborn lamb after carotid body denervation. Respir. Physiol. 60, 109–119.

Bureau, M.A., Cote, A., Blanchard, P.W., Hobbs, S., Foulon, P., and Dalle, D. (1986) Exponential and diphasic ventilatory response to hypoxia in conscious lambs. J. Appl. Physiol. 61, 836–842.

Calabrese, R.L. (1979) The roles of endogenous membrane properties and synaptic interaction in generating the heartbeat rhythm of the leech, Hirudo medicinalis. J. Exp. Biol. 82, 163–176.

Chidekel, A.S., Friedman, J.E., and Haddad, G.G. (1997) Anoxia-induced neuronal injury: role of Na^+ entry and Na^+-dependent transport. Exp. Neurol. 146, 403–413.

Chow, E., and Haddad, G.G. (1998) Differential effects of anoxia an glutamate on cultured neocortical neurons. Exp. Neurol. 150, 52–59.

Clarke, J.A., deBurgh, Daly M., and Ead, H.W. (1990) Comparison of the size of the vascular compartment of the carotid body of the fetal, neonatal and adult cat. Acta Anat. 138, 166–174.

Cole, S., Lindenberg, L.B., Gallioto, F.M., Howe, P.E., DeGraff, C., Davis, J.M., et al. (1979) Ultrastructural abnormalities of the carotid body in sudden infant death syndrome. Pediatrics 63, 13–17.

Dekin, M.S., and Haddad, G.G. (1990) Membrane and cellular properties in oscillating networks: implications for respiration. J. Appl. Physiol. 69, 809–821.

Donnelly, D.F., and Haddad, G.G. (1986) Respiratory changes induced by prolonged laryngeal stimulation in unanesthetized piglets. J. Appl. Physiol. 61, 1018–1024.

Donnelly, D.F., and Haddad, G.G. (1990) Prolonged apnea and impaired survival in piglets after sinus and aortic nerve section. J. Appl. Physiol. 68, 1048–1052.

Eden, G.J., and Hanson, M.A. (1987) Maturation of the respiratory response to acute hypoxia in the newborn rat. J. Physiol. (Lond.) 392, 1–9.

Farber, J.P. (1985) Motor responses to positive pressure breathing in the developing opossum. J. Appl. Physiol. 58, 1489–1495.

Farber, J.P. (1988) Medullary inspiratory activity during opossum development. Am J. Physiol. R578–R584.

Farber, J.P. (1989) Medullary expiratory activity during opossum development. J. Appl. Physiol. 66, 1606–1612.

Farber, J.P. (1990) Effects on breathing of rostral pons glutamate injection during opossum development. J. Appl. Physiol. 69, 189–195.

Feldman, J.L., Smith, J.C., Ellenberger, H.H., Connelly, C.A., Greer, J.J., Lindsay, A.D., et al. (1990) Neurogenesis of respiratory rhythm and pattern: emerging concepts. Am. J. Physiol. 259, 879–886.

Friedman, J.E., Chow, E.J., and Haddad, G.G. (1998) State of actin filaments is changed by anoxia in cultured rat neocortical neurons. Neuroscience 82(2), 421–427.

Fung, M.L., and Haddad, G.G. (1997) Anoxia-induced depolarization in CA1 hippocampal neurons: role of Na^+-dependent mechanisms. Brain Res. 762, 97–102.

Funk, G.D., and Feldman, J.L. (1995) Generation of respiratory rhythm and pattern in mammals: insights from developmental studies. Curr. Opin. Neurobiol. 6, 778–785.

Feldman, J.L., Smith, J.C., and Liu, G. (1991) Respiratory pattern generation in mammals: in vitro en bloc analyses. Curr. Opin. Neurobiol. 1, 590–594.

Ganfornina, M.D., and Lopez-Barneo, J. (1991) Single K^+ channels in membrane patches of arterial chemoreceptor cells are modulated by O_2 tension. Proc. Natl. Acad. Sci. 88, 2927–2930.

Ge, Q., and Feldman, J.L. (1998) AMPA receptor activation and phosphatase inhibition affect neonatal rat respiratory rhythm generation. J. Physiol. 509(1), 255–266.

Getting, P.A. (1981) Mechanisms of pattern generation underlying swimming in Tritonia. I. Neuronal network formed by monosynaptic connections. J. Neurophysiol. 46, 65–79.

Getting, P.A. (1983a) Mechanisms of pattern generation underlying swimming in Tritonia. II. Network reconstruction. J. Neurophysiol. 49, 1017–1035.

Getting, P.A. (1983b) Mechanisms of pattern generation underlying swimming in Tritonia. III. Intrinsic and synaptic mechanisms for delayed excitation. J. Neurophysiol. 49, 1036–1050.
Getting, P.A. (1988) Comparative analysis of invertebrate central pattern generators. In: Neural control of rhythmic movements in vertebrates, Cohen, A.H., Rossignol, S., and Grillner, S. (eds.) pp. 101–127. John Wiley and Sons, New York.
Gonzalez, C., Almaraz, L., Obeso, A., and Rigual, R. (1992) Oxygen and acid chemoreception in the carotid body chemoreceptors. TINS. 15, 146–153.
Grillner, S. (1985) Neurobiological bases of rhythmic motor acts in vertebrates. Science 228, 143–149.
Haddad, G.G., and Donnelly, D.F. (1988) The interaction of chemoreceptors and baroreceptors with the central nervous system. Ann. N.Y. Acad. Sci. 533, 221–227.
Haddad, G.G., and Getting, P.A. (1989) Repetitive firing properties of neurons in the ventral region of the nucleus tractus solitarius. In vitro studies in the adult and neonatal rat. J. Neurophysiol. 62, 1213–1224.
Haddad, G.G., Donnelly, D.F., and Getting, P.A. (1990) Biophysical membrane properties of hypoglossal neurons in vitro: intracellular studies in adult and neonatal rats. J. Appl. Physiol. 69, 1509–1517.
Haddad, G.G. (1991) Cellular and membrane properties of brainstem neurons in early life. In: Developmental neurobiology of breathing, Haddad, G.G., and Farber, J.P. (eds.) pp. 155–175. Marcel Dekker, New York.
Haddad, G.G., and Jiang, C. (1993) O_2 deprivation in the central nervous system: on mechanisms of neuronal response, differential sensitivity, and injury. Prog. Neurobiol. 40, 277–318.
Haddad, G.G., and Jiang, C. (1994) Mechanisms of neuronal survival during hypoxia: ATP-sensitive K^+ channels. Biol. Neonate 65, 160–165.
Haddad, G.G., and Jiang, C. (1997) O_2–sensing mechanisms in excitable cells: role of plasma membrane K^+ channels. Annu. Rev. Physiol. 59, 23–43.
Haddad, G.G., Sun, Y.-A. Wyman, R.J., and Xu, T. (1997) Genetic basis of tolerance to O_2 deprivation in Drosophila melanogaster. Proc. Natl. Acad. Sci. USA 94, 10809–10812.
Hanson, M.A., Kumar, P., and McCooke, H.B. (1987) Post-natal resetting of carotid chemoreceptor sensitivity in the lamb. J. Physiol. (Lond.) 382, 57P.
Hofer, M.A. (1984) Lethal respiratory disturbance in neonatal rats after arterial chemoreceptor denervation. Life Sci. 34, 489–496.
Hofer, M.A. (1985) Sleep-wake state organization in infant rats with episodic respiratory disturbance following sinoaortic denervation. Sleep 8, 40–48.
Jiang, C., and Haddad, G.G. (1991) The effect of anoxia on intracellular and extracellular K^+ in hypoglossal neurons in vitro. J. Neurophysiol. 66, 103–111.
Jiang, C., Cummins, T.R., and Haddad, G.G. (1994) Membrane ionic currents and properties of freshly dissociated rat brainstem neurons. Exp. Brain Res. 100, 407–420.
Jiang, C., Sigworth, F.J., and Haddad, G.G. (1994) Oxygen deprivation activates an ATP-inhibitable K^+ channel in substantia nigra neurons. J. Neurosci. 14(9), 5590–5602.
Jiang, C., and Haddad, G.G. (1994) Oxygen deprivation inhibits a K^+ channel independently of cytosolic factors in rat central neurons. J. Physiol. 481.1, 15–26.
Jiang, C., and Haddad, G.G. (1997) Modulation of K^+ channels by intracellular ATP in human neocortical neurons. J. Neurophysiol. 77, 93–102.
Kholwadwala, D., and Donnelly, D.F. (1992) Maturation of carotid chemoreceptor sensitivity to hypoxia: in vitro studies in the newborn rat. J. Physiol. (Lond.) 453, 461–473.
LeGallois, J.J.C. (1813) Experiments on the principle of life. In: Pulmonary and respiratory physiology, part II, Comroe, Jr., J.H. (ed.) pp. 12–16. Dowden, Hutchinson, & Ross, Pennsylvania.
Ma, E., and Haddad, G.G. (1997) Anoxia regulates gene expression in the central nervous system of *Drosophila melanogaster*. Mol. Brain Res. 46, 325–328.
Ma, E., Xu, T., and Haddad, G.G. (1999) Gene regulation by O_2 deprivation: an anoxia-regulated novel gene in *Drosophila melanogaster*. Mol. Brain Res. 63, 217–224.
Mironov, S.L., and Richter, D.W. (1998) L-type Ca^{2+} channels in inspiratory neurones of mice and their modulation by hypoxia. J. Physiol. (Lond.) 512(1), 75–87.
Mironov, S.L., and Richter, D.W. (1999) Cytoskeleton mediates inhibition of the fast Na^+ current in respiratory brainstem neurons during hypoxia. Eur. J Neurosci. 11(5), 1831–1834.

Moore, P.J., Clarke, J.A., Hanson, M.A., deBurgh Daly, M., and Ead, W. (1991) Quantitative studies of the vasculature of the carotid body in fetal and newborn sheep. J. Dev. Physiol. 15. 211–214.
Naeye, R.L., Fisher, R., Ryser, M., and Whalen, P. (1976) Carotid body in sudden infant death syndrome. Science 191, 567–569.
O'Reilly, J.P., Jiang, C., and Haddad, G.G. (1995) Major differences in response to graded hypoxia between hypoglossal and neocortical neurons. Brain Res. 683, 179–186.
O'Reilly, J.P., Cummins, T.R., and Haddad, G.G. (1997) Oxygen deprivation inhibits Na^+ current in rat hippocampal neurones. J. Physiol. (Lond.) 503.3, 479–488.
Peers, C. (1990) Effect of lowered extracellular pH on Ca^{2+}-dependent K^+ currents in type I cells from the neonatal rat carotid body. J. Physiol. (Lond.) 422, 381–395.
Perrin, D.G., Cutz, E., Becker, L.E., Bryan, A.C., Madapallimatum, A., and Sole, M.J. (1984) Sudden infant death syndrome: increased carotid body dopamine and noradrenaline content. Lancet 2(8402), 535–537.
Pokorski, M., and Lahiri, S. (1981) Effects of naloxone on carotid body chemoreception and ventilation in the cat. J. Appl. Physiol. 51, 1533–1538.
Rekling, J.C., and Feldman, J.L. (1998) PreBotzinger complex and pacemaker neurons: hypothesized site and kernel for respiratory rhythm generation. Annu. Rev. Physiol. 60, 385–405.
Richter, D.W., Ballanyi, K., and Schwarzacher, S. (1992) Mechanisms of respiratory rhythm generation. Curr. Opin. Neurobiol. 2, 788–793.
Robertson, R.M., and Pearson, K.G. (1985) Neural circuits in the flight system of the locust. J. Neurophysiol. 53, 110–128.
Selverston, A.I., Russell, D.F., Miller, J.P., and King D. (1976) The stomatogastric nervous system: structure and function of a small neural network. Prog. Neurobiol. 7, 215–290.
Sieck, G.C., and Fournier, M. (1991) Developmental aspects of diaphragm muscle cells. In: Developmental neurobiology of breathing, lung biology in health and disease, Haddad, G.G., and Farber, J.P. (eds.) pp. 375–428. Marcel Dekker, New York.
Smith, J.C., Ellenberger, H., Ballanyi, K., Richter, D.W., and Feldman, J.L. (1991) Pre-Botzinger complex: a brainstem region that may generate respiratory rhythm in mammals. Science 254, 726–729.
Tankersley, C.G. (1999) Genetic control of ventilation: What are we learning from murine models? Curr. Opin. Pulm. Med. 5, 344–348.
Watchko, J.F., Mayock, D.E., Standaert, T.A., and Woodrum, D.E. (1991) The ventilatory pump: neonatal and developmental issues. Adv. Pediatr. 38, 109–134.
Xia, Y., and Haddad, G.G. (1994) Voltage-sensitive Na^+ channels increase in number in newborn rat brain after in utero hypoxia. Brain Res. 635, 339–344.

Recommended Readings

Altose, M.D., and Kawakami, Y. (eds.) (1999) Control of breathing in health and disease. Marcel Dekker, New York.
Caruana-Montaldo, B., Gleeson, K., and Zwillich, C.W. (2000) The control of breathing in clinical practice. Chest 117, 205–225.
Dahan, A. (1998) Physiology and pharmacology of cardio-respiratory control. Kluwer Academic, Dordrecht.
Hughson, R.L., Cunningham, D.A., and Duffin, J. (eds.) (1998) Advances in modeling and control of ventilation. Plenum, New York.
Stocks, J. (1999) Respiratory physiology during early life. Monaldi Arch. Chest Dis. 54, 358–364.
Tankersley, C.G. (1999) Genetic control of ventilation: what are we learning from murine models? Curr. Opin. Pulm. Med. 5, 344–348.

7
Arterial Chemoreceptors

Constancio González, Laura Almaraz, Ana Obeso, and Ricardo Rigual

Cell respiration is the process by which oxygen (O_2) is delivered to cells to oxidize nutrients and to obtain the energy necessary for all their functions. Carbon dioxide (CO_2) and water are the main byproducts of cell respiration.

In pluricellular organisms, cells obtain O_2 from the internal milieu that surrounds them. In higher terrestrial and aerial animals, the delivery of O_2 from the atmosphere to the internal milieu is provided by the respiratory (ventilatory) and by the circulatory systems. The process involves the following steps: pumping of O_2-rich air from the atmosphere to the alveoli, diffusion of O_2 from the alveoli to the blood, binding of O_2 to hemoglobin, circulation of the O_2-rich blood to the capillaries of the whole organism, dissociation of O_2 from hemoglobin, and diffusion of O_2 from blood to the internal fluid and to the cell interior and mitochondria, where cell respiration occurs. Carbon dioxide follows the opposite path from mitochondria to the atmosphere.

From this ample conception of respiration it emerges that hypoxia, defined as any situation in which the utilization of O_2 by the cells is insufficient to maintain their normal function, may result from (1) an inadequate provisioning of O_2 to the blood (hypoxic hypoxia), (2) a decreased capacity of blood to bind O_2 (anemic hypoxia), (3) an insufficient rate of circulation of blood (stagnant hypoxia), and (4) an impaired ability of the cells to respire, to use O_2 (histotoxic hypoxia). In a healthy animal, the only type of hypoxia that may occur is a hypoxic hypoxia due to a decrease of PO_2 in the atmosphere, as it occurs at high altitude.

Evolution has endowed higher animals with regulating mechanisms, that using as effectors the respiratory and/or the circulatory systems, tend to prevent any type of hypoxia. The arterial chemoreceptors, comprising the aortic bodies and, in particular, the carotid bodies (CB), are the commanders of these regulating mechanisms. The arterial chemoreceptors "sense" blood PO_2 (the CBs) and blood O_2 content (the aortic bodies), and initiate a series of reflex responses directed to the respiratory and to the circulatory systems with the net result of an increase in air pumping into the lungs and blood pumping into the tissues. By this means, the arterial chemoreceptors (mostly the CBs) subserve an important adaptative role, which makes possible for higher animals, including humans, to survive and to perform adequately (i.e., to adapt) at high altitudes with barometric pressures below 350 mmHg. In addition, the CBs have an important homeostatic function regarding hydrogen ions (H^+): they sense blood PCO_2 and $[H^+]$, and, when their concentration is above normal, respiratory and circulatory reflexes are initiated to discharge the excess of H^+ as CO_2 into the atmosphere.

The CBs are secondary sensory receptors whose chemoreceptor or type I cells sense arterial PO_2, PCO_2 and pH (PaO_2, $PaCO_2$ and pH). Chemoreceptor cells release neurotransmitters in resting conditions (normal blood gases and pH); this release increases when PaO_2 or pH falls, or when $PaCO_2$ increases. Released neurotransmitters set the discharge (action potential frequency) in the sensory fibers of the carotid sinus nerve (CSN; a branch of the IX cranial nerve), whose central projections terminate in the brainstem. CSN discharges are integrated in the brainstem, leading to a proportionate reflex ventilatory and cardiovascular response. In successive sections of this chapter, we will summarize the structural organization of the CB, the functions of the CB, and the mechanisms of the CB functions. A brief account of the aortic bodies will also be included. The mechanisms of integration of the CSN discharges in the brainstem and the efferent pathways of the chemoreceptor reflexes can be found in other chapters of this book.

Structural Organization of the Carotid Body

Microscopic Organization of the Carotid Body

The mammalian CB is a small paired organ located on a branch of the carotid artery near its bifurcation. Figure 7.1A presents the location of the CB in the rabbit. Since its discovery at the middle of the eighteenth century, the CB had been considered a small ganglion and a gland; however, in 1928 De Castro showed that those fibers of the CSN that innervate the CB parenchyma were sensory, and not secretomotor. He had earlier shown that the sympathetic fibers that reach the CB via the gangliomerular nerves, innervate the CB vessels rather than the CB cells. De Castro concluded that the CB was a specialized sensory organ, not a ganglion or a gland. In 1930, Heymans and co-workers demonstrated that the CB senses PaO_2, $PaCO_2$, and pH, and that the hyperventilation induced by a decrease in PaO_2 or pH or by an increase in $PaCO_2$ was initiated at the CB. For this work, Heymans was awarded the Nobel Prize in 1938.

The parenchyma of the CB is organized in clusters (Fig. 7.1B) separated from each other by connective tissue, which converges on the surface of the organ to form a capsule. The clusters are formed by two cell types, chemoreceptor cells (or type I cells) and sustentacular or type II cells. Chemoreceptor cells are more numerous than sustentacular cells, have a clear and round nucleus, and are located towards the center of the clusters; type II cells surround chemoreceptor cells and have a disc-shaped nucleus with dense chromatin (Figure 7.1B). The CB is a very vascularized organ with 25–30%

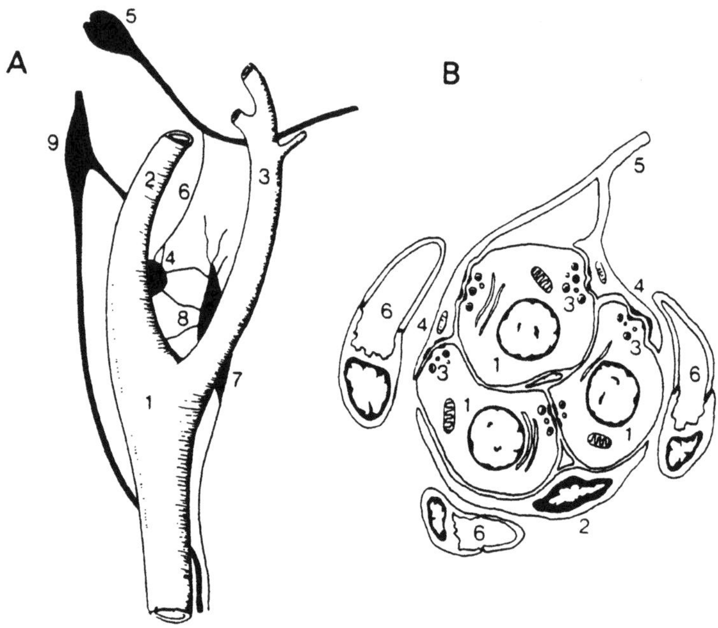

FIGURE 7.1. (A) A schematic drawing showing the anatomical relationships of the carotid body in the rabbit. The common carotid artery (1) divides to give internal (2) and external (3) carotid arteries. The carotid body (4) is attached to the internal carotid artery near its origin. It is innervated by sensory fibers that have their soma in the petrosal ganglion (5) and reach the carotid body via the carotid sinus nerve (6) and by sympathetic fibers that having their origin in the superior cervical ganglion (7) reach the carotid body by the ganglioglomerular nerves (8). The nodose ganglion (9) is lateral to the internal carotid artery. (B) Schematic drawing of a cellular cluster of the rabbit carotid body. Chemoreceptor cells (1) are partially surrounded by sustentacular cells (2). The cytoplasm of chemoreceptor cells has an heterogeneous population of synaptic vesicles (3) frequently located facing the sensory nerve endings (4) that originate from branching fibers of the carotid sinus nerve (5). A dense net of capillaries (6) surrounds the cell clusters (*Source*: from Gonzalez et al. Trends Neurosci. 15, 146–153, 1992, with permission).

of the surface of histological sections occupied by vessel lumen. A net of small vessels, including wide fenestrated capillaries, traverse the connective tissue and establish a close relationship with the cell clusters before reaching a venous plexus in the CB's surface.

Bundles of the CSN sensory nerve fibers penetrate the CB and synapse with chemoreceptor cells; functionally these fibers are chemoreceptor fibers. The sympathetic fibers that enter the CB innervate the blood vessels. Some parasympathetic fibers originate from neurons scattered along the CSN nerve also innervate CB blood vessels. Part of the CSN sensory fibers never enter the CB, but they reach the carotid sinus, and functionally constitute the carotid baroreceptors.

Chemoreceptor cells are round or ovoid and have a diameter of ~10 μm and occasional fingerlike processes. In addition to the general cell organelles, they have abundant catecholamine-containing (CA; dopamine, DA, and norepinephrine, NE) dense-core vesicles, similar to those of adrenomedullary cells. Dense-core vesicles accumulate in the periphery of the cytoplasm, sometimes facing the nerve endings. Chemoreceptor cells also possess smaller clear-core vesicles that appear to contain acetylcholine (ACh). The sensory nerve endings that synapse with chemoreceptor cells are variable in form and size, and contain clear-core and dense-core vesicles. These synapses are ultrastructurally of two types: most have the direction of cells to sensory nerve endings, being the morphological correlate of the sensory (chemosensory) function of the CB; the rest has the opposite direction (i.e., of nerve endings to cells), and represent pathways for feedback control of chemoreceptor cells.

This account of the structural organization of the CB allows us to define the functional ensemble of its elements. The capillaries minimize the diffusion distance between blood and chemoreceptor cells due to their close contact with cell clusters, so that the chemoreceptor can detect promptly any variation in PaO_2, $PaCO_2$, and pH. Upon detection, chemoreceptor cells transduce variations in blood parameters into a neurosecretory response and the neurotransmitters contained in the vesicles are released. The frequency of action potentials in the CSN chemosensory fibers varies accordingly. Modulation of chemoreceptor cell function can be achieved by activation of synaptic transmission directed from the sensory nerve endings to chemoreceptor cells, as well as reflexely via the sympathetic and parasympathetic innervation of the CB blood vessels. As might be expected, modifications of the CB blood flow affect the stimuli reaching chemoreceptor cells, and hence their response at a given PaO_2 or $PaCO_2$/pH. Type II cells are supporting (glial-like) structures to chemoreceptor cells and sensory nerve endings.

Histochemistry and Immunohistochemistry of the Carotid Body

If the account of the structural organization enables us to envision the functional relationship of the different elements of the CB, the histochemical–immunohistochemical data would be helpful to understand how the CB elements communicate with each other.

Chemoreceptor cells originate from the neural crest and contain several neurotransmitters. Catecholamines were demonstrated in chemoreceptor cells on the basis of chromaffin and fluorescent reactions, whereas more recent immunocytochemical studies have shown the enzymes involved in their synthesis. In the rat and rabbit CB, only a minority of chemoreceptor cells are positive to dopamine beta hydroxylase, the enzyme required for the synthesis of NE, whereas all of them contain tyrosine hydroxylase. Thus, most of chemoreceptor cells in these species are dopaminergic, and only a small percentage would be noradrenergic. Consistent with these immunocytochemical findings, chemical analyses show a DA/NE ratio of nearly 5/1 in the CB of these two species. On the contrary, in the cat CB most chemoreceptor cells are dopamine beta hydroxylase positive and DA/NE ratios of nearly 1/1 have been reported; those findings would imply that most chemoreceptor cells contain both CA. In general, the CB of all studied species, including humans, have a high concentration of CA with DA as the dominant CA in most instances. CA-degrading enzymes have been shown in the CB of several species. Other two biogenic amines, serotonin and ACh, have also been identified in the CB of several mammals.

Chemoreceptor cells exhibit opioid-like immunoreactivity with opioid peptides being co-localized with CA in dense-core vesicles; biochemical analyses show that most of the opioid activity in the CB is in the form of Met-enkephalin and Leu-enkephalin. Substance P, tachykinin A, cholecystokinin, galanin, neurotensin, bombesin and atrial natriuretic peptide have been found in variable percentages of chemoreceptor cells and with several patterns of coexistence depending on the species. The significance of these neurotransmitters will be discussed in a later section.

Radioreceptor binding and *in situ* hybridization studies have shown DA-2 dopamine receptors in chemosensory nerve endings as well as in chemoreceptor cells of the CB in several species. Similarly, northern blot analysis and reverse transcription-polymerase chain reaction have demonstrated the expression of DA-1 receptors in the CB of the rabbit, rat and cat, with some additional pharmacological evidence indicating that they are located in blood vessels. Radioreceptor binding and autoradiographic studies have also shown nicotinic and muscarinic receptors in chemoreceptor cells, and electrophysiological and neurochemical findings have corroborated these findings. Thus, nicotinic agents activate inward currents and induce release of CA, and muscarinic agonists tend to inhibit the release of CA induced by hypoxia in chemoreceptor cells. However, nicotine receptors are not restricted to chemoreceptor cells, as immunocytochemical studies have identified α-7 subunits of the nicotine receptors in tyrosine hydroxylase positive chemosensory nerve endings of the CSN. Finally, several lines of evidence indicate that chemoreceptor cells also contain α_2 and β adrenergic and δ-opioid receptors, as agonists and/or antagonists of these receptors modify release of CA and/Ca^{2+} currents in these cells. But it cannot be excluded that other CB structures might also contain these receptors.

Sensory fibers that innervate chemoreceptor cells contain DA, tachykinin A, substance P, calcitonin-gene related peptide and neuropeptide Y with different patterns of incidence and coexistence in different species. Immunocytochemical techniques have also revealed that these sensory nerve endings are not mature at birth nor contain the neurotransmitters that they exhibit in the adult animals. It should be stressed that the poor sensitivity of the CBs of neonatal animals to hypoxia parallels this incomplete maturation of the sensory nerve endings. Sympathetic fibers innervating CB blood vessels contain CA, and some of them are neuropeptide Y- and/or vasoactive intestinal peptide-positive. Parasympathetic fibers, which also innervate blood vessels, are positive to nitric oxide synthase and producing vasodilatation can negatively modulate responses of the CB to hypoxia. These nitrergic fibers form part of the so called efferent pathway to the CB.

The location of carbonic anhydrase was important to define as the structure that sensed $PaCO_2$ in the CB. In the middle eighties it was debated whether the acidic (PCO_2/pH) stimulus was detected in chemoreceptor cells or in the chemosensory nerve endings. Carbonic anhydrase was considered a key component of the PCO_2 sensing machinery because the inhibitors of the enzyme eliminate the transient component of the response to high $PaCO_2$, and therefore, the structure that possessed the enzyme would be the one detecting this stimulus. A combination of histochemical techniques and axonal flow experiments located the enzyme in chemoreceptor cells.

Functions of the Carotid Body

Carotid body function can be studied at two levels. Some physiologists have sought the cellular level to learn how the chemoreceptor cells "sense" blood gases or release neurotransmitters, or how the action potentials in the CSN are generated. Others have sought the systemic level of the CB function, and have tried to define the responses of the ventilatory and cardiovascular system when the CBs are activated either by natural stimuli or by drugs. With a greater emphasis in the cellular level, we will cover both aspects. With Fig. 7.2 in mind, we shall examine in this section the outputs of the CB at all levels, including the release of neurotransmitters from chemoreceptor cells, the activity in the CSN, and the ventilatory and cardiocirculatory responses elicited by the stimulation of the CB. The mechanisms of transduction and neurotransmission, that is,

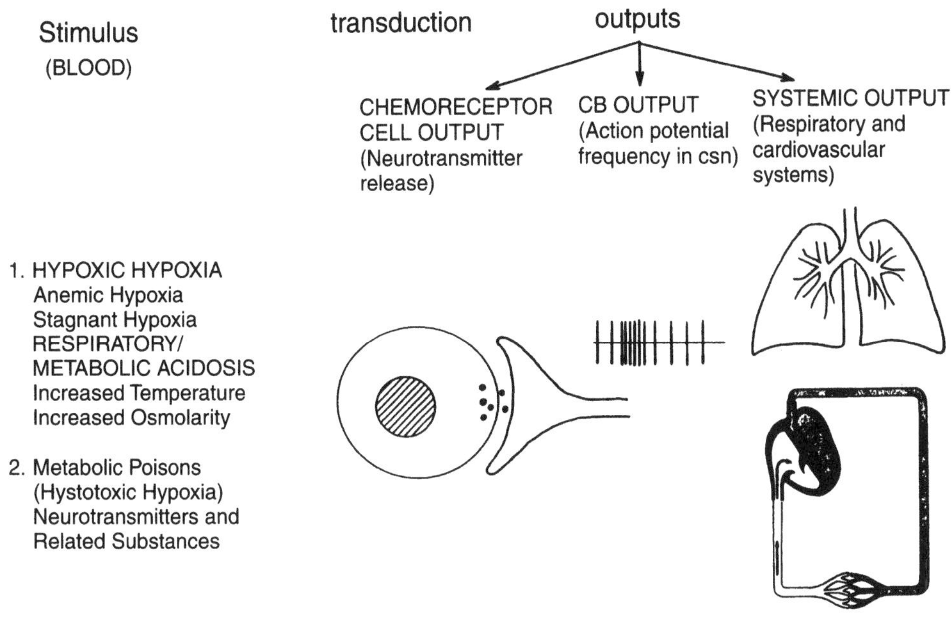

FIGURE 7.2. Functional organization of the carotid body. Physiologically, stimuli reach the carotid body by the blood vessels. Chemoreceptor cells transduce the stimuli into a neurosecretory response (the release of neurotransmitters) that constitutes the output of chemoreceptor cells. The released neurotransmitters set the action potential frequency in the carotid sinus nerve (whole CB output). The discharges of the carotid sinus nerve are integrated at the brain stem and an adequate reflex ventilatory and cardiovascular response is generated.

the mechanisms of the detection of the stimuli and its transformation into a neurosecretory response and the mechanisms used by neurotransmitters to generate the action potentials in the CSN will be discussed in a later section. Finally, a small section will be devoted to summarize the CB function in situations of chronic hypoxia.

Release of Neurotransmitters from Chemoreceptor Cells

Given the variety of neurotransmitters in chemoreceptor cells, the patterns of release of the different neurotransmitters in response to different stimuli, or at different intensities of a given stimulus, can be variable. Such patterns of co-release have not yet been explored. However, with independence of the possible patterns of the release, the neurotransmitters must appear in the synaptic space, and hence in the bathing solutions when the CB is superfused *in vitro*. Under such conditions quantification of neurotransmitters present in the bathing solutions provides a measure of the output of the chemoreceptor cells. That is, any stimulus that modifies the resting release of neurotransmitters is assumed as having been detected by the cells.

In the late 1930s, it was found that prostigmine, which is an inhibitor of ACh-esterase, activated ventilation in a manner similar to injected ACh, suggesting that chemoreceptor cells release ACh. Eyzaguirre and co-workers showed later that chemoreceptor cells in cat CB release a substance that is degraded by ACh-esterase and exerts nicotinic effects.

The release of DA from chemoreceptor cells in response to hypoxic stimulation was inferred from several *in vivo* studies which demonstrated a decrease in DA levels in the CB after an episode of acute hypoxia. In the late 1970s Gonzalez and Fidone measured the actual release of DA under low PO_2 conditions *in vitro*, and later provided a detailed characterization of the release response. They found that the DA released was proportional to the intensity of the hypoxia (Fig. 7.3A), and that it was simultaneously paralleled by the CSN action potential frequency recorded. The release of DA and CSN discharge elicited by low PO_2 were greatly reduced in Ca^{2+}-free solutions; antagonists of voltage-dependent Ca^{2+} channels inhibited low PO_2-induced release of DA.

Any inferences regarding the release of CA based on the measurement of endogenous levels after exposure of animals to hypercapnia (respiratory acidosis) could be misleading. Acidosis releases CA, but activates tyrosine hydroxylase, and thereby the synthesis of CA; in other words, an increase in CA levels in the CB, rather than a decrease, may occur during acidosis. The release of CA from the CB during acidic stimulation has been demonstrated *in vitro* (Fig. 7.3B). As in the case of hypoxic stimulation, a close relationship exists between decrease of external pH and the release of DA or CSN discharge. Low external pH-induced release of DA is also a Ca^{2+} dependent process, but antagonists of Ca^{2+} channels fail to block or reduce it. For a given extracellular pH, the release response was greater if the superfusing solution mimicked a respiratory acidosis rather than a metabolic acidosis. Superfusion of the CB with isohydric hypercapnic solutions or with solutions containing weak acids at a pH of 7.40 also augmented DA release and CSN discharge. Finally, inhibitors of carbonic anhydrase reduced in parallel the release of DA and the activity in the CSN elicited by hypercapnic stimulation.

Hypoxic and acidic stimuli release NE, albeit in smaller amounts than DA. In fact, in a recent work performed with rat CB it was found that hypoxia releases preferentially DA over NE by a factor of nearly 10 times. On the other hand, acidosis released DA and NE in proportions paralleling their tissue content, i.e., DA was released in amounts about 5 times larger than NE.

Metabolic poisons (e.g., cyanide, which produces histotoxic hypoxia, and dinitrophenol, a protonophore that acidifies intracellularly and uncouples oxidative phosphorylation), which represent the most classical stimulating agents for the CB

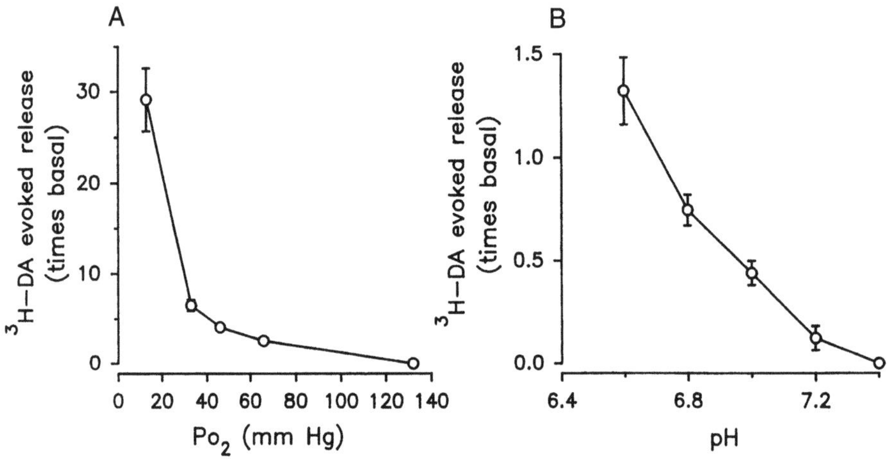

FIGURE 7.3. Release of dopamine from chemoreceptor cells in response to low PO_2 (A) and low pH (B) stimulation. Note the greater intensity of the response during low PO_2 stimulation.

chemoreceptors, also elicit a parallel increase in CA release and in the CSN nerve activity.

The effects of several neurotransmitter-related agents on the release of DA have also been investigated. For example, nicotinic agonists produced release of CA and parallel increases in the CSN discharge of the rabbit and cat CB. Unlike hypoxia, nicotinic agonists released NE preferentially in the CB of the rabbit, rat and cat. This would suggest that the location of nicotinic receptors is in NE-containing chemoreceptor cells, or a specific coupling of nicotinic receptors to the mechanisms releasing NE-containing granules. Muscarinic agonists behave as inhibitory agents in most species by producing a parallel decrease in the release of CA and CSN action potential frequency elicited by hypoxia. Isoproterenol, a β-adrenergic agonist, augmented the basal and stimulus-induced release of CA, the effect of which was inhibited by propranolol, a β-adrenergic antagonist. On the contrary, α_2-agonists inhibited low PO_2-induced release of CA, and α_2-antagonists augmented the low PO_2-induced release of DA. It should be pointed out that the effects of the adrenergic agents on the release of DA are paralleled by their actions on CSN discharge. DA-2 dopaminergic blockers augmented the resting and low PO_2-induced release of DA in the rabbit and cat CB, indicating that released DA is acting on DA-2 autoreceptors to inhibit via feedback its own release. Selective δ-opioid agonists inhibited the release of DA induced by several stimuli including low PO_2, and naloxone, an opioid antagonist, increased low PO_2-induced release. There is also a parallelism between the δ-opioid receptor mediated inhibition of the release of DA and CSN discharge. There are no studies aimed at characterizing the differential release of DA and NE in response to adrenergic, dopaminergic and opiod agents.

In summarizing the data on the release of CA, several aspects should be emphasized: first, that the rate of the release of DA parallels the intensity of stimulation and the frequency of discharge in the CSN. Second, that the pharmacological data indicate that chemoreceptor cells possess receptors for most of the neurotransmitters present in the CB, and third, that the modulation of the release of DA exerted by antagonists of the neurotransmitters indicates that those neurotransmitters are released in the CB in sufficient amounts to regulate the function of chemoreceptor cells.

The release of the other neurotransmitters present in chemoreceptor cells has been less studied. Acute hypoxic stimulation *in vivo* (8% O_2, 3 h) reduced at the same extent (40%) DA and opioid activity levels in the rabbit CB, suggesting co-release of both neurotransmitters, as it should be expected from their co-storage in the same secretory vesicles. Hypoxia also released substance P from the rabbit CB in a Ca^{2+}-dependent manner, but the precise origin of the released substance P (chemoreceptor cells or sensory nerve endings) was not established. Recently, it has also been shown that the cat CB releases acetylcholine in basal conditions and that the amount released increases during hypoxic stimulation. It is also known that adenosine appears in the solution bathing the CB when perfused with normoxic solutions, its amount increasing on lowering the bathing PO_2. The origin of this adenosine is dual: part of it would represent a final degradation product of ATP in the general metabolism of the cell, and the rest, would originate extracellularly from the ATP co-released with other neurotransmitters and degraded by extracellular nucleotidase.

Activity in the Carotid Sinus Nerve

In the following section we will consider the characteristics of chemosensory discharge during resting conditions and during the application of different types of stimulation, the categories of which are depicted in Fig. 7.2.

Resting Carotid Sinus Nerve Discharge

In a pentobarbital anesthetized cat with normal blood gases and pH, CSN fiber discharge varies with the type of fiber studied. Chemoreceptor A-fibers discharge between 1 and 15 impulses/second with most units firing at 2–5 impulses/second; C-fibers discharge but rarely fire more than 4–5 impulses/second with most firing less than 2 impulses/second.

Single nerve fiber resting activity is random because interspike intervals follow a Poisson distribution; however, if the resting discharge of a single fiber is averaged over several respiratory cycles, an oscillatory discharge pattern becomes evident. The oscillations in CSN discharge have their origin in the periodic nature of breathing that produces oscillations in alveolar PO_2 and PCO_2 that are transmitted to arterial blood and detected by the CB. Relevant questions on CSN discharge oscillations include their amplitude, frequency and physiological significance.

The amplitude of the oscillations in relation to the mean CSN frequency in resting, air-breathing cats is about 0.5. The amplitude of $PaCO_2$ and PaO_2 oscillations during the respiratory cycle averages around 2 mmHg and 5 mmHg, respectively, and contribute similarly to the genesis of the CSN discharge oscillations in normoxia and moderate hypoxia. The frequency of oscillations in CSN discharge matches that of breathing, but the amplitude of the CSN discharge oscillations decreases as the breathing frequency increases. In the cat, the CB can follow $PaCO_2$ oscillations of up to 72/minute.

The question now is: What is the significance of CSN discharge oscillations?. The CB drive (CSN input to the brainstem) is an important signal for the generation and control of the breathing pattern under resting conditions, as evidenced by the fact that CB resection or CSN denervation causes a decrease in resting minute ventilation and PaO_2, and an increase in $PaCO_2$. Taking into account that in resting conditions the amplitude of the CSN discharge oscillations is about 0.5, then, the oscillating nature of the CSN discharge could be a signal as important as the mean discharge. In fact, ventilation is reduced when blood gas oscillations at the CB are eliminated by interposing a mixing chamber in the common carotid artery. It also appears that CSN discharge oscillations are important signals in the genesis of exercise hyperventilation. The CB contributes importantly to the hyperventilation seen in exercise (see below Fig. 7.6.A), but the stimulus to the CB has not been identified because in exercise of up to moderate severity mean arterial blood gases and pH remain normal. Then, the possibility exists that the oscillations in the CSN discharge increase in amplitude and represent the CB drive to respiration during exercise. Consistent with this suggestion, experimental evidence demonstrates that increased venous CO_2 loads and increased K^+ concentrations in blood such as those seen in exercise are able to increase the amplitude of the oscillations, and thereby their contribution to the respiratory drive.

Carotid Sinus Nerve Discharge in Response to Hypoxia

A sudden step down in PaO_2 produces an increase in the CSN discharge that starts within about 1 second and peaks at about 3 seconds. The peak level of discharge is maintained as long as PaO_2 is low (i.e., it does not adapt) and, on returning to normal PaO_2, the discharge returns to the basal level within 2–3 seconds with little or no undershoot. Figure 7.4.A shows that the relationship between PaO_2 and chemoreceptor fiber discharges in acute hypoxia is almost exponential; at very low PaO_2 (<10 mmHg) the discharges may tend to level-off or to decrease.

After a few hours of exposure to hypoxia (sustained hypoxic hypoxia) there is acclimatization, i.e., the hyperventilation starts to increase even when the level of

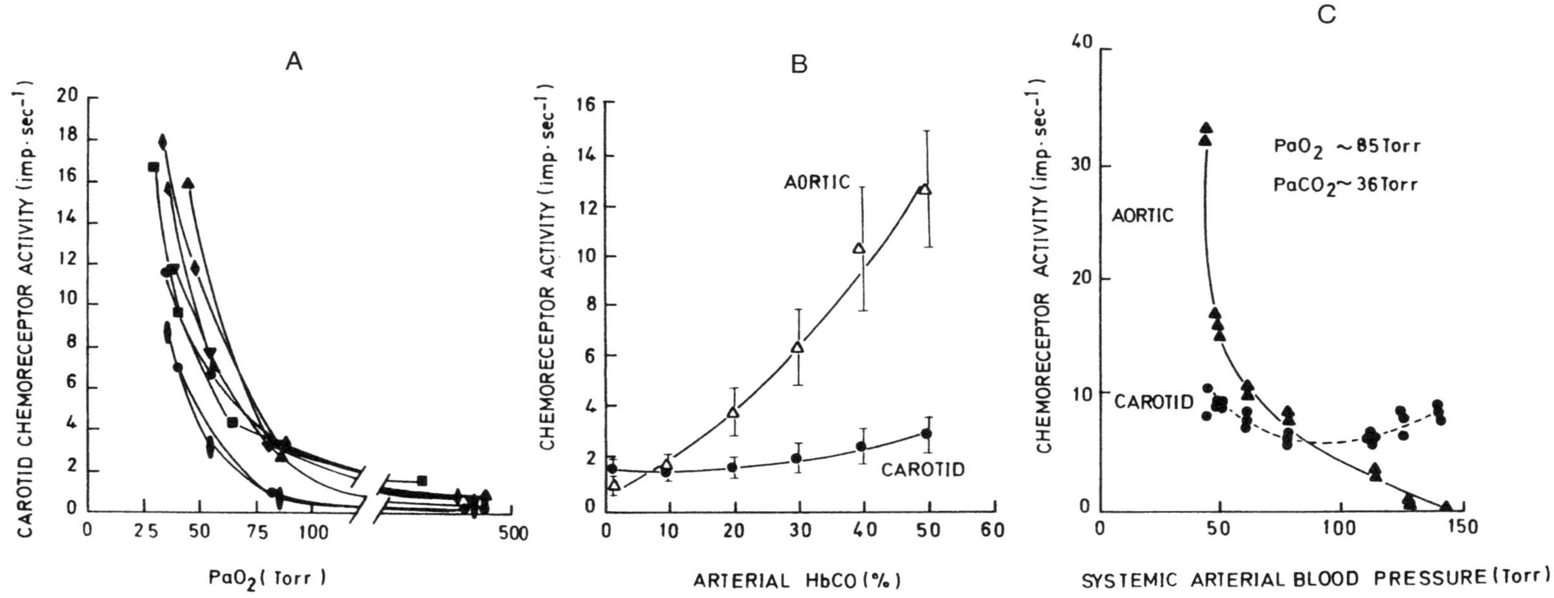

FIGURE 7.4. Carotid sinus nerve action potential frequency during acute hypoxic stimulation in the cat. (A) Steady state activity as a function of PaO_2 in 8 experiments (*Source*: from Mulligan et al. J. Appl. Physiol. 51, 438–446, 1981, with permission). (B) Comparison of the chemoreceptor response to anemic hypoxia (carboxyhemoglobinemia) in the carotid and aortic bodies (*Source*: from Lahiri et al. J. Appl. Physiol. 50, 580–586, 1981, with permission). (C) Comparison of the chemoreceptor response to stagnant hypoxia (hypotension) in the carotid and aortic bodies (*Source*: from Lahiri et al. J. Appl. Physiol. 48, 781–788, 1980, with permission).

hypoxia is maintained. When CSN discharge was recorded continuously during 5–6 h of sustained hypoxia, the same acclimatization type of response appeared, and selective perfusion of the CB with hypoxic blood produced acclimatization both in the CSN discharge and in ventilation. It has also been shown that animals that have been exposed to hypoxia for 48–60 hours and were tested for hypoxic ventilatory responses or CSN activity, exhibited augmented responses in comparison to their corresponding non-exposed controls. These observations are in agreement with classical studies on ventilation in humans during hypoxic exposure, and in addition, they indicate that the CBs are responsible for the acclimatization.

The data shown in Fig. 7.4A indicate that there is no true threshold for the CB response to low PaO_2 because the CSN is already firing at a PaO_2 5 times higher than the normal PaO_2 of 100 mmHg. It is common, however, to talk about the "physiological" threshold for CSN activation. This threshold corresponds to a PaO_2 of 70–75 mmHg. At this PaO_2 the slope of the curve defining the relationship between PaO_2 and CSN discharges (and ventilation) starts to increase quasi exponentially (Fig. 7.4A). At a PaO_2 of ~75 mmHg, the O_2 content in arterial blood is ~95% of the normal. This implies that the CB triggers a hyperventilation before real tissue hypoxia starts to appear, and therefore, that the CB has a tissue hypoxia-preventing role. The lack of adaptation of the CSN response to hypoxia described above corresponds to the required needs of the organism under physiological conditions of sustained hypoxia, such as high altitude.

The response of the CB to anemic and stagnant hypoxia is physiologically unimportant. Very intense anemia and hypotension, which are hardly compatible with life, are needed to elicit an increase in the CSN discharge or ventilation as far as PaO_2 is normal. On the contrary, the aortic chemoreceptors respond to moderate intensities of anemic and stagnant hypoxia (Figs. 7.4B, C), but the ventilatory effects of aortic chemoreceptor stimulation are negligible. The reasons for the different behavior of the CB and aortic bodies is not known. It has been suggested that the much lower blood flow to the aortic bodies in comparison to the CB, would cause a moderate reduction in arterial O_2 content (anemic hypoxia) or in O_2 delivery (stagnant hypoxia) thus producing marked hypoxia and excitation of the aortic bodies without affecting the CB. In turn, this suggestion implies that the aortic bodies detect rates of O_2 delivery, while the CBs detect preferentially changes in PaO_2, independently, to some extent, of the total O_2 delivery. The CB responses to histotoxic hypoxia will be considered later under metabolic poisons.

Carotid Sinus Nerve Discharge in Response to High PCO_2 and Low pH

Discharge in the CSN increases almost linearly with $PaCO_2$ (or decreasing pH) but only up to certain levels; at very high $PaCO_2$ (>100 mmHg) or very low pH (<6.9), the discharge tends to plateau. As Fig. 7.5A shows, the slope of the lines relating $PaCO_2$ and CSN discharge increases as PaO_2 decreases, indicating a positive interaction between both stimuli.

The CSN response to a step rise in $PaCO_2$ (respiratory acidosis) has a latency of less than 1 second, and a time to peak discharge of around 2 seconds. From this fast onset, discharge adapts within about 30 s at a lower sustained level. The fast component of the $PaCO_2$ response disappears after inhibition of carbonic anhydrase, while the steady-state response is preserved (Fig. 7.5B). The loss of this component after the inhibition of the carbonic anhydrase reflects the elimination of the fast swing in $[H^+]$ produced by the fast hydration of CO_2 in the presence of the enzyme, indicating that the response to PCO_2 (in fact to H^+) is generated in chemoreceptor cells where the enzyme is

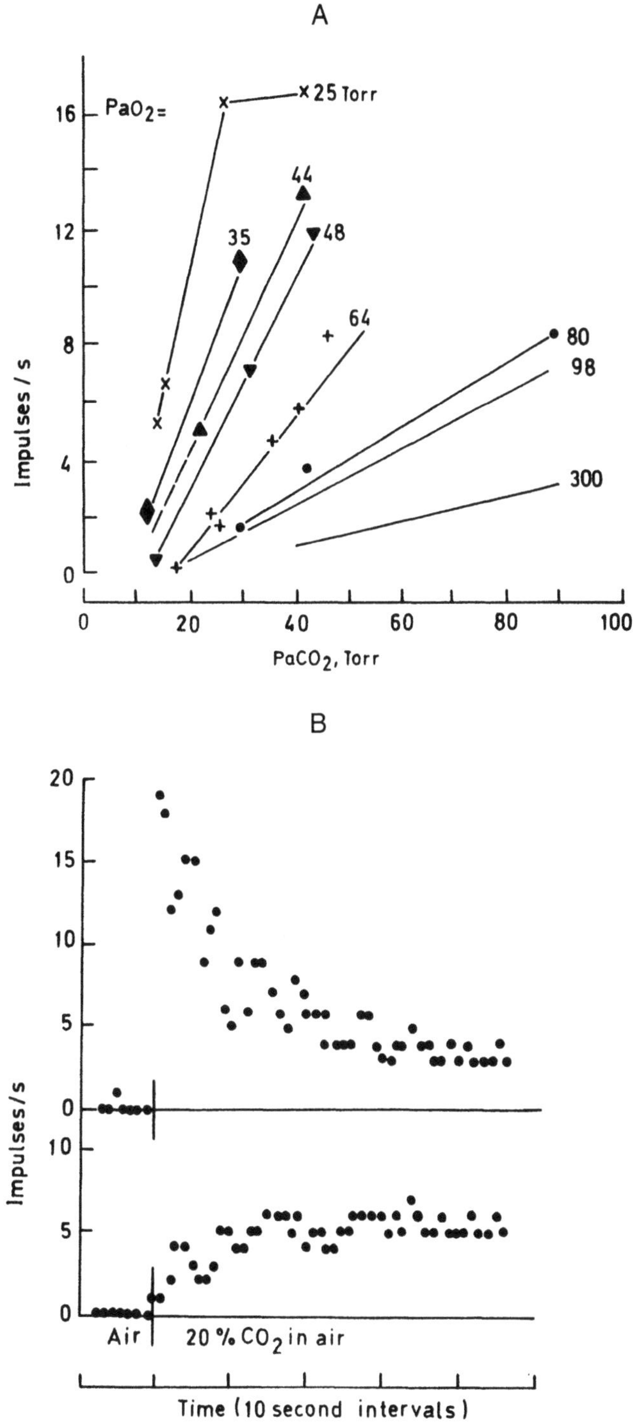

FIGURE 7.5. (A) Cat carotid sinus nerve responses to $PaCO_2$ at different PaO_2. Note the low intensity of the responses to PCO_2 in normoxic and hyperoxic conditions (*Source*: from Lahiri and Delaney, Resp. Physiol. 24, 249–266, 1975, with permission). (B) Effect of sudden CO_2 breathing on the level of discharge of a single carotid sinus nerve fiber in the cat before (top) and after carbonic anhydrase inhibition (bottom) (*Source*: from Black et al. Resp. Physiol. 13, 36–49, 1971, with permission).

located. The response to low pH at normal $PaCO_2$ (metabolic acidosis) rises slowly and does not show the initial adaptation, so that the time to steady discharge is attained in around 30 s. For a given pH drop, the steady-activity is similar for both types of acidosis. Isohydric hypercapnia (proportional rise of PCO_2 and bicarbonate) produces a peak CSN discharge comparable to that seen in hypercapnic acidification, but the adaptation may be more marked.

Carotid Sinus Nerve Discharge in Response to Temperature and Osmolarity

The CB is very sensitive to changes in temperature. For example, a change of 5°C (36 to 41°C) doubles, or more than doubles CSN discharges. This sensitivity may be relevant for the control of ventilation during exercise (body core temperature can increase 1–4°C) and during fever. Then, the combined effect of greater blood gas oscillations, increased plasma K^+, and increased temperature can represent the stimulus for the CB during exercise.

The CB chemoreceptors also respond to changes in blood osmolarity. It has been suggested that the increases in CSN activity seen on moderate increases in blood osmolarity might contribute to the hypothalamic release of antidiuretic hormone.

Carotid Sinus Nerve Response to Metabolic Poisons

All blockers of the respiratory chain produce histotoxic hypoxia and most of them have been tested on CB function. Many inhibitors of oxidative phosphorylation, including uncoupling agents and blockers of the mitochondrial ATPase, have similarly been tested on different aspects of the CB function including CSN discharge. Interest in the actions of metabolic poisons on chemoreceptor function is linked to the metabolic hypothesis of low PO_2 chemosensory transduction. According to this hypothesis, low PO_2, alike metabolic poisons, would activate the CB by decreasing ATP levels in chemoreceptor cells. The decrease in ATP would then trigger the release of the neurotransmitters (see last section).

CSN discharge or ventilation, *in vivo*, increase linearly with increasing doses of CN^- from 2 to 30 $mg{\cdot}Kg^{-1}$, and *in vitro*, threshold CN^- concentration for DA release and increase in CSN activity is around 10^{-5} M. Carbon monoxide (at >300 mmHg) inhibits cytochrome oxidase (as CN^- does) and also increases CSN discharge.

Dinitrophenol increases CSN discharge and release of DA with a threshold of 2.5×10^{-5} M *in vitro*. At concentrations 5 to 10 higher than threshold, DNP increases glucose consumption and does not affect the ATP levels of the CB, indicating that a decrease in ATP levels is not necessary to activate chemoreceptor cells. Thus, DNP must activate chemoreceptor cells by mechanisms other than decreasing cellular ATP.

Oligomycin increases CSN discharge and prevents the response to low PO_2, CN^- and uncouplers, leaving intact the response to high PCO_2. These actions of oligomycin were explained, taking into account only its action on ATP synthesis. However, oligomycin inhibits the Na^+-K^+ pump and Ca^{2+}-activated K^+-channels at concentrations comparable to those required to inhibit the synthesis of ATP.

Carotid Sinus Nerve Responses to Neurotransmitters, Their Agonists and Antagonists

Interpretation of the action of neurotransmitters on CSN discharge on the basis of their bath application or intravascular injection is difficult, because neurotransmitters can act on several sets of receptors with different affinities. For example presynaptic and extrasynaptic receptors have higher affinity for neurotransmitters than postsynaptic receptors, and then the response to "the lower physiological doses" of a neurotransmitter might reflect a presynaptic or extrasynaptic effect rather than a postsynaptic effect.

Acetylcholine, acting on nicotinic receptors, is a powerful stimulant of the CSN activity, whereas acting on muscarinic receptors is inhibitory. The net effect of ACh on the CSN discharge in a given species depends on the ratio of nicotinic to muscarinic receptors; thus, in the case of the cat CB, where nicotinic receptors predominate, ACh produces excitation. In the rabbit, however, the opposite is true. Since cholinergic receptors are located in chemoreceptor cells, sensory nerve endings and blood vessels (muscarinic receptors), the effects of ACh on CSN discharge should represent an integrated effect of its action on the different sets of receptors.

In pioneer work, it was reported that DA increased CSN discharge in the dog *in vivo*; later it was shown that this was true at high doses, and that at low doses DA decreased CSN discharge. In the cat and rabbit CB *in vivo*, DA or its agonists inhibit the resting and the stimulus-induced CSN discharge at low doses, but again high doses produce excitation. Dopamine also reduced ventilation (or phrenic output) in several species including man. Consistent with these findings, dopaminergic antagonists increased CSN discharge and, ventilation. However, most of the studies that have generated the preceeding data are qualitative, as single or few doses of the dopaminergic agents were tested, and, therefore, they should be interpreted with caution. For example, in one study in humans using a single dose of droperidol (a DA-2 dopaminergic blocker, ~35 mg.Kg^{-1}), it was found that the drug increased the hypoxic ventilatory response and prevented the inhibitory effects of DA infusion. It was concluded that droperidol acts by blocking the action of endogenously released DA (without further consideration of the level of inhibition), and, therefore, DA is an inhibitory transmitter. In another study carried out using the cat, it was found that droperidol at low doses (ca. 8–20 mg·Kg^{-1}) increased basal CSN discharge and augmented the response to a 30 second asphyctic test; at higher doses (ca. 40–120 mg·Kg^{-1}). However, it produced a short lasting (about 40 seconds) increase in resting discharge that was followed by an inhibition of the CSN discharge and an abolition of the response to the asphyctic test. In comparing both studies, it appears that at low doses droperidol blocks the action of DA at some level in the CB and produces CSN excitation, whereas at higher doses it blocks the DA action at an additional level and abolishes CSN discharge.

Studies with dose-response curves have shown that high doses of DA in the cat produced either excitation or a short-lasting inhibition followed by a sustained excitation. Then, in the cat CB there are at least two sets of DA receptors with different accessibility and/or affinity for DA. Those more accessible or with higher affinity for DA mediate the inhibitory action of DA; those with lower accessibility or affinity mediate the excitatory actions.

In superfused preparations, where blood vessels are not functional, the excitatory action of DA in the cat is more frequently observed, and in the rabbit DA is excitatory always. In this preparation of the cat CB, haloperidol (0.13–5.32 mM) reduced or abolished spontaneous and ACh-induced CSN discharge. Finally, it has been reported that cat and rabbit CBs isolated from animals reserpinized for 24 h (reserpine depletes CA stores) exhibit a very low resting CSN discharge and reduced responses to low PO_2 and high PCO_2. DA is excitatory at all doses in reserpinized cats. Whether endogenous DA is an excitatory or inhibitory transmitter is a question we shall address in the final section.

Injection of norepinephrine and epinephrine into the blood stream increases ventilation in several species including humans. Both also produce an increase in CSN discharge (preceded by a transient inhibition) in the cat, rabbit, and dog. Pharmacological analyses of these responses indicate that β and α_2 receptors mediate the excitatory and inhibitory actions, respectively.

Adenosine in the cat increases CSN *in vivo* and *in vivo*, and produces hyperventilation in the rat. Agonists, inhibitors of transport and blockers of degradation of adenosine mimic the excitatory action of adenosine.

Efferent Control of Carotid Sinus Nerve Discharge

Many years ago it was demonstrated that the afferent activity recorded from a thin filament of the CSN was higher when the entire nerve was sectioned, implying that efferent inhibitory impulses to the CB are travelling down the CSN. Electrical stimulation of the distal stump of the CSN while recording from a thin filament of it mimics the efferent inhibition. However, many aspects of the efferent control of the CB remained elusive: what is the anatomical origin and identity of the fibers mediating the efferent inhibition? And what is the mechanism of the efferent inhibition? And what is its significance? Recent experiments indicate that two different sets of fibers, both of them nitrergic, mediate the efferent inhibition. One set comprises the parasympathetic fibers originating in the neurons scattered along the CSN, which cause the inhibition of the afferent CSN activity, thus producing vasodilatation via muscarinic and nitrergic mechanisms. The other set are the nitrergic sensory fibers which upon activation would increase their release of NO to inhibit directly chemoreceptor cells, apparently via cGMP dependent and independent mechanisms. In addition, the dopaminergic sensory fibers release dopamine upon activation; this dopamine might contribute to the efferent inhibition acting on the dopaminergic receptors of chemoreceptor cells. In sum, an anatomically defined reflex loop with parasympathetic fibers in its efferent arm and the efferent function of some (or all) afferent fibers conform to a system known in the literature as the efferent pathway, probably aimed to control acutely and chronically the responsiviness of the CB chemoreceptors to hypoxia. A second reflex loop with the ganglioglomerular sympathetic fibers as efferent arms would also contribute to modulate chemosensory sensitivity mostly via vascular mechanisms.

Reflex Responses Originating at the Carotid Body: Systemic Functions of the Carotid Body

The main targets of the CB reflexes are the respiratory and the cardiovascular systems, which act in concert to provide O_2 to the cells of the entire organism in accordance with their varying demands. The mechanisms involved in the control of each system are multiple, existing complex interactions among them which are not always well defined. There are certain facts that if considered in isolation are devoid of a logical purpose for the performance of the whole organism. For example, the most significant direct effects of CB stimulation on the cardiovascular system are bradycardia, decreased cardiac contractility and hypertension. These cardiovascular effects seem to render wasteful the increased rate of air pumping as they would tend to pose difficulties for O_2 distribution throughout the organism. However, other regulating mechanisms override the direct effects of the CB on the cardiovascular system so as to achieve purposefulness, i.e. accompanying an increase in the rated O_2-rich air pumping into the lungs, there is an increase at rate of O_2 delivery to the tissues by the circulatory system.*

* O_2-rich air has a relative meaning. Even when PO_2 in the breathing atmosphere may be well below normal, it is O_2-rich in comparison to the CO_2-rich gas present in the lungs. Hyperventilation diminishes PCO_2 in the alveoli and allows PO_2 to increase to values near those seen in the atmosphere.

It appears that the direct effects of the CB stimulation on the cardiovascular system tend to protect the myocardium during hypoxia.

Systemic Functions of the Carotid Bodies with Normal Blood Gases and pH

Under resting conditions at sea level, the CB discharges at low frequency and with an oscillatory pattern dictated by the breathing frequency itself. This basal input to the central respiratory controller contributes to the genesis and maintenance of an adequate resting level of ventilation. The oscillating discharge of the CSN may also contribute to the genesis of resting respiration-related variations in the cardiovascular system, as exemplified by an increased heart frequency and a fall in blood pressure during inspiration.

In exercise of up to moderately severe intensity (i.e., below the anaerobic threshold) blood gases and pH are maintained normal in spite of the increased rate of O_2 consumption and CO_2 production. But in exercise of severe intensity a metabolic acidosis develops. The genesis of the ventilatory and the circulatory responses during exercise is multifactorial. Although a detailed treatment of this topic will be made in chapter 17, we should mention that the input from the CB contributes significantly (about 30%) to the hyperventilation seen in exercise of up to moderately severe intensity (Fig. 7.6A). However, in more severe exercise, the acidosis would be expected to stimulate the CB, and, therefore, its contribution to the hyperventilation would be greater.

Systemic Functions of the CB During Hypoxia

The CB responds to the four types of hypoxia, but the response to hypoxic hypoxia is physiologically the most relevant. In humans, as in many other species, the contribution of aortic chemoreceptors to low PO_2 hyperventilation is negligible, and a centrally-mediated ventilatory depression appears during hypoxia after sectioning both CSN.

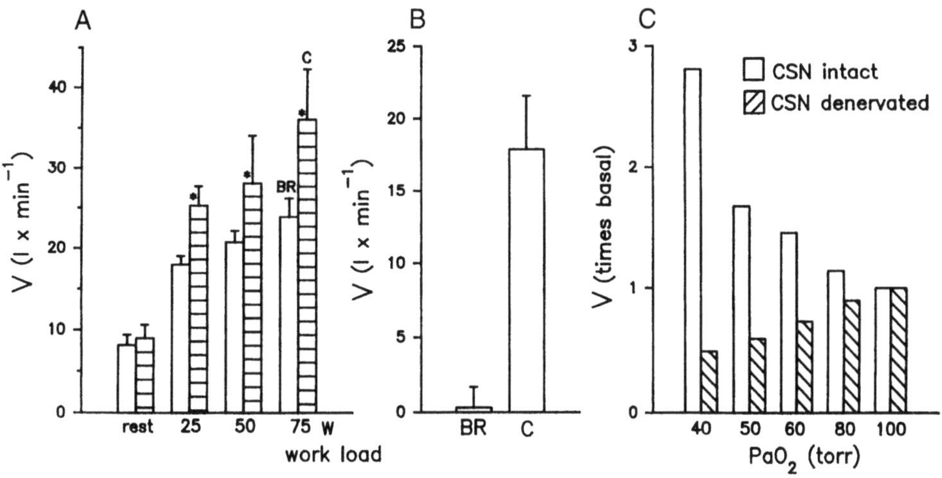

FIGURE 7.6. (A) Ventilatory responses to exercise of increasing intensity in control (C) and in bilaterally carotid body resected humans (BR). Note that the ventilatory response is smaller at all intensities in glomectomized subjects. (B) Ventilatory response to acute hypoxia (PaO_2, 40 mmHg) in control (C) and in bilaterally carotid body resected humans (BR). (C) Schematic drawing of the relation between PaO_2 and ventilation in mammals with intact (control) and carotid sinus nerve denervated carotid bodies. (*Source*: A and B are taken from Honda, Jap. J. Physiol. 35, 535–544, 1985, with permission).

This means that the CBs mediate all the hyperventilation seen during physiologic hypoxic hypoxia at high altitude and during the pathologic hypoxic hypoxias seen in lung pathology. Figures 7.6B and C show in a quantitative manner the effects of hypoxic hypoxia on ventilation in different experimental conditions, including CB resected humans and CSN denervated animals (compare with Figs. 7.3A and 7.4A). The increase in minute ventilation is achieved by an increase in tidal volume and in respiratory frequency, achieved mostly by a reduction in the expiratory time.

The reported cardiovascular effects of hypoxia vary with species, with the intensity of hypoxia, and with the experimental protocols used, including whether or not the animal were anaesthetized and the type and dose of anesthesia. In addition, experimental attempts to individualize the effect of any given input on the cardiovascular system quite commonly alter other inputs. For example, denervation of the CB to determine the contribution of this chemoreceptor to the control of cardiovascular system may result in removal of the carotid baroreceptors, which play a critical role in the control of the cardiovascular system. Taking these facts into consideration, the consensus appears to be that acute hypoxic hypoxia in subprimate animals produces tachycardia, minor changes in blood pressure, increase in cardiac output and reduction of total peripheral resistances; bradycardia and increase in peripheral resistances may appear in intense hypoxias. In primates and humans, tachycardia, increased cardiac output and peripheral vasodilatation are seen at all levels of hypoxia. Therefore, in short-term hypoxia the ratio cardiac output/O_2 consumption increases.

It is possible to partially dissect out the contribution of the different systems to the genesis of those responses by making adequate experimental protocols. For example, a selective stimulation of the CB (by selectively perfusing the carotid sinus region with hypoxic blood) under controlled ventilation produces peripheral vasoconstriction, bradycardia and decreased left ventricle inotropism. On the contrary, selective stimulation of the aortic bodies produces tachycardia and increased heart contractility, constriction of the capacitance vessels, and increased peripheral resistances. If the animals are now allowed to breath spontaneously, hyperventilation follows CB stimulation, and hyperventilation, by changing the thoracic pressures, affects venous return and cardiac contractility, and by activating the lung inflation reflex, modifies further the cardiovascular function. Overall, the lung-originated mechanisms override the effects of the CB with the net result of tachycardia and a tendency to peripheral vasodilatation. Finally, when the whole animal is made hypoxic, as it happens naturally, the direct effects of hypoxia on the heart and blood vessels add to the previous mechanisms of control. Hypoxic hypoxia of mild intensity may tend to increase contractility, but moderate to intense hypoxias would tend to produce a negative inotropic effect; hypoxia produces vasodilation in all vascular beds, except for the pulmonary circulation where low alveolar PO_2 causes vasoconstriction.* Thus, the addition of these single mechanisms generates the response to acute hypoxia described above, that is, tachycardia, increase in cardiac output and peripheral vasodilation in primates which facilitates the distribution of O_2 to the body tissues. After hypoxic acclimatization the cardiac output returns to near control values at sea level and the ratio cardiac output/O_2 consumption normalizes. The gain in the CB chemoreflex that occurs during acclimatization would tend to protect the heart by keeping its work at normal level.

The CBs do not respond to anemic and stagnant hypoxia compatible with life, but the aortic bodies respond to moderate intensities of both types of hypoxia. The venti-

* This pulmonary vasoconstriction improves the ventilation perfusion ratio and seems to improve the gas exchange.

latory effects of aortic bodies stimulation, and hence those of anemic hypoxia and hypotension are negligible,* but the cardiovascular effects are significant. The tachycardia and positive inotropism, the constriction of capacitance vessels, and the increased peripheral resistances, that result from the aortic bodies stimulation, tend to prevent tissue hypoxia occurring during anemic and stagnant hypoxia.

Systemic Functions of the Carotid Bodies During Hypercapnia and Acidosis

The CB responses to high PCO_2 and low pH play an homeostatic role, which in conjunction with the responses of central chemoreceptors (see Chapter 15), allow organisms to pump out CO_2 at a rate matching production and to maintain the organism pH. Exercise is the only physiological situation associated with an increase in CO_2 production; the CBs play an important role in the exercise hyperventilation and hence in maintaining $PaCO_2$ (see earlier).

Contrary to exercise, where blood gases and pH are maintained normal, there are pathological situations, respiratory and metabolic acidosis, in which $PaCO_2$ and/or $[H^+]$ increase; in these situations there is also hyperventilation. The CB has a lower threshold for acidic stimuli than central chemoreceptors, and therefore the hyperventilation seen in response to mild respiratory acidosis originates mostly at the CB, but at more intense respiratory acidosis the CB chemoreceptors contribute by about 30–40% to the hyperventilatory response (Fig. 7.7). The contribution of the CB to the hyperventilation seen in metabolic acidosis, especially in the initial periods, seems to be greater. H^+ diffuse slowly to the cerebrospinal fluid, and the CB-triggered hyperventilation

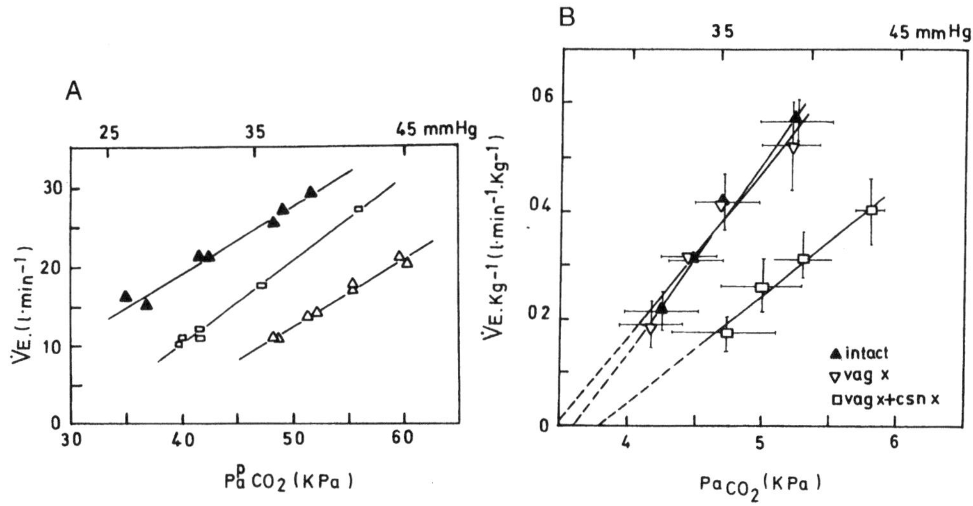

FIGURE 7.7. Effects of $PaCO_2$ on ventilation in the cat. (A) Ventilation as a function of the peripheral $PaCO_2$ at three different levels of central $PaCO_2$ (3.85 kPa, open triangles; 4.68 kPa, open squares; 5.73 kPa, filled triangles) (*Source*: from Heeringa et al. Respir. Physiol. 37, 365–379, 1979, with permission). (B) Effects of $PaCO_2$ on ventilation in four cats before (filled triangles), after vagotomy (open triangles) and after subsequent carotid sinus nerve neurotomy (squares) (*Source*: from Berkenbosch et al. Respir. Physiol. 37, 381–390, 1979, with permission).

*Note that hyperventilation would not add significant amounts of O_2 to the circulating blood during anemic or stagnant hypoxia due to the shape of the hemoglobin dissociation curve.

decreases $PaCO_2$ and forces CO_2 to diffuse from the cerebrospinal fluid to blood. Normohydria or even alcalosis of the cerebrospinal fluid occurs, and thereby lack of stimulation of the central chemoreceptors. If the disbalance lasts beyond 24–48h, the pH in blood and cerebrospinal fluid tend to equilibrate, and both chemoreceptors are stimulated.

The effects of hypercapnia on the cardiovascular system are comparable to those of hypoxia. This is so partly because the activated neurogenic mechanisms are the same (CSN and stretch receptors from the lung), and partly because CO_2, like low PO_2, produces direct vasodilatation. However, acidosis of either type produces an important activation of the sympathetics, including secretion of CA from the adrenal medulla, that tend to offset the vasodilatation produced by the inflation reflex and by the CO_2 itself.

Mechanisms of the Carotid Body Chemoreception

The responses of the CB to natural and pharmacological stimuli measured as the release of neurotransmitters, as discharges in the CSN, and as ventilatory and cardiovascular responses have already been described. The events occurring in chemoreceptor cells from the arrival of the stimuli to the release of neurotransmitters (i.e., the transduction process), and the role of released neurotransmitters in the genesis of discharges in the CSN (i.e., the transmission process) will now be considered.

Sensory Transduction in Chemoreceptor Cells

PO_2 Transduction in Chemoreceptor Cells

Chemoreceptor cells must have some mechanism operationally working as a "molecular" sensor that detects PO_2, and when PO_2 decreases, this sensor must trigger a chain of events that culminates in the release of neurotransmitters. Before we inquire into the nature of the O_2-sensor, however, we must make it clear as to what is meant by low PO_2. From Fig. 7.4A, it is evident that an important PO_2 should be that found in the CB tissue when PaO_2 is about 70–75mmHg; at that tissue PO_2 the O_2-sensor must start signaling the drop in PO_2. Even if the absolute values for the CB tissue PO_2 are in dispute, we are inclined to accept those reported by Whalen and co-workers who found that at normal PaO_2 of 100mmHg the CB tissue PO_2 is ~70mmHg, and at a PaO_2 of ~70–75mmHg, the CB tissue PO_2 is ~50–55mmHg. Thus, this O_2 pressure of 50–55mmHg is an important parameter in the functioning of the O_2-sensor.

There are three proposals on the nature of the molecular mechanisms acting as the putative O_2-sensor. To some authors, the cytochrome oxidase in the mitochondria of chemoreceptor cells would be "special", and would lead to a decrease in chemoreceptor cell phosphate potential ([ATP]/[ADP][AMP][P_i]) when the PaO_2 falls below 70mmHg (or CB tissue PO_2 falls below 50mmHg). According to this proposal, some component of the phosphate potential equation (such as a decrease in ATP) would be the signal that activates the release of neurotransmitters. However, experimental supporting data are not available. In the rabbit, CB hypoxia does not reduce ATP levels.

Other authors have proposed that a haem-linked NADPH oxidase would act as the O_2-sensor. According to this proposal when PO_2 drops, the oxidase-mediated production of $O_2^{\bullet-}$, and H_2O_2 in chemoreceptor cells would decline. The decrease in reactive oxygen species (ROS) would alter the redox state of glutathione, and this in turn would affect the disulfide bonds in ionic channel proteins, leading ultimately to the inhibition of K^+ currents that is known to occur during hypoxia, depolarization of chemoreceptor cells, activation of voltage-operated Ca^{2+} channels, influx of Ca^{2+} into cells and the release of neurotransmitters. This attractive proposal, corning from the models used to explain pulmonary hypoxic vasoconstriction, does not seem to be supported by recent findings including the following observations: (a) diphenileneiodonium (DPI), an inhibitor of NADPH oxidase, is able to promote the release of neurotransmitters from chemoreceptor cells in a dose-dependent manner in normoxia, but two other inhibitors of the oxidase (phenylarsine oxide and neopterin) were not; (b) the release induced by DPI was only partially dependent on extracellular Ca^{2+}, while hypoxia-induced release was completely Ca^{2+}-dependent; (c) none of the three inhibitors at maximal doses were able to alter the release of neurotransmitters induced by hypoxia; (d) hypoxia does not decrease the production of ROS, at least in PC12 cells, and in the CB it does not augment the levels of reduced glutathione; and (e) knockout mice for the oxidase exhibit normal chemoreceptor responses to hypoxia and normal pulmonary hypoxic vasoconstriction. In recent years models for O_2-sensing have been proposed which imply an increase rather than a decrease of ROS during hypoxia. The observation that scavengers of ROS do not alter the response of chemoreceptor cells to hypoxia casts serious doubt on the real significance of ROS levels as messengers mediating O_2-sensing in the CB.

Finally, other authors have proposed that the O_2-sensor could be a hemoglobin-like protein present in the plasma membrane of chemoreceptor cells: on decreasing PO_2, the hemoglobin-like O_2-sensor would be desaturated, this being the trigger signal for certain ionic conductances to change. This model was derived from the experimental observations that follow. The release of DA, which is taken as the final step in the low PO_2 transduction cascade, requires the entry of Ca^{2+} into the cells. Ca^{2+} enters the cells via voltage-dependent Ca^{2+} channels, and that voltage-dependent Na^+ channels were also activated during hypoxia, because blockers of both types of channels inhibit the release of DA elicited by hypoxia. Therefore, low PO_2 must depolarize the cells as a necessary step to activate these voltage-dependent channels. It was later found that chemoreceptor cells possess several types of K^+ currents (a transient current in the rabbit, a Ca^{2+}-dependent current in the rat) that are inhibited reversibly by low PO_2, so that on decreasing CB tissue PO_2 the overall membrane permeability to K^+ would decrease and the membrane depolarization required to activate voltage-dependent Na^+ and Ca^{2+} channels would follow. The involvement of a hemoprotein in O_2-sensing is indicated by the additional observation showing that carbon monoxide (CO), at ≈70 mmHg, prevents the action of low PO_2 on K^+ currents. The location of the O_2-sensor in the plasma membrane is supported by the observation that low PO_2 decreases the opening probability of K^+ channels in isolated patches of chemoreceptor cell membrane. This proposal (Fig. 7.8A), although consistent with most of the known facts on chemoreceptor cell function, has been criticized on the following basis. The channels originally described as O_2-regulated are voltage-dependent and have apparent activation thresholds of around −40 mV. Because the normoxic resting membrane potential of chemoreceptor cells is about −55 mV, it would appear that the O_2-sensitive channels should be closed in normoxia, and therefore hypoxia could not decrease their opening probability. However, in rabbit chemoreceptor cells hypoxic depolarization seems to be mediated, at least in part, by voltage-dependent K^+ channels of the Kv4 family as viral gene transfer of a dominant-negative Kv4 construct causes permanent cell depolarization and reduces markedly the O_2-sensitive transient K^+ current and the hypoxic depolarization. In the case of the rat chemoreceptor cells, there are data in favor and against the participation of maxi-K^+ channels in the maintenance of resting membrane

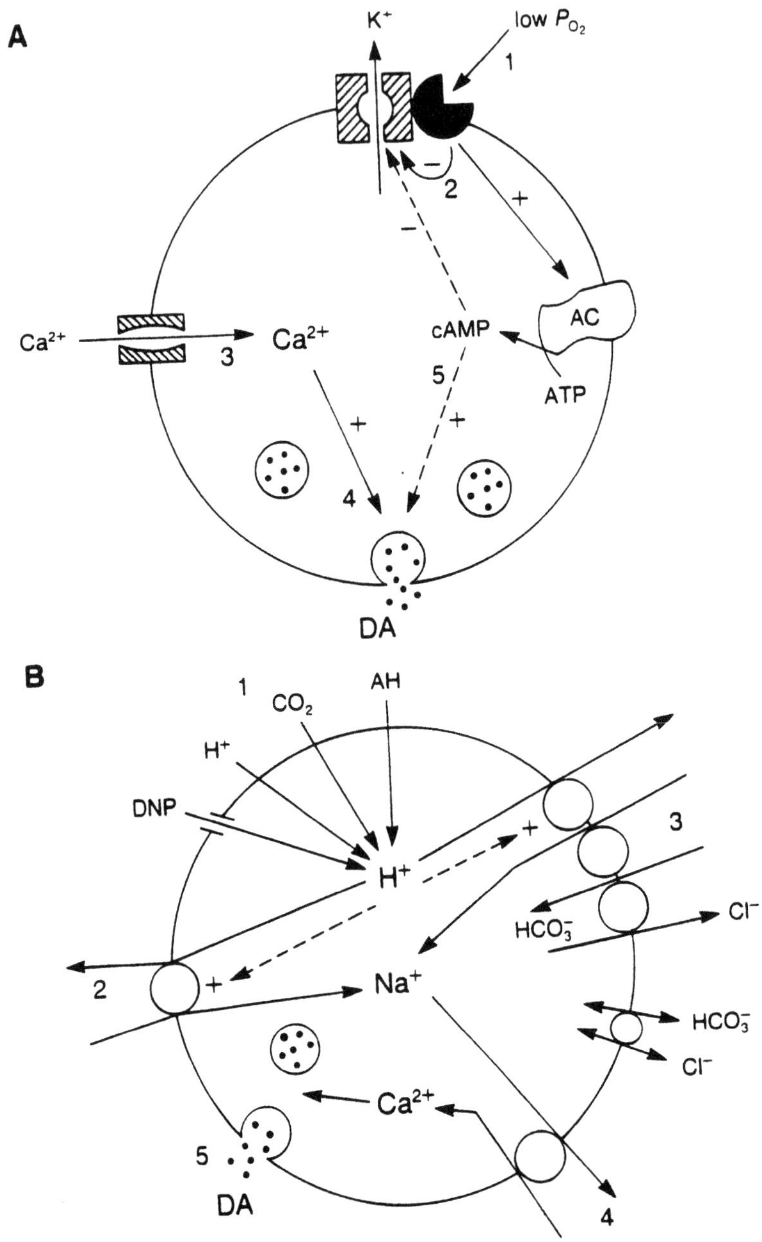

FIGURE 7.8. Proposed models for hypoxic (A) and acidic (B) transduction in chemoreceptor cells of the carotid body. (A) A decrease in the PO_2 at the level of the O_2 sensor would generate a signal that inhibits the O_2-sensitive K^+ channel (1) and activates an adenylate cyclase (2). The inhibition of the K^+ current would depolarize the cell leading to Ca^{2+} channel activation (3), entry of Ca^{2+} and the release of neurotransmitters (4). Increased cAMP levels modulate low PO_2 effects at the level of the K^+ channels, and Ca^{2+} actions at the exocytotic machinery. (B) Several maneuvers that produce an increase in $[H^+]_i$ are depicted in (1). The increased $[H^+]_i$ stimulates (dashed arrows) the Na^+-dependent H^+-extruding mechanism (2,3) causing an increase of influx of Na^+ into chemoreceptor cells and an increase in $[Na^+]_i$. The increase in $[Na^+]_i$ drives the entry of Ca^{2+} through the Na^+/Ca^{2+} exchanger (4) and the exocytosis of neurotransmitters follow (5) (*Source*: from Gonzalez et al. Trends Neurosci. 15, 146–153, 1992, with permission).

potential, and therefore in the initiation of cell depolarization and in the triggering of the release response elicited by hypoxia. In addition, rat chemoreceptor cells possess a TASK-like background K^+ channel which is also sensitive to PO_2 and which will be responsible for or contribute to the initial hypoxic depolarization of chemoreceptor cells. O_2-sensitive K^+ currents have also been found in smooth muscle cells of the pulmonary arteries that contract in response to low PO_2

Acidic Stimuli Transduction in Chemoreceptor Cells

The parameter sensed by chemoreceptor cells is intracellular pH, and not extracellular pH or CO_2 as molecular CO_2. The evidence supporting this statement includes: (1) superfusion of the CBs with solutions at high PCO_2 and normal pH_e or with weak acid-containing solutions at normal pH_e excite chemoreceptor cells, induce release of DA and increase CSN discharge; and (2) carbonic anhydrase accelerates the CO_2 hydration, and thereby the rate of the decrease in intracellular pH; as a consequence, if the sensed signal is a decrease in intracellular pH, the inhibition of carbonic anhydrase must slow the response to CO_2, an expectation that has been experimentally corroborated. Low extracellular pH solutions also acidify intracellularly because H^+ diffuses across plasma membranes, albeit at a slower rate than CO_2 and weak acids.

The obvious question now is how an increase in $[H^+]_i$ in chemoreceptor cells triggers the release of neurotransmitters. In the rabbit CB, the release of catecholamines induced by high PCO_2/low pH and by weak acids, while being Ca^{2+} and Na^+-dependent and inhibitable by blockers of proton extruding exchangers, it is insensitive to dihydropyridines and to tetrodotoxin, blockers of voltage-dependent Ca^{2+} and Na^+ channels, respectively. These findings prompted the proposal that ion exchange mechanisms involving first, the extrusion of H^+ and the concomitant uptake of Na^+ and then, the reversal of the normal operation of the Na^+/Ca^{2+} antiporter, which raises internal Ca^{2+}, would be responsible for the transmitter release (Figure 7.8B). Interestingly enough, in rabbit chemoreceptor cells, Na^+, K^+ and Ca^{2+} currents are inhibited by acidic stimuli with an almost identical pK. Contrary to that, in the chemoreceptor cells of the rat, the two K^+ channels (maxi-K^+ and TASK) sensitive to PO_2 are also exquisitely sensitive to acid, and acidic stimuli raise intracellular Ca^{2+} through membrane depolarization and voltage-gated Ca^{2+} entry.

Second Messengers and Chemoreceptor Cell Function

In rods and cones, taste cells and olfactory receptors cyclic nucleotides are necessary for the initiation of the transduction process. In chemoreceptor cells they do not seem to play such a role, because low PO_2 inhibits K^+ channels in isolated membrane patches bathed with balanced saline. However, cAMP and PGE_2 have important roles in modulating the responses both, to low PO_2 and to acidic stimuli. Low PO_2 and acidic stimuli increase cAMP levels in the CB, and the nucleotide potentiates the release response evoked by both types of stimuli. Cyclic-AMP inhibits the same K^+ current that low PO_2, explaining the nucleotide potentiation of the low PO_2 induced release response. Prostaglandin E_2 is released from the CB in resting conditions (normal PO_2 and PCO_2/pH), and its release increases during low PO_2 and high PCO_2/low pH stimulation. The release of DA induced by low PO_2 is greatly attenuated by bath application of PGE_2 and potentiated by inhibitors of PG synthesis. Because PGE_2 inhibits Ca^{2+} currents in chemoreceptor cells, the inhibition of Ca^{2+} entry can explain the reduction of the DA release response. Phospholipase C-derived second messengers do not appear to play a significant role in the minute to minute control of chemoreception, because acute hypoxia does not alter inositol 1,4,5-triphosphate levels in the CB.

Neurotransmission Between Chemoreceptor Cells and Sensory Nerve Endings

Neurotransmission at the chemoreceptor cell-sensory nerve ending synapse has been, historically, a very debated issue. Several neurotransmitters including ACh, NE, substance P and DA, have been proposed to play the role of the primary transmitter. That is, the transmitter responsible for transformation of the transduction process in chemoreceptor cells into action potentials in the sensory nerve endings of the CSN. The existence of a primary transmitter implies that the other transmitters are secondary ones, having a modulating role in the action of the primary neurotransmitter. An absolute requirement for a candidate to be the primary neurotransmitter is that the sensory nerve endings of the CSN (the postsynaptic element) must possess specific receptors for it. But it is the most commonly secondary neurotransmitters that have receptors in the presynaptic structure (chemoreceptor cells) that accomplish their role modulating the release of the primary transmitter. However, the primary transmitter may also have receptors presynaptically, and secondary neurotransmitters may have receptors postsynaptically. This complexity provides synapses with many degrees of freedom to transmit particular codes of information.

It follows from these considerations that only those neurotransmitters possessing receptors in the sensory nerve endings can be considered real candidates for the role of the primary neurotransmitter. Dopamine, ACh and substance P appear to satisfy this requirement; in addition, sensory nerve endings also possess P2X2 and P2X3 for ATP which is known to be co-stored and co-released with other neurotransmitters. Future research might demonstrate postsynaptic receptors for other neurotransmitters.

In previous sections we have shown that DA is present in the presynaptic element and is released in proportion to the intensity of the stimuli and to the action potential frequency in the CSN. However, DA exogenously applied does not always mimic the action of natural stimuli at the level of the CSN. As discussed previously, the variable effects of exogenous DA on CSN discharge can be explained on the basis of different accessibility of DA to the dopaminergic receptors and/or on the basis of the different affinity of the several sets of CB dopaminergic receptors for DA. Therefore, the lack of mimicry between exogenous DA and natural stimuli does not exclude DA as the primary neurotransmitter. In fact, Donnelly and coworkers have provided further support for DA as the primary transmitter by finding that the postnatal resetting of chemoreceptor function (i.e., the increase in the capacity of the CB to produce hyperventilation in response to mild hypoxia) was paralleled by the ability of chemoreceptor cells to release DA in response to hypoxic stimulus. However, recent experiments have demonstrated that spontaneous e.p.s.ps and action potentials, as well as hypoxia elicited action potentials, recorded in petrosal neurons apposed to chemoreceptor cells in cocultures of both cell types, are inhibited by Ca^{2+}-free solutions and by blockers of nicotinic and/or P2 purinergic receptors. These findings indicate that the electrical activity recorded in the petrosal neurons has a synaptic origin and that acetylcholine and ATP mediate the synaptic transmission in this preparation. If these results are verified in intact preparations, they would imply that ACh acting in conjunction with ATP should be considered the primary transmitters in the CB chemoreceptor synapse.

The Carotid Body During Chronic Hypoxia

When the CB is subjected to chronic hypoxia, such as occurs under certain physiological (i.e., high altitude) and pathological (i.e., sudden infant death syndrome; SIDS) conditions, important changes occur in the structure, neurotransmitter content and function of the organ. Permanent high altitude residents exhibit hypertrophy and hyperplasia of CB type I cells. Both among animals and humans living permanently at high altitude, there is a frequency of chemodectomas (CB tumors) which greatly exceeds that found at sea level. Even if there is no clinical tumor, such organs have an increased concentration of catecholamines. Permanent residents at high altitude exhibit a blunted ventilatory response to further hypoxic challenges, but in most cases normal responsiveness is regained upon returning to sea level. In SIDS, CB morphology is also altered, but in a complex manner, while the overall dopamine concentration of the organ is increased. Although there are additional alterations in the organization of certain nuclei of the brainstem related to the control of respiration, it has been suggested that the primary alteration in the CB may be pathogenic in this fatal disorder. This contention seems to be supported by epidemiological and experimental observations. Thus, the clinical history of the infants dying of SIDS manifests very frequent episodes of respiratory infections usually associated with episodes of hypoxia in the initial months of life. On the experimental side, it is known that the maintenance of newborn animals in hypoxic atmospheres is associated with an incomplete development of the CB and chemoreceptor cell function.

By contrast, subjects living at sea level and ascending to high altitude develop a hyperreactivity to hypoxia (known as acclimatization to hypoxia; see earlier) in a period of hours to a few days in such a manner that a further hypoxic challenge produces a greater hyperventilation than that measured in the same subjects before ascending to high altitude. Experimental evidence indicates that acclimatization is multifactorial in its origin involving up and down regulation of membrane mechanisms (e.g., ion channels and receptors for neurotransmitters), neurotransmitter levels (e.g., catecholamines, endothelin, substance P, and NO) and trophic factors (e.g., vascular growth factors, endothelin, and angiotensin); the gain of the efferent pathway and both in its dopaminergic and nitrergic components are also upregulated during acclimatization.

Summary

The CB is a secondary sensory receptor that is activated when PaO_2 decreases or when $PaCO_2$ or $[H^+]$ increase. The primary sensory elements in the CB are chemoreceptor cells. These cells contain a great number of neurotransmitters, which are released during natural stimulation. The concerted action of these neurotransmitters fixes the level of action potential frequency in the CSN according to the intensity of the stimulation. The action potential frequency in the CSN at normal blood gas pressures and pH is very low. If PaO_2 decreases, CSN action potential frequency increases with a very small slope until PaO_2 drops to near 75 mmHg, and thereafter CSN discharge increases exponentially; minute ventilation follows, CSN discharge. If $PaCO_2$ increases and/or the blood pH decreases, CSN action potential frequency increases linearly with PCO_2 or with the increase in blood $[H^+]$. Ventilation parallels the increase in the CSN discharge.

The cascade of stimuli transduction by chemoreceptor cells is not completely defined in a step by step manner, but it appears that chemoreceptor cells possess in their membrane an O_2-sensor. The sensor could be a hemoprotein, which on desaturation by a drop in PaO_2, causes specific types of K^+ channels to decrease their opening probability. As a result of that, chemoreceptor cells are depolarized, Ca^{2+} enters the cells via voltage-operated channels and the release of neurotransmitters follows. An increase in $PaCO_2$ or in $[H^+]$, lead to an increase in $[H^+]$ in chemoreceptor cells. Chemoreceptor cells respond by extruding H^+ in exchange for external Na^+, resulting increased in $[Na^+]_i$.

The increased $[Na^+]_i$ triggers the sodium/calcium exchanger to pump Na^+ out in exchange for Ca^{2+}; Ca^{2+} enters the cells and exocytosis of neurotransmitters is activated.

Acknowledgments. This work was supported by Spanish DGICYT Grant PB 97/0400.

References

Acker, H., and Xue, D. (1995) Mechanisms of O_2 sensing in the carotid body in comparison with other O_2-sensing cells. News. Physiol. Sci. 10, 211–215.

Almaraz, L., Perez-Garcia, M.T., Gomez-Niño, A., and Gonzalez, C. (1997) Mechanisms of α_2 adrenoceptor-mediated inhibition in rabbit carotid body. Am. J. Physiol. 272, C628–C637.

Archer, S.L., Huang, J., Henry, T., Peterson, D., and Weir, E.K. (1993) A redox-based O_2 sensor in rat pulmonary vasculature. Circ. Res. 73, 1100–1112.

Archer, S.L., Reeve, H.L., Michelakis, E., Puttagunta, L., Waite, R., Nelson, D.P., Dinauer, M.C., and Weir, E.K. (1999) O2 sensing is preserved in mice lacking the gp91 phox subunit of NADPH oxidase. Proc. Natl. Acad. Sci. USA 96, 7944–7949.

Archer, S.L., Will, J.A., and Weir, E.K. (1986) Redox status in the control of pulmonary vascular tone. Herz 11, 127–141.

Bairam, A., Frenette, J., Dauphin, C., Carroll, J.L., and Khandjian, E.W. (1998) Expression of dopamine D1-receptor mRNA in the carotid body of adult rabbits, cats and rats. Neurosci. Res. 31, 147–54.

Berkenbosch, A., Van Dissel, J., Olievier, C.N., De Goede, J., and Heeringa, J. (1979) The contribution of the peripheral chemoreceptors to the ventilatory response to CO_2 in anaesthetized cats during hyperoxia. Resp. Physiol. 37, 381–390.

Biscoe, T.J., and Duchen, R.M. (1990) Monitoring PO_2 by the carotid chemoreceptor. News Physiol. Sci. 5, 229–233.

Biscoe, T.J., Purves, M.J., and Sampson, S.R. (1970) The frequency of nerve impulses in single carotid body chemoreceptor afferent fibers recorded in vivo with intact circulation. J. Physiol. 208, 121–131.

Black, A.M.S., McCloskey, D.I., and Torrance, R.W. (1971) The responses of carotid body chemoreceptors in the cat to sudden changes of hypercapnic and hypoxic stimuli. Resp. Physiol. 13, 36–49.

Buckler, K.J. (1997) A novel oxygen-sensitive potassium current in rat carotid body type I cells. J. Physiol. 498, 649–662, 1997.

Buckler, K.J., and Vaughan-Jones, R.D. (1998) Effects of mitochondrial uncouplers on intracellular calcium, pH and membrane potential in rat carotid body type I cells. J. Physiol. 513, 819–833.

Buckler, K.J., Williams, B.A., and Honore, E. (2000) An oxygen-, acid- and anaesthetic-sensitive TASK-like background potassium channel in rat arterial chemoreceptor cells. J. Physiol 525, 135–142.

De Castro, F. (1928) Sur la structure et l'innervation du sinus carotidien de l'homme et des mammifères: Nouveaux faits sur l'innervation et la fonction du glomus caroticum. Trab. Lab. Invest. Biol. Univ. Madrid. 25, 330–380.

Dvorakova, M., Hohler, B., Vollerthun, R., Fischbach, T., and Kummer, W. (2000) Macrophages: a major source of cytochrome b558 in the rat carotid body. Brain Res. 852, 349–354.

Eyzaguirre, C., Koyano, H., and Taylor, J.R. (1965) Presence of acetylcholine and transmitter release from carotid body chemoreceptors. J. Physiol. 178, 463–476.

Fidone, S., Gonzalez, C., and Yoshizaki, K. (1982) Effects of low oxygen on the release of dopamine from the rabbit carotid body in vitro. J. Physiol. 333, 93–110.

Fitzgerald, R.S., Shirahata, M., and Wang, H.Y. (1999) Acetylcholine release from cat carotid bodies. Brain Res. 841, 53–61.

Gonzalez, C., Almaraz, L., Obeso, A., and Rigual, R. (1992) Oxygen and acid chemoreception in the carotid body chemoreceptors. Trends Neurosci. 15, 146–153.

Gonzalez, C., Almaraz, L., Obeso, A., and Rigual, R. (1994) Carotid body chemoreceptors: From natural stimuli to sensory discharges. Physiol. Rev. 74, 829–898.

Hanson, M.A. (1998) Role of chemoreceptors in effects of chronic hypoxia. Comp. Biochem. Physiol. A Mol. Integr. Physiol. 119, 695–703.

Heeringa, J., Berkenbosch, A., De Goede, J., and Olievier, C.N. (1979) Relative contribution of central and peripheral chamoreceptors to the ventilatory response to CO_2 during hyperoxia. Resp. Physiol. 37, 365–379.

Heymans, C., Bouckaert, J.J., and Dautrebande, L. (1930) Sinus carotidien et réflexes respiratoires,II. Influences respiratoires réflexes de l'acidose, de l'alcalose, de l'anhydride carbonique, de l'ion hydrogene et de l'anoxémie: Sinus carotidiens et e'changes respiratoires dans les poumons et au dela des poumons. Arch. Int. Pharmacodyn. Ther. 39, 400–408.

Honda, H. (1985) Role of carotid chemoreceptors in control of breathing at rest and in exercise: studies on humans subjects with bilateral carotid body resection. Jap. J. Physiol. 35, 535–544.

Honda, H. (1992) Respiratory and circulatory activities in carotid body-resected humans. J. Appl. Physiol. 73, 1–8.

Hornbein, T.F. (1968) The relation between stimulus to chemoreceptors and their response. In: Arterial Chemoreceptors, Torrance, R.W. (ed.) Blackwell Scientific Publications, Oxford, U.K. pp. 65–78.

Hohler, B., Lange, B., Holzapfel, B., Goldenberg, A., Hanze, J., Sell, A., Testan, H., Moller, W., and Kummer, W. (1999) Hypoxic upregulation of tyrosine hydroxylase gene expression is paralleled, but not induced, by increased generation of reactive oxygen species in PC12 cells. FEBS Lett 457, 53–56.

Kim, D.K., Oh, E.K., Summers, B.A., Prabhakar, N.R., and Kumar, G.K. (2001) Release of substance P by low oxygen in the rabbit carotid body: evidence for the involvement of calcium channels. Brain Res. 892, 359–69.

Lahiri, S., and Delaney, R.G. (1975) Stimulus interaction in the responses of carotid body chemoreceptor single afferent fibers. Respir. Physiol. 24, 249–266.

Lahiri, S., Mulligan, E., Nishino, T., Mokashi, A., and Davies, R.O. (1981) Relative responses of aortic body and carotid body chemoreceptors to carboxyhemoglobinemia. J. Appl. Physiol. 50, 580–586.

Lahiri, S., Nishino, T., Mokashi, A., and Mulligan, E. (1980) Relative responses of aortic body and carotid body chemoreceptors to hypotension. J. Appl. Physiol. 48, 781–788.

Leitner, L.M., and Roumy, M. (1985) Effects of dopamine superfusion on the activity of rabbit carotid chemoreceptors in vitro. Neuroscience 16, 431–438.

Leitner, L.M., and Roumy, M. (1986) Chemoreceptor response to hypoxia and hypercapnia in catecholamine depleted rabbit and cat carotid bodies in vitro. Pflügers Arch. Eur. J. Physiol. 406, 419–423.

Lopez-Barneo, J., Lopez-Lopez, J.R., Ureña, J., and Gonzalez, C. (1988) Chemotransduction in the carotid body: K^+ current modulated by Po_2 in type I chemoreceptor cells. Science 241, 580–582.

McQueen, D.S. (1983) Pharmacological aspects of putative transmitters in the carotid body. In: *Physiology of the Peripheral Arterial Chemoreceptors*, Acker, H., and O'Regan, R.G. (eds.) Elsevier Science Publishers, Amsterdam, pp. 149–195.

Monteiro, E.C., and Ribeiro, J.A. (2000) Adenosine-dopamine interactions and ventilation mediated through carotid body chemoreceptors. Adv. Exp. Med. Biol. 475, 671–684.

Mulligan, E., Lahiri, S., and Storey, B.T. (1981) Carotid body O_2 chemoreception and mitocondrial oxidative phosphorylation. J. Appl. Physiol. 51, 438–446.

Nolan, W.F., Donnelly, D.F., Smith, E.J., and Dutton, R.E. (1985) Haloperidol-induced suppression of carotid chemoreception in vitro. J. Appl. Physiol. 59, 814–820.

Obeso A., Gomez-Niño, M.A., Almaraz, L., Dinger, B., Fidone, S., and Gonzalez, C. (1997) Evidence for two types of nicotinic receptors in the cat carotid body chemoreceptor cells. Brain Res. 754, 298–302.

Obeso, A., Gomez-Niño, A., and Gonzalez, C. (1999) NADPH oxidase inhibition does not interfere with low PO2 transduction in rat and rabbit CB chemoreceptor cells. Am. J. Physiol. 276, C593–C601.

Obeso, A., Gonzalez, C., Rigual, R., Dinger, B., and Fidone, S. (1993) Effect of low O_2 on glucose uptake in rabbit carotid body. J. Appl. Physiol. 74, 2387–2393.

Peers, C. (1990) Hypoxic suppression of K^+ currents in type I carotid body cells: selective effect on the Ca^{2+}-activated K^+ current. Neurosci. Lett. 119, 253–256.
Perez-Garcia, M.T., Lopez-Lopez, J.R., and Gonzalez, C. (1999) Kvβ1.2 subunit coexpression in HEK293 cells confers O_2 sensitivity to Kv4.2 but not to Shaker channels. J. Gen. Physiol. 13, 897–907.
Perez-Garcia, M.T., Lopez-Lopez, J.R., Riesco, A.M., Hoppe, U., Gonzalez, C., Marban, E., and Johns, D.C. (2000) Supression of transient outward K^+ currents in chemoreceptor cells of the rabbit carotid body by viral gene transfer of inducible dominant negative Kv4.3 constructs. J. Neurosci. 20, 5689–5695.
Ponte, J., and Purves, M.J. (1974) Frequency response of carotid body chemoreceptors in the cat to changes of $PaCO_2$, PaO_2, and pHa. J. Appl. Physiol. 37, 635–647.
Rigual, R., Almaraz, L., Gonzalez, C., and Donnelly, D.F. (2000) Developmental changes in chemoreceptor nerve activity and catecholamine secretion in rabbit carotid body: possible role of Na+ and Ca2+ currents. Pflugers Arch. Eur. J. Physiol. 439, 463–70.
Rigual, R., Lopez-Lopez, J.R., and Gonzalez, C. (1991) Release of dopamine and chemoreceptor discharge induced by low pH and high PCO_2 stimulation of the cat carotid body. J. Physiol. 433, 519–531.
Rigual, R., Cachero, M.T.G., Rocher, A., and Gonzalez, C. (1999) Hypoxia inhibits the synthesis of phosphoinositides in the rabbit carotid body. Pflügers Arch. Eur. J. Physiol. 437, 839–845.
Rocher, A., Obeso, A., Gonzalez, C., and Herreros, B. (1991) Ionic mechanisms for the transduction of acidic stimuli in rabbit carotid body glomus cells. J. Physiol.433, 533–548.
Sanz-Alfayate, G., Obeso, A., Agapito, M.T., and González, C. (2001) Reduced to oxidized glutathione ratios and oxygen sensing in calf and rabbit carotid body chemoreceptor cells. J. Physiol. 537, 209–220.
Shirahata, M., Ishizawa , Y., Rudisill, M., Schofield, B., and Fitzgerald, R.S. (1998) Presence of nicotinic acetylcholine receptors in cat carotid body afferent system. Brain Res. 814, 213–217.
Verna, A. (1979) Ultrastructure of the carotid body in the mammals. Int. Rev. Cytol. 60, 271–330.
Verna, A., Talib, N., Roumy, M., and Pradet, A. (1990) Effects of metabolic inhibitors and hypoxia on the ATP, ADP and AMP content of the rabbit carotid body in vitro: the metabolic hypothesis in question. Neurosci. Lett. 116, 156–161.
Vicario, I., Rigual, R., Obeso, A., and Gonzalez, C. (2000) Characterization of the synthesis and release of catecholamine in the rat carotid body *in vitro*. Am. J. Physiol. 278, C490–C499.
Wang, Z.Z., Stensaas, L.J., Dinger, B., and Fidone, S.J. (1992) The co-existence of biogenic amines and neuropeptides in the type I cells of the cat carotid body. Neuroscience 47, 473–480.
Wang, Z.Z., Stensaas, L.J., Dinger, B., and Fidone, S.J. (1995) Nitric oxide mediates chemoreceptor inhibition in the carotid body. Neuroscience 65, 217–229.
Whalen, W.J., and Nair, P. (1983) Oxidative metabolism and tissue Po_2 of the carotid body. In: *Physiology of the Peripheral Arterial Chemoreceptors*, Acker, H., and O'Regan, R.G. (eds.) Elsevier Science Publishers, Amsterdam, pp. 117–132.
Zapata, P. (1975) Effects of dopamine on carotid chemo- and baroreceptors in vitro. J. Physiol. 244, 235–251.
Zhang, M., Zhong, H., Vollmer, C., and Nurse, C.A. (2000) Co-release of ATP and ACh mediates hypoxic signalling at rat carotid body chemoreceptors. J. Physiol. 525, 143–58.

Recommended Readings

De Burgh Daly, M. (1997) Peripheral Arterial Chemoreceptors and Respiratory-Cardiovacular Integration. Oxford University Press, Oxford.
Gonzalez, C. (ed.) (1997) *The carotid body chemoreceptors.* Springer-Verlag, Heidelberg.

8
The Respiratory Muscles in Health and Disease

André De Troyer

The mechanical action of any skeletal muscle is essentially determined by the anatomy of the muscle and by the structures it has to displace when it contracts. The respiratory muscles are structurally and functionally skeletal muscles, and their vital task is to displace the chest wall rhythmically to pump gas in and out of the lungs. This chapter therefore starts with a discussion of the basic mechanical structure of the chest wall. It next analyzes the actions of the muscles that displace the chest wall. For the sake of clarity, the functions of the diaphragm, the muscles of the rib cage, and the muscles of the abdominal wall will be analyzed sequentially; however, because all these muscles normally work together in a coordinated manner, the most critical aspects of their mechanical interactions are also emphasized. Finally, some disorders are considered in which the respiratory displacements of the chest wall are abnormal due to a particular distribution of muscle weakness or chronic hyperinflation.

The Chest Wall

Although the chest wall is a complex structure, it can be thought of as consisting of two compartments: the rib cage and the abdomen (Konno and Mead, 1967). These two compartments, which are separated from each other by the diaphragm (Fig. 8.1), are mechanically arranged in parallel. Expansion of the lungs, therefore, can be accommodated by expansion of either the rib cage or the abdomen, or by both compartments simultaneously.

The displacements of the rib cage during breathing are essentially related to the motion of the ribs, and this motion occurs primarily through a rotation around the axes defined by the articulations of the ribs with the vertebral bodies and transverse processes, as shown in Fig. 8.2 (Jordanoglou, 1970; Wilson et al., 1987). Indeed, each rib is fixed dorsally by these two vertebral joints that together form a hinge. The axis of this hinge, however, is oriented laterally, dorsally, and caudally. In addition, the ribs are curved and slope caudally and ventrally from their costotransverse articulations, such that their ventral ends and the costal cartilages are more caudal than their dorsal part. When the ribs are displaced in the cranial direction, therefore, their ventral ends move laterally and ventrally as well as cranially, the cartilages rotate cranially around the chondrosternal junctions, and the sternum is displaced ventrally. As a result, there is usually an increase in both the lateral and the dorsoventral diameter of the rib cage (Fig. 8.2). On the other hand, an axial displacement of the ribs in the caudal direction is usually associated with a decrease in rib cage diameters. The muscles that elevate the ribs as their primary action therefore have an inspiratory effect on the rib cage, whereas the muscles that lower the ribs have an expiratory effect on the rib cage. It must be appreciated, however, that although the ribs move predominantly by a rotation, there is some misfit at the surfaces of the costovertebral and chondrosternal joints. The long cartilages of ribs 8, 9, and 10 even articulate with one another by little synovial cavities, rather than with the sternum. Hence, although the upper ribs tend to move as a unit with the sternum, the lower ribs have some freedom to move independently (De Troyer and Decramer, 1985; De Troyer, Estenne, and Vincken, 1986). Deformations of the rib cage may therefore occur under the influence of muscle contraction both in animals and in humans.

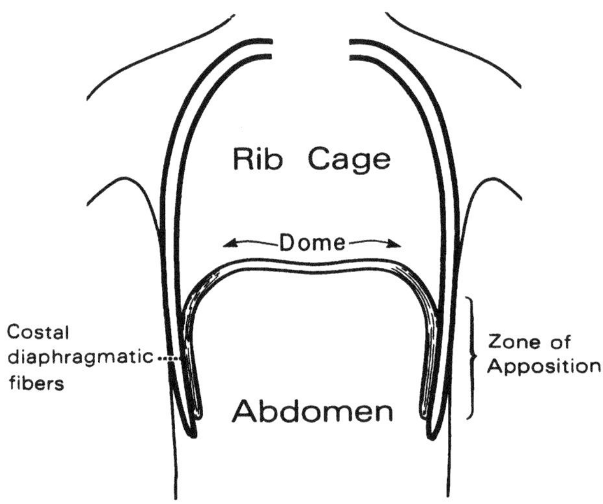

FIGURE 8.1. Frontal section of the chest wall at end-expiration. Note the cranial orientation of the costal diaphragmatic fibers and their apposition to the inner aspect of the lower rib cage (zone of apposition). (*Source*: Reprinted, with permission, from De Troyer, A., and Loring, S.H. [1995] Actions of the respiratory muscles. In: The thorax, second ed., Roussos, C. (ed.) pp. 535–563. Marcel Dekker, New York.)

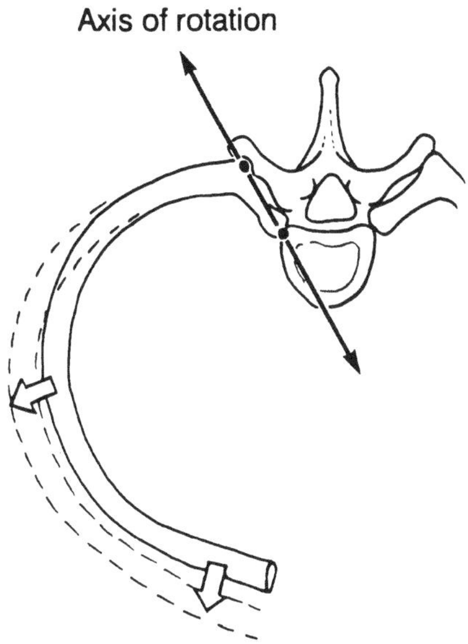

FIGURE 8.2. Diagram of a typical thoracic vertebra and a rib (viewed from above). The rib articulates both with the body and the transverse process of the vertebra (closed circles) and moves essentially through a rotation around the axis defined by these articulations. From these articulations, however, the rib slopes downward and ventrally. When it becomes more horizontal in inspiration (dotted line), it therefore causes an increase in both the anteroposterior and the transverse diameter of the rib cage (open arrows). (*Source*: Reprinted, with permission, from De Troyer, A. [1996] Mechanics of the chest wall muscles. In: Neural control of the respiratory muscles, Miller, A.D., Bishop, B., and Bianchi, A.L. (eds.) pp. 59–73. CRC Press, Boca Raton.)

The respiratory displacements of the abdominal compartment are more straightforward because if one neglects the 100–300ml of abdominal gas volume, its content is virtually incompressible. As a corollary, the abdomen behaves as a liquid-filled container, and any local inward displacement of its boundaries results in an equal outward displacement elsewhere. Furthermore, many of these boundaries (e.g., the spine dorsally, the pelvis caudally, and the iliac crests laterally) are virtually fixed. Thus, the parts of the abdominal container that can be displaced are essentially limited to the ventral abdominal wall and the diaphragm. When the diaphragm contracts during inspiration (see later), therefore, its descent usually results in an outward displacement of the ventral abdominal wall; conversely, when the abdominal muscles contract, they generally cause an inward displacement of the belly wall that results in a cranial motion of the diaphragm into the thoracic cavity.

The Diaphragm

The diaphragm is anatomically unique among skeletal muscles in that its muscle fibers radiate from a central tendinous structure (the central tendon) to insert peripherally into skeletal structures. The crural (or vertebral) portion of the diaphragmatic muscle inserts on the ventrolateral aspect of the first three lumbar vertebrae and on the aponeurotic arcuate ligaments; the costal portion inserts on the xiphoid process of the sternum and the upper margins of the lower six ribs. From their insertions, the costal fibers run cranially so that they are directly apposed to the inner aspect of the lower rib cage (Fig. 8.1). In standing humans at rest, this so-called zone of apposition of the diaphragm to the rib cage (Mead, 1979) is about 6–9cm in height in the midaxillary line and occupies 25–30% of the total internal surface area of the rib cage. Although the older literature has suggested the possibility of an intercostal motor innervation of some portions of the diaphragm, it is now clearly established that its only motor supply is through the phrenic nerves that originate in the third, fourth, and fifth cervical segments in humans.

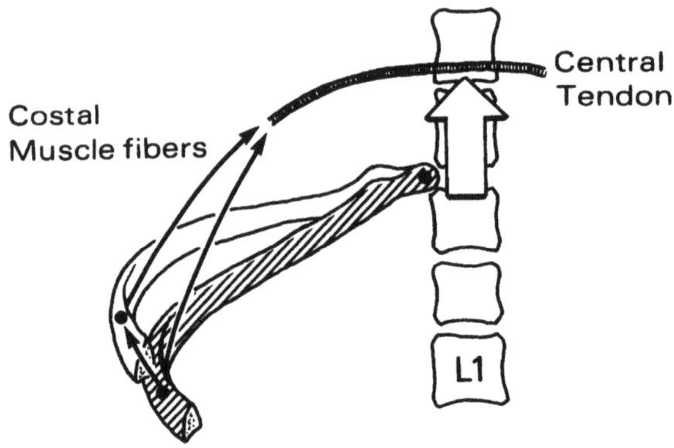

FIGURE 8.3. Insertional component of diaphragmatic action. During inspiration, as the fibers of the costal diaphragm contract, they exert a force on the lower ribs (arrow). If the abdominal visceral mass effectively opposes the descent of the diaphragmatic dome (open arrow), this force is oriented cranially. As a result, the lower ribs are lifted and rotate outward. (*Source*: Reprinted, with permission, from De Troyer, A. [1996] Mechanics of the chest wall muscles. In: Neural control of the respiratory muscles, Miller, A.D., Bishop, B., and Bianchi, A.L. (eds.) pp. 59–73. CRC Press, Boca Raton.)

Actions of the Diaphragm

As the muscle fibers of the diaphragm are activated during inspiration, they develop tension and shorten. As a result, the axial length of the apposed diaphragm diminishes, and the dome of the diaphragm, which corresponds primarily to the central tendon, descends relative to the costal insertions of the muscle. The dome remains relatively constant in size and shape during breathing, but its descent has two effects. First, it expands the thoracic cavity along the craniocaudal axis. Hence, pleural pressure falls, and lung volume increases or alveolar pressure falls depending on whether the airways are open or closed. Second, it produces a caudal displacement of the abdominal visceral mass and an increase in abdominal pressure that, in turn, pushes the ventral abdominal wall outwards.

In addition, because the muscle fibers of the costal diaphragm insert onto the upper margins of the lower six ribs, they also apply a force on these ribs when they contract. In fact, this force is equal to the force exerted on the central tendon, and under normal circumstances it is directed cranially due to the cranial orientation of the fibers (Fig. 8.3). It therefore has the effect of lifting the ribs and rotating them outward. The fall in pleural pressure and the increase in abdominal pressure induced by diaphragmatic contraction, however, act simultaneously on the rib cage, which probably explains why the action of the diaphragm on the rib cage has been controversial for so long.

Action of the Diaphragm on the Rib Cage

When the diaphragm in anesthetized dogs is activated selectively by electrical stimulation of the phrenic nerves, the upper ribs move caudally and the cross-sectional area of the upper portion of the rib cage decreases (D'Angelo and Sant'Ambrogio, 1974). In contrast, the lower ribs move cranially and the cross-sectional area of the lower portion of the rib cage increases. When a bilateral pneumothorax is subsequently introduced so that the fall in pleural pressure is eliminated, isolated contraction of the diaphragm causes a greater expansion of the lower rib cage, but the dimensions of the upper rib cage then remain unchanged. Thus, when contracting alone, the canine diaphragm has two opposing effects on the rib cage. On the one hand, it has an expiratory action on the upper rib cage, and the fact that this action is abolished by a pneumothorax indicates that it is due to the fall in pleural pressure. On the other, it has also an inspiratory action on the lower rib cage. Measurements of chest wall motion during phrenic nerve pacing in humans with transection of the upper cervical cord (Danon et al., 1979; Strohl et al., 1984) and during resting breathing in subjects who use their diaphragm exclusively because of a traumatic transection of the lower cervical cord (Mortola and Sant'Ambrogio, 1978; Estenne and De Troyer, 1985) have shown that the diaphragm in humans, as in the dog, has both an expiratory action on the upper rib cage and an inspiratory action on the lower rib cage.

Theoretical and experimental work has confirmed that the inspiratory action of the diaphragm on the lower rib cage results in part from the force the muscle applies on the ribs by way of its insertions; this force is conventionally referred to as the *insertional force* (De Troyer et al., 1982; Loring and Mead, 1982). This inspiratory action of the diaphragm, however, is also related to its apposition to the rib cage. The zone of apposition makes the lower rib cage, in effect, part of the abdominal container, and measurements in dogs and rabbits have established that during breathing the changes in pressure in the pleural recess between the apposed diaphragm and the rib cage are almost equal to the changes in abdominal pressure (Urmey et al., 1988). Pressure in this pleural recess rises, rather than falls, during inspiration, thus indicating that the rise in abdominal pressure is truly transmitted through the apposed diaphragm to expand the lower rib cage. This mechanism of diaphragmatic action has been called the *appositional force.*

Although the insertional and appositional forces make the normal diaphragm expand the lower rib cage, it should be appreciated that this action of the diaphragm is largely determined by the resistance provided by the abdominal contents to diaphragmatic descent (Fig. 8.3). If this resistance is high (i.e., if abdominal compliance is low), the dome of the diaphragm descends less, so that the zone of apposition remains significant throughout inspiration and the rise in abdominal pressure is greater. For a given diaphragmatic contraction, therefore, the appositional force is greater and the expansion of the lower rib cage is increased. On the other hand, if the resistance provided by the abdominal contents is small (i.e., if the abdomen is very compliant), the dome of the diaphragm descends more easily, the zone of apposition decreases more, and the rise in abdominal pressure is smaller. As a result, the inspiratory action of the diaphragm on the rib cage is reduced. If the resistance provided by the abdominal contents were eliminated, the zone of apposition would disappear in the course of inspiration and the contracting diaphragmatic muscle fibers would become oriented transversely inward at their insertions onto the ribs. The insertional force would then have an expiratory, rather than inspiratory, action on the lower rib cage. Indeed, when a dog is eviscerated, the diaphragm causes a decrease, rather than an increase, in lower rib cage dimensions (Duchenne, 1867; D'Angelo and Sant'Ambrogio, 1974; De Troyer et al., 1982).

Influence of Lung Volume

The balance between pleural pressure and the insertional and appositional forces of the diaphragm is also markedly affected by changes in lung volume. As lung volume decreases from functional residual capacity (FRC) to residual volume (RV), the zone of apposition increases and the fraction of the rib cage exposed to pleural pressure decreases. As a result, the appositional force increases and the effect of pleural pressure diminishes; therefore, the inspiratory action of the diaphragm on the rib cage is enhanced. On the other hand, as lung volume increases above FRC, the zone of apposition decreases and a larger fraction of the rib cage becomes exposed to pleural pressure. The inspiratory action of the diaphragm on he rib cage is therefore diminished (D'Angelo and Sant'Ambrogio, 1974; De Troyer et al., 1982; Loring and Mead, 1982). When lung volume approaches total lung capacity (TLC), the zone of apposition all but disappears, and the diaphragmatic muscle fibers become oriented transversely inward as well as cranially. As in the eviscerated animal, the insertional force of the diaphragm is then expiratory, rather than inspiratory, in direction.

The Muscles of the Rib Cage

The Intercostal Muscles

The intercostal muscles are two thin layers of muscle occupying each of the intercostal spaces. They are termed *external* and *internal* because of their surface relations, with the external being superficial to the internal. The external intercostals extend from the tubercles of the ribs dorsally to the costochondral junctions ventrally, and their fibers are oriented obliquely caudad and ventrally from the rib above to the rib below. In contrast, the internal intercostals extend from the angles of the ribs dorsally to the sternocostal junctions ventrally, and their fibers run obliquely caudad and dorsally from the rib above to the rib below. Thus, although the intercostal spaces in their lateral portion contain two layers of intercostal muscle running approximately at right angles to each other, they contain a single muscle layer in their ventral and dorsal portions. On the ventral side, between the sternum and the chondrocostal junctions, the only fibers are those of the internal intercostal muscles, and they are conventionally called the *parasternal intercostals*. On the dorsal side, from the angles of the ribs to the vertebrae, the only fibers come from the external intercostal muscles. These latter, however, are duplicated by a spindle-shaped muscle running in each interspace from the tip of the transverse process of the vertebra cranially to the angle of the rib caudally; this muscle is the *levator costae*. All the intercostal muscles are innervated by the intercostal nerves.

Actions on the Ribs

The actions of the intercostal muscles on the ribs are conventionally regarded according to the theory proposed by Hamberger (1749). This theory is based on geometrical considerations and is illustrated in Fig. 8.4. When an intercostal muscle contracts in one interspace, it pulls the upper rib down and the lower rib up. As the fibers of the external intercostal slope obliquely caudad and ventrally from the rib above to the one below, however, their lower insertion is more distant from the center of rotation of the ribs (i.e., the vertebral articulations) than the upper one. When this muscle contracts, the torque acting on the lower rib is therefore greater than that acting on the upper rib; consequently, its net effect is to raise the ribs. The orientation of the levator costae is similar to that of the external intercostal, and its action is also to raise the ribs. In contrast, the fibers of the internal intercostal run obliquely caudad and dorsally from the rib above to the one below. Their lower insertion, therefore, is less distant from the center of rotation of the ribs than the upper one. As a result, when this muscle contracts, the torque acting on the lower rib is less than that acting on the upper rib, and its net effect is to lower the ribs. Hamberger finally concluded that although the

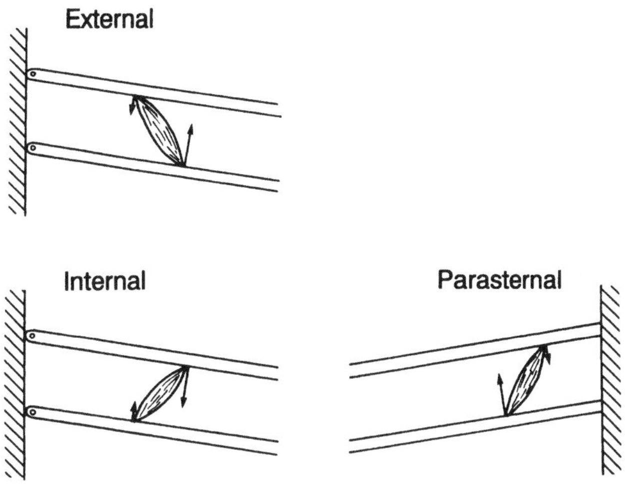

FIGURE 8.4. Diagram illustrating the actions of the intercostal muscles on the ribs, as proposed by Hamberger (1749). The hatched area in the left panels represents the spine (dorsal view); the hatched area in the lower right panel represents the sternum (ventral view). The two bars oriented obliquely represent two adjacent ribs. The external and internal intercostal muscles are depicted as single bundles, and the torques acting on the ribs during contraction of these muscles are represented by arrows. (*Source*: Reprinted, with permission, from De Troyer, A. [1995] Respiratory muscle function. In: Respiratory medicine, second ed. Brewis, R.A.L., Corrin, B., Geddes, D.M., and Gibson, G.J. (eds.) pp. 125–133. W.B. Saunders, London.)

parasternal intercostals are part of the internal intercostal layer, their action should be referred to the sternum rather than to the vertebral column. Their contraction, therefore, should raise the ribs.

Several of these conclusions have received direct experimental support. When the parasternal intercostal muscles in the dog are selectively activated by electrical stimulation, they produce a cranial displacement of the ribs into which they insert and an increase in lung volume (De Troyer and Kelly, 1982). Measurements of the changes in length of the canine intercostal muscles have also shown that, like the diaphragm, the parasternal intercostals in all interspaces shorten during passive inflation; therefore, they have a clear-cut inspiratory mechanical advantage (Decramer and De Troyer, 1984; De Troyer, Legrand, and Wilson, 1996). Stimulation of the levator costae in a single intercostal space similarly causes a cranial displacement of the rib into which it inserts (De Troyer and Farkas, 1989). When either the external or the internal interosseous intercostal muscle in a single interspace is selectively stimulated in the dog, however, there is a mutual approximation of the adjacent ribs, but the cranial displacement of the rib below is always greater than the caudal displacement of the rib above (De Troyer, Kelly, and Zin, 1983; Ninane, Gorini, and Estenne, 1991). In addition, whereas the Hamberger mechanism would predict that inflation of the relaxed respiratory system above FRC produces shortening of all the external intercostals and lengthening of all the internal interosseous intercostals, measurements in dogs have shown that the changes in length of these muscles are variable and largely determined by the location of the muscles along the rostrocaudal axis of the rib cage. Thus, in the rostral interspaces, these two sets of intercostal muscles tend to shorten during passive inflation, whereas they both tend to lengthen in the caudal interspaces (Decramer, Kelly, and De Troyer, 1986).

The Hamberger theory is thus incomplete. Saumarez (1986) and Wilson and De Troyer (1993) have emphasized the two major reasons why this theory cannot entirely explain the actions of the intercostal muscles. First, the Hamberger model is planar, whereas the real ribs are curved. As a result, the mechanical advantage of the external and internal intercostal muscles, as reflected by their changes in length during passive inflation, varies as a function of the position of the muscle fibers along the rib. Thus, because of the curvature of the ribs, these changes in muscle length are greatest in the dorsal region of the rib cage, decrease progressively as one moves around the cage, and are reversed as one approaches the sternum. The second reason is that the Hamberger theory is based on the idea that all the ribs rotate by equal amounts around parallel axes; therefore, the distance between adjacent ribs remains constant. In fact, the radii of curvature of the different ribs are different, increasing from the top downward; their rotation compliances are different as well (Wilson and De Troyer, 1993; De Troyer, Legrand, and Wilson 1996).

Respiratory Function of the Intercostal Muscles

Regardless of the limitations of the Hamberger theory, a number of electromyographic studies in dogs (De Troyer and Kelly, 1982; De Troyer and Farkas, 1993), cats (Greer and Martin, 1990), and baboons (De Troyer and Farkas, 1994) have clearly established that the parasternal intercostals are electrically active during the inspiratory phase of the breathing cycle (Fig. 8.5). Electromyographic recordings from intercostal muscles and nerves in these animals have also established that the external intercostal and levator costae muscles are active only during inspiration (Fig. 8.5), whereas the internal interosseous intercostals are active only during expiration (Sears, 1964; Bainton, Kirkwood and Sears, 1978; Hilaire, Nicholls and Sears, 1983; De Troyer and Ninane, 1986a; De Troyer and Farkas, 1989; Greer and Martin, 1990). It is of interest that the inspiratory activation of the external intercostals takes place predominantly in the dorsal region of the rostral interspaces (Baintain, Kirkwood, and Sears, 1978; De Troyer

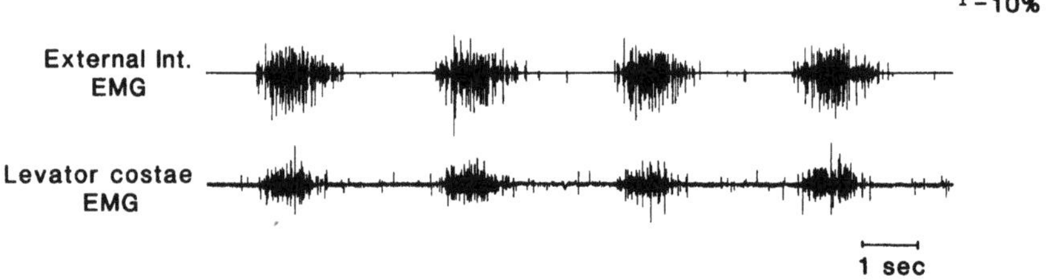

FIGURE 8.5. Pattern of electrical activation of the parasternal intercostal, external intercostal, and levator costae muscles during resting breathing in the baboon. The traces of lung volume (increase upward) and parasternal intercostal length (decrease downward) are also shown. Note that the three muscles (recorded in the third intercostal space) are active during inspiration; the parasternal intercostal also shortens in phase with inspiration.

and Ninane, 1986a; Greer and Martin, 1990) where the muscles are thickest and have the greatest inspiratory mechanical advantage (Wilson and De Troyer, 1993). These features correspond to an inspiratory action on the rib cage; indeed, when the diaphragm and parasternal intercostals in dogs are denervated so that the external intercostals and levator costae are the only muscles active during inspiration, the ribs continue to move cranially (De Troyer and Farkas, 1989). On the other hand, expiratory activation of the internal interosseous intercostals is predominant in the caudal interspaces (De Troyer and Ninane, 1986a), where the muscles have a greater expiratory mechanical advantage (Wilson and De Troyer, 1993). This pattern corresponds to an expiratory action on the rib cage and the lung (Loring and Woodbridge, 1991).

Although the parasternal intercostals, the external intercostals in the rostral interspaces, and the levator costae contract together during inspiration and contribute to the inspiratory cranial displacement of the ribs, there is substantial evidence that the parasternal intercostals play a larger role than do the external intercostals and levator costae during resting breathing in anesthetized animals. In anesthetized dogs and cats, the cranial motion of the ribs occurs together with a caudal displacement of the sternum (Da Silva et al., 1977; De Troyer and Kelly, 1982; De Troyer and Decramer, 1985). This pattern of motion is due to the action of the parasternal intercostals. Indeed, selective stimulation of these muscles causes the sternum to move caudally (De Troyer and Kelly, 1982), whereas both the external intercostals and the levator costae displace the sternum cranially (De Troyer and Farkas, 1989). Of more importance, when the canine parasternal intercostals are denervated in all interspaces, the inspiratory cranial motion of the ribs is reduced by 60%, yet the external intercostal and levator costae inspiratory activities are markedly increased (De Troyer, 1991a; De Troyer and Yuehua, 1994). These increases in inspiratory activity presumably reduce the decrease in cranial rib motion that results from the denervation of the parasternals. In contrast, when the canine external intercostals in all interspaces are severed and the parasternal intercostals are left intact, the inspiratory cranial displacement of the ribs decreases by only 10%, although the parasternal inspiratory activity is unchanged (De Troyer, 1991a).

Normal humans breathing at rest also have inspiratory activity in the parasternal intercostals and in the external intercostals of the most rostral interspaces (Taylor, 1960; Delhez, 1974). Although it is difficult to compare the amounts of activity recorded in different muscles, activity in the human external intercostals also appears to be less consistent and to involve fewer motor units than activity in the parasternal intercostals (Taylor, 1960; Delhez, 1974; Whitelaw and Feroah, 1989). This suggests that in humans as in quadrupeds, the contribution of the parasternal intercostals to resting breathing is greater than that of the external intercostals. In contrast to the parasternal intercostals, the external intercostals and levator costae are abundantly supplied with muscle spindles (Duron, Jung-Caillol, and Marlot, 1978); therefore, they might constitute a reserve, "load-compensating" system (De Troyer, 1991b).

Nonrespiratory Function of the Intercostal Muscles

The insertions and orientations of the external intercostals suggest that contraction of these muscles on one side of the sternum might also rotate the ribs in a transverse plane so that the upper ribs would move forward while the lower ribs would move backward (Fig. 8.6). In contast, contraction of the internal intercostals on one side of the sternum would displace the upper ribs backward and the lower ribs forward. These muscles, therefore, would be ideally suited to twist the rib cage.

Measurements of the changes in muscle length during passive rotations of the thorax in anesthetized dogs have supported this idea (Decramer, Kelly, and De Troyer, 1986).

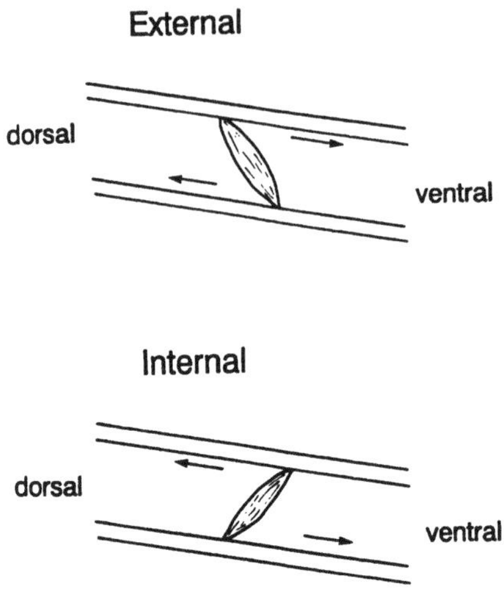

FIGURE 8.6. Diagram illustrating the actions of the intercostal muscles during rotations of the trunk. Lateral view of an intercostal space on the right side of the chest. The two bars oriented obliquely represent two adjacent ribs. The external and internal intercostal muscles are depicted as single bundles, and the arrows indicate the component of tension vector acting along the ribs. (*Source*: Reprinted, with permission, from De Troyer, A. [1996] Mechanics of the chest wall muscles. In: Neural control of the respiratory muscles, Miller, A.D., Bishop, B., and Bianchi, A.L. (eds.) pp. 59–73. CRC Press, Boca Raton.)

When the animal's trunk was twisted to the left, the external intercostals on the right side of the chest and the internal interosseous intercostals on the left side shortened considerably. At the same time, the external intercostals on the left side and the internal intercostals on the right side lengthened. The opposite pattern was seen when the animal's trunk was passively rotated to the right, with a marked shortening of the right internal and left external intercostals and a lengthening of the left internals and right externals. Thus, the length of these muscles changed in the way expected if they were producing the rotations, and indeed electromyographic studies in normal humans have recently demonstrated that the external intercostals on the right side of the chest are active when the trunk is rotated to the left, whereas they are silent when the trunk is rotated to the right (Whitelaw et al., 1992). On the other hand, the internal intercostals on the right side of the chest are active only when the trunk is rotated to the right. Active use of these muscles during such postural movements is also consistent with their abundant supply of muscle spindles.

The Triangularis Sterni

The triangularis sterni, also called *transversus thoracis*, is a flat muscle that lies deep to the sternum and the parasternal intercostals. As shown in Fig. 8.7, its fibers originate from the dorsal aspect of the caudal half of the sternum and insert into the inner surface of the chondrocostal junctions of ribs 3 to 7. The muscle receives its motor supply from the intercostal nerves, and in the dog, its selective stimulation causes a caudal displacement of the ribs with a cranial motion of the sternum and a decrease in lung volume (De Troyer and Ninane, 1986b).

Although this muscle has long been overlooked, it has an important respiratory function in mammalian quadrupeds. In the dog and in the cat, it invariably contracts during the expiratory phase of the breathing cycle (De Troyer and Ninane, 1986b; Hwang, Zhou, and St John, 1989), and, in so doing, it pulls the ribs caudally and deflates the rib cage below its neutral (resting) position. As a result, when the muscle relaxes at the end of expiration, there is a passive rib cage expansion and an increase in lung volume that precedes the onset of inspiratory muscle contraction. In these animals, the triangularis sterni thus shares the work of breathing with the inspiratory muscles and helps

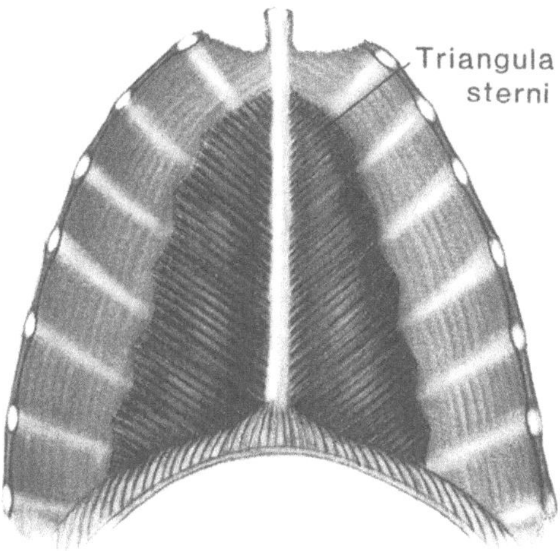

FIGURE 8.7. Dorsal aspect of the ventral wall of the rib cage in the dog, illustrating the insertions of the triangularis sterni muscle. Note that the rostral portion of the sternum and the first two ribs have been reflected; the triangularis sterni has no insertions there. (*Source*: Reprinted, with permission, from De Troyer, A., and Loring, S.H. [1995] Actions of the respiratory muscles. In: The thorax, second ed. Roussos, C. (ed.) pp. 535–563. Marcel Dekker, New York.)

the parasternal intercostals produce the rhythmic inspiratory expansion of the rib cage (De Troyer and Ninane, 1986b).

In contrast to quadrupeds, the triangularis sterni in normal humans is usually inactive during resting breathing (De Troyer et al., 1987); however, it invariably contracts during voluntary or involuntary expiratory efforts such as coughing, laughing, and speech. In fact, normal humans cannot produce expiratory efforts without contracting the triangularis sterni. The muscle then presumably acts in concert with the internal interosseous intercostals to deflate the rib cage and increase pleural pressure.

The Scalenes

The scalenes in humans comprise three muscle bundles that run from the transverse processes of the lower five cervical vertebrae to the upper surface of the first two ribs. When these muscles are selectively activated by electrical stimulation in dogs, they produce a marked cranial displacement of the ribs and sternum, and cause an increase in the rib cage anteroposterior diameter. Although the scalenes have traditionally been considered as "accessory" muscles of inspiration, electromyographic studies with concentric needle electrodes have established that in normal humans they invariably contract in concert with the diaphragm and the parasternal intercostals during inspiration (Raper et al., 1966; Delhez, 1974; De Troyer and Estenne, 1984).

There is no clinical setting that causes paralysis of all the inspiratory muscles without also affecting the scalenes; therefore, the isolated action of these muscles on the human rib cage cannot be precisely defined. Several observations, however, indicate that contraction of the scalenes is an important determinant of the motion of the sternum and the upper ribs during resting breathing. First, the sternum in resting humans moves cranially during inspiration. In contrast, the scalenes in the dog are not active during breathing (De Troyer, Cappello, and Brichant, 1994), and the inspiratory contraction of the parasternal intercostals causes the sternum to move caudally, rather than cranially (De Troyer and Kelly, 1982). Second, when normal subjects attempt to inspire with the diaphragm alone, there is a marked, selective decrease in scalene activity associated with either less inspiratory increase or a paradoxical decrease in anteroposterior diameter of the upper rib cage (De Troyer and Estenne, 1984). Third, the inward inspiratory displacement of the upper rib cage characteristic of quadriplegia (see later) is usually not observed when scalene function is preserved after the lower cervical cord transection (Estenne and De Troyer, 1985). Because the scalenes are innervated from the lower five cervical segments, persistent inspiratory contraction is frequently seen in subjects with a transection at the C_7 level or below. In such subjects, the anteroposterior diameter of the upper rib cage tends to remain constant or to increase slightly during inspiration.

The Sternocleidomastoids and Other Accessory Muscles of Inspiration

Many additional muscles (e.g., the pectoralis minor, the trapezius, the erector spinae, the serrati, and the sternocleidomastoids) can elevate the ribs when they contract. These muscles, however, run between the shoulder girdle and the rib cage, between the spine and the shoulder girdle, or between the head and the rib cage. They therefore have primarily postural functions. In healthy individuals, they contract only during increased inspiratory efforts; in contrast to the scalenes, they are thus real "accessory" muscles of inspiration.

Of all these muscles, only the sternocleidomastoids have been thoroughly studied. These descend from the mastoid process to the ventral surface of the manubrium sterni and the medial third of the clavicle, and their action in humans has been inferred from measurements of chest wall motion in subjects with transection of the upper cervical cord. Indeed, in such patients, the diaphragm and the intercostal, scalene, and abdominal muscles are paralyzed, but the sternocleidomastoids (the motor innervation of which largely depends on the eleventh cranial nerve) are spared and contract forcefully during unassisted inspiration (Danon et al., 1979; De Troyer, Estenne, and Vincken, 1986). When breathing spontaneously, these patients show a marked inspiratory cranial displacement of the sternum and a large inspiratory expansion of the upper rib cage, particularly in its anteroposterior diameter. These patients, however, also have an inspiratory decrease in the transverse diameter of the lower rib cage (Fig. 8.8).

The Abdominal Muscles

The four abdominal muscles that have a significant respiratory function in humans constitute the ventrolateral wall of the abdomen. The *rectus abdominis* is the most ventral of these muscles. It originates from the ventral aspect of the sternum and the fifth, sixth, and seventh costal cartilages, and it runs caudally along the whole length of the abdominal wall to insert into the pubis. This muscle is enclosed in a sheath formed by the aponeuroses of the other three muscles. The most superficial of these is the *external oblique*, which originates by fleshy digitations from the external surface of the lower eight ribs, well above the costal margin, and directly covers the lower ribs and intercostal muscles. Its fibers radiate caudally to the iliac crest and inguinal ligament and medially to the linea alba. The *internal oblique* lies deep to the external oblique. Its fibers arise from the iliac crest and inguinal ligament, and they diverge to insert on the costal margin and an aponeurosis contributing to the rectus sheath down to the pubis.

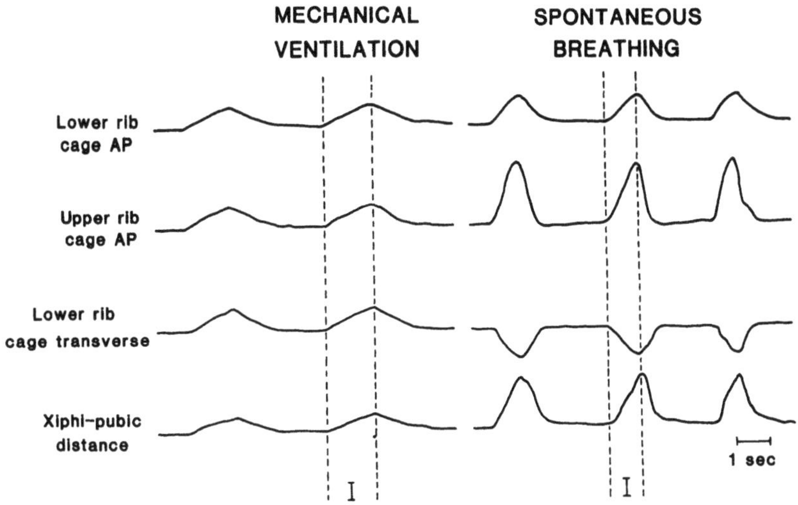

FIGURE 8.8. Pattern of rib cage motion during mechanical ventilation (left) and during spontaneous breathing (right) in a quadriplegic subject with a traumatic transection of the upper cervical cord (C_1). Each panel shows, from top to bottom, the respiratory changes in anteroposterior (AP) diameter of the lower rib cage, the changes in AP diameter of the upper rib cage, the changes in transverse diameter of the lower rib cage, and the changes in xiphi-pubic distance. Upward deflections correspond to an increase in diameter or an increase in xiphi-pubic distance (i.e., a cranial displacement of the sternum). I indicates the duration of inspiration. Note that during mechanical ventilation, all rib cage diameters and the xiphi-pubic distance increase in phase. During spontaneous inspiration, however, the xiphi-pubic distance and the upper rib cage AP diameter increase more than the lower rib cage AP diameter, and the lower rib cage transverse diameter decreases. (*Source*: Reprinted, with permission, from De Troyer, A. [1996] Mechanics of the chest wall muscles. In: Neural control of the respiratory muscles, Miller A.D., Bishop, B., and Bianchi, A.L. (eds.) pp. 59–73. CRC Press, Boca Raton.)

The *transversus abdominis* is the deepest of the muscles of the lateral abdominal wall. It arises from the inner surface of the lower six ribs, where it interdigitates with the costal insertions of the diaphragm. From this origin and from the lumbar fascia, the iliac crest, and the inguinal ligament, its fibers run circumferentially around the abdominal visceral mass and terminate ventrally in the rectus sheath.

Actions of the Abdominal Muscles

These four muscles have important functions as flexors (rectus abdominis) and rotators (external oblique, internal oblique) of the trunk, but as respiratory muscles, they have two principal actions. First, as they contract, they pull the abdominal wall inward and produce an increase in abdominal pressure. This causes the diaphragm to move cranially into the thoracic cavity, and this motion, in turn, results in an increase in pleural pressure and a decrease in lung volume. Second, these four muscles displace the rib cage due to their insertions on the ribs. These insertions would suggest that the action of all abdominal muscles is to pull the lower ribs caudally and to deflate the cage, another expiratory action. Measurements of rib cage motion during electrical stimulation of the four abdominal muscles in dogs have shown, however, that the rise in abdominal pressure produced by these muscles also confers on them an inspiratory action on the rib cage (De Troyer et al., 1983; D'Angelo, Prandi, and Bellemare, 1994).

Indeed, the zone of apposition of the diaphragm to the rib cage (see Fig. 8.1) allows the rise in abdominal pressure to be transmitted to the lower rib cage and to expand it. In addition, by forcing the diaphragm cranially, the rise in abdominal pressure causes passive diaphragmatic tension. This passive tension also tends to raise the lower ribs and to expand the lower rib cage in the same way as does an active diaphragmatic contraction ("insertional" force of the diaphragm).

The action of the abdominal muscles on the rib cage is thus determined by the balance between the insertional, expiratory force of the muscles and the inspiratory force related to the rise in abdominal pressure. Isolated contraction of the external oblique in humans produces a small caudal displacement of the sternum and a large decrease in the rib cage transverse diameter, but the rectus abdominis, although causing a marked caudal displacement of the sternum and a large decrease in the anteroposterior diameter of the rib cage, also produces a small increase in the rib cage transverse diameter (Mier et al., 1985). The isolated actions of the internal oblique and transversus abdominis muscles on the human rib cage are not known, but the anatomical arrangement of the transversus suggests that among the abdominal muscles, this muscle has the smallest insertional, expiratory action on the ribs and the greatest effect on abdominal pressure. Isolated contraction of the transversus should therefore produce little or no expiratory rib cage displacement.

Respiratory Function of the Abdominal Muscles

Irrespective of their actions on the rib cage, the abdominal muscles are primarily expiratory muscles through their action on the diaphragm and the lung, and they play important roles in activities such as coughing and speaking. When they contract rhythmically in phase with expiration and reduce lung volume to less than the neutral position of the respiratory system, however, their relaxation at end-expiration may promote a passive descent of the diaphragm and induce an increase in lung volume before the onset of inspiratory muscle contraction. The abdominal muscles, therefore, may also be considered as accessory muscles of inspiration.

This inspiratory action of the abdominal muscles takes place all the time in quadrupeds. In dogs placed in the head-up or the prone posture, the relaxation of the abdominal muscles at end-expiration accounts for up to 40–60% of the tidal volume (Farkas, Estenne, and De Troyer, 1989; Farkas and Schroeder, 1990). Adult humans do not utilize such a breathing strategy at rest. Phasic expiratory contraction of the abdominal muscles, however, also occurs in healthy subjects whenever the demand placed on the inspiratory muscles is abnormally increased (e.g., during exercise or during CO_2-induced hyperpnea). It is noteworthy that in these conditions, the transversus muscle is recruited well before activity can be recorded from either the rectus or the external oblique (De Troyer et al., 1990; Abe et al., 1996). In view of the actions of these muscles, this differential recruitment also supports the idea that the effect of the abdominal muscles on abdominal pressure is more important to the act of breathing than their action on the rib cage.

There is a second mechanism by which the abdominal muscles can assist inspiration. When adopting the standing posture most normal human subjects develop tonic abdominal muscle activity unrelated to the phases of the breathing cycle (Floyd and Silver, 1950). Studies in subjects with transection of the upper cervical cord, in whom bilateral pacing of the phrenic nerves allows the degree of diaphragmatic activation to be maintained constant, have clearly illustrated the effect of this tonic abdominal contraction on inspiration (Danon et al., 1979; Strohl et al., 1984). When the subjects were supine, the unassisted paced diaphragm was able to generate an adequate tidal volume.

When the subjects were tilted head up or moved to the seated posture, however, the weight of the abdominal viscera and the absence of abdominal muscle activity caused the belly wall to protrude. The tidal volume produced by pacing in this posture was markedly reduced relative to the supine posture, but the reduction was significantly diminished when a pneumatic cuff was inflated around the abdomen to mimic the tonic abdominal muscle contraction. Thus, by contracting throughout the breathing cycle in the standing posture, the abdominal muscles make the diaphragm longer at the onset of inspiration and prevent it from shortening excessively during inspiration; in accordance with the length-tension characteristics of the muscle, its ability to generate pressure is therefore increased.

Respiratory Muscle Function in Diseases

Quadriplegia

As pointed out previously, the particular distribution of muscle paralysis in subjects with traumatic transection of the lower cervical cord causes distinct abnormalities in the pattern of chest wall motion during breathing (Fig. 8.9). Because diaphragmatic function is preserved, the expansion of the abdomen during inspiration is associated with an expansion of the lower rib cage; however, whereas the entire rib cage expands synchronously and uniformly in healthy subjects, the lower rib cage expands predominantly over its lateral walls where the area of apposed diaphragm is greater (greater appositional force) in quadriplegic subjects (Estenne and De Troyer, 1985). In addition, the paralysis of the rib cage inspiratory muscles, particularly the scalenes and the parasternal intercostals, is such that many quadriplegic subjects breathing at rest have an inspiratory decrease (paradoxical motion) of the anteroposterior diameter of the upper rib cage (Mortola and Sant'Ambrogio, 1978; Estenne and De Troyer, 1985).

Quadriplegic subjects also have complete paralysis of all the well-recognized muscles of expiration (e.g., abdominal muscles, internal intercostals, triangularis sterni). As a result, the expiratory reserve volume (ERV) is markedly reduced, and RV is usually greater than normal. The peak pleural pressures developed during cough are also less than normal, such that the efficiency of cough and the clearance of bronchial secretions are severely impaired. Studies have demonstrated, however, that most quadriplegic subjects have residual expiratory muscle function due to the action of the clavicular portion of the pectoralis major (De Troyer, Estenne, and Heilporn, 1986; Estenne and De Troyer, 1990). In subjects with transection at the C_6 segment or below, this muscle bundle invariably contracts during voluntary expiration and during cough, and its insertions on the humerus and the medial half of the clavicle make it displace the manubrium sterni and the upper ribs in the caudal direction when it contracts on both sides of the chest (De Troyer, Estenne, and Heilporn, 1986). In so doing, it produces collapse of the upper rib cage and partial emptying of the lung. In a number of subjects, vigorous contraction of the clavicular portion of the pectoralis major may even induce dynamic compression of the intrathoracic airways (Estenne et al., 1994), thus indicating that cough in this setting is not necessarily a passive phenomenon as conventionally thought.

Diaphragmatic Paralysis

Paralysis or severe weakness of both hemidiaphragms is usually seen in the context of generalized respiratory muscle weakness; however, in occasional patients the

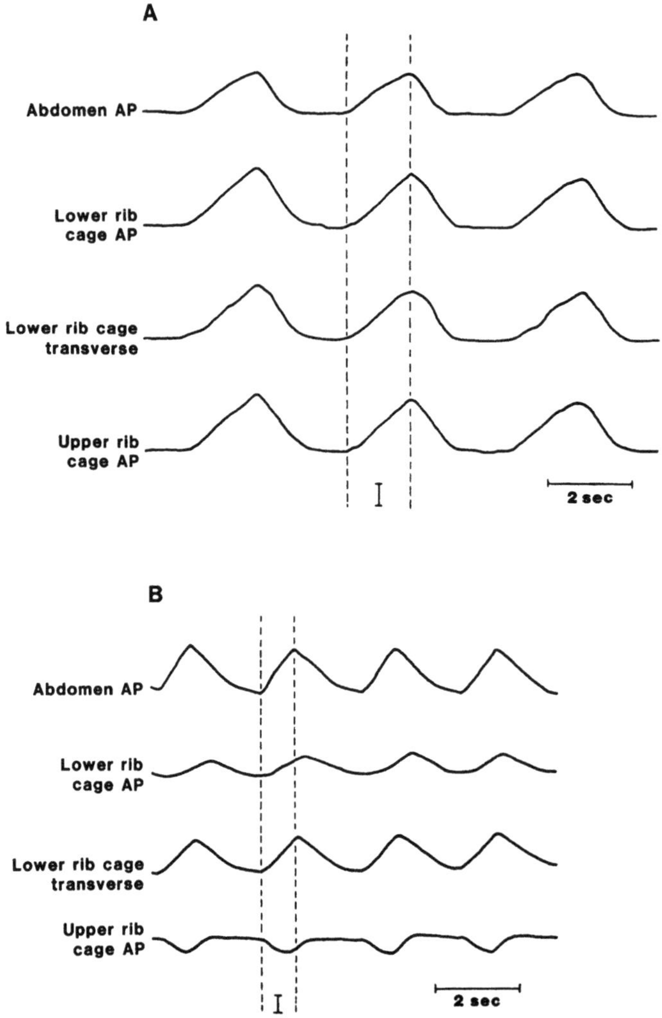

FIGURE 8.9. Pattern of chest wall motion in a healthy subject (panel A) and a C_5 quadriplegic patient (panel B) breathing at rest in the seated posture. The respiratory changes in anteroposterior (AP) diameter of the abdomen, lower rib cage, and upper rib cage are shown, as well as the changes in transverse diameter of the lower rib cage. Same conventions as in Fig. 8.8. (*Source*: Reprinted, with permission, from De Troyer, A., and Estenne, M. [1995] The respiratory system in neuromuscular disorders. In: The thorax, second ed. Roussos, C. (ed.) pp. 2177–2212. Marcel Dekker, New York.)

diaphragm is specifically or disproportionately affected. Selective paralysis of the diaphragm results in a compensatory increase in the activation of the inspiratory rib cage muscles, so that the inspiratory expansion of the rib cage compartment of the chest wall is accentuated (De Troyer and Kelly, 1982). In addition, whereas the simultaneous contraction of the diaphragm and the rib cage inspiratory muscles causes a rise in abdominal pressure associated with a fall in pleural pressure in healthy subjects, in the presence of diaphragmatic paralysis the fall in pleural pressure is transmitted through the flaccid diaphragm such that abdominal pressure falls as well. As a

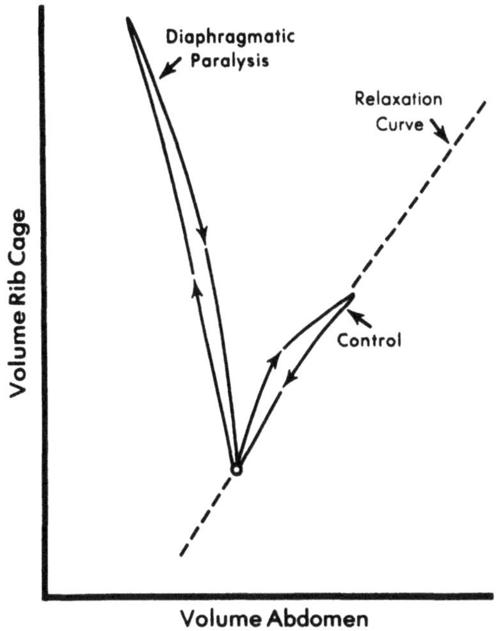

FIGURE 8.10. Pattern of chest wall motion during resting breathing in a supine anesthetized dog before (control) and after bilateral section of the phrenic nerves in the neck (diaphragmatic paralysis). The changes in abdominal cross-section are on the abscissa (increase rightward), and the changes in rib cage cross-section are on the ordinate (increase upward). The broken line represents the relaxation curve of the thoracoabdominal system, and the solid loops represent tidal volume cycles; arrows indicate the direction of the loops, and the open circle corresponds to end expiration. During breathing in the control condition, the chest wall moves on its relaxed configuration. After induction of diaphragmatic paralysis, however, the rib cage expands markedly during inspiration, but the abdomen moves paradoxically inward (*Source*: Adapted from De Troyer, A., and Kelly, S. [1982] J. Appl. Physiol., with permission.)

result, the abdomen moves paradoxically inward, opposing the inflation of the lung (Fig. 8.10).

Some patients also compensate for diaphragmatic paralysis by contracting the abdominal muscles during expiration, thus displacing the abdomen inward and the diaphragm cranially into the thorax. Relaxation of the abdominal muscles at the onset of inspiration may therefore result in outward abdominal motion and passive descent of the diaphragm (Newsom et al., 1976; Kreitzer et al., 1978). Such a contraction of the abdominal muscles during expiration seems to be particularly frequent in the erect patient. When present it may remove the inspiratory inward motion of the abdomen that is the cardinal sign of diaphragmatic paralysis on clinical examination; however, this does not occur in the supine posture, where the abdominal muscles usually remain relaxed during the whole respiratory cycle.

Chronic Obstructive Pulmonary Disease

Measurements of thoracoabdominal motion during breathing have shown that patients with chronic obstructive pulmonary disease (COPD) and hyperinflation have relatively greater expansion of the rib cage and smaller expansion of the abdomen than healthy subjects (Sharp et al., 1977; Martinez, Couser, and Celli, 1990). The normal inspiratory positive swing of abdominal pressure is also attenuated, whereas the fall in pleural pressure is greater than normal due to the increased airflow resistance and the reduced dynamic pulmonary compliance. In patients with severe disease, abdominal pressure may even become negative during inspiration, and the abdomen may move paradoxically inward, as if the diaphragm were paralyzed (Ashutosh et al., 1975; Sharp et al., 1977; Martinez, Couser, and Celli, 1990). This altered pattern has led to the widespread belief that these patients have relatively more use of the rib cage inspiratory muscles and less use of the diaphragm than healthy subjects, possibly due to diaphragmatic fatigue (Cohen et al., 1982; Martinez, Couser, and Celli, 1990).

In agreement with this idea, the scalenes and parasternal intercostals feel tense on palpation in many patients with severe COPD. Electromyographic studies using concentric needle electrodes have also shown that although most of these patients do not contract the sternocleidomastoids when breathing at rest (De Troyer et al., 1994), they have increased firing frequencies in the parasternal intercostal and scalene motor units compared with normal subjects (Gandevia et al., 1996). The patients also have a greater number of active motor units in these two muscles; however, diaphragmatic motor units also demonstrate substantial increases in firing frequencies during resting breathing, thus indicating that COPD is associated with an increase in neural drive both to the rib cage inspiratory muscles and to the diaphragm (De Troyer et al., 1997). This suggests that the altered thoracoabdominal motion seen during inspiration in these patients results from mechanical factors alone.

The diaphragm in such patients is characteristically flat and low compared with normal subjects, and the zone of apposition is reduced in size. Irrespective of the degree of neural activation, the ability of the diaphragmatic dome to descend is therefore impaired; hence, the rise in abdominal pressure and the outward displacement of the abdominal wall must be reduced. In some patients with severe hyperinflation, the zone of apposition has virtually disappeared, and the normal curvature of the diaphragm is even reversed, with its concavity facing upward rather than downward. The muscle fibers at their insertions on the ribs then run transversely inward rather than cranially. In this situation, contraction of the diaphragm cannot result in any descent of the dome. Instead, the vigorous contraction of the rib cage inspiratory muscles, which results in a greater than normal elevation of the ribs, will tend to pull the diaphragm cranially and to displace the ventral abdominal wall inward. Contraction of this flat diaphragm, however, produces the so-called Hoover's sign (i.e., an inspiratory decrease in the transverse diameter of the lower rib cage) (Gilmartin and Gibson, 1986).

Patients with severe COPD also have a greater than normal neural drive to the abdominal muscles. Indeed, in contrast to normal subjects, many resting patients with severe COPD have phasic expiratory contraction of the abdominal muscles, particularly the transversus abdominis (Ninane et al., 1992). This causes a decrease in abdominal dimensions associated with an increase in abdominal pressure, and produces a paradoxical increase in the anteroposterior diameter of the lower rib cage in occasional patients (Gilmartin and Gibson, 1986). Although this expiratory muscle contraction might be considered as a compensation for the relatively ineffective diaphragm, its usefulness is uncertain. As previously pointed out, expiratory contraction of the abdominal muscles is a natural component of the response of the normal respiratory system to increased stimulation. In the absence of expiratory flow limitation, this expiratory muscle contraction is appropriate because it allows the work of breathing to be shared between the inspiratory and expiratory muscles. In patients with severe COPD and airflow limitation, however, this "automatic" response to increased ventilatory stimulation might be present during resting breathing without producing significant deflation of the respiratory system (Ninane et al., 1992).

Conclusions

Although the diaphragm is the main respiratory muscle in humans, it is not the only important contracting muscle. Thus, the expansion of the cranial half of the rib cage is accomplished by the scalenes and the inspiratory intercostals, particularly those of the parasternal area. Additional muscles (e.g., the transversus abdominis and the triangularis sterni) are also frequently involved in the act of breathing, particularly when the ventilatory requirements are increased. These muscles are usually considered to be expiratory because they displace the respiratory system below its resting volume. By relaxing at end-expiration, however, they also cause an increase in lung volume, thereby reducing the load on the inspiratory muscles. Moving the chest wall during breathing is thus a complex, integrated process that involves many muscles, and the control mechanisms that promote coordinated use of these different muscles are critically important to maintaining alveolar ventilation within acceptable limits. These mechanisms already play an important role in healthy subjects, but they become absolutely essential to life in conditions where the diaphragm is less effective or paralyzed.

References

Abe, T., Kusuhara, N., Yoshimura, N., Tomita, T., and Easton, P.A. (1996) Differential respiratory activity-of four abdominal muscles in humans. J. Appl. Physiol. 80, 1379.

Ashutosh, K., Gilbert, R., Auchincloss, J.H., Jr., and Peppi, D. (1975) Asynchronous breathing movements in patients with chronic obstructive pulmonary disease. Chest 67, 553.

Bainton, C.R., Kirkwood, P.A., and Sears, T.A. (1978) On the transmission of the stimulating effects of carbon dioxide to the muscles of respiration. J. Physiol. (Lond.) 280, 249.

Cohen, C.A., Zagelbaum, G., Gross, D., Roussos, C., and Macklem, P.T. (1982) Clinical manifestations of inspiratory muscle fatigue. Am. J. Med. 73, 308.

D'Angelo, E., and Sant'Ambrogio, G. (1974) Direct action of contracting diaphragm on the rib cage in rabbits and dogs. J. Appl. Physiol. 36, 715.

D'Angelo, E., Prandi, E., and Bellemare, F. (1994) Mechanics of the abdominal muscles in rabbits and dogs. Respir. Physiol. 97, 275.

Danon, J., Druz, W.S., Goldberg, N.B., and Sharp, J.T. (1979) Function of the isolated paced diaphragm and the cervical accessory muscles in C_1 quadriplegics. Am. Rev. Respir. Dis. 119, 909.

Da Silva, K.M.C., Sayers, B.M.A., Sears, T.A., and Stagg, D.T. (1977) The changes in configuration of the rib cage and abdomen during breathing in the anesthetized cat. J. Physiol. (Lond.) 266, 499.

Decramer, M., and De Troyer, A. (1984) Respiratory changes in parasternal intercostal length. J. Appl. Physiol. 57, 1254.

Decramer, M., Kelly, S., and De Troyer, A. (1986) Respiratory and postural changes in intercostal muscle length in supine dogs. J. Appl. Physiol. 60, 1686.

Delhez, L. (1974) Contribution électromyographique à l'étude de la mécanique et du contrôle nerveux des mouvements respiratoires de l'homme. Vaillant-Carmanne, Liège, Belgium.

De Troyer, A. (1991a) Inspiratory elevation of the ribs in the dog: primary role of the parasternals. J. Appl. Physiol. 70, 1447.

De Troyer, A. (1991b) Differential control of the inspiratory intercostal muscles during airway occlusion in the dog. J. Physiol. (Lond.) 439, 73.
De Troyer, A., Cappello, M., and Brichant, J.F. (1994) Do the canine scalene and sternomastoid muscles play a role in breathing. J. Appl. Physiol. 76, 242.
De Troyer, A., and Decramer, M. (1985) Mechanical coupling between the ribs and sternum in the dog. Respir. Physiol. 59, 27.
De Troyer, A., and Estenne, M. (1984) Coordination between rib cage muscles and diaphragm during quiet breathing in humans. J. Appl. Physiol. 57, 899.
De Troyer, A., Estenne, M., and Heilporn, A. (1986) Mechanism of active expiration in tetraplegic subjects. N. Engl. J. Med. 314, 740.
De Troyer, A., Estenne, M., Ninane, V., Van Gansbeke, D., and Gorini, M. (1990) Transversus abdominis muscle function in humans. J. Appl. Physiol. 68, 1010.
De Troyer, A., Estenne, M., and Vincken, W. (1986) Rib cage motion and muscle use in high tetraplegics. Am. Rev. Respir. Dis. 133, 1115.
De Troyer, A., and Farkas, G.A. (1989) Inspiratory function of the levator costae and external intercostal muscles in the dog. J. Appl. Physiol. 67, 2614.
De Troyer, A., and Farkas, G.A. (1993) Mechanics of the parasternal intercostals in prone dogs: statics and dynamics. J. Appl. Physiol. 74, 2757.
De Troyer, A., and Farkas, G.A. (1994) Contribution of the rib cage inspiratory muscles to breathing in baboons. Respir. Physiol. 97, 135.
De Troyer, A., and Kelly, S. (1982) Chest wall mechanics in dogs with acute diaphragm paralysis. J. Appl. Physiol. 53, 373.
De Troyer, A., Kelly, S., and Zin, W.A. (1983) Mechanical action of the intercostal muscles on the ribs. Science 220, 87.
De Troyer, A., Legrand, A., and Wilson, T.A. (1996) Rostrocaudal gradient of mechanical advantage in the parasternal intercostal muscles of the dog. J. Physiol. (Lond.) 495, 239.
De Troyer, A., Leeper, J.B., Mc Kenzie, D.K., and Gandevia, S.C. (1997) Neural drive to the diaphragm in patients with severe COPD. Am. J. Respir. Crit. Care Med. 155, 1335.
De Troyer, A., and Ninane, V. (1986a) Respiratory function of intercostal muscles in supine dog: an electromyographic study. J. Appl. Physiol. 60, 1692.
De Troyer, A., and Ninane, V. (1986b) Triangularis sterni: a primary muscle of breathing in the dog. J. Appl. Physiol. 60, 14.
De Troyer, A., Ninane, V., Gilmartin, J.J., Lemerre, C., and Estenne, M. (1987) Triangularis sterni muscle use in supine humans. J. Appl. Physiol. 62, 919.
De Troyer, A., Peche, R., Yernault, J.C., and Estenne, M. (1994) Neck muscle activity in patients with severe chronic obstructive pulmonary disease. Am. J. Respir. Crit. Care Med. 150, 41.
De Troyer, A., Sampson, M., Sigrist, S., and Kelly, S. (1983) How the abdominal muscles act on the rib cage. J. Appl. Physiol. 54, 465.
De Troyer, A., Sampson, M., Sigrist, S., and Macklem, P.T. (1982) Action of costal and crural parts of the diaphragm on the rib cage in dog. J. Appl. Physiol. 53, 30.
De Troyer, A., and Yuehua, C. (1994) Intercostal muscle compensation for parasternal paralysis in the dog. Central and proprioceptive mechanisms. J. Physiol. (Lond.) 479, 149.
Duchenne, G.B. (1867) *Physiologie des mouvements.* Baillière, Paris.
Duron, B., Jung-Caillol, M.C., and Marlot, D. (1978) Myelinated nerve fiber supply and muscle spindles in the respiratory muscles of cat: quantitative study. Anat. Embryol. 152, 171.
Estenne, M., and De Troyer, A. (1985) Relationship between respiratory muscle electromyogram and rib cage motion in tetraplegia. Am. Rev. Respir. Dis. 132, 53.
Estenne, M., and De Troyer, A. (1990) Cough in tetraplegic subjects: an active process. Ann. Intern. Med. 112, 22.
Estenne, M., van Muylem, A., Gorini, M., Kinnear, W., Heilporn, A., and De Troyer, A. (1994) Evidence of dynamic airway compression during cough in tetraplegic subjects. Am. Rev. Respir. Dis. 150, 1081.
Farkas, G.A., Estenne, M., and De Troyer, A. (1989) Expiratory muscle contribution to tidal volume in head-up dogs. J. Appl. Physiol. 67, 1438.
Farkas, G.A., and Schroeder, M.A. (1990) Mechanical role of expiratory muscles during breathing in prone anesthetized dogs. J. Appl. Physiol. 69, 2137.

Floyd, W.F., and Silver, P.H.S. (1950) Electromyographic study of patterns of activity of the anterior abdominal wall muscles in man. J. Anat. 84, 132.
Gandevia, S.C., Leeper, J.B., Mc Kenzie, D.K., and De Troyer, A. (1996) Discharge frequencies of parasternal intercostal and scalene motor units during breathing in normal and COPD subjects. Am. J. Respir. Crit. Care Med. 153, 622.
Gilmartin, J.J., and Gibson, G.J. (1986) Mechanisms of paradoxical rib cage motion in patients with chronic obstructive pulmonary disease. Am. Rev. Respir. Dis. 134, 684.
Greer, J.J., and Martin, T.P. (1990) Distribution of muscle fiber types and EMG activity in cat intercostal muscles. J. Appl. Physiol. 69, 1208.
Hamberger, G.E. (1749) *De Respirationis Mechanismo et usu genuino.* Jena, Germany.
Hilaire, G.G., Nicholls, J.G., and Sears, T.A. (1983) Central and proprioceptive influences on the activity of the levator costae motoneurones in the cat. J. Physiol. (Lond.) 342, 527.
Hwang, J.C., Zhou, D., and St John, W.M. (1989) Characterization of expiratory intercostal activity to triangularis sterni in cats. J. Appl. Physiol. 67, 1518.
Jordanoglou, J. (1970) Vector analysis of rib movement. Respir. Physiol. 10, 109.
Konno, K., and Mead, J. (1967) Measurement of the separate volume changes of rib cage and abdomen during breathing. J. Appl. Physiol. 22, 407.
Kreitzer, S.M., Feldman, N.T., Saunders, N.A., and Ingram, R.H. (1978) Bilateral diaphragmatic paralysis with hypercapnic respiratory failure. A physiologic assessment. Am. J. Med. 65, 89.
Loring, S.H., and Mead, J. (1982) Action of the diaphragm on the rib cage inferred from a force-balance analysis. J. Appl. Physiol. 53, 756.
Loring, S.H., and Woodbridge, J.A. (1991) Intercostal muscle action inferred from finite-element analysis. J. Appl. Physiol. 70, 2712.
Martinez, F.J., Couser, J.I., and Celli, B.R. (1990) Factors influencing ventilatory muscle recruitment in patients with chronic airflow obstruction. Am. Rev. Respir. Dis. 142, 276.
Mead, J. (1979) Functional significance of the area of apposition of diaphragm to rib cage. Am. Rev. Respir. Dis. 119, 31.
Mier, A., Brophy, C., Estenne, M., Moxham, J., Green, M., and De Troyer, A. (1985) Action of abdominal muscles on rib cage in humans. J. Appl. Physiol. 58, 1438.
Mortola, J.P., and Sant'Ambrogio, G. (1978) Motion of the rib cage and the abdomen in tetraplegic patients. Clin. Sci. Mol. Med. 54, 25.
Newsom Davis, J., Goldman, M., Loh, L., and Casson, M. (1976) Diaphragm function and alveolar hypoventilation. Q. J. Med. 45, 87.
Ninane, V., Gorini, M., and Estenne, M. (1991) Action of intercostal muscles on the lung in dogs. J. Appl. Physiol. 70, 2388.
Ninane, V., Rypens, F., Yernault, J.C., and De Troyer, A. (1992) Abdominal muscle use during breathing in patients with chronic airflow obstruction. Am. Rev. Respir. Dis. 146, 16.
Raper, A.J., Thompson, W.T., Jr., Shapiro, W., and Patterson, J.L., Jr. (1966) Scalene and sternomastoid muscle function. J. Appl. Physiol. 21, 497.
Saumarez, R.C. (1986) An analysis of action of intercostal muscles in human upper rib cage. J. Appl. Physiol. 60, 690.
Sears, T.A. (1964) Efferent discharges in alpha and fusimotor fibres of intercostal nerves of the cat. J. Physiol. (Lond.) 174, 295.
Sharp, J.T., Goldberg, N.B., Druz, W.S., Fishman, H., and Danon, J. (1977) Thoracoabdominal motion in chronic obstructive pulmonary disease. Am. Rev. Respir. Dis. 115, 47.
Strohl, K.P., Mead, J., Banzett, R.B., Lehr, J., Loring, S.H., and O'Cain, C.F. (1984) Effect of posture on upper and lower rib cage motion and tidal volume during diaphragm pacing. Am. Rev. Respir. Dis. 130, 320.
Taylor, A. (1960) The contribution of the intercostal muscles to the effort of respiration in man. J. Physiol. (Lond.) 151, 390.
Urmey, W.F., De Troyer, A., Kelly, S.B., and Loring, S.H. (1988) Pleural pressure increases during inspiration in the zone of apposition of diaphragm to rib cage. J. Appl. Physiol. 65, 2207.
Whitelaw, W.A., and Feroah, T. (1989) Patterns of intercostal muscle activity in humans. J. Appl. Physiol. 67, 2087.

Whitelaw, W.A., Ford, G.T., Rimmer, K.P., and De Troyer, A. (1992) Intercostal muscles are used during rotation of the thorax in humans. J. Appl. Physiol. 72, 1940.
Wilson, T.A., and De Troyer, A. (1993) Respiratory effect of the intercostal muscles in the dog. J. Appl. Physiol. 75, 2636.
Wilson, T.A., Rehder, K., Krayer, S., Hoffman, E.A., Whitney, C.G., and Rodarte, J.R. (1987) Geometry and respiratory displacement of human ribs. J. Appl. Physiol. 62, 1872.

Recommended Readings

Aubier, M., and Viires, N. (1998) Calcium ATPase and respiratory muscle Function. Europ. Resp. J. 11, 758–766.
Celli, B. (1998). The diaphragm and respiratory muscles. Chest Surg. Cl. N. Amer. 8, 207–224.
Dumas, J.P., Bardou, M., Goirand, F., and Dumas, M. (1998) Assessment of respiratory muscle function and strength. Postgrad. Med. J. 74, 208–215.
Gandevia, S.C., Allen, G.M., Butler, J.E., Gorman, R.B., and McKenzie, D.K. (1998) Human respiratory muscles: sensations, reflexes and fatiguability. Clin. Exptl. Pharmacol. Physiol. 25, 757–763.
Hussain, S.N. (1998) Respiratory muscle dysfunction in sepsis. Mol. Cell. Biochem. 179, 125–134.
Supinski, G. (1998) Free radical induced respiratory muscle dysfunction. Mol. Cell. Biochem. 179, 99–110.

9
Mechanics of Respiration

BERNHARD F. MULLER

Altered pulmonary mechanics leads to many of the pulmonary symptoms encountered by physicians. An understanding of the ideas discussed in this chapter is essential for comprehension of many of the concepts that are presented in this volume. The material is meant to provide basic physiological information in an intuitively understandable way; an intuitive grasp of the complex and dynamic interrelationships that exist in the normal and diseased lung can make deeper understanding easier, and lead to successful diagnostic and therapeutic efforts.

The mechanical characteristics of the lung can be divided into those that are measurable principally with the lung at rest, and those that become important when air is actively moving into and out of the lung. Those with the lung at rest are called *static* characteristics; those with the lung in motion are called *dynamic* characteristics.

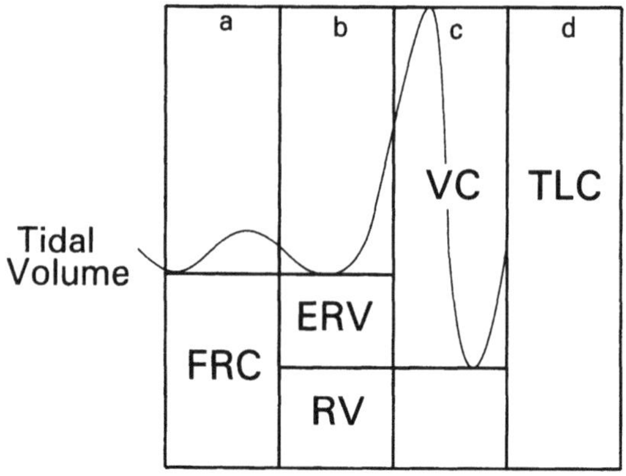

FIGURE 9.1. Schematic depiction of the physiologic lung volumes. VC = vital capacity; FRC = functional residual capacity; ERV = expiratory reserve volume; RV = residual volume; TLC = total lung capacity. (See text for details.)

Static Characteristics

The lung can be divided conceptually into air containing volumes that describe the state of the lung under different conditions of inflation and have little to do with anatomic subdivisions. In Fig. 9.1, each large rectangular block represents the volume of the lung when it is fully inflated at total lung capacity (TLC). The sinusoidal curve represents the volume change during quiet breathing, called the tidal volume (V_T). It is then the volume of air exhaled per one respiratory cycle. To determine minute volume or total ventilation (V_E), the average V_T is multiplied by the breathing frequency. Block (A) of Fig. 9.1 depicts the functional residual capacity (FRC) as the volume of air remaining in the lung at the end of a normal tidal expiration. Block (B) shows that the FRC can be further subdivided into the expiratory reserve volume (ERV) and the residual volume (RV). The RV is that volume remaining in the lungs after a maximum effort to expel as much air as possible from the lungs. Block (C) depicts the vital capacity (VC) as the difference between TLC and RV. Measurement of each of these static lung volumes reveals different aspects of the status of the respiratory system. Note, however, that lung volumes vary among healthy individuals. With the exception of the residual volume, all volumes are measured with a respirometer.

Total Lung Capacity

This is the volume of gas that the lung contains at the time of maximal voluntary inspiration. The inspiratory muscles become shorter as they inflate the chest, thus reducing the maximum force they can exert. At the same time, the force required to inflate the lung against the elastic recoil pressure gets greater. (Note: The lungs are somewhat like rubber balloons in that a force is required to inflate them.) Near TLC, the inspiratory muscles also have to work actively against the chest wall, which has a natural volume somewhat smaller than TLC. At TLC, then, the force required to inflate the lungs and chest wall becomes greater than the force that the inspiratory muscles can generate, and expansion ceases. Thus, TLC is determined by the balance between inspiratory muscle strength and the elastic properties of the lungs and chest wall. The normal TLC

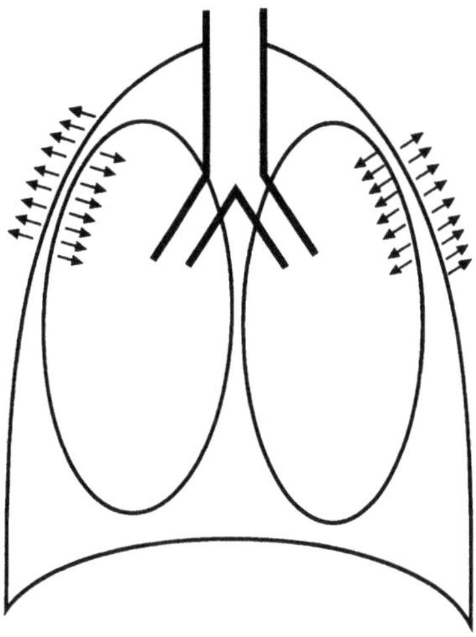

FIGURE 9.2. The thorax at FRC. The tendency of the lungs to collapse exactly balances the tendency of the chest wall to expand. The negative intrapleural pressure keeps the pleural surfaces together.

range is 5–7 L. Variations in measured TLC reflect lungs that are abnormal in size, or stiffness, and also reflect neuromuscular disorders that weaken the muscles of inspiration. An abnormally large TLC is often found in emphysema, a disorder in which the lungs are very compliant and easily inflated. On the other hand, in diffuse interstitial fibrosis, the lungs are stiff, and TLC is generally reduced. Likewise, in neuromuscular disease, TLC is reduced.

Functional Residual Capacity

FRC is the volume of gas that the lungs contain at the end of a normal inspiration–expiration cycle, when the respiratory muscles are at rest (i.e., 2.5 L or 50% of TLC). When the respiratory muscles are relaxed, and in the absence of the lungs' collapsing force, or in a surgical patient with the chest open, the chest wall assumes a "natural volume" that is somewhat larger than FRC, but slightly smaller than TLC. Because the pleural space binds the lung and the chest wall together, the volume of the lungs at FRC is determined by the balance between the lung's propensity to deflate, and the chest wall's resistance to being made smaller than its natural volume (see Fig. 9.2). A very stiff lung, such as is found in pulmonary fibrosis, or a very compliant chest wall, as in multiple rib fractures, will lead to a reduced FRC; a compliant lung as found in emphysema will lead to an increase in the FRC.

Neuromuscular disease, on the other hand, usually has a minimal effect on the FRC because muscular effort generally plays little role in determining the size of the FRC. Exceptions to this rule do exist, such as in asthma, where tonic muscular activity increases FRC, as well as in long-standing neuromuscular disease, where repeated aspiration pneumonia has led to scarred shrunken lungs.

Residual Volume

The RV is the volume of gas remaining in the lungs at the end of a maximal effort to expel all of the air possible. A normal range is 1–2 L or 25% of TLC. In young persons (i.e., less than 20 or 25 years old), the volume of the RV is determined by the balance of opposing forces between the strength of the expiratory muscles on the one hand, and the force needed to collapse the chest wall on the other. For example, neuromuscular weakness will lead to an increase in the RV because the chest wall cannot be compressed as much as normal by the weak muscles. In older individuals, at small lung volumes, small airways collapse, trapping air behind them. This phenomenon, which occurs at progressively higher lung volumes as we age, determines the volume of the lungs at RV rather than a balance of forces.

Vital Capacity

VC is the volume of gas contained in the largest breath possible from full inspiration to full expiration; that is, the difference between TLC and RV (4–6 L). Almost any lung disease can reduce the size of the VC, and the degree of reduction can often help estimate the degree of disability to be expected from that lung disease.

Expiratory Reserve Volume

The ERV is the volume of gas that can be expelled voluntarily from the lungs following a normal inspiratory–expiratory breath; that is, the difference between FRC and RV (2.5 L–1 to 2 L). The ERV is typically reduced in morbid obesity. The RV is determined by the closure of small airways, and the FRC is reduced by the layers of fat on the chest and the abdominal wall pushing up the diaphragm. So, as the obesity worsens, the FRC is reduced while the RV remains fixed; the ERV becomes smaller.

Anatomical Dead Space Volume

This space (V_D) is defined as the volume of inspired air that does not participate in respiratory gas exchange. The nonexchanging air is the air remaining in the upper conducting airways; that is, in the nasopharynx, larynx, conducting airways (i.e., trachea, bronchi, and bronchioles), and unperfused alveolae at the end of inspiration. Note, however, that the inspired air is warmed and humidified during passage in the conducting airways. Note, too, that the presence of ventilated unperfused alveolae means that it is a component of total dead space; hence, we can write:

$$V_D \text{ physiol} = V_D \text{ anat} + V_D \text{ alv}$$

where the physiological dead space is the sum of the anatomical and alveolar dead spaces. Measurement of V_D is carried out by a single inspiration of 100% O_2, followed by measuring the concentration of N_2 in the exhalation as a function of exhaled volume. A plot of exhaled volume versus exhaled N_2 concentration (FEN_2) is sigmoidal in shape. The initial phase represents air from the anatomic dead space, which is free of N_2. This space is distinguished from the alveolar space by drawing a vertical line through the midpoint of the curve (the transition phase) and extrapolating the points back to the abscissa. In an adult male, the value of V_D anat is in the 150–180 ml range.

In sum, then, the air entering the alveoli during inspiration includes the dead space air from the preceding breath and a portion of the new air. This does not initially alter alveolar P_{O_2} and P_{CO_2} until the new or fresh air reaches the alveoli. It is the fresh air (of the tidal volume) that raises alveolar P_{O_2} and lowers alveolar P_{CO_2}. It is the last dead space part of the fresh air that remains in the conducting airways at the end of normal inspiration. Thus, this fresh air is wasted.

Lung Distensibility

The lungs and chest wall are likened to two interconnecting balloons, one inside the other. The three pressures in such a system are the alveolar pressures (P_A), the intrapleural pressure, PP_L, and the atmospheric (outside) pressure, P_O. The difference between P_A and P_O represents the transmural respiratory system pressure, whereas the difference between PP_L and P_O represents the transmural chest wall pressure. As for the intrapleural pressure, it is a negative pressure that holds the visceral pleura (outside lining of the lungs) and parietal pleural (inside lining of the chest wall) closely together but separately by a film of pleural fluid.

The lungs, chest wall tissue, and the respiratory muscles possess elastic properties that are like those of a balloon. That is, an inflated lung has the tendency to recoil which can be stopped by having a distending pressure difference between P_A and PP_L. Such a pressure or force is provided by the chest wall and the respiratory muscles (see p. 174 and Fig. 9.12).

Compliance and Elastance

Imagine a bellows such as that depicted in Fig. 9.3. Blowing air into these bellows will increase their volume, stretching the springiness that returns the bellow to the closed state when air is released from them. As more air is forced into the bellows, they are stretched and the pressure inside as measured by the manometer rises (in contrast to the pressure decrease that occurs when the bellows are inflated by the more traditional use of the handles). As the air is released from the bellows, the pressure drops until the bellows are fully contracted; a small amount of air remains inside the residual volume.

If the volume of the bellows is changed repeatedly by a small amount, and the change in the pressure that occurs across the wall of the bellows is measured, a series of volume and pressure changes can be plotted, as shown in Fig. 9.4.

Compliance (C)

The slope of this volume pressure line, viz., the change of volume (Δ), divided by the change of pressure (Δ) is termed the *compliance* of the system; that is, compliance, C, is given by the expression $C = \Delta V/\Delta P$ in units of liters per centimeter of water (absolute volume). The dashed line in Fig. 9.4 represents a bellows with a lower compliance than that represented by the solid line. A compliance for the lungs can be measured in a similar manner.

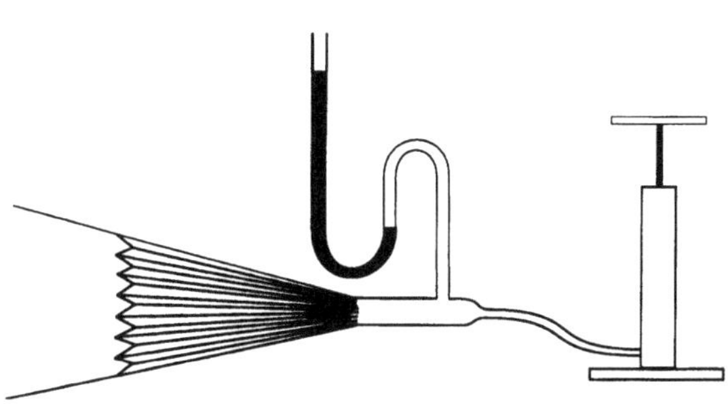

FIGURE 9.3. The bellows as a lung analogy. (See text for details.)

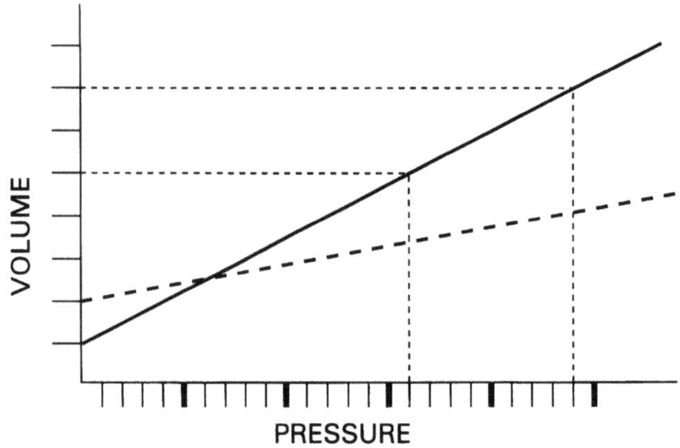

FIGURE 9.4. The volume pressure curves of two different compliant systems. (See text for details.)

Elastance (E)

The reciprocal of the compliance is called the *elastance*; that is, $E = \Delta P/\Delta V$. It is important to realize that elastic behavior and compliant behavior are merely opposite sides of the same coin, and that a highly elastic structure has a low compliance, whereas a highly compliant structure has a low elastance.

Elastic Recoil Pressure

ERP refers to the specific pressure between the inside (P_A) and the outside of the lung (PALV) at the volume in question, whereas the term *elastance* refers to the reciprocal of the compliance. The elastic recoil pressure of the unit depicted by the solid line in Fig. 9.4 at 7 L volume is 24 pressure units. The elastic recoil pressure of this same unit at 5 L volume is 16 pressure units. The elastance, then, is 24 – 16 = 8 pressure units change divided by 7 – 5 = 2 liters change: 4 pressure units/L. The compliance is 0.25 L change/pressure unit change. The dimensions of compliance are volume divided by pressure; in pulmonary physiology, volume is measured in liters and pressure in centimeters of water; therefore, the dimensions of compliance would be liters per centimeter of water.

It is useful to remember that the term *elastic* refers to the tendency of an object to restore itself to its initial state after being deformed, like an elastic (rubber) band. A tennis ball is thus more elastic than a foam rubber ball, and the foam rubber ball is more compliant than the tennis ball. Diseases in which collagen is laid down in the lung (e.g., sarcoidosis) lead to an increase in the elastic recoil or elastance of the lung with a concomitant decrease in the compliance. Emphysema, with its loss of elastic fibers, leads to a loss of elastance and an increase in the compliance of the lung.

Parallel and Series

The terms *parallel* and *series* describe two ways of arranging interacting structures. In a parallel arrangement, additional structures can all be filled without having to exert additional pressure (e.g., the two lungs are in parallel). In a series arrangement, the structures are so arranged that the pressure needed to fill them increases with the addition of additional structures (e.g., the lung and the chest wall are in series). The lung and chest wall combination requires more pressure to expand than either the chest wall or the lung alone would require.

Parallel

The system in Fig. 9.5(A) illustrates compliant structures (depicted schematically as balloons) arranged in parallel. A parallel arrangement of compliant or elastic structures is an arrangement in which the compliance of the combined structures is equal to the sum of the individual compliances of each part. The pressure in the two balloons/compliances clearly must be equal; otherwise, air would just flow from the higher pressure to the lower pressure device until the pressures are equal. Now, assume the compliances of the two devices are 2 and 3, as indicated in Fig. 9.5(A). If five additional liters are added to the system, the gas would distribute itself according to the compliances of the individual devices and the equation:

$$\text{Volume change} = \text{Pressure change} * C.$$

Thus, the compliance 3 device would receive 3 L, and the compliance 2 device would receive 2 L (note, carefully, that the 5 L distributed itself in this way as a result of the fact that the pressures in the two balloons are equal). According to the foregoing equation, the pressure in each device would rise by 1 pressure unit, and the pressure in the entire system would also rise by 1 unit. The total compliance of the entire system consisting of the two units in parallel can be calculated from the compliance equation as a volume change of 5 divided by the pressure change of 1 for a total compliance of 5. This is the same value we would obtain by adding the compliances of the component parts of the system. The two lungs are in parallel, and the total pulmonary compliance is equal to the sum of the compliances of each lung alone.

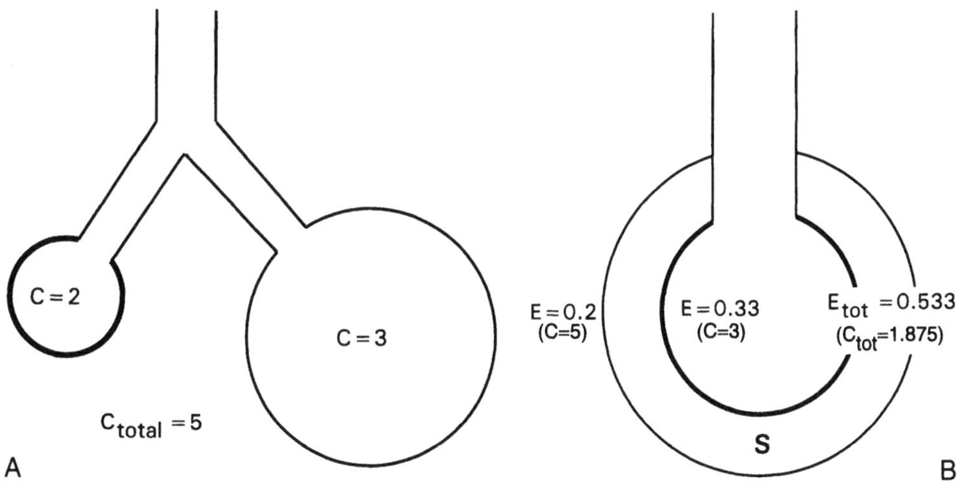

FIGURE 9.5. (A) Parallel lung units. The total compliance is equal to the sum of the compliances of the individual units. Examples of parallel lung units are: the two lungs are in parallel with one another, and the upper and lower lobes. (B) Lung units in series. The total elastance is equal to the sum of the elastances of each individual unit. An example of series units is: the lungs and the chest wall are in series with one another.

Series

The structures shown in Fig. 10.5(B) are effectively in series; in a series arrangement the elastances (the reciprocal of the compliances) of each structure must be added together to calculate the total elastance of the system. Assuming that the compliance of the outer is 5, its elastance is 0.2; the compliance of the inner structure is 3, and its elastance is 0.333. The total elastance then is 0.333 + 0.20, or 0.533. The overall compliance is then 1/0.533 or 1.875 [it might be amusing to calculate the total elastance of the combined structures in Fig. 9.5(A); the answer is 0.2]. The pressure across the inner structure is given by the inside pressure minus the pressure in space, S. The pressure across the outer structure is given by the pressure in space, S, minus the outside pressure. The total pressure across both structures is the pressure difference between the inside of the inner structure and the outside of the outer one. If we add 15 L of air to the inside balloon, both structures must stretch by that volume. In other words, the pressure across the outside structure increases by 15/5 pressure units, and the pressure across the inside structure increases by 15/3 pressure units. The total pressure across both structures increases by 15/5 + 15/3, or 8 pressure units; therefore, the compliance of the entire system is given by 15/8, or 1.875, and the elastance is 8/15 or 0.533. Note that this is a smaller value of compliance than either structure alone has. The lungs and the chest wall are effectively in series; therefore, the elastances of the lungs and the chest wall add, making the total elastance greater than the elastance of either one alone. Of course, the compliance of the lung/chest wall system is smaller than the compliance of either the lungs or chest wall alone.

Surface Tension

General Considerations

When different phases such as water and air are in contact, an interface is found between them. This is a surface whose properties are unlike those of water or air—the bulk phases. The surface differs from the bulk phase because it represents a boundary whose outside is not a bulk phase, but its inside is (anisotropy). In other words, the surface is not surrounded by a bulk phase of water molecules.

Now, the force in the surface tries to limit the surface area to a minimum. This force called the *surface tension* is either reduced or slightly increased by adding solute to the bulk solution. A solute that decreases surface tension concentrates at the surface (Gibbs absorption isotherm). A textbook example are the surfactants in the lungs (Chapter 3) where they increase the surface area available in the alveolar-capillary membrane for gas exchange. The surface tension of water in air at 298 K is about 72 $dyn\,cm^{-1}$ or $m\,N\,m^{-1}$.

Special Considerations

If an excised lung is filled with air, then a volume pressure curve can be constructed, as indicated by the dashed line in Fig. 9.6. The slope of this volume pressure curve is the compliance of the lung. Notice, however, that the slope is not constant; therefore, the compliance of the lung varies with its volume. Pulmonary physiologists usually measure the compliance of the lung at FRC to avoid measurement variation due to volume change alone. Notice also that when the lung is filled with saline rather than with air, as indicated by the solid line in Fig. 9.6, the slope of the curve (the compli-

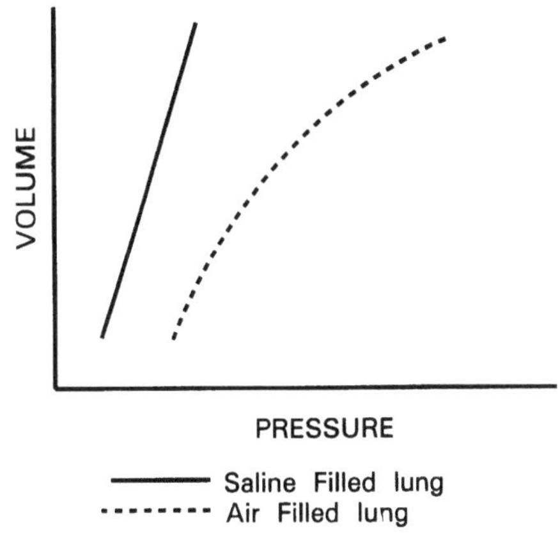

FIGURE 9.6. Volume pressure curves for air-filled and saline-filled lungs. (See text for details.)

ance) increases markedly. This increase of compliance in the saline-filled lung is taken to mean that the compliance (and elastance) in the air-filled lung is determined to a large extent by the presence of an air–water interface. The alveoli are lined by alveolar fluid, which is mainly water. Filling the alveoli completely with water eliminates the interface, and decreases the elastance (increasing the compliance) of the lung.

Compliance or distensibility is readily estimated by perusing plots of the recoil pressure–volume relation for the ventilated lung. The relation is expressed by the slope of the curve (i.e., $\Delta V/\Delta P$). A decline in the slope occurs as lung volume approaches total lung capacity. The reason for this is that elastin reaches its limit of distensibility. A textbook example is the patient with interstitial fibrosis (sarcoidosis or idiopathic pulmonary fibrosis). The lung is less compliant and stiff as the result of fibrosis and loss of elastin. In contrast, the patient with emphysema has a highly compliant and distensible lung resulting from progressive dilatation and obliteration of the alveolar walls without the occurrence of fibrosis.

The Law of LaPlace

The LaPlace relationship describes how the pressure inside a spherical container such as an alveolus may be influenced by surface tension. LaPlace's law states that the pressure inside an elastic hollow sphere is equal to twice the tension in the wall divided by the radius of the sphere (Fig. 9.7). That is,

$$\Delta P = 2T/r$$

where ΔP = the pressure difference across the container surface
r = the radius
T = the tension in the wall

In a soap bubble, as in a rubber balloon, the tension in the wall generates the pressure inside. In a soap bubble, as in an alveolus at small volume, the tension in the wall (and hence, the pressure inside the alveolus) is caused mainly by the surface tension of the air–liquid interface. The value of T is independent of bubble size because surface tension is dependent only on the chemical nature of the fluid, not on how much it is stretched. Therefore, pressure inside such a sphere is inversely proportional to its size. Given the two unequal sized air-filled alveoli shown in Fig. 9.8, it follows that the smaller has a greater pressure inside than the larger one, and that it would empty its contents into the larger alveolus. If all small alveoli tended to empty into larger

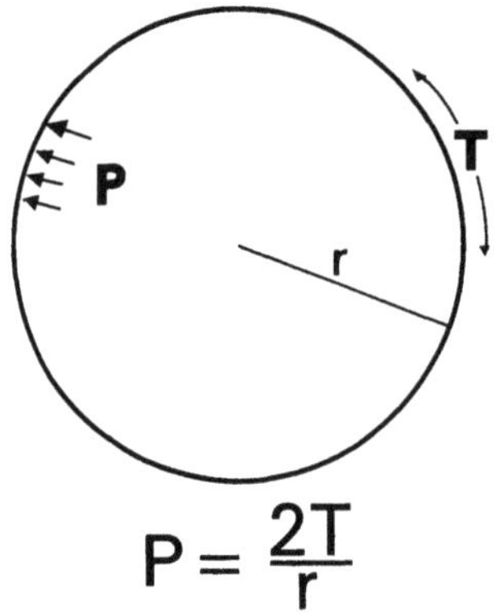

FIGURE 9.7. Schematic depiction of the LaPlace relationship in a sphere.

ones, the lung would eventually consist only of a very few distended alveoli, with the majority in the collapsed state, which is a situation incompatible with physiologic respiration. Such an unstable situation would not occur in a system that consists of interconnected rubber balloons because the tension in the wall increases as the balloon gets larger in a rubber balloon. Thus, the tendency would be to increase the pressure inside and to counteract the tendency of the smaller balloon to empty into the larger. In the lung, however, an unstable situation does not occur at the end of expiration; smaller alveoli have a reduced surface tension and do not collapse into the larger alveoli. This is because alveolar stability is due to the presence of surfactant.

As described by Harwood and his colleagues in Chapter 3, surfactant is a unique chemical that is synthesized and secreted by the Type-II alveolar cells. Surfactant lowers the surface tension of the alveolar fluid much as a detergent might, reducing the effort required to inhale. Surfactant also has the unique properties of changing its surface tension characteristics as the surface is stretched (Fig. 9.9). The dashed line represents the surface tension of a water (liquid)–air interface. Adding a detergent to the water decreases the surface tension, as is indicated by the dotted line. The solid line represents the surface tension of a water–surfactant solution, and, as can be seen, it varies with the area of the alveolus. As an alveolus increases in size, the internal pressure tends to decrease according to the LaPlace relationship; however, when the alveolar fluid contains surfactant, the surface tension is no longer constant; rather, it rises as the surface area of the alveolus is increased, thereby opposing the effect of the increasing radius; the internal pressure no longer goes down. In other words, the pressure inside small alveoli is no longer greater than that in larger alveoli; hence, the lung

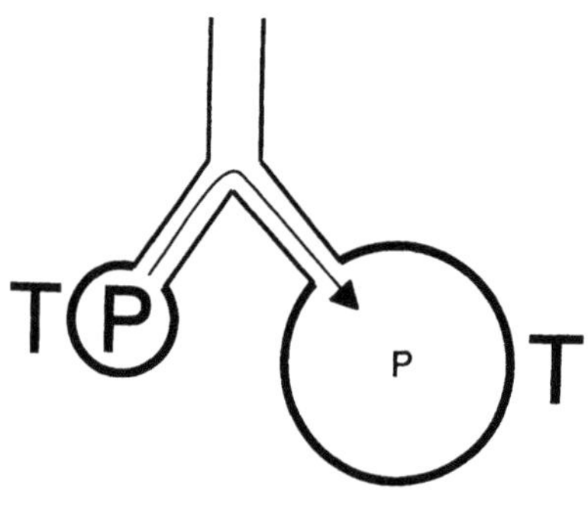

FIGURE 9.8. Air filled alveoli of unequal size. In the absence of surfactant, the smaller tends to empty into the larger.

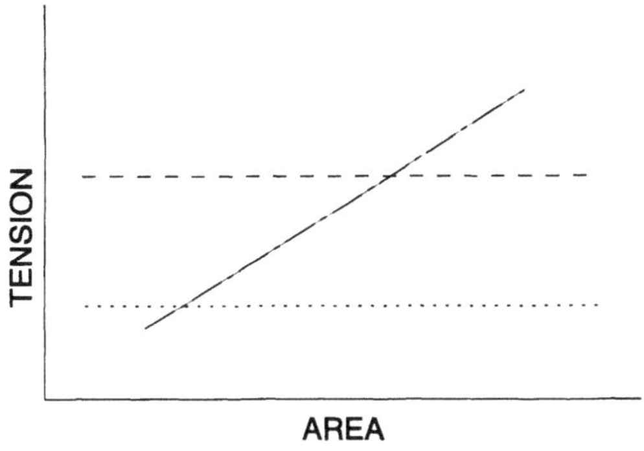

FIGURE 9.9. Surface tension vs area plots for water (dashed line), detergent solution (dotted line), and surfactant solution (dash dot line).

is stable (Fig. 9.10). In addition, surfactant is able to maintain dryness of the alveoli by preventing fluid transudation from capillaries to alveoli.

Hysteresis

The phenomenon exhibited by the lung in which the inflation–deflation volume–pressure curves are not identical is called *hysteresis* (Fig. 9.11). The graphs of volume–pressure curves shown up to this point in the discussion have included only the deflation branch of the complete curve. In clinical practice, only this portion is taken into account when measuring compliance.

Surfactant also displays hysteresis whereby the area–pressure relationships during stretching and shrinking are not alike. As it turns out, surfactant is responsible for almost all the hysteresis the intact lung exhibits. This is borne out by the fact that little or no hysteresis is seen in a saline filled lung. It is a moot point whether or not the hysteresis associated with surfactant is a significant contributor to lung stability; however, one could argue conceptually that hysteresis might be beneficial. For an alveolus trying to empty into another would find itself on the lower pressure deflation part of the curve, whereas the alveolus about to be inflated would find itself on the higher pressure inflation branch. Thus, the alveolus trying to empty would be at a lower pressure than the alveolus trying to fill.

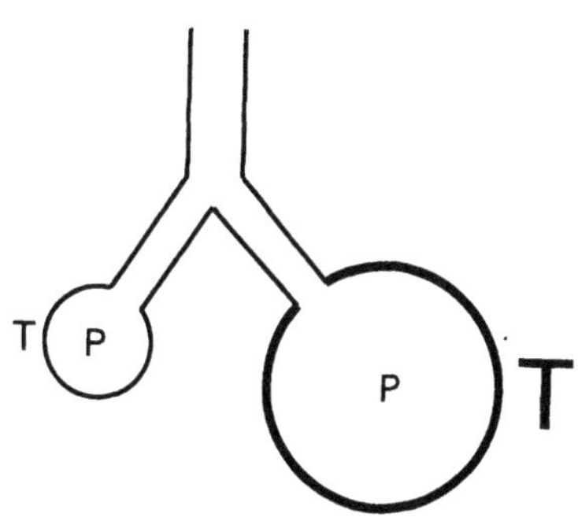

FIGURE 9.10. The increasing surface tension with size characteristic of surfactant helps to stabilize alveoli of unequal size.

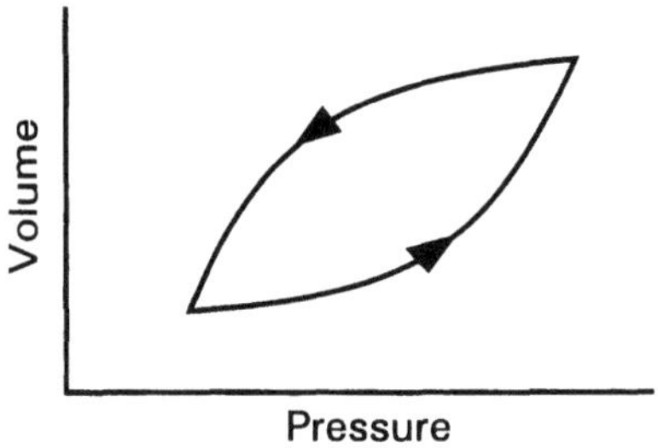

FIGURE 9.11. Volume pressure curve of normal lung. The phenomenon of differing curves on inflation and deflation is called *hysteresis*.

Respiratory Distress Syndrome of the Newborn

This syndrome, which is known as *hyaline membrane disease*, is marked by poor alveolar–capillary membrane gas exchange and ventilatory failure caused by a deficiency in surfactant. The mere fact that surfactant replacement therapy in these prematurely born infants dramatically reduces the mortality rate is indirect but compelling evidence that the syndrome is due to surfactant deficiency. The hallmarks of surfactant deficiency include stiff and hard-to-inflate lungs, increased work of breathing, widespread microatelectasis, alveoli filled with transudate, and hypoxemia. The resulting pulmonary vasoconstriction and increase in pulmonary resistance leads to hypoperfusion and, hence, to capillary damage and necrosis of the alveoli. Fluid, in turn, leaks into both the interstitial and alveolar spaces, and hyaline membrane is formed. The hypoxemia is rendered worse through right-to-left shunting (patency of the foramen ovale and ductus anteriosus), which results from pulmonary collapse and elevated pulmonary artery pressure.

Chest Pressures

The various pressures in the thorax that relate to lung expansion are depicted in Fig. 9.12. The relationship between these pressures is given by the expression:

$$\text{Palv} = \text{Pel} + \text{PP}_l$$

where Palv = the pressure difference between the alveolus and the body surface
Pel = the elastic recoil pressure of the lung
PP_l = the pleural pressure referred to the body surface (atmospheric)

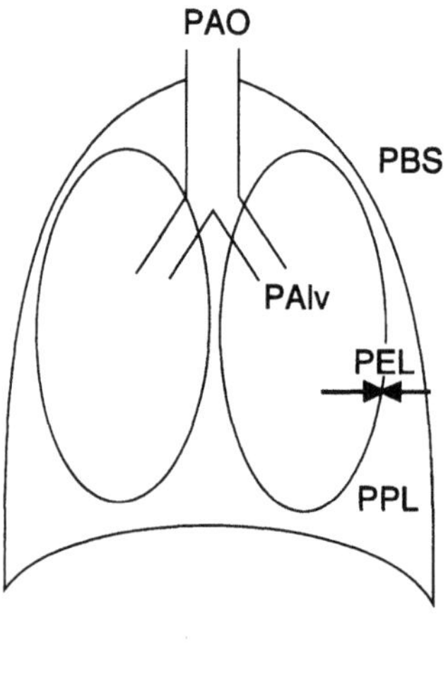

FIGURE 9.12. Important thoracic pressures. PBS = barometric pressure at the body surface; PAO = the pressure at the airway opening; Palv = the pressure inside the alveolus, and is usually expressed as a pressure difference from the body surface; PEl = the pressure difference between the inside and outside of the lung (in this case, the pressure outside the lung is equal to the pleural pressure); PPl = the pleural pressure, and is usually expressed as a pressure difference from the body surface. (See text for explanation.)

Pel is measured as the pressure difference between the inside of the alveolus and the outside of the lung. Because the elastic tissue of the lung together with surface tension forces make the intra-alveolar pressure higher than the surrounding pressure, Pel is always a positive number. PP_l is usually negative during quiet breathing, but can become positive during forced expiration.

Dynamic Characteristics

Ohm's Law

The flow of air or water through a pipe is determined by a variant of Ohm's law which states that the flow is proportional to the driving pressure, and inversely proportional to the resistance to flow:

$$V = (P_1 - P_2)/R$$

where V = flow, P_1 = the pressure at the inflow to the pipe, P_2 = the pressure at the outflow, and R = resistance [Fig. 9.13(a)].

The driving pressure is the pressure difference between the inlet to the resistor and the outlet. In pulmonary physiology, driving pressure is usually measured in cm water and the flow rate in liters per second. The units of resistance are centimeters of water per liters per second. Flow always occurs from a region of greater pressure to one of lower pressure and is proportional to the pressure difference. If both ends of the resistor are at the same pressure, no flow will occur. The converse also is true: with gravity ignored, an unblocked pipe with no flow has a pressure drop of 0cm water.

Poiseuille's Law

While Ohm's law is very useful in understanding and describing the function of the airways, it is something of a simplification. The actual cause of the resistance referred to in Ohm's Law is based in thermodynamics. During conditions of laminar flow (e.g., in airways less than 2mm in diameter), which is generally slow and smooth, the resistance to flow is influenced by the length and size of the pipe and the viscosity of the fluid. That is,

$$R = Kl\eta/r^4$$

where R = the resistance, K = a proportionality constant, l = the length of the pipe, η = the viscosity of the fluid, and r = the radius of the pipe.

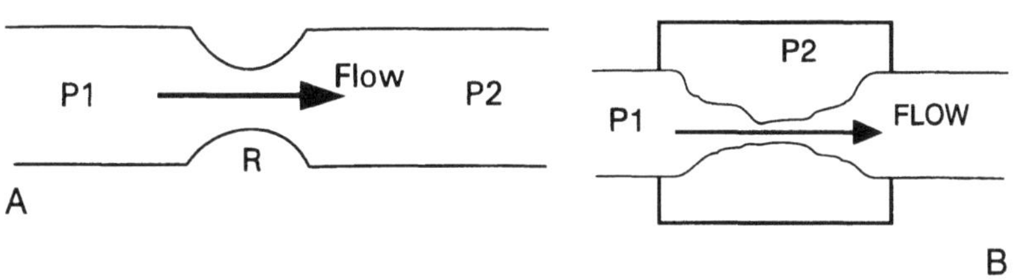

FIGURE 9.13. (A) Schematic representation of the factors determining flow in a pipe. R is resistance; and P_1 and P_2 are the pressures at the inlet and outlet of the pipe, respectively. (B) In a Starling Resistor, flow is determined by the pressure difference between the inlet pressure (P_1) and the pressure in the surrounding chamber (P_2).

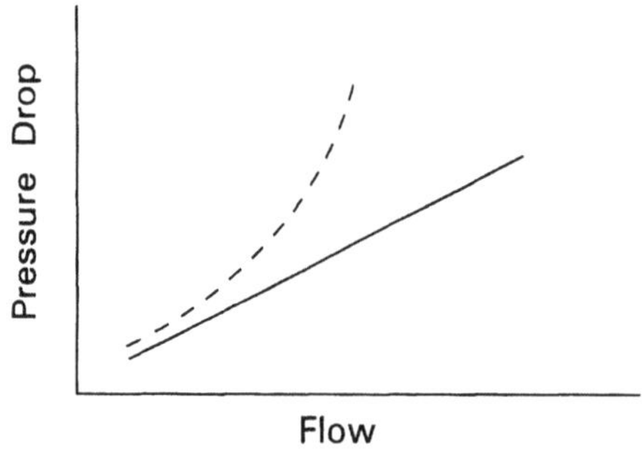

FIGURE 9.14. When flow is laminar, the flow rate is directly proportional to the pressure difference between the inlet and outlet of a pipe (solid line.) When flow is turbulent, a greater pressure is required for a similar flow (dashed line.)

The fact that the radius of the pipe is raised to the fourth power means that small changes in the diameter of the airways lead to large changes in the resistance to flow. For example, halving the diameter leads to a 16-fold increase in the resistance to flow. A similar mechanism occurs in blood vessels; the body uses this characteristic to regulate blood flow and air flow to the appropriate tissues.

Turbulent Flow

Under conditions of turbulent flow (e.g., in the trachea and upper bronchi), which is the actual state of affairs in most regions of the lungs, the resistance becomes even more dependent on the radius of the pipe, and somewhat dependent on the flow rate as well. As shown in Fig. 9.14, when flow is turbulent the pressure required to sustain a certain flow rate rises much faster than under laminar conditions. Even small amounts of bronchospasm can lead to large increases in the work of breathing. Bear in mind that the driving pressure, ΔP, for turbulent flow is considerably larger than for laminar flow.

Starling Resistors

The Starling Resistor consists of a collapsible tube placed inside a pressurizable chamber [Fig. 9.13(b)]. As the pressure inside the chamber is increased, the tube is compressed, and the resistance to flow through the tube is increased. It can be shown that in the case of a starling Resistor, the pressure drop of Ohm's law is between the inlet pressure, and the pressure inside the chamber pressing off the tube. If the inlet pressure is much greater than the chamber pressure, a large flow can occur. As the chamber pressure is increased, the total flow is decreased proportionately, until the chamber pressure is made equal to or greater than the inlet pressure, when flow is shut off entirely. Many bodily systems are influenced by the Starling Resistor effect: Blood flow is modified by tissue pressure that acts like the chamber pressure; air flow is modified by pleural pressure, which also acts like the chamber pressure in a Starling Resistor.

Series Resistors

If two resistors are taken and connected in series so that the outflow from one flows into the inflow of the next one (Fig. 9.15), the total resistance of the system is equal to

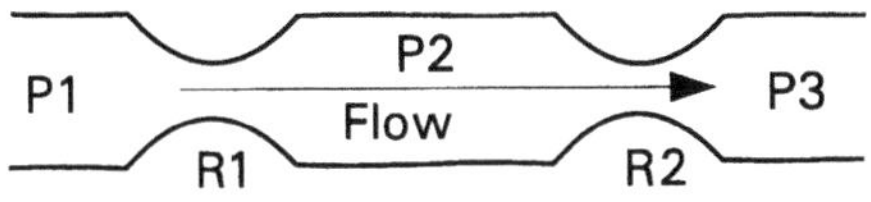

FIGURE 9.15. The total resistance to flow is equal to $R_1 + R_2$. See text for further explanation.

the sum of the two resistances. If we now add a third resistor, the total resistance is equal to the sum of all three. This is a situation similar to the upper airway, where the resistance of the nose, pharynx, and the trachea are in series, and add to one another to form part of the total respiratory resistance. It is important to note that the flow through each of the resistors in a series arrangement is exactly the same as through all the others, even if the resistances are different. This becomes intuitively obvious when we recognize that the total output of the first resistor in line must pass through the second, and that no air is added or lost en route. Because the flow through all the resistors is the same, we can calculate the pressure drop across them (Fig. 9.16). We assume P_4 to be 0, or ambient. Because flow = $(P_3 - P_4)/R$ (Ohm's Law), and then, if flow = 30, and R = 8, we obtain $30 = (P_3 - 0)/8$; hence, P_3 is calculated to be 240.

If we go through a similar exercise, we find that P_2 is 60 pressure units greater than P_3, or 300 pressure units. P_1 is 150 units greater than P_2 and is therefore 450 pressure units. Furthermore, the total pressure drop through the pipes in Fig. 9.16 is $P_1 - P_4$ or 450 pressure units. The flow is given as 30, so by Ohm's law again, the total resistance is 450/30 or 15 units. This corresponds to 5 + 2 + 8, as described earlier.

The flow through series resistors is the same in each resistor, but the pressure drop across each resistor varies proportionately to the resistance in the resistor; the largest pressure drop being across the largest resistor, and the smallest across the smallest resistor. Because that portion of the airway with the largest resistance will dissipate the greatest pressure drop (and patients with sleep apnea can have a very high upper airway resistance), a very high respiratory effort may be required to generate enough pressure to pass sufficient air through the upper airway for respiration. Notice also how the airway pressure diminishes downstream. In the airway, during forced expiration, the pressure is greatest in the alveolus, and becomes progressively less toward the mouth, until it is 0 or ambient at the mouth.

Parallel Resistors

Imagine a large swimming pool, and it is spring, so the thing must be filled. It will take a very long time to fill it from the garden hose because the resistance of the hose and the pressure available will not generate a large enough flow rate. Is there a way to

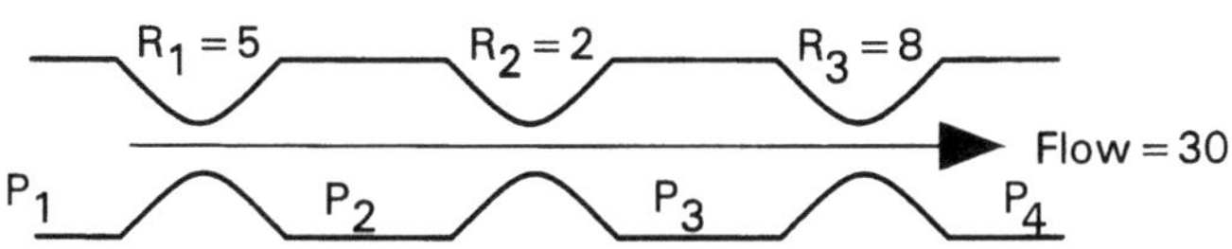

FIGURE 9.16. Three resistors in series. (See text for explanation.)

speed up the process? One that comes to mind is to connect hoses to several outside faucets at a time, or better yet, attach several hoses to faucets on the homes of all your neighbors! Each additional hose you attach will increase the flow rate available to fill your pool. The hoses are now connected in parallel, which is a connection that increases flow, in contrast to the series connection (all the hoses connected end to end), which serves to decrease flow if pressure remains constant.

Just as the reciprocal of elastance is compliance, the reciprocal of resistance is conductance, for which the symbol is G (because C is already taken up by the symbol for compliance). All resistors are also conductors, and all conductors are also resistors; however, a good resistor is a poor conductor, and a good conductor is a poor resistor. In a series connection, the resistances of the individual units are added together to calculate the total resistance. In a parallel connection, however, conductances of the individual units are added together to calculate the total conductance. In Fig. 9.17, the two pipes are connected in parallel. The resistance through the upper branch is R_1, and the conductance through that branch is $G_1 = 1/R_1$. Flow through the upper branch is given by $Flow_1 = G_1 * (P_1 - P_2)$. The total flow through both pipes is simply the sum of the two flows, viz.,

$$\text{Total Flow} = \text{Flow}_1 + \text{Flow}_2$$

$$\text{Total Flow} = (P_1 - P_2) * G_1 + (P_1 + P_2) * G_2$$

$$\text{Total Flow} = (P_1 - P_2) * (G_1 + G_2)$$

Because the total conductance is given by

$$G_{total} = \text{Total Flow}/(P_1 - P_2)$$

$$G_{total} = G_1 + G_2$$

The total conductance in a parallel system is then simply the total of all the conductances, just like the total resistance in a series system is equal to the sum of all the series resistances.

When conductors are connected in parallel (Fig. 9.17), the conductances add, and the total resistance decreases. Note that in a parallel system, the flow through each branch is proportional to the conductance in that branch. The resistance of the individual small airways is very high because they have such a small radius. There are very many of them, however, and they are all functionally in parallel, so the total airways resistance is small enough that reasonable pressures generated by normal muscles can move sufficient air for respiration.

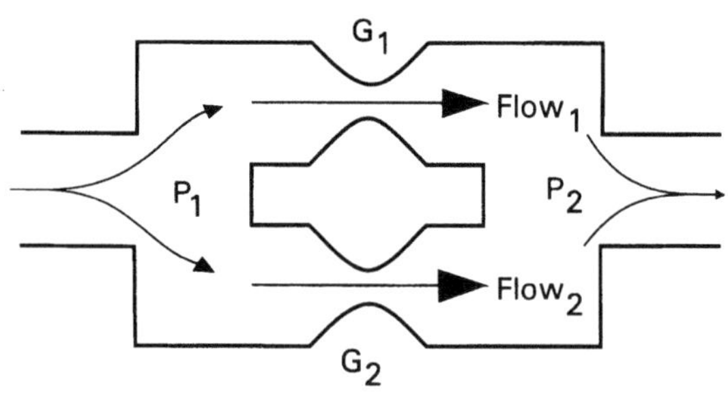

FIGURE 9.17. Two conductors in parallel. G_1 and G_2 are equal to $1/R_1$ and $1/R_2$ respectively. (See text for explanation.)

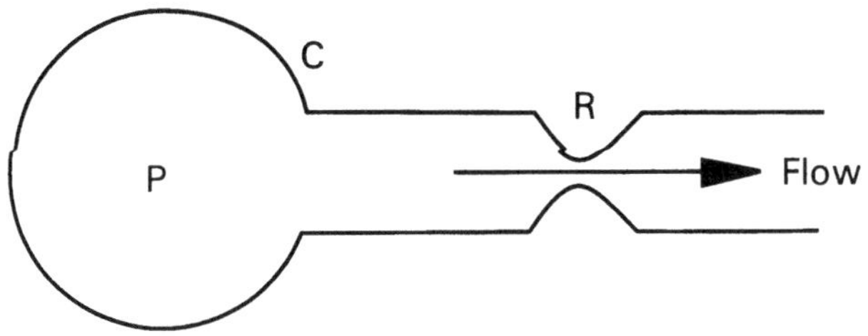

FIGURE 9.18. A compliant structure emptying through a resistor.

Time Constant

Imagine a balloon emptying through a resistor (Fig. 9.18). How long will it take for the balloon to empty? At first, the balloon will empty with a flow rate given by the pressure difference between the inside of the balloon and the outside, divided by the resistance. As the balloon empties, however, the pressure inside it decreases according to the change in volume divided by the compliance. The smaller pressure will cause a smaller flow rate through the resistor, so, as the balloon gets smaller, the flow decreases as well. In theory, at least, the balloon can never completely empty itself. Systems such as this have an exponential filling and emptying curve, and can best be described using a number called the *time constant* (Fig. 9.19), which is given by

$$\text{Time Constant} = RC$$

where R = resistance and C = compliance

The time constant, appropriately enough, has the units of time: The units of resistance are pressure/(volume/time); compliance is volume/pressure. Multiplying the two gives a product of time. In a time interval equal to one time constant, the balloon will empty to 37% of its initial volume. In another such interval it will empty to 37% of the volume at the end of the first interval, or 13.5% of the initial volume, and so forth. Lung units that have a long time constant will fill and empty slowly, and units that have

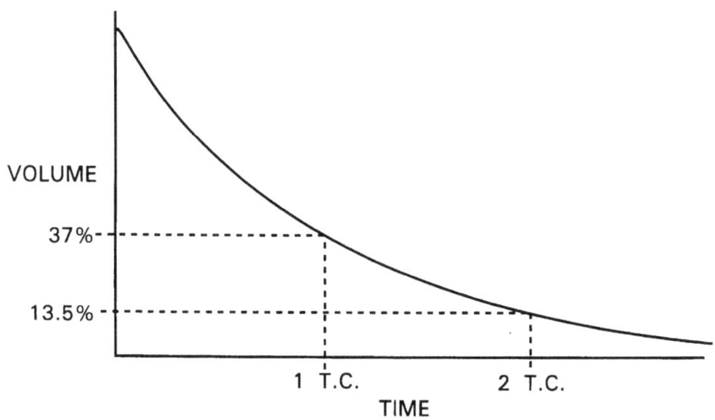

FIGURE 9.19. Exponential volume–time curve of the unit in Fig. 9.18 showing two time constant intervals.

a short time constant will fill and empty quickly. In diffuse interstitial fibrosis, the lungs and their subunits are stiff (i.e., they have a low compliance). Because the resistance is usually not changed greatly in this condition, the time constant of the lung units will be small, indicating rapid emptying and filling. By contrast, in emphysema the lungs are very compliant due to loss of elastic tissue. Airway resistance is often near normal. Due to the greatly increased compliance of all the individual lung units, however, their time constants are long. In chronic bronchitis, the elastic tissue is nearly normal, but the resistance of the smaller airways is greatly increased; therefore, the time constants of these lung units are once again increased, leading to slow filling and emptying. If these effects happened uniformly, there would be a minimal increase in the A–a gradient. As we will see, however, the damaged lung is rarely uniformly damaged, and the nonuniformity leads to problems.

Distribution of Ventilation

Even in normal lungs, ventilation is not uniformly distributed, but favors the bases of the lungs. There are both static as well as dynamic reasons for this nonuniformity. During full inspiration [see Fig. 9.20(A)], all the alveoli in the lung are fully distended; however, at maximal expiration [Fig. 10.20(B)], the alveoli at the bases are fully deflated, while those at the apex are still partially inflated. The volume change, hence ventilation, of the basal alveoli is thus greater than that at the apex. It is not fully understood why this happens, but a pleural pressure gradient caused by gravity, the shape of the chest wall, and regional compliance variations all probably contribute to the cause.

In terms of dynamics, variation in time constants can cause scattered regional areas of nonuniform ventilation, and is an important cause of hypoxemia seen in diffuse lung disease. In addition, respiratory muscles play a role in where ventilation goes. If inspiration is limited to the diaphragm, the bases are preferentially ventilated. If the accessory muscles of the shoulder girdle are used, then more of the ventilation goes to the apex of the lung.

Flow Limitation

If a normal individual undertakes a series of forced expiratory maneuvers measured by a spirometer, with each maneuver being done at a greater effort while flow rate measurements are made at specific lung volumes, then a flow rate plateau occurs after

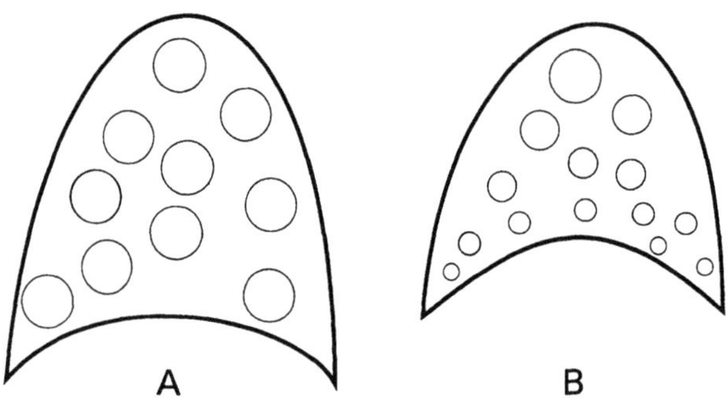

FIGURE 9.20. The lung units at the bases change volume to a greater degree than units at the apices during a respiratory cycle. A represents the lung at TLC, and B the lung below FRC.

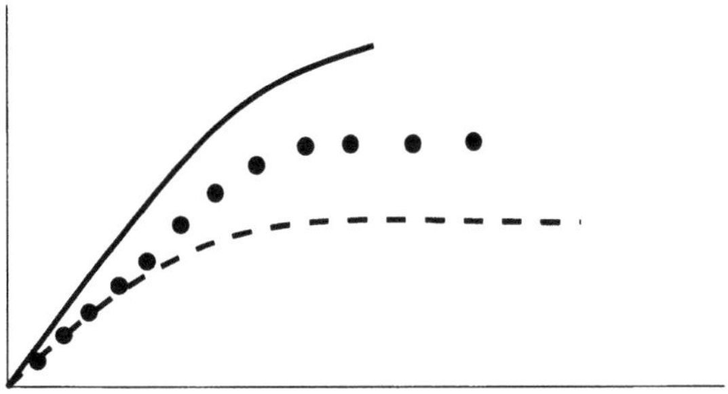

FIGURE 9.21. Isovolume flow/pressure curves. Pleural pressure varies with expiratory effort. The dotted curve is measured near FRC, the solid curve near TLC, and the dash curve near RV.

a certain amount of effort (Fig. 9.21). At low levels of effort, successive breaths result in an increase in flow rate as effort is increased. Above a certain level of effort, however, increases in effort do not result in an increase in flow rate at any but the largest lung volumes in normal people. This maximum effort independent flow rate is called V_{max}, and several models have been proposed to explain its existence. In patients with various kinds of obstructive lung disease, the airflow has a similar plateau effect, but at lower flow rates than in normals.

The EPP Concept

In 1967, Mead et al. proposed the equal pressure point (EPP) model to explain the plateau phenomenon (Fig. 9.22). The model assumes a collapsible airway and an elastic source of pressure (the alveolus). The EPP is the point in the airway where the pressure inside the airway equals the pleural pressure. Remember that the alveolar pressure equals the pleural pressure plus the elastic recoil pressure ($P_{alv} = PP_l + P_{el}$). During forced expiration, the pleural pressure is positive, and the alveolar pressure is greater than the pleural pressure. The pressure at the mouth is defined as 0. Inside the

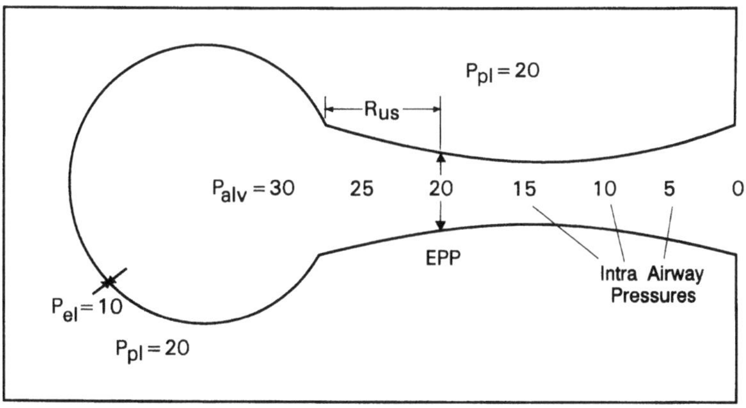

FIGURE 9.22. The EPP model for airflow limitation. The intraairway pressures diminish progressively from 30 at the alveolar end of the airway to 0 at the mouth. The EPP is the point where intraairway pressure equals the pleural pressure. Rus is the resistance of the airway segment from the alveolus to the EPP.

airway, there is a gradient from the alveolar pressure to the mouth pressure, a location exists where the pressure is the same as the pleural pressure, which is called the EPP. The model provides a simple and elegant equation, viz.,

$$V_{max} = Pel/Rus$$

where Vmax = the maximum flow achievable at a given lung volume, Pel = the elastic recoil pressure of a lung segment, and Rus = the resistance of the "upstream" segment of airway between the alveolus and the EPP.

The model suffers, however, from the fact that the EPP does not necessarily correspond to a segment of the airway where the actual flow limitation is occurring, so it does not explain physically what is really happening in the airway. The equation does suggest, however, that both a loss of elastic recoil, as well as an increase in upstream resistance, can have deleterious effects on V_{max}, a fact that is easily observable in patients with emphysema and chronic bronchitis.

The Waterfall Model

The Waterfall model of Pride et al. (1967) presented in Fig. 9.23 attempts to explain what is actually happening. It does so by analyzing the area of collapse (i.e., "the flow limiting segment") in terms of how rigid the airway is, and the internal and external pressures acting upon it. This collapsing segment is a kind of Starling Resistor and equations become complex. Analysis is compounded by the problem of knowing just how rigid the airway is during forced expiration. Furthermore, the static pressures invoked ignore the Bernoulli phenomenon, which tends to reduce intraairway pressures below those predicted on the basis of airway resistance alone. The Bernoulli phenomenon is the physical property of fluid flow dictating that as flow rate goes up, pressure within the fluid goes down. It is the basis of perfume atomizers, and most respiratory therapy nebulizers.

The Wavespeed Model

This is a second model that has the benefits of actually explaining what is going on, together with equations that are relatively simple. The concept itself, however, is somewhat more complex (Fig. 9.24). In brief, the Wavespeed Model states that the maximum

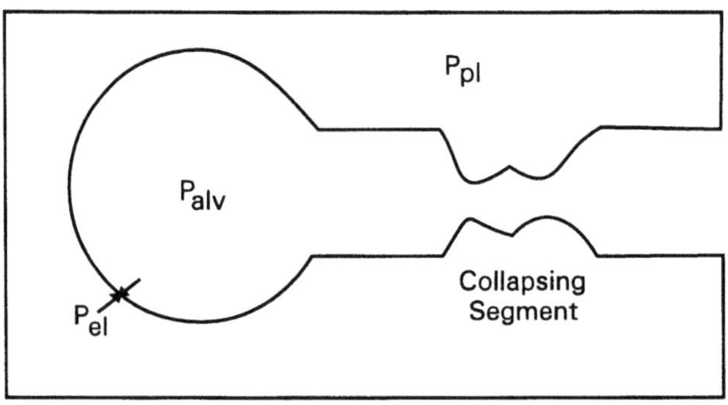

FIGURE 9.23. The Waterfall model of airflow limitation. At some point in the airway, the extramural pressure on the airway is sufficiently greater than the intramural pressure to cause collapse of the airway and consequent airflow limitation.

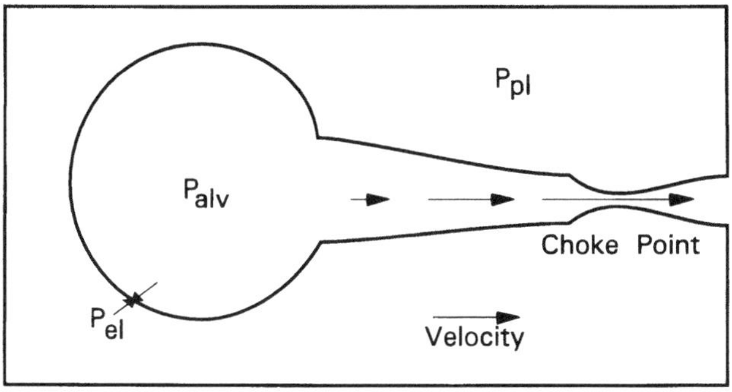

FIGURE 9.24. The Wavespeed model. As the cumulative cross-sectional area of the airways decreases toward the mouth, the particle speed of the air increases, until, at some point in the airway, the speed of the air particles equals the speed of wave propagation in the airway. At that point, the speed of airflow can increase no further.

flow rate occurs when the speed of the gas particles through the airway equals the speed of wave propagation at any point in that airway. The speed of wave propagation is the maximum speed at which air molecules can get out of each others' way, and in a collapsible tube, is considerably slower than in open air. The more easily collapsible the tube, the slower the speed at which a pressure wave will propagate through the tube. The slowing of a pressure wave in a collapsible tube is due at least in part to the (partial) collapse of the tube at a place where the resistive pressure drop plus the Bernoulli pressure drop cause the tube to collapse at a point (called the *choke points*) where the pressure across the tube overwhelms the inherent stiffness of the tube.

Blood Flow

Because ventilation is distributed nonuniformly, with more air normally going to the bases of the lungs than the apices, there would be significant mismatching of blood flow and ventilation if blood flow were not similarly nonuniformly distributed. There is fortunately a gradient of blood flow from the top of the lung to the bottom very similar to that which occurs in ventilation. This gradient of blood flow is due largely to the effects of gravity.

It is useful to divide the lung into three perfusion zones as first described by West (Fig. 9.25). The capillaries at the apex of the lung are well above the level of the heart in the upright posture, and the systolic pressure of the pulmonary system is not great enough to pump blood that high. This region is called *zone 1*, and the capillaries are not usually perfused in the upright posture.

At midlung levels, the pulmonary systolic pressure is great enough to fill the capillaries, but the pulmonary venous pressure is not high enough to fill the venous end of the capillaries. The capillaries act as if they were Starling Resistors with the alveolar pressure as the force compressing the venous end of the capillaries. Blood flow is determined by the difference between arterial pressure at that point in the lung and the alveolar pressure. Because arterial pressure is greater lower in the lung due to gravity, and is pulsatile, the blood flow in this zone is greater lower in the lung, and is pulsatile. In zone three, both pulmonary venous and arterial pressures are greater than alveolar pressure, and the capillaries are filled continuously. A fourth zone appears at the base of the lung at less than full lung inflation. In this zone, perfusion decreases from the

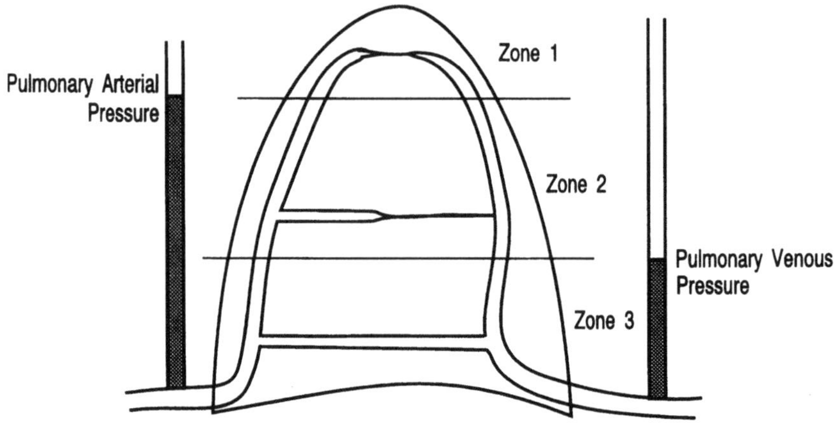

FIGURE 9.25. Blood flow through the three zones of the lung. In zone 1, the pressure generated by the right ventricle is less than the hydrostatic pressure needed to counteract gravity. In zone 2, the systolic pressure of the right ventricle is greater than the hydrostatic pressure, but the left atrial pressure is less than the hydrostatic pressure. The flow is pulsatile. In zone 3, both right ventricular and right atrial pressures are greater than the hydrostatic pressure.

top of the zone to the bottom due to the relative compression of the lung tissue and vessels by gravity with concomitant compression of the extraalveolar vessels.

It is noteworthy that the pulmonary arterioles have only a small amount of smooth muscle in their walls, and are not very active in regulating blood flow under normal circumstances. However, the stimulus of alveolar hypoxia is a powerful vasoconstrictor, and this vasoconstriction serves to deflect blood away from under ventilated alveoli in normal persons.

Alveolar–Capillary Diffusion

Diffusion that refers to the process in which respiratory gasses pass through the alveolar walls into the red blood cell (Fig. 9.26) is described by Fick's first law. The law states that the flux J in moles of a substance (e.g., gas) through a tissue unit area in unit time is directly proportional to the concentration gradient of the substance across the tissue. This is described by

$$J = -D(dC/dx)$$

where C = the molar concentration of the substance
x = a measure of distance from a reference point along the diffusion path
D = the diffusion coefficient of the substance

The path through which respiratory gasses must pass includes the alveolar lining fluid, the alveolar lining cells, the alveolar basement membrane, the capillary basement membrane, the capillary endothelial cell, the blood plasma, the erythrocyte cell wall, and the erythrocyte cytoplasm. The rate at which gasses pass through from the alveolus to the capillary is dependent on several factors: the partial pressure difference between the alveolus and the capillary for the gas in question, the chemical nature of the gas and its solubility in the interstitium, the linear distance between the capillary and the alveolus, and, finally, the surface area available for diffusion. It is generally thought that only the surface area of the interface between the alveolus and capillary

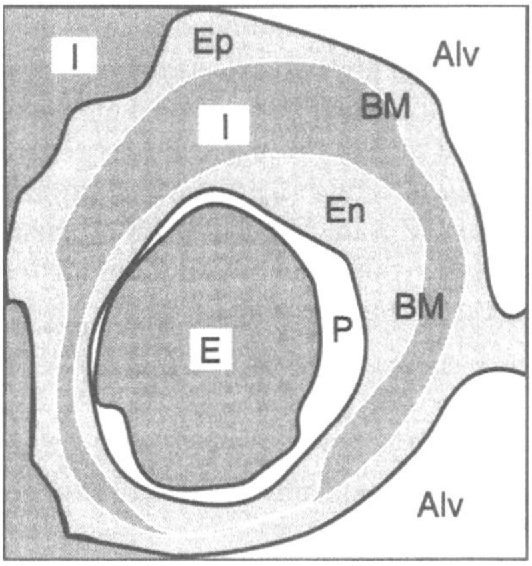

FIGURE 9.26. The diffusion path. The layers of gasses must traverse between the alveolus and the erythrocyte. E = Erythrocyte; Alv = alveolus; P = blood plasma; Ep = the alveolar epithelium; En = the capillary endothelium; BM = basement membrane of the capillary endothelium, and the alveolar epithelium; I = Interstitium (collagen, ground substance, etc.). Adapted from an electron micrograph.

is the important factor in limiting diffusion of gasses in human illness. In such diseases as emphysema, a very considerable loss of surface area occurs due to the loss of alveolar walls, and hence very profound disturbances in diffusion can exist. Loss of surface area also occurs in fibrosing disorders because of the replacement of alveolar–capillary tissue with collagen.

Hypoxemia

The result of deranged pulmonary physiology is often hypoxemia or an abnormally low partial pressure of oxygen in the blood.

A useful concept to an understanding of arterial oxygen levels is the idea of the alveolar to arterial oxygen gradient. The alveolar–arterial oxygen gradient (A–a gradient) is simply the difference between the oxygen tension in an average alveolus and that in the end-capillary blood. In normal young persons, the gradient can be close to zero and up to 10 mmHg pressure difference. With increasing age, the gradient increases slightly so that a normal 70 year old might have a gradient of 15 or 20. This is calculated from the arterial oxygen tension as measured by arterial blood gasses, and the simplified alveolar air equation:

$$P_{AO_2} = P_{IO_2} - (P_{aCO_2}/R)$$

where P_{IO_2} can be calculated as:

$$P_{IO_2} = (P_{atm} - 47) * F_{IO_2}$$

where Patm = ambient atmospheric pressure (760 at sea level)
47 = the vapor pressure of water at body temperature
F_{IO_2} = the fraction of inspired oxygen, 0.21 on room air, and 1.0 on 100% oxygen

For all practical purposes, P_{IO_2} can be taken to be about 150. If we apply the preceding equations, we can then calculate the alveolar P_{AO_2} as follows:

$$P_{AO_2} = 150 - (40/0.8)$$

$$P_{AO_2} = 100$$

at sea level on room air. The A–a normally is less than 10, so the normal young individual will have an arterial oxygen tension of greater than 90 mmHg. The normal A–a gradient is not 0 because of the physiologic shunt. Deoxygenated blood is admitted to the left atrium through the Thebesian veins, which drain the myocardium directly into the left atrium, and through the bronchial veins that likewise drain directly into the left atrium. This deoxygenated blood mixes with fully oxygenated blood from the lungs in the left atrium, and the resultant mixture has a lower hemoglobin saturation than blood directly from the lungs.

Arterial Hypoxemia and Its Causes

Arterial hypoxemia means a reduction in arterial Po_2 below normal. The causes of hypoxemia are:

Hypoventilation and low P_{IO_2}	Shunt
V/Q mismatch	Diffusion limitation

Hypoventilation and Low P_{IO_2}

Hypoventilation is defined as ventilation insufficient to maintain a normal P_{ACO_2} in the range of 38–42 mmHg at sea level. As the concentration of carbon dioxide in the blood rises, the concentration in the alveoli also rises. This elevated alveolar P_{ACO_2} effectively displaces oxygen from the alveolar air, leading to a low alveolar, and, hence low arterial oxygen tension. Hypoventilation can be voluntary, or due to such disease states as severe chronic obstructive pulmonary disease (COPD), neuromuscular disease, or drug overdose. Supplemental oxygen administration will correct the hypoxemia due to hypoventilation because the alveolar O_2 is increased; of course, the hypercarbia is not corrected by oxygen. With regard to a low P_{IO_2}, it occurs principally at high altitude, or during conditions where the oxygen percentage is reduced below the usual 20.9%. Oxygen administration reverses this hypoxemia.

Ventilation–Perfusion (V/Q) Mismatch

The most important cause of hypoxemia in human disease is V/Q mismatch (Fig. 9.27). In the lung, the perfusion of blood and the ventilation of alveoli are normally fairly well matched. In some disease states, however, the lung becomes variably, nonuniform and some areas of the lung become relatively underperfused, others are relatively underventilated. Areas that are underperfused contribute small amounts of well-oxygenated blood to the circulation, and are not responsible for hypoxemia. For all the areas that are underperfused, however, there are areas that are correspondingly poorly ventilated. In these alveoli, there is insufficient oxygen being introduced into the alveoli because of the inadequate local ventilation to properly oxygenate the blood circulating there. The result is the contribution of an amount of poorly oxygenated blood to the systemic circulation. The overall ventilation to the entire lung can usually be increased sufficiently not to lead to hypercarbia in this situation; however, the fact that the hemoglobin in the well-ventilated alveoli cannot be saturated more than 100%, no matter how much that alveolus is hyperventilated, means that there is a limit to how much oxygen the well-ventilated alveoli can contribute. This well-oxygenated blood from well-ventilated alveoli is then diluted in the pulmonary veins by the poorly oxygenated blood from the poorly ventilated alveoli, so that the resulting arterial blood has a lower than normal overall saturation.

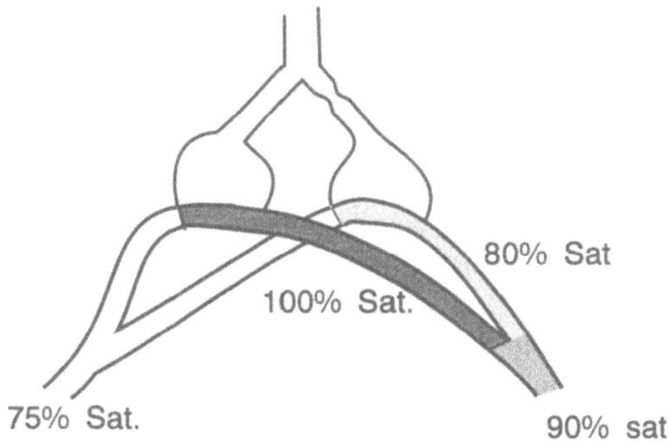

FIGURE 9.27. Schematic depiction of a V/Q mismatch. Due to the constriction to the right-hand alveolus, the share of total ventilation it receives is reduced. The left-hand alveolus receives additional ventilation, but due to the plateau in the hemoglobin saturation curve, the blood flowing past this alveolus, being already 100% saturated, can carry no additional oxygen. The net result is reduced saturation in the pulmonary vein.

Shunt

A situation in which blood bypasses any ventilated alveoli, and contributes venous blood (deoxygenated) to the arterial side of the circulation is called a *shunt* (Fig. 9.28). The shunt represents the extreme case of low ventilation–perfusion ratio. The question then is, At what level of low ventilation does it cease being a V/Q defect and become a shunt? In practical terms, when local ventilation is so low that 100% oxygen administration no longer provides enough oxygen to the affected capillaries to oxygenate the blood in them, it is considered to be a shunt. By this concept, then, V/Q mismatch hypoxemia is correctable by 100% oxygen administration, whereas shunt hypoxemia is not.

In normal individuals, up to about 5% of the cardiac output is shunted away from alveolar capillaries, predominantly via the cardiac Thebesian veins, and the bronchial veins. A large variety of disease states can significantly increase the amount of shunted blood, including alveolar edema (e.g., in pulmonary edema and adult respiratory distress syndrome), alveolar filling diseases (e.g., lobar pneumonia), and collapse of lung (atelectasis) due to endobronchial obstruction. Right-to-left shunts of the heart due to atrial or ventricular septal defects constitute other causes of arterial hypoxemia.

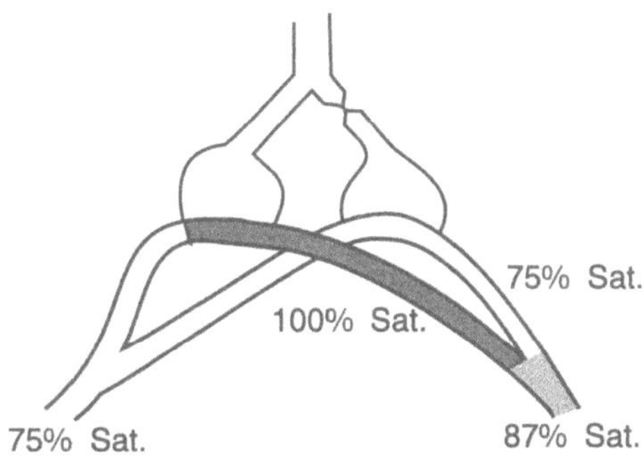

FIGURE 9.28. Schematic depiction of a shunt (note that the right lung unit is totally obstructed). The desaturated blood from the shunt dilutes the fully saturated blood from the ventilated alveolus. Additional oxygen does not get to the unventilated alveolus, and the blood from the well-ventilated one is already fully saturated.

Diffusion Limitation

A diffusion limitation or defect exists if the diffusion path is so poor (i.e., due to small surface area) that oxygen cannot diffuse into the blood stream rapidly enough to meet metabolic demand. This never occurs in normal humans at rest, but it could be a factor under such extreme conditions as mountain climbing. In disease states, diffusion limitation rarely contributes materially to hypoxemia at rest, but it can become significant during exercise. Measuring oxygen saturation during exercise is a very sensitive test for early diffusion limitation. Oxygen administration reverses the hypoxemia of diffusion limitation by increasing the diffusion pressure gradient from alveolus to capillary, thus increasing the amount of gas that can diffuse per unit time.

Conclusion

The static and dynamic characteristics of the lungs are differently deranged by various diseases. Understanding how different diseases affect these characteristics can help make a diagnosis, and suggest treatment. The end result of most pathologic processes in the lungs is an increase in the A–a gradient due to V/Q mismatch. If the defect proves to be due to a shunt as demonstrated by a relative resistance-to-oxygen administration, another set of diagnoses suggests itself. An understanding of the physiology of the lung is clinically useful in both the diagnosis, as well as the treatment of a wide variety of lung disorders.

Recommended Readings

Farhi, L.E., and Tenney, S.M. (eds.) (1987) Handbook of physiology. Section 3, the respiratory system. Volume IV, gas exchange. American Physiological Society, Bethesda, MD.

Hlastala, M.P., and Berger, J.I. (1996) Physiology of respiration. Oxford University Press, Oxford.

Levitzky, M. (1999) Pulmonary physiology. McGraw Hill and Co., New York.

Lumb, A., and Nunn, J.F. (2000) Nunn's applied respiratory physiology. Oxford University Press, Oxford.

Pierce R.J., and Worsnop, C.J. (1999) Upper airway function and dysfunction in respiration. Clin. Expt./Pharmacol. Physiol. 26, 1–10.

West, J.B. (2000) Respiratory physiology. Lippincott—Williams and Wilkins, Philadelphia.

10 The Physiology and Pathophysiology of Gas Exchange

Brian J. Whipp

The cells of the body require a continuous exchange of energy to subserve their functions. This is supplied in almost all cases by the high, free energy of hydrolysis of the terminal phosphate bond(s) of adenosine triphosphate (ATP). In most cells ATP is present only in low concentrations—approximately 5 mM. This concentration is maintained relatively constant, despite large variations of the rate of energy exchange, through mechanisms that produce ATP at rates commensurate with its utilization within the cell. An increased ATP production rate can be mediated from three sources:

1. Breakdown of creatine phosphate in the creatine kinase reaction (i.e., creatine phosphate may therefore be considered to be a labile store of high-energy phosphate).
2. Oxidative phosphorylation in which hydrogen atoms derived from ingested substrates are processed through the mitochondrial electron transport chain as protons and electrons (Fig. 10.1), with the terminal oxidation at cytochrome oxidase being provided by molecular oxygen.
3. The production of ATP through anaerobic glycolysis in the cell cytoplasm.

The major source of ATP production is oxidative phosphorylation. This demands an increased flow of oxygen into the cell to maintain cellular oxygen partial pressures at

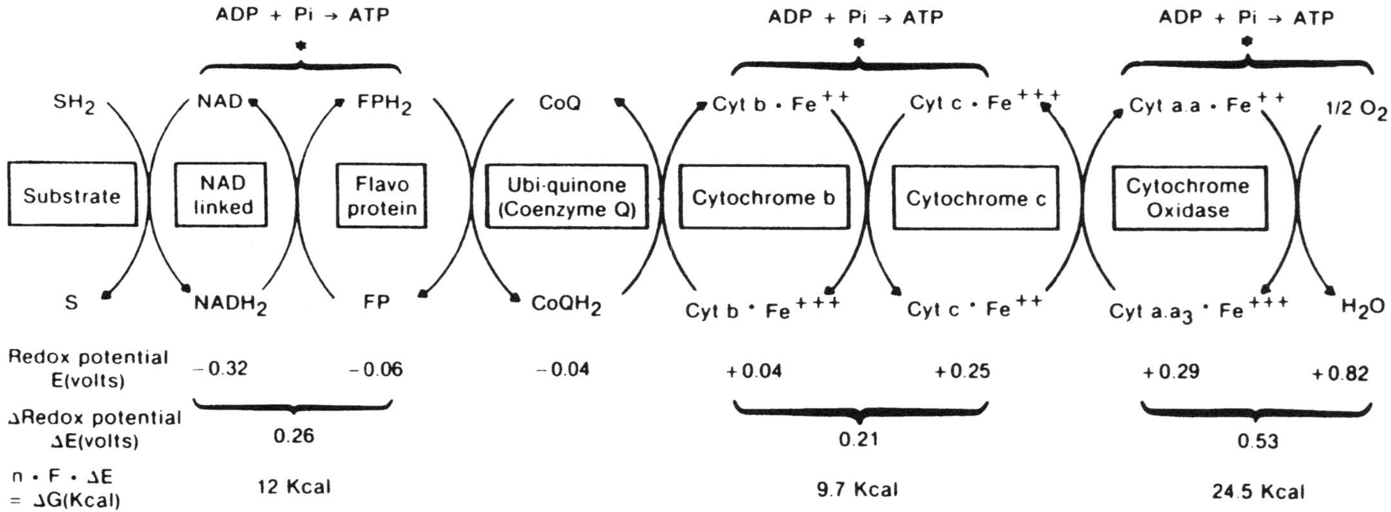

FIGURE 10.1. Schematic representation of the processing of protons and electrons through the electron-transport chain, resulting in ATP production. The sites of sufficient free energy change to support the proton transfer across the mitochondrial membrane, establishing the transmembrane proton gradient that generates ATP, are represented by the brackets. (Taken from: Whipp, B.J. (1994) The bioenergetic and gas-exchange basis of exercise testing. *Clin. Chest Med.* 15, 173–192.)

TABLE 10.1. Oxygen and carbon dioxide demands of substrate oxidation.

	$\dot{V}O_2$* (L/min)	$\dot{V}CO_2$ (L/min)	RQ	~P:O_2	~P:CO_2	O_2:~P	CO_2:~P
Glycogen	1.0	1.0	1.0	6.00	6.00	0.17	0.17
Palmitate	1.0	0.7	0.7	5.65	8.13	0.18	0.12
Glycogen/palmitate	1.0	1.43	1.43	1.06	0.74	0.94	1.42

* Values at a particular VO_2. For a particular work rate or high-energy phosphate utilization rate, less O_2 is needed when glycogen is the substrate than when palmitate is used. (Taken from: Whipp, B. J. (1994) The bioenergetic and gas-exchange basis of exercise testing. *Clin. Chest Med.* 15, 173–192)

levels commensurate with the mitochondrial utilization. One may therefore consider the mitochondria to be "sinks" for oxygen, and the lung to be the "source."

Cellular metabolism, however, also yields molecular CO_2 at a rate dependent both on the metabolic rate for O_2 and on the substrate being utilized (Table 10.1). This CO_2 must also be cleared from the lungs at rates commensurate with its production rate, if the body's partial pressure of CO_2, and, hence, its hydrogen ion concentration, is to be maintained within relatively narrow limits.

This presents an immediate physiological control "problem." On those occasions when the demands for O_2 and CO_2 exchange differ, the lung cannot simultaneously meet the demands for both gases. The resolution of this "dilemma" (considered later) involves the body's ventilatory control strategies.

The cells of the body require *numbers* of molecules of oxygen to subserve the demands for oxidative phosphorylation. That is, oxidative phosphorylation yields 3 ATP from one atom of O_2 utilized in the electron transport chain, or 2 ATP if the hydrogen ions are transferred to FAD rather than NAD (Table 10.1). The convention of pulmonary physiology, however, is to consider *volumes* of gases that are exchanged, and the gas pressures that are consequent to the exchange. It is important, therefore, that consideration be given to the interrelationships among the pressure, volume, temperature, and number of molecules of a given gas. These will now be considered.

The Gas Laws

Avogadro's Law

The same number of molecules of any gas will occupy the same volume, as long as the pressures and temperatures are the same. For example, 1 mole of an ideal gas contains 6.02×10^{23} molecules (Avogadro's number). This will occupy 22.4L at standard temperature and pressure, where the standard temperature is 0°C or 273K and the standard pressure is 1 atmosphere. It should be noted that the 22.4L described in this relationship refers to an ideal gas (i.e., one in which there is no interaction among the molecules, for example as a result of polarity). Of the relevant respiratory gases, CO_2 is most disparate in this regard, with a molar volume at STPD of 22.26L compared with 22.39L for O_2.

Boyle's Law

The pressure and volume of a given number of molecules of gas at constant temperature are inversely related. That is, if a given volume of gas, V_1, expresses a pressure, P_1, then it will exhibit a new pressure, P_2, if it is compressed, for example, to a new volume V_2, as described by the equation,

$$P_1 \cdot V_1 = P_2 \cdot V_2$$

Charles' Law

If the pressure of a gas is maintained constant, then the volume of the gas is directly proportional to its absolute temperature (i.e., degrees Kelvin). That is, when the temperature of a gas at a volume V_1 is raised from T_1 to T_2, the volume will increase to a new value, V_2; i.e.,

$$V_1/V_2 = T_1/T_2$$

This law allows gas volumes that are conventionally expressed at different temperatures to be readily related. For example, the metabolic exchange of gases in the body is typically expressed at standard temperature (0°C or 270 K), whereas the pulmonary ventilation is normally expressed at body temperature (this is commonly, but not always 37°C or 310 K).

The General Gas Law

These three laws may therefore be combined into a single representation of P, V, and T for a given number of moles of gas, i.e.,

$$(P \cdot V)/T = N \cdot R$$

where R = the gas constant.

It is important to remember, however, that even though this formulation is perfectly adequate for general respiratory exchange purposes, extreme precision would demand corrections both for the effects of interactions among the gas molecules and the space that they occupy (i.e., the van der Waals effect).

Graham's Law

The rate at which a gas diffuses through a gas mixture is inversely proportional to the square root of its molecular weight (MW). Lighter molecules travel faster than heavy ones and hence diffuse more rapidly. As a result, oxygen will diffuse faster than CO_2, i.e.,

$$\text{the rate of oxygen diffusion/rate of } CO_2 \text{ diffusion}$$
$$= \sqrt{MW_{CO_2}}/\sqrt{MW_{O_2}} = \sqrt{44}/\sqrt{32} = 6.6/5.6 = 1.18$$

Dalton's Law

The pressure exerted by a mixture of gases is equal to the sum of the individual (or partial) pressures exerted by each gas. If one considers dry atmospheric air, for example, its total or barometric pressure P_B is a result of the sum of the pressures of its constituent gases

$$P_B = P_{O_2} + P_{CO_2} + P_{N_2}$$

Here we have disregarded the small effect of inert gases other than nitrogen (e.g., argon, neon, krypton, etc.), as these are commonly "lumped" with nitrogen because they are normally present at such low concentrations. As a result the partial pressure of gas X is a manifestation of the fractional concentration of *that* gas in the total, i.e.,

$$P_X = F_X \cdot P_B$$

Taking dry atmospheric air to contain 20.93% oxygen, 0.04% CO_2 and 79.03% N_2, then the partial pressures exerted by these gases, with a total barometric pressure of 760 mmHg, will be:

$$P_{O_2} = 20.93/100 \times 760 = 159\,\text{mmHg}$$

$$P_{CO_2} = 0.04/100 \times 760 = 0.3\,\text{mmHg}$$

$$P_{N_2} = 79.03/100 \times 760 = 601\,\text{mmHg}$$

When the air is not dry, however, then the partial pressure of water vapor will necessarily reduce the pressures exerted by the other gases if the pressure remains the same (i.e., atmospheric). Here it is useful to consider air that is inhaled, which becomes fully saturated with water by the time it enters the bronchial tree. Because the partial pressure of water vapor at the normal body temperature 37°C is 47 mmHg, the total pressures exerted by O_2, CO_2, and N_2 will now be equal to the total pressure minus that of water vapor (i.e., 760 – 47 mmHg). The P_{O_2} of the saturated tracheal air is therefore $20.93/100 \cdot 713 = 149$ mmHg. This means there is a reduction of partial pressure of oxygen from 159 mmHg for the dry inspired air to 149 mmHg for the air saturated with water vapor at a body temperature of 37°C despite no O_2 having been exchanged.

Alveolar Gas Partial Pressures

It is apparent from the foregoing discussion that in the steady state (i.e., the condition in which the body gas stores are stable) the amount of oxygen utilized by the cells per minute will be equivalent to that loaded into the blood from the lungs in the same interval of time.

That is, as long as care is taken to consider the conditions under which the gas exchange is expressed, then the average metabolic rate of the body can be determined from the rate at which that metabolic gas is exchanged at the lung. For example, consider the metabolic rate of CO_2. In the steady state this must be equal to the amount of CO_2 exchanged per unit time across the lung ($\dot{V}_{CO_2}$). In turn, this is equal to the amount of CO_2 that is exhaled from the alveoli in 1 minute minus the amount of CO_2 that was inhaled into the alveoli during that minute. This may be symbolically represented as:

$$\dot{V}_{CO_2} = \left[\dot{V}A_{(exp)} \times FA_{CO_2}\right] - \left[\dot{V}A_{(insp)} \times FI_{CO_2}\right]$$

where $\dot{V}A_{(insp)}$ and $\dot{V}A_{(exp)}$ = the inspired and expired alveolar ventilations
FI_{CO_2} and FA_{CO_2} = the inspired and alveolar fractional CO_2 concentrations, respectively.

Because FI_{CO_2} is effectively zero, then

$$\dot{V}_{CO_2} = \dot{V}A_{(exp)} \cdot FA_{CO_2}$$

On the other hand, of course, this can be represented as

$$FA_{CO_2} = \dot{V}_{CO_2}/\dot{V}A$$

or

$$PA_{CO_2}/[P_B - 47] = \dot{V}_{CO_2}/\dot{V}A$$

This equation deserves attention. For example, one might ask the question, How might alveolar ventilation ($\dot{V}_A$) and the body's CO_2 production rate in the steady state influence the alveolar CO_2 partial pressure? This equation demonstrates that in order for alveolar P_{CO_2} to be regulated at a constant level, with increased rates of metabolic CO_2 production, then alveolar ventilation must increase and it must increase in exact proportion to that of $\dot{V}_{CO_2}$ (Fig. 10.2). That is, regulation of alveolar P_{CO_2} requires proportional changes of alveolar ventilation and CO_2 production. Note, however, that if P_{CO_2} is regulated at a lower than normal level, then a given increase in $\dot{V}_{CO_2}$ will require a greater increase in alveolar ventilation. This is because a given volume of gas at a lower partial pressure will contain fewer CO_2 molecules. Similarly, if P_{CO_2} is regulated at a higher than normal level, a given increase in metabolic rate will require less of an increase in alveolar ventilation. In this case, a given volume of alveolar gas will contain a greater number of CO_2 molecules if the P_{CO_2} is high.

On the other hand, one might ask: What is the relationship between alveolar P_{CO_2} and the amount that the subject breathes (i.e., $\dot{V}_A$) if metabolic rate (i.e. $\dot{V}_{CO_2}$) remains constant? The equation shows that P_{CO_2} is inversely related to $\dot{V}_A$ in a precise hyperbolic fashion (as shown in Fig. 10.3); that is, a given level of pulmonary CO_2 exchange ($\dot{V}_{CO_2}$) can be accomplished by an effectively infinite combination of $\dot{V}_A$ and $P_{A_{CO_2}}$. A low $\dot{V}_A$ would result in a high P_{CO_2} or a high $\dot{V}_A$ would result in a low P_{CO_2} with both, however, yielding the same $\dot{V}_{CO_2}$. This relationship, of course, remains hyperbolic if considered at different metabolic rates, the difference being that the position of the hyperbola is altered as shown in Fig. 10.3.

Similar relationships may of course be developed for oxygen exchange. That is, in the steady state the body's metabolic rate for oxygen ($\dot{V}_{O_2}$) will equal the amount of oxygen loaded into the alveolus per minute minus the amount that is exhaled from the alveolus per minute, with the remainder being loaded into the blood

$$\dot{V}_{O_2} = \left[\dot{V}_{A(insp)} \times F_{I_{O_2}}\right] - \left[\dot{V}_{A(exp)} \times F_{A_{O_2}}\right]$$

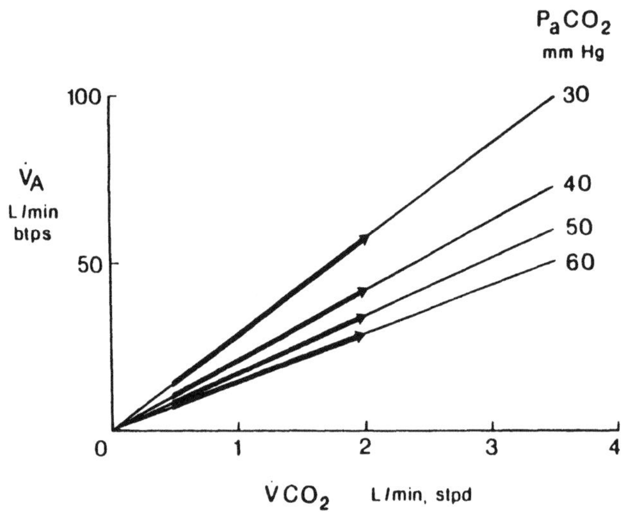

FIGURE 10.2. The change in alveolar ventilation required to regulate arterial P_{CO_2} at different values as a function of increasing metabolic rate for CO_2 exchange.

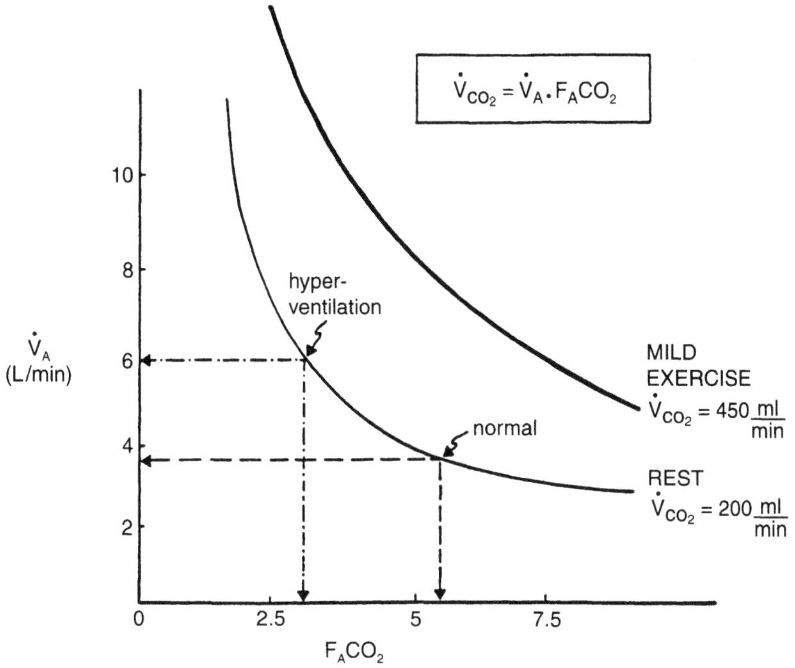

FIGURE 10.3. The "metabolic hyperbola." Note that a particular level of CO_2 output can be attained with an infinite variation of alveolar ventilation and alveolar CO_2 fractional concentration. The position of the hyperbola is shifted as shown by increasing metabolic rate.

where FI_{O_2} and FA_{O_2} are the inspired and alveolar fractional O_2 concentrations, respectively. As a result,

$$[PI_{O_2}{*} - PA_{O_2}]/[P_B - 47] = \dot{V}_{O_2}/\dot{V}A$$

The asterisk in this equation represents the fact that, in its present form, the equation is an approximation because an additional term is required here to account for small differences between inspired and expired nitrogen concentration that result from the volume exhaled being normally slightly less than the volume inhaled; i.e., the typical western diet has a respiratory quotient (R.Q. = $\dot{V}_{CO_2}/\dot{V}_{O_2}$) which is typically less than one. Therefore, a slightly smaller volume of CO_2 is excreted than the volume of oxygen which is consumed.

With reasoning equivalent for that described for P_{CO_2} earlier, you will note that the extraction of oxygen [i.e., $(PI_{O_2} - PA_{O_2})$] depends upon the relationship between alveolar ventilation to metabolic rate. Regulating alveolar P_{O_2} therefore demands that alveolar ventilation increase in proportion to $\dot{V}_{O_2}$ when metabolic rate changes. At a given level of metabolic rate (i.e., $\dot{V}_{O_2}$), the inspired-minus-alveolar oxygen pressure difference will similarly be inversely related to the alveolar ventilation. That is, the higher the alveolar ventilation at a given metabolic rate, the smaller will be the value of $(PI_{O_2} - PA_{O_2})$. Since PI_{O_2}, which is the inspired oxygen partial pressure, is constant, it then follows that increases of alveolar ventilation out of proportion to that of metabolic rate will elevate the alveolar O_2 partial pressure (i.e., narrow the difference between the inspired and the alveolar level).

For simplicity of understanding these concepts, we have presented the foregoing relationships as if the gases were measured under the same conditions. It is, however, conventional to express metabolic rate (i.e., $\dot{V}_{CO_2}$ and $\dot{V}_{O_2}$) under conditions of standard temperature and pressure, and dry. That is, we must keep track of the actual number of moles of gas produced or consumed. For ventilation, however, it is typically more important to know the actual volume that is breathed in a given situation. Consequently, $\dot{V}A$ is conventionally expressed under conditions of body temperature and pressure, saturated. A correction must therefore be applied to relate these quantities to the same conditions, i.e.,

$$\dot{V}A_{(BTPS)} = \dot{V}A_{(STPD)} \cdot (760/713) \cdot (310/273)$$

that is,

$$\dot{V}A_{(BTPS)} = 1.21 \cdot \dot{V}A_{(STPD)}$$

As a result,

$$FA_{CO_2} = \left[1.21\dot{V}_{CO_2(STPD)}\right] / \dot{V}A_{(BTPS)}$$

And as the partial pressure of CO_2 is typically of more interest than its fractional concentration, we need to multiply both sides of this equation by ($P_B - 47$) (i.e., 713) to yield

$$PA_{CO_2} = \left[863 \cdot \dot{V}_{CO_2(STPD)}\right] / \dot{V}A_{(BTPS)}$$

We shall now follow the convention of assuming that arterial P_{CO_2} (PA_{CO_2}) is equal to alveolar P_{CO_2} (i.e., $Pa_{CO_2} = PA_{CO_2}$) (the reality and consequences of this assumption will be discussed later). Further consideration of the interrelationships among these ariables requires that we distinguish between certain terms.

Hyperpnea and Hyperventilation

Hyperpnea refers to any condition in which the level of ventilation is increased above that expected for a normal resting subject. Hyperventilation, on the other hand, expresses a condition in which ventilation is increased *relative* to the current levels of CO_2 output. As a result, alveolar and, hence, arterial P_{CO_2} are low with hyperventilation. In fact, a low arterial P_{CO_2} is the *sine qua non* of the condition.

Hypopnea and Hypoventilation

Hypopnea refers to any decrease in ventilation relative to the normal resting condition. *Hypoventilation*, on the other hand, is a condition in which ventilation is low relative to the current metabolic rate of CO_2 output. As a result, alveolar and arterial P_{CO_2} are high in hypoventilation.

In order to ensure that there is a clear distinction between these terms, consider the following question. Is it possible for a subject to be hyperpneic and hypoventilating at the same time? The answer is clearly, *yes*. This reflects a condition in which the absolute level of ventilation is increased but is not increased sufficiently in proportion to the increased CO_2 output. As a result, the absolute level of ventilation is high, but it is associated with a high P_{CO_2}. Hence, although the subject is *hyperpneic*, the subject is also *hypoventilating*. This is not unusual for subjects with high airway resistance when they exercise.

Tachypnea refers to rapid breathing that is often, but not necessarily, associated with shallow breathing. *Bradypnea* is slow breathing that is often, but not necessarily always, being associated with deep breathing. *Apnea* is a total cessation of breathing.

Dead Space Ventilation

Of the inspired volume of atmospheric air, only that fraction that reaches the "alveolus," and hence its exchange surface with the blood in the pulmonary capillaries, can contribute to gas exchange. The part of the inspirate that remains in the conducting airways does not take part in gas exchange with the pulmonary capillary blood. This volume, from the entrance of the gas into the body to the site of alveolar gas exchange (i.e., beginning at the respiratory bronchiolar level), is termed the *anatomical dead space* [$V_{D(anat)}$]. As a result, the volume of the total inspired tidal volume (V_T) that reaches the alveoli (V_A) is given by

$$V_A = V_T - V_{D(anat)}$$

The alveolar ventilation $\dot{V}_A$ is given by

$$\dot{V}_A = V_T \cdot f - V_{D(anat)} \cdot f$$

or

$$\dot{V}_A = \dot{V}_E - \dot{V}_{D(anat)}$$

where f = the breathing frequency and $\dot{V}_E$ = the total minute ventilation

It may readily be seen, therefore, that the alveolar ventilation depends both on the volume of the dead space and the breathing frequency of the subject. For example, consider a subject with a minute ventilation of 8 L/minute and a breathing frequency of 10 (i.e., $\dot{V}_E$ = 8000 ml/minute; f = 10 per minute). If we consider $V_{D(anat)}$ to be 150 ml, then

$$\dot{V}_A = 8000 - [150 \times 10]\,\text{ml/minute} = 6500\,\text{ml/minute}$$

Were the anatomical dead space to be increased to a level of 300 ml, then

$$\dot{V}_A = 8000 - [300 \times 10] = 5000\,\text{ml/minute}$$

The alveolar ventilation would consequently be less despite the same total ventilation. Thus, the ability of the ventilation to effect gas exchange would be reduced.

A second important consideration is the pattern of breathing. Here we consider the total ventilation to be the same as in the initial condition (i.e., 8000 ml/minute), and the anatomical dead space to be the same at 150 ml. However, here the subject achieves the minute ventilation with a tidal volume that is halved and a frequency that is doubled (i.e., V_T = 400 ml; f = 20 per minute). Then,

$$\dot{V}_A = 8000 - [150 \times 20] = 5000\,\text{ml/minute}$$

Note that in this case the alveolar ventilation is reduced, despite both minute ventilation and the size of the anatomical dead space remaining the same. It should also be noted that in these examples we have assumed for simplicity that the anatomical dead space remains constant independent of the tidal volume. But this is not the case because during inspiration the more negative intrapleural pressure at high lung volumes increases the transmural pressure across the walls of the intrapulmonary airways, causing them to be distended. As a result, the size of the anatomical dead space increases somewhat as tidal volumes get larger.

It follows from the foregoing example that the subject ought not breathe too rapidly because this leads to high resistive work of breathing and it creates a large dead space ventilation. In contrast, very deep breaths would lead to high elastic work of breathing. Subjects normally tend to select an optimum pattern that minimizes the work of breathing while providing appropriate alveolar gas exchange. It is important to keep

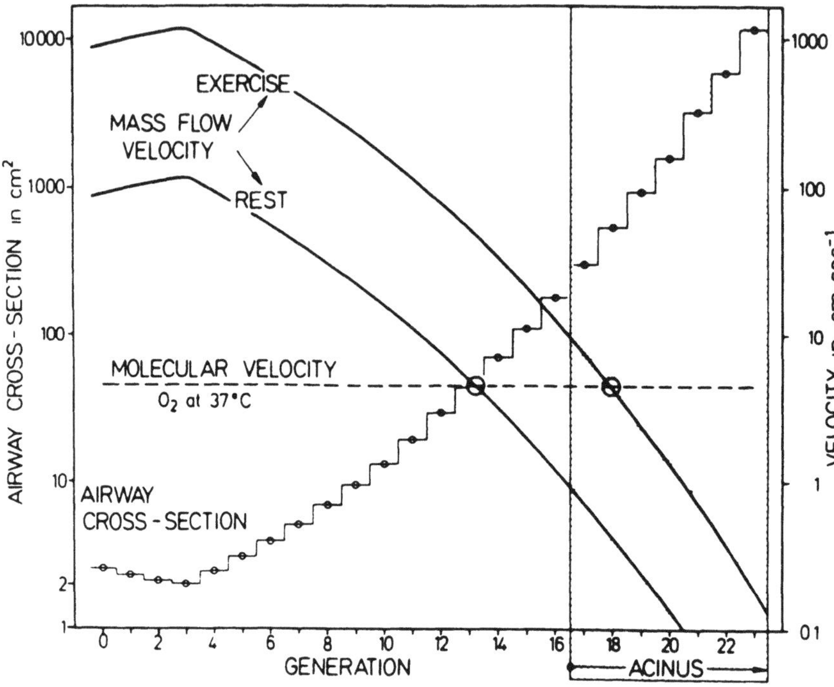

FIGURE 10.4. Characterization of the influence of the airway cross-sectional area on the linear velocity of the inspired gas. Note that the velocity resulting from convective flow falls below that of the diffusional flow, shown for oxygen as the horizontal dashed line. The point at which this occurs penetrates deeper into the lung with increasing levels of exercise. (Taken from: Weibel, E.R. (1984) *The Pathway for Oxygen*, p. 285. Harvard University Press, Cambridge, MA.)

in mind that only 50% or so of the total anatomical dead space is intrathoracic, and hence amenable to this distension as pleural pressure changes. Furthermore, even though the model of dead space in the alveoli is often in the form of a tube with a bubble at the end, representing the dead space and alveolus, respectively, the anatomical dead space is, of course, much more complex in that the conducting airways are of different lengths, and, hence, of different volumes to different regions of the lungs. In addition, there is not a sharp interface between gas in the dead space containing atmospheric air at the end of an inspiration and gas in the alveoli. This is because, even though airflow through the conducting airways is predominantly via convection, the linear velocity of the inspired gas progressively decreases as the gas penetrates further and further into the airways as a result of the increased cross-sectional area. This is depicted in Fig. 10.4. As a result, the gas flow into and across the alveolus is predominantly a diffusional process. The depth of convective penetration of the inspired gas into the lung (i.e., down to what has been termed the *stationary interface*) varies with the airflow. As a result, even though anatomical dead space may be relatively easy to define structurally, it is by no means as easy to quantify; however, there are two basic techniques for estimating the volume of this anatomical or series dead space.

Fowler Technique

Were the lung to be a relatively simple structure in which the anatomical dead space had a single, discrete separation from the alveolar volume with no diffusional gas flux

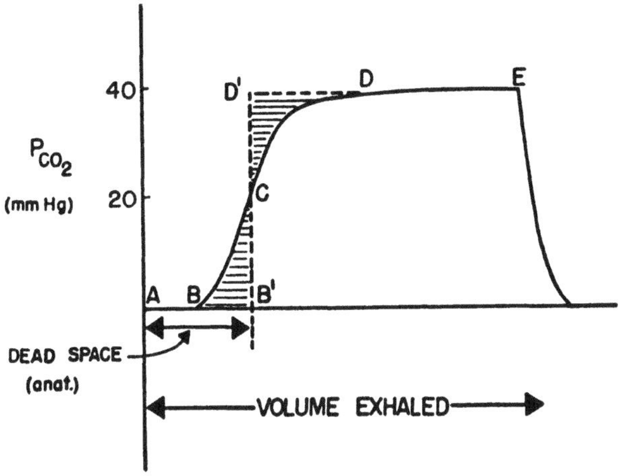

FIGURE 10.5. Schematic representation of the profile of expired P_{CO_2} during an exhaled tidal volume. The actual profile measured is given by the sequence A → B → C → D → E. The anatomical deadspace is estimated from the volume exhaled to the point at which the area B, B′, C is exactly equal to the area C, D′, D.

across the interface, then atmospheric air following an inspiration would fill the anatomical dead space and alveolar gas would occupy the alveoli. The profile of P_{CO_2} during the subsequent exhalation would appear as shown in Fig. 10.5. The partial pressure for CO_2 would effectively be zero initially as the anatomical dead space is cleared. The gas from the alveoli would then be expressed with an abrupt transition between the phases. The anatomical dead space would simply be the volume exhaled up to point B′.

In reality, even though the initial part of the expirate is derived exclusively from the dead space and the final part *exclusively* from the alveoli, there is an intermediate transitional region in which some alveoli with short conducting airways would begin to empty before all the dead space is cleared. In addition, there would be some diffusional mixing across the interface between the dead space and alveolar compartments. Rather than appearing as shown by the dashed line in Fig. 11.5, the expired P_{CO_2} profile would therefore be similar to that shown by the solid line in Fig. 10.5.

The useful technique devised by Fowler provides a geometrical means of functionally delaying the influence of the alveoli which on average emptied "too early," and bringing forward the influence of the anatomical dead space regions which, on average, emptied "too late". This is achieved practically as shown in Fig. 10.5. A vertical line is drawn through the slewed interface between the dead space and the alveolar compartments such that the area (B − B′ − C) = area (C − D′ − D). In other words, the technique "creates" a condition in which the lung apparently empties with independent and sequential dead space and alveolar compartments.

CO_2 Proportion Technique (Bohr)

This technique is based upon the reasonable assumption that the volume of CO_2 exhaled in the breath originates entirely from the alveolar compartment; that is, none originates in the anatomical dead space because this was filled with atmospheric air at the start of the expiration. As a result, the total amount of CO_2 that is collected in a mixed expired tidal volume will equal the CO_2 volume that was evolved from the alveolar compartment, i.e.,

$$V_T \cdot F\bar{E}_{CO_2} = V_A \cdot F_{A_{CO_2}}$$

where $F\bar{E}_{CO_2}$ = the mixed expired CO_2 fraction

As $V_A = V_T - V_{D(anat)}$, this yields

$$V_T \cdot F\bar{E}_{CO_2} = [V_T \cdot F_{A_{CO_2}}] - [V_{D(anat)} \cdot F_{A_{CO_2}}]$$

i.e.,

$$V_{D(anat)}/V_T = [F_{A_{CO_2}} - F\bar{E}_{CO_2}]/F_{A_{CO_2}}$$

or,

$$V_{D(anat)}/V_T = [P_{A_{CO_2}} - F\bar{E}_{CO_2}]/P_{A_{CO_2}}$$

Although this relationship is conceptually simple, it is in reality extremely difficult to determine. This is because there is not a simple means of establishing $P_{A_{CO_2}}$. The values in different alveoli are highly variable throughout the lung, even in normal subjects (as discussed later). In some patients with lung disease, some alveoli may be ventilated, but not perfused. The P_{CO_2} in these units will therefore be close to that of atmospheric air because they do not participate in gas exchange. The volume of the expirate that originated in those alveoli that do not participate in gas exchange is termed the *alveolar dead space*. As a result, because it is the effectiveness of gas exchange that is of interest, the influence of the total dead space becomes meaningful [i.e., the sum of both the anatomical and the alveolar dead space ($V_{D(alv)}$)]. This is termed the *physiological dead space* ($V_{D(physiol)}$); i.e.,

$$V_{D(physiol)} = V_{D(anat)} + V_{D(alv)}$$

One then makes the simplifying assumption that the arterial P_{CO_2} will be in equilibrium with the P_{CO_2} in the perfused alveoli only. This allows us to replace the unknown $P_{A_{CO_2}}$ with the readily determined $P_{A_{CO_2}}$. As a result, the mixed expired P_{CO_2} will differ from the arterial P_{CO_2} because of the effects of both the anatomical and the alveolar dead spaces. Thus, the physiological dead space may be determined as

$$V_{D(physiol)}/V_T = [Pa_{CO_2} - P\bar{E}_{CO_2}]/Pa_{CO_2}$$

In normal subjects, $V_{D(physiol)}$ is approximately equal to $V_{D(anat)}$, and accounts for about 25–30% of the tidal volume at rest. In patients with lung disease (e.g., pulmonary vascular occlusive disease or obstructive lung disease), $V_{D(physiol)}$ is usually appreciably larger than $V_{D(anat)}$ and can account for 50% or more of the tidal volume.

Diffusion

Although diffusion would be an ineffective process for appropriate rates of gas transfer if the diffusion pathway were to be long, the average alveolar diameter is only on the order of 100 μ, and so diffusion equilibrium throughout the alveolus takes place rapidly. In fact, Comroe had estimated that, even if the diffusion distance were as large as 0.5 mm, diffusion of the gas in the alveolar phase would be 80% complete within 0.002 seconds. Because the blood flowing through the lungs is exposed to the gas exchange surface of the pulmonary capillary bed for 0.75 to 1 seconds, it is clear that alveolar gas diffusion does not normally limit gas transfer into the blood. In a disease such as pulmonary emphysema, however, the diffusion distance can be much larger because some alveolar septa are destroyed and larger single air sacs are formed where numerous alveoli were originally located. In these abnormal units, therefore, diffusion within the air space might be expected to impair efficiency of oxygenation of the blood; however, such an effect has proven difficult to demonstrate, even in patients with chronic obstructive lung disease.

The most essential component of the gas transfer, however, is the uptake of O_2 across the alveolar–capillary membrane into the blood and CO_2 evolution in the reverse direction. The rate of gas exchange (i.e., volume per unit time) from alveolus to blood depends on the following factors.

Diffusion Constant (d)[1] *of the Gas*

This depends on both the rate of diffusion in the gas phase as well as the solubility (S) of the gas in the liquid phase; i.e.,

$$d \propto [S/\sqrt{MW}]$$

Although CO_2 diffuses about 20% more slowly than O_2, i.e., the solubility[2] of CO_2 is 0.59, whereas O_2 has a solubility of 0.024. The relative rates of gas transfer across the alveolar–capillary membrane are therefore conventionally presented as:

$$\begin{aligned}&\text{rate of } CO_2 \text{ transfer/rate of } O_2 \text{ transfer}\\&= [S_{CO_2}/S_{O_2}] \times [\sqrt{MW_{O_2}}/\sqrt{MW_{CO_2}}\\&= [0.59/0.024] \times [5.6/6.6]\\&= 20/1\end{aligned}$$

Geometry of Alveolar–Capillary Interface

The greater the surface area for diffusion, the greater is the net rate of gas transfer. For the lung this is the area (A) of the alveolar surface in contact with functioning pulmonary capillaries; however, the diffusion rate is inversely related to the diffusion distance (T).

Thus, the geometric factor affecting diffusion is A/T. It should be remembered that the total diffusion pathway for O_2 is: alveolar diffusion; diffusion through alveolar surface lining (surfactant) layer; diffusion through alveolar–capillary membrane; diffusion through plasma; diffusion through red cell (erythrocyte) membrane; diffusion through intracellular fluid of the red cell; and, finally, chemical combination with hemoglobin.

Rate of Gas Transfer

This also depends upon the pressure gradient across the capillary bed. Thus, by combining (a), (b) and (c), we see that the net rate of gas transfer of volume per unit time ($\dot{V}$)—or its "conductance"—is given by:

$$\dot{V} = d \times A/T \times (P_1 - P_2)$$

---

[1] Lower case d is used to symbolize this diffusion constant. The upper case D will be reserved for the variable "diffusing capacity" of the lung.

[2] In this case solubility is usually expressed as milliliters of gas dissolved per milliliter of fluid at a pressure of 1 atmosphere.

Diffusion—Uptake of O_2 into Blood

As mixed venous blood flows from the pulmonary artery into the beginning of the pulmonary capillary bed, it has a normal P_{O_2} at rest of about 40 mmHg. It is brought into close apposition with alveolar air, which has a P_{O_2} of 100 mmHg or more. That is, even though the mixed venous P_{O_2} is relatively stable, the alveolar P_{O_2} varies with the phase of the breathing cycle, being higher at end-inspiration and lower at end-expiration as a result of O_2 being transferred into the capillary blood. This continued gas exchange during expiration results in the end-tidal P_{O_2} being lower than the mean alveolar value. End-tidal P_{CO_2} is likewise higher than mean alveolar (and arterial) P_{CO_2}, and during exercise this difference can be as much as 6 mmHg.

The average gradient between mixed venous and alveolar P_{O_2} is 60 mmHg or more at the entrance to the pulmonary capillary bed. As a result, O_2 is taken up into the pulmonary capillary blood until it is in equilibrium with the alveolar air. This equilibrium normally takes 0.25–0.3 seconds; however, as pulmonary capillary blood normally takes an average of 0.8 seconds to traverse the capillary bed (at rest), there is ample time available for diffusion equilibrium.

The pattern of increase in pulmonary capillary P_{O_2} from its mixed venous value to its equilibrium with alveolar air is complex; the instantaneous pressure gradient for P_{O_2} that defines the instantaneous rate of O_2 transfer is itself dependent upon the characteristics of O_2 reaction with red cell hemoglobin (see later). The general time course of pulmonary capillary P_{O_2}, however, is given by the solid curve in Fig. 10.6. It should be noted that the initial rate of uptake is rapid because of the large P_{O_2} gradient from alveolus to capillary. As the gradient is progressively reduced, the rate of transfer decreases until P_{O_2} reaches an equilibrium.

It is important, however, to consider in some greater detail the factors that influence the rate at which P_{O_2} increases as the blood traverses the length of the pulmonary capillary. It is intuitively obvious that the greater the diffusive flux of the gas, the greater the quantity of that gas that flows into the blood per unit time will be. The

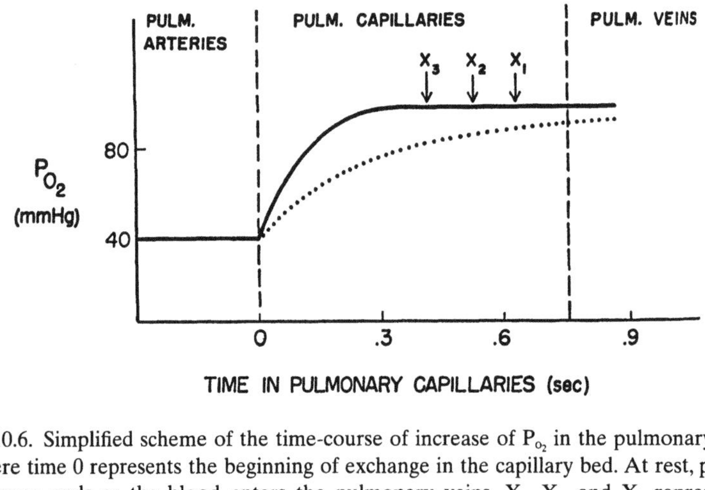

FIGURE 10.6. Simplified scheme of the time-course of increase of P_{O_2} in the pulmonary capillary bed, where time 0 represents the beginning of exchange in the capillary bed. At rest, pulmonary gas-exchange ends as the blood enters the pulmonary veins. X_1, X_2, and X_3 represent three progressively increased levels of muscular exercise that shortens the transit-time in the capillary bed. The dotted line represents the time-course through a markedly thickened alveolar capillary interface.

consequent increase in pulmonary capillary P_{O_2}, however, will depend both upon the slope of the blood "dissociation curve" for the gas (β) (i.e., the relationship between content change) and partial pressure change) and also upon the rate at which the blood flows through the capillary bed ($\dot{Q}c$). These factors can be combined into what Piiper and Scheid have termed an *equilibrium coefficient*, $D/\dot{Q}c.\beta$. This determines both the instantaneous rate of rise of the gas partial pressure as the blood traverses the pulmonary capillary as well as whether equilibrium will have been achieved between the gas (PA) and the blood (Pa) phases when the blood passes into the pulmonary veins, such that

$$P_A - Pa = (P_A - P\bar{v}) \cdot e^{D/\dot{Q}c.\beta}$$

This equation is rigorous for gases for which β is independent of partial pressure (i.e., inert gases and, to a good approximation, for O_2 in the hypoxic and hyperoxic ranges, where the dissociation slopes may be considered to be functionally linear, albeit with different slopes of β values. Throughout the normal physiological range for O_2, β changes with partial pressure (and other factors such as pH, P_{CO_2}, temperature, intraerythrocytic 2,3-diphosphoglycerate levels, etc.) and requires a more complex consideration of the profile of P_{O_2} increase.

Consideration of this equation, however, is instructive because it demonstrates that pulmonary diffusion limitation for a particular gas depends not simply on D but on $D/\dot{Q}c.\beta$, i.e., the "diffusive-to-perfusive conductance" ratio. It is clear, therefore, that conditions in which D is small and $\dot{Q}c$ and/or β large predispose to diffusion limitation rather than perfusion limitation of gas uptake. This is schematized in Fig. 10.7 for various gases. The foregoing discussion, however, has not taken into consideration other factors that complicate precise characterization of the diffusion process in the lungs; for example, any incomplete mixing of gas in the lung distal to the "stationary interface" (i.e., where diffusive flow takes over from convective flow in the airway); regional differences in thickness and composition; chemical reaction rates in the blood (e.g., O_2 binding with hemoglobin or $CO_2 \leftrightarrow HCO_3^-$ exchange, in which the enzyme carbonic anhydrase plays such an important role); the hematocrit itself—it is, for example, the erythrocyte "contact area" with the pulmonary capillary surface that dominates the "surface area" for diffusion, the O_2 solubility of plasma being so low; the pulsatility of $\dot{V}c$ over the cardiac cycle; and regional inhomogeneities of $D/\dot{Q}c.\beta$.

The time available for gas transfer into the blood is commonly quantified as the mean transit time (t_{tr}). This will depend both on the capillary volume (Vc) and the flow through it (i.e., the capacitance-to-conductance ratio). As the pulmonary blood flow ($\dot{Q}p$) is functionally equal to the cardiac output:

$$t_{tr}(sec) = Vc(ml)/\dot{Q}p(ml/sec)$$

There will be a distribution of transit times around the mean, however, with the smallest values being most susceptible to diffusional impairments: This will be exacerbated by lung disease.

If cardiac output is increased (e.g., during muscular exercise), the transit time through the pulmonary capillary bed will naturally be shortened. However if, for example, cardiac output is doubled, capillary transit time is not proportionally halved because the lung has great capacity to recruit previously nonperfused, or poorly perfused, capillaries and also has some capacity to distend the already perfused capillaries. This increases the pulmonary capillary blood volume. The reduction in transit time that does occur still does not typically reach the levels at which O_2 transfer is impaired in normal, healthy subjects at sea level, as shown by the symbols X_1, X_2, and X_3, which represent three progressively higher cardiac outputs (see Fig. 10.6). Some highly

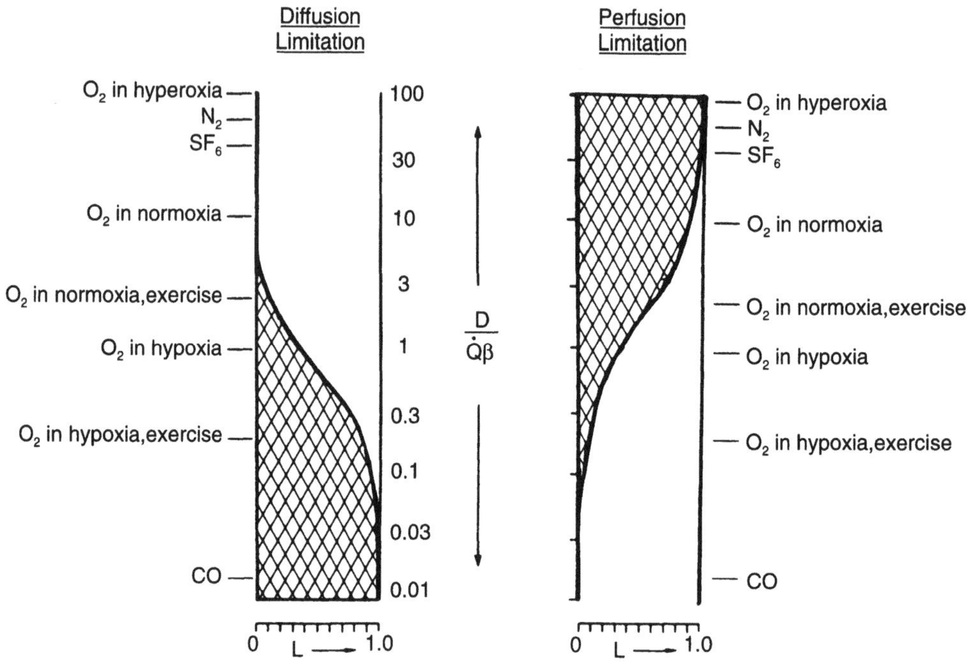

FIGURE 10.7. Influence of different values of the diffusive-to-perfusive conductances ($D/\dot{Q}\beta$). Note the diffusion limitation is associated with low values of this ratio, whereas perfusion limitation is the consequence of a high ratio. L_{diff} and L_{perf} represent the dependence of limitation by diffusion and perfusion, respectively; the "Diffusion limitation" and "Perfusion limitation" lines at the left represent the respective ranges in which $L > 5\%$. (*Source*: Modified from Piiper and Scheid.) (Taken from: Piiper, J., and Scheid, P. (1980) Blood-gas equilibration in the lungs. In: *Pulmonary Gas Exchange*, Vol. I. West, J.B. (ed), pp. 131–171. Academic Press, New York, NY.)

trained athletes, however, are capable of such high levels of pulmonary blood flow that arterial hypoxemia does result from this mechanism (i.e., pulmonary capillary volume recruitment is inadequate for the pulmonary blood flow recruitment).

Insufficient time for O_2 equilibrium, however, can impair O_2 transport across the alveolar–capillary bed under at least the following three conditions.

Increased Diffusion Path Length

If the alveoli are partially filled by some exudate (e.g., the proteinaceous material that is extruded into the alveolar space in the disease *pulmonary alveolar proteinosis*), or by pulmonary edema fluid, the path length for O_2 diffusion can be appreciably increased. This leads to a slower rate of rise of pulmonary capillary P_{O_2} toward its equilibrium value with alveolar P_{O_2}. Under these conditions, the time for equilibrium may be increased from the normal 0.3–0.6 seconds or even, in extreme cases, to values greater than the 0.8 seconds normally available for diffusion equilibrium. When the time needed for O_2 equilibrium exceeds the capillary transit time, the pulmonary venous blood and therefore arterial blood will clearly have low P_{O_2} and be hypoxemic. Even if there is just sufficient time for equilibrium at rest, however, increases in cardiac output leading to decreased pulmonary capillary transit time will tend to cause hypoxemia, which becomes progressively greater as cardiac output is progressively increased.

Reduction of Functioning Pulmonary Capillary Bed

Any condition that leads to fewer functioning pulmonary capillaries (e.g., blockade of pulmonary vessels with emboli), will cause the mean transit time of blood in the capillary bed to be decreased. In this case the same cardiac output, or pulmonary blood flow, will traverse a smaller volume in less time. Thus, the residence time of the blood in the pulmonary capillaries approaches that needed for equilibrium. Exercise can lead to insufficient time for equilibrium in such cases and arterial hypoxemia ensues. It should be noted that the actual structure of the diffusion path length may be unaltered in those regions through which the diffusion occurs; rather, the hypoxemia simply results from lack of sufficient time in the pulmonary capillary bed for diffusion equilibrium.

Reduced Alveolar P_{O_2}

When the driving pressure for O_2 transfer is reduced (e.g., low alveolar P_{O_2} due to a sojourn at high altitude or when a subject is given a low O_2 concentration to breathe), the rise time of pulmonary capillary P_{CO_2} will be slowed. Again, this can lead to the pulmonary blood leaving the capillary bed before it has come into equilibrium with the alveolar gas, especially during exercise.

The most striking condition for diffusion impairment of O_2 transfer is therefore a combination of a low alveolar P_{O_2} and exercise in a subject with pulmonary disease such as in those examples mentioned previously. In contrast, CO_2 retention caused by diffusion impairment is less common, even though the impairment is sufficient to cause severe hypoxemia.

Diffusing Capacity (or Pulmonary Gas Transfer Index)

The pulmonary diffusing capacity (DL) is defined as the volume of gas taken up into the pulmonary capillary blood from the alveoli per unit time ($\dot{V}$) divided by the pressure gradient for that gas across the alveolar–capillary interface ($P_1 - P_2$); that is,

$$\mathrm{D_L} = [\dot{V}/(P_1 - P_2)]$$

If we recall that the general formula for gas transfer across the lung is

$$\dot{V} = d \cdot A/T \cdot (P_1 - P_2)$$

then, it is apparent that DL really takes both the specific diffusing characteristics of the gas as well as the area and thickness of the intervening diffusing pathway into account.

For O_2 the diffusing capacity is given by the formula

$$\mathrm{D_{L_{O_2}}} = \dot{V}_{O_2}/(P_{A_{O_2}} - P\bar{c}_{O_2})$$

where $P\bar{c}_{O_2}$ = the mean pulmonary capillary P_{O_2}.

We can readily measure O_2 uptake in a given time and establish an estimate of mean alveolar P_{O_2}. Mean capillary P_{O_2}, however, is extremely difficult to measure because it represents an integrated mean value of the instantaneous increments of P_{O_2} of capillary blood as this rises from mixed venous to its arterial value. As stated earlier, this has a complex time course that is influenced by the O_2 in solution and by its reaction and binding with hemoglobin of the red cells.

To circumvent these inherent difficulties, the diffusing capacity for carbon monoxide (CO) is usually determined. Carbon monoxide is considered an appropriate gas for this use because of the following properties:

1. CO follows the same pulmonary transfer route as O_2 and binds to hemoglobin on the same site.

2. Except in areas where CO pollutes the air or in subjects who smoke tobacco, the CO concentration in venous blood entering the pulmonary capillary bed is effectively zero.

3. The affinity of CO for hemoglobin is about 210 times greater than that of O_2. This means that if small quantities[3] are administered to a subject, the CO in the capillary blood will be preferentially bound to the hemoglobin; consequently, the P_{CO} in the capillary blood will be disappearingly small and in practice is considered to be zero. Hence,

$$D_{L_{CO}} = \dot{V}_{CO}(\text{ml/min})/P_{A_{CO}}(\text{mmHg})$$

where V_{CO} is the rate of CO uptake.

4. $D_{L_{O_2}}$ can be readily determined from $D_{L_{CO}}$ by referring to the solubilities and molecular weights of the gases:

$$\begin{aligned} D_{L_{O_2}}/D_{L_{CO}} &= S_{O_2}/S_{CO} \times \sqrt{MW_{CO}}/\sqrt{MW_{O_2}} \\ &= 0.024/0.018 \times \sqrt{28}/\sqrt{32} \\ &= 1.23 \end{aligned}$$

i.e.,

$$D_{L_{O2}} = 1.23 \cdot D_{L_{CO}}$$

Several tests are used to determine $D_{L_{CO}}$, but a common procedure is to inhale a gas mixture containing a low concentration (~0.3%) CO and then to hold the breath for 10 seconds. While the breath is being held, CO enters the blood; obviously, the more CO that enters in this time, the greater the diffusion capacity for the gas.

The volume of CO taken up by the blood during breath-holding is computed ($\dot{V}_{CO}$) and the mean alveolar CO during the time is measured. Hence $D_{L_{CO}}$ is computed. The normal value for $D_{L_{CO}}$ by this test is about 25 ml/min/mmHg in resting subjects.

Factors that Affect $D_{L_{CO}}$

In addition to the thickness and area of the surface, several other factors affect $D_{L_{CO}}$.

Hemoglobin (Hb) Concentration

The greater the Hb concentration, the greater the number of sites for combination with CO per unit blood volume and thus the greater the rate of CO uptake. Hence, polycythemia leads to high $D_{L_{CO}}$, whereas anemia leads to low $D_{L_{CO}}$.

[3] As CO can be a lethal gas in somewhat higher concentrations (it binds to the transport site of O_2 on the hemoglobin and hence impairs O_2 delivery to the tissues), only a very small amount is inhaled in these tests (e.g., 0.3% CO).

Pulmonary Blood Volume

Increased volume of blood in the capillary bed increases the effective surface area for diffusion, especially if there is recruitment of pulmonary capillaries, and makes more Hb molecules available for uptake of CO. Thus, exercise or recumbency increases $D_{L_{CO}}$ by increasing pulmonary blood volume. The $D_{L_{CO}}$, however, does not increase simply due to increases in pulmonary blood flow per se because so few of the available binding sites for CO on the Hb are filled during this test. As a result, so many are still available for binding that it does not alter the rate of CO uptake if new red cells replace the original ones in the pulmonary capillary bed with increased blood flow.[4] The uptake rate of CO, therefore, would be maintained for a long period even though the blood flow were to stop. Most conditions in which pulmonary blood flow increases, however, lead to a simultaneous increase in pulmonary blood volume, or alteration of the perfusion pattern, so that $D_{L_{CO}}$ is, in fact, affected by the indirect effects of the increased flow.

Body Size

The $D_{L_{CO}}$ increases with body size, presumably owing to the larger pulmonary surface area available for diffusion.

Altered Alveolar P_{O_2}

If O_2 concentration is high, there is greater competition for Hb-binding sites. This leads to reduced $D_{L_{CO}}$; similarly, low P_{O_2} increases $D_{L_{CO}}$. As the membrane component of $D_{L_{CO}}$ is presumably unaltered by these variations in P_{O_2}, it is possible to perform $D_{L_{CO}}$ tests at various levels of P_{O_2} and determine the separate effects of the membrane and the Hb reaction components. For example, the total resistance to diffusion of the gas is made up of the sum of the membrane resistance and the resistance offered to the Hb binding. Recalling that the resistance to gas flow is the inverse of its conductance, then,

$$1/D_L = 1/D_M + 1/\theta \cdot Vc$$

where D_L = the total diffusing capacity
D_M = the membrane component (this includes alveolar–capillary and red cell membranes)
Vc = the volume of blood in the pulmonary capillaries
θ = the reaction rate coefficient for the chemical combination of Hb with gas

If $D_{L_{CO}}$ is measured at various levels of P_{O_2} [i.e., altering the reaction rate coefficient (θ) and assuming that it is all that is altered], then plotting $1/D_L$ against $1/\theta$ results in a straight line relationship (Fig. 10.8) with a slope of $1/Vc$ (hence, we can determine pulmonary capillary blood volume) and intercept at $1/D_M$. This technique is therefore a refinement of the commonly used $D_{L_{CO}}$ tests because it allows us to determine whether or not reduced $D_{L_{CO}}$ is due predominantly to impaired membrane diffusion.

[4] This is not true for O_2.

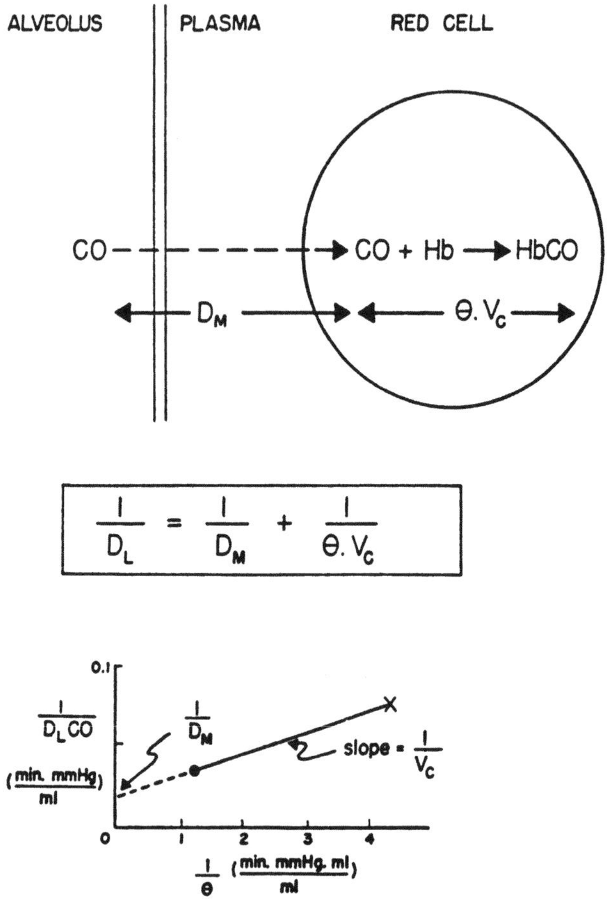

FIGURE 10.8. Representation of a technique that allows the pulmonary diffusing capacity to be partitioned into its membrane and red-cell components. Variations of θ (the reaction rate of oxygen with hemoglobin) are usually achieved by varying the inspired oxygen partial-pressure.

Mechanisms of Hypoxemia and Hypercapnia

In humans breathing normal air at sea level, arterial hypoxemia (i.e., reduced arterial O_2 pressure) can result from four different causes, or any combination of the four. These are: (1) alveolar hypoventilation; (2) diffusion abnormality; (3) right-to-left shunt; and (4) ventilation–perfusion ($\dot{V}A/\dot{Q}$) inequality.

Alveolar Hypoventilation

The fundamental relationship between the rate of metabolic O_2 utilization ($\dot{V}O_2$) and alveolar ventilation ($\dot{V}A$) defines the level of alveolar P_{O_2}; that is,

$$\dot{V}_{O_2} = \dot{V}A(FI_{O_2} - FA_{O_2})$$

or,

$$863 \cdot \dot{V}_{O_2(STPD)} / \dot{V}A_{(BTPS)} = PI_{O_2} - PA_{O_2}$$

It is clear, therefore, that if alveolar ventilation is inappropriately low for the current metabolic O_2 requirements, the difference between inspired and alveolar P_{O_2} will increase; this means that alveolar P_{O_2} will fall. As a result, arterial P_{O_2} is reduced by hypoventilation. The difference between alveolar and arterial P_{O_2} [i.e., $(A\text{–}a)P_{O_2}$], however, remains unchanged.

If alveolar ventilation is too low for the rate of metabolic CO_2 production, $P_{A_{CO_2}}$ and Pa_{CO_2} will rise similarly, as described by:

$$P_{A_{CO_2}} = 863 \cdot \left[\dot{V}CO_{2(STPD)} / \dot{V}_{A(BTPS)}\right]$$

Thus, alveolar ventilation leads to reduced arterial P_{O_2} (hypoxemia) and increased arterial P_{CO_2} (hypercapnia). The fall in Pa_{CO_2} is related to the rise of Pa_{CO_2} by the gas exchange ratio (R) as shown by the following:

$$R = \left(\dot{V}_{CO_2} / \dot{V}_{O_2}\right) \simeq \dot{V}A \cdot FA_{CO_2} / \dot{V}A(FI_{O_2} - FA_{O_2})^{5}$$

which reduces to

$$R \simeq PA_{CO_2} / (PI_{O_2} - PA_{O_2})$$

Thus, when R = 1, the change in PA_{CO_2} and PA_{O_2} is the same; the change in Pa_{CO_2} is therefore equal to the change in Pa_{O_2}; However, as R is normally about 0.8 at rest, Pa_{CO_2} will normally increase by about 8 mmHg for each 10 mmHg decrease in Pa_{O_2} due to hypoventilation.

It is important to recognize that alveolar hypoventilation can occur although the lung itself is normal such as in diseases affecting the medullary respiratory center, or respiratory neuromuscular diseases. Hypoventilation, however, is also seen in severe obstructive airways disease with high work of breathing.

Diffusion Abnormality

As described earlier, impaired diffusion from alveolus to pulmonary capillary blood can lead to arterial hypoxemia. This is exacerbated by high cardiac output states such as muscular exercise. As O_2 is loaded into the alveolus at an appropriate rate in this condition but fails to be loaded adequately into the blood, $(A\text{–}a)P_{O_2}$ will be widened.

The CO_2 retention from diffusion impairment is less common. Any increase in Pa_{CO_2} that might ensue is normally corrected by ventilatory control mechanisms, which are normally highly sensitive to Pa_{CO_2}. In contrast, ventilatory stimulation from low arterial P_{O_2} only becomes appreciable after Pa_{O_2} falls below approximately 60 mmHg. Thus, for moderate hypoxemia from diffusion impairment, Pa_{O_2} is decreased, $(A\text{–}a)P_{O_2}$ is widened, and Pa_{CO_2} can be normal. When more severe hypoxemia leads to hypoxic stimulation of the carotid bodies, hyperventilation ensues: Pa_{O_2} is low, $(A\text{–}a)P_{O_2}$ is widened, but now Pa_{CO_2} is low.

Right-to-Left Shunt

In normal subjects the bronchial circulation (to the larger airways) and thebesian vessels (from the myocardium) drain the venous blood into the arterialized blood, "downstream" of the pulmonary capillary bed. This venous blood consequently reduces mixed arterial blood P_{O_2} (Pa_{O_2}) to values less than those of the pulmonary end-capillary

[5] For clarity, the normal small difference between inspired and expired alveolar ventilation is disregarded here.

blood; however, because the normal right-to-left shunt is only a few percent of the cardiac output, this reduction in Pa_{O_2} is only a few millimeters of mercury.

Cardiac and/or pulmonary disease can lead to marked increases in the fraction of the cardiac output that does not undergo gas exchange in the lung. Atelectatic (collapsed) regions of the lung that still have pulmonary blood flow and abnormal pulmonary arteriopulmonary venous fistulas contribute to intrapulmonary right-to-left shunts. Intracardiac right-to-left shunts can also develop (e.g., through atrial or ventricular septal defects) if associated with abnormally high blood pressures in the right side of the heart. The reduction in arterial P_{O_2} from right-to-left shunts depends on several factors. Also include the shunt fraction of the cardiac output ($\dot{Q}s/\dot{Q}t$), mixed venous O_2 content ($C\bar{v}_{O_2}$), and HbO_2 dissociation curve. If one is prepared to make some assumptions, such as pulmonary end-capillary blood has a uniform composition, and all blood that is "shunted" has a mixed venous O_2 content, then the shunt fraction of cardiac output can be readily determined. We know that arterial O_2 flow (per unit time) is made up of a pulmonary capillary component and a shunt component:

$$\text{cardiac output} \times Ca_{O_2} = (\text{pulmonary blood flow} \times \text{end-capillary } O_2 \text{ content}) + (\text{shunt blood flow} \times C\bar{v}_{O_2})$$

$$\therefore \dot{Q}t \cdot Ca_{O_2} = (\dot{Q}p \cdot Cc'_{O_2}) \times (\dot{Q}s \cdot C\bar{v}_{O_2})$$

but

$$\dot{Q}p = \dot{Q}t - \dot{Q}s$$

$$\therefore \dot{Q}t \cdot Ca_{O_2} = (\dot{Q}t \cdot Cc'_{O_2}) - (\dot{Q}s \cdot Cc'_{O_2}) + (\dot{Q}s \cdot C\bar{v}_{O_2})$$

or,

$$\dot{Q}s/\dot{Q}t = (Cc'_{O_2} - Ca_{O_2})/(Cc'_{O_2} - C\bar{v}_{O_2})$$

In this equation, the pulmonary end-capillary O_2 content (Cc'_{O_2}) is determined by assuming that end-capillary P_{O_2} is equal to the "ideal" alveolar $P_{O_2}(PA_{O_2})$. The true mean alveolar P_{O_2} clearly would be extremely difficult to measure in a lung with maldistribution of $\dot{V}A/\dot{Q}$. PA_{O_2} can be determined from the following equation:

$$PA_{O_2} = PI_{O_2} - Pa_{CO_2}/R + [Pa_{CO_2} \cdot FI_{O_2} \cdot (1-R)/R]$$

Because the term in the square brackets normally accounts for a few millimeters of mercury (and is zero when R = 1), it is often neglected in estimating PA_{O_2}; that is,

$$PA_{O_2} = PI_{O_2} - (Pa_{CO_2}/R)$$

As CO_2 in the shunted blood is higher than in the pulmonary end-capillary blood, right-to-left shunt also tends toward CO_2 retention. This is rarely observed, however, because of the normally small venous-to-arterial P_{CO_2} difference (about 6 mmHg) compared with O_2 (around 60 mmHg), and also because the mechanisms of ventilatory control normally respond to increased P_{CO_2} with an increase of ventilation. As stated earlier, if P_{O_2} falls appreciably because of the right-to-left shunt, hypoxemia will stimulate ventilation by its effect on the peripheral chemoreceptors: this leads to hyperventilation and reduced Pa_{CO_2}. However, as the alveolar and end-capillary P_{CO_2} will be less than the arterial P_{CO_2} as a consequence of the shunt, this will influence the computation of the dead space fraction of the breath, such that it will appear as if the dead space is inappropriately large even when this is not the case. In summary, therefore, moderate right-to-left shunt leads to reduced Pa_{O_2}, widened $(A\text{–}a)P_{O_2}$, but relatively normal Pa_{CO_2}. Severe right-to-left shunts cause markedly reduced Pa_{O_2}, wide $(A\text{–}a)P_{O_2}$, and also reduced Pa_{CO_2}.

The magnitude of the shunt can be determined by giving the subject 100% O_2 to breathe until the lung becomes nitrogen free (and assuming that the O_2 does not change the magnitude of the shunt). In the ideal lung, the arterial P_{O_2} would be expected to

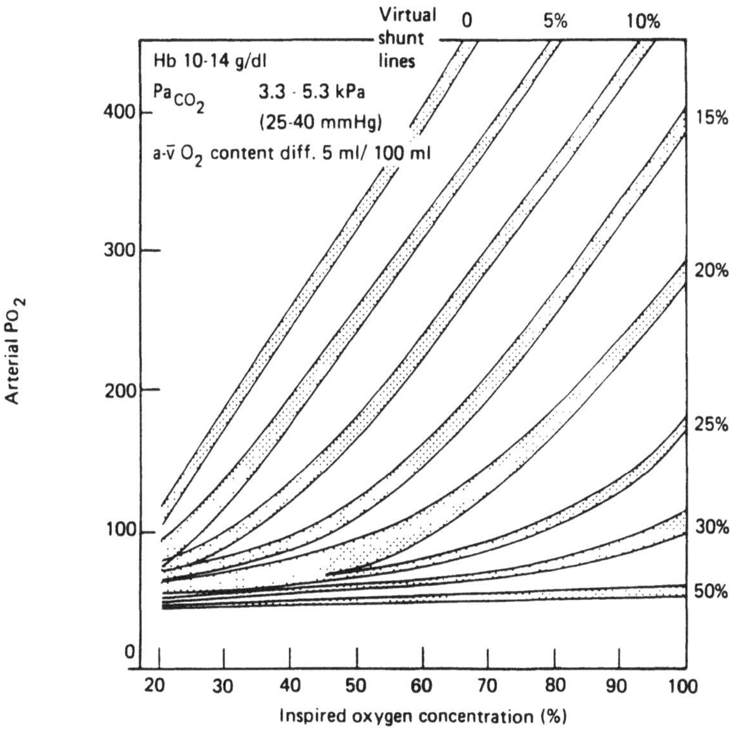

FIGURE 10.9. The effect of different inspired oxygen concentrations on the measured arterial P_{O_2} for different levels of constant right-to-left shunt. The right-hand axis reflects the commonly used influence of 100% oxygen breathing on the arterial P_{O_2} at different levels of right-to-left shunt. This graph assumes an arterial venous oxygen content difference of 5 ml/100 ml (i.e., consistent with the resting condition). (*Source*: Reproduced from Nunn.)

increase to an "expected" value of roughly 650 mmHg [i.e., equal to the PA_{O_2} of (760 − 40 mmHg PA_{CO_2} − 47 mmHg water vapor)]. When the arteriovenous difference for O_2 is normal at 5 ml/100 ml, each 5% shunt will reduce Pa_{O_2} by approximately 100 mmHg below the expected value. Under these conditions if the Pa_{O_2} increased to only 450 mmHg (i.e., 200 mmHg below the "expected" 650 mmHg), then there would be a 10% right-to-left shunt. This useful "rule of thumb" is only valid while the arterial hemoglobin is fully saturated, the shunted blood only affects the quantity of blood dissolved, and with an arteriovenous O_2 content difference of 5 ml/100 ml. Figure 10.9 depicts this relationship for various levels of hyperoxic inspirates. One might therefore consider the question: At greater than what value of right-to-left shunt is it no longer possible to maintain an arterial P_{O_2} of 100 mmHg even breathing 100% O_2? As shown in Fig. 10.9, the answer is approximately 30%.

$\dot{V}A/\dot{Q}$ *Inequality*

Under conditions in which overall, or average, $\dot{V}A$ is approximately equal to overall, or average, $\dot{Q}$(i.e., a normal average $\dot{V}A/\dot{Q}$), the lung may still have regions of high, normal, and low $\dot{V}A/\dot{Q}$ interspersed. The consequence of such a distribution is that blood from the high $\dot{V}A/\dot{Q}$ region will reflect hyperventilation, with high P_{O_2} and low P_{CO_2}; that blood from the normal $\dot{V}\dot{Q}$ region will have normal values for P_{O_2} and P_{CO_2}; and that blood from the low $\dot{V}A/\dot{Q}$ region will reflect hypoventilation, with low P_{O_2} and high P_{CO_2}. The overall arterial P_{O_2} and P_{CO_2} will therefore be a consequence of the aver-

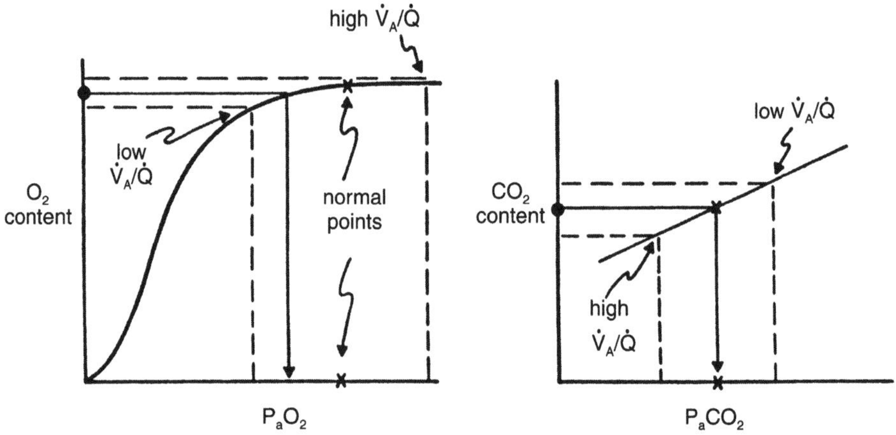

FIGURE 10.10. The influence of altered ventilation-to-perfusion ratios on the mean arterial partial pressures of O_2 and CO_2. The nonlinear O_2 dissociation curve results in obligatory hypoxemia (arrowhead) compared with X, the relatively linear CO_2 dissociation curve that obviates this effect. (See text for further discussion.)

aging of the total contents of the gas from each "stream," in proportion to the blood flow from each region. The blood dissociation curve for O_2, is nonlinear however. The result is that low $\dot{V}A/\dot{Q}$ regions lead to both low P_{O_2} and low O_2 content, whereas high $\dot{V}A/\dot{Q}$ regions lead to high P_{O_2}, with only small increases in O_2 content above the normal value (i.e., the O_2 dissociation curve is relatively flat in this range). Mixing blood from low $\dot{V}A/\dot{Q}$ regions with blood of high $\dot{V}A/\dot{Q}$ regions will therefore result in an average P_{O_2} that is "weighted" toward the low $\dot{V}A/\dot{Q}$ blood value (Fig. 10.10). The resulting value for the mean P_{O_2} will depend on the volumes of blood from each "region" that stream into the mixed arterial blood per unit time. Thus, the high $\dot{V}A/\dot{Q}$ regions (even if their hemoglobin is completely saturated) tend to be functionally unable to "compensate" for the regions of low $\dot{V}A/\dot{Q}$. The result is that even under conditions in which the overall $\dot{V}/\dot{Q}$ is normal, uneven distribution of $\dot{V}A$ with respect to $\dot{Q}$ results in arterial hypoxemia.

Maldistribution of $\dot{V}A/\dot{Q}$ also impairs the lung's ability to clear CO_2 despite the relatively linear CO_2 dissociation curve (Fig. 10.10). This is illustrated in Fig. 10.11. For example, suppose that the lung initially operates as an "ideal" gas exchanger—such that the arterial O_2 and CO_2 values are at points "x" (Fig. 10.10). Now, if we assume that the lung develops uneven $\dot{V}A/\dot{Q}$ through maldistribution of ventilation (but for simplicity we shall consider the perfusion pattern to remain normal) such that the Pa_{O_2} of the hypoventilated region is reduced below normal by the same amount that the Pa_{O_2} of the hyperventilated region is raised above the normal value. When blood from these regions mix, the average of the O_2 contents will necessarily result in a reduced mean Pa_{O_2}, as shown by the arrows that originate at the dot (•) on the content scale (Fig. 10.10).[6] Applying the same model to CO_2, however, we see that the resulting mixed arterial P_{CO_2} (as shown by the arrow), is at or close to normal because of the relatively linear CO_2 dissociation curve (Fig. 10.10). It is important to consider, however, how the lung with such maldistribution of ventilation effects such changes of P_{CO_2}. The

[6] Effects such as those induced by Haldane and Bohr effects are neglected in this example.

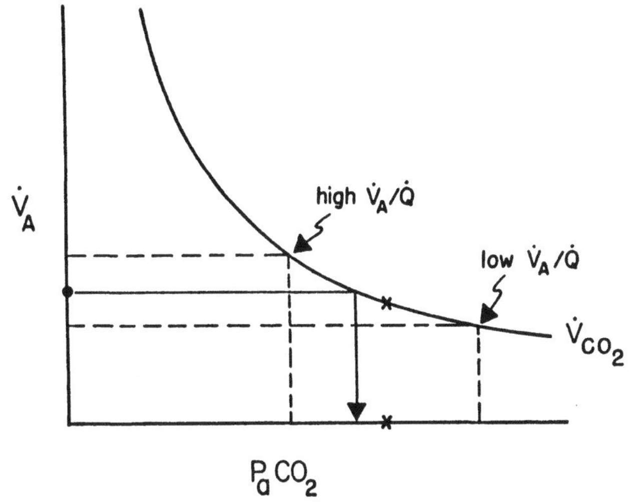

FIGURE 10.11. Schematic representation to show that in order to maintain arterial P_{CO_2} constant in the face of variations of ventilation-to-perfusion ratio, the alveolar ventilation to the high $\dot{V}A/\dot{Q}$ region must be appreciably higher than the decrement in alveolar ventilation to the low $\dot{V}A/\dot{Q}$ region. As a result, the mean alveolar P_{CO_2} (represented by the vertical arrowhead) will be low relative to the arterial value. (Taken from: Nunn, J.F. (1987) *Applied Respiratory Physiology*, p. 171. Butterworths, London, U.K.)

hyperventilated and the hypoventilated lung regions in this example are shown on the graphical representation (Fig 10.11) of the equation

$$\dot{V}CO_2 = \dot{V}A \cdot FA_{CO_2}$$

or

$$\dot{V}CO_2 = \dot{V}A \cdot PA_{CO_2}/(PB - 47)$$

Notice that the *increment* in alveolar ventilation required to maintain P_{CO_2} of the *hyper*ventilated region is much greater than the *decrement* of alveolar ventilation required to maintain P_{CO_2} of the *hypo*ventilated region. As a result, although the mixed arterial blood can be normal (points "x" in Fig. 10.11) in such a condition, the total alveolar ventilation must be increased to maintain it (symbol • in Fig. 10.11). A further consequence is that mean alveolar P_{CO_2} (arrow in Fig. 10.11) will be lower than if the lung exchanged gas in an "ideal" manner. As a result, maldistribution of $\dot{V}A/\dot{Q}$ results in both a widened $(A–a)P_{O_2}$ and a widened $(a–A)P_{CO_2}$.

The pattern of arterial blood and alveolar gas tensions in $\dot{V}A/\dot{Q}$ maldistribution is such that with mild or moderate maldistribution Pa_{O_2} is low, $(A–a)P_{O_2}$ is widened, and Pa_{CO_2} can be normal or low, depending on the degree of ventilatory stimulation consequent to the hypoxemia (or other factors such as regional lung distortion). In severe $\dot{V}A/\dot{Q}$ impairment, associated with severe airways obstruction, hypoventilation can ensue leading to increased Pa_{CO_2} that serves to reduce Pa_{O_2} even more.

It should be noted that of the four independent causes of arterial hypoxemia at sea level, only hypoventilation is inevitably associated with CO_2 retention. It is also the only cause that does not lead to increased $(A–a)P_{O_2}$ (Table 10.2). The $(A–a)P_{O_2}$ might therefore be considered the best single indicator of the efficiency of the lung for gas exchange, as (1) diffusion impairment, (2) right-to-left shunt, and (3) maldistribution of $\dot{V}A/\dot{Q}$ all widen the gradient.

TABLE 10.2. Alveolar and arterial blood gas tensions typically associated with causes of hypoxemia.

Cause of hypoxemia	Pa_{O_2}	$(A\text{–}a)P_{O_2}$	Pa_{CO_2}	$(a\text{–}A)P_{CO_2}$
Alveolar hypoventilation	↓	Normal	↑	Normal
Diffusion impairment	↓	↑	Normal or ↓	Normal or ↑
Right-to-left shunt	↓	↑	Normal or ↓	↑
Maldistribution of $\dot{V}_A/\dot{Q}$	↓	↑	Normal, ↓ or ↑	↑

It is clear that maldistribution of alveolar ventilation ($\dot{V}_A$) with respect to its perfusion ($\dot{Q}$) predisposes to arterial hypoxemia, and that an obligatory widening of the alveolar-to-arterial P_{O_2} difference is one thing while quantifying its effects is another. Wagner and West and their colleagues have pioneered a technique that has been generally accepted to do precisely that. The technique works on the principle that if a mixture of inert gases (e.g., dissolved in saline) is infused at a constant rate into the inflow of the pulmonary capillary bed, then in the steady state, the amount "lost" into the alveolar gas (i.e., what does not remain in the pulmonary venous blood) will be a function only of the partition coefficient (λ) for the individual gas and the $\dot{V}_A/\dot{Q}$ of the particular lung unit, such that:

$$\frac{Pa}{P\bar{v}} = \frac{\lambda}{\lambda + \dot{V}_A/\dot{Q}}$$

Note that when λ is small (as for a gas such as sulfur hexafluoride), there will be relatively little remaining in the end-capillary or arterial blood, whereas when λ is large (as for acetone) more will be retained in the arterial blood. The greater the $\dot{V}_A/\dot{Q}$ value, however, the less gas will be retained in the arterial blood for a given mixed venous value, regardless of the value of λ.

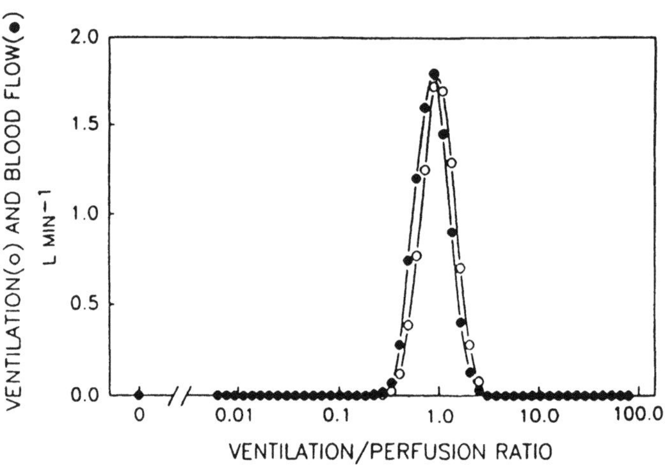

FIGURE 10.12. Characterization of the distribution of ventilation (open circles) and pulmonary blood flow (solid circles) in a young, healthy subject in the upright position. Note that there is very little mismatching of ventilation to perfusion, and no discernible right-to-left shunt (i.e., a ventilation-to-perfusion ratio of zero). (Taken from: West, J.B., and Wagner, P.D. (1991) Ventilation-perfusion relationships. In: *The Lung: Scientific Foundations*. Crystal, R.G., and West, J.B. (eds.), pp. 1289–1305. Raven Press, New York, NY.)

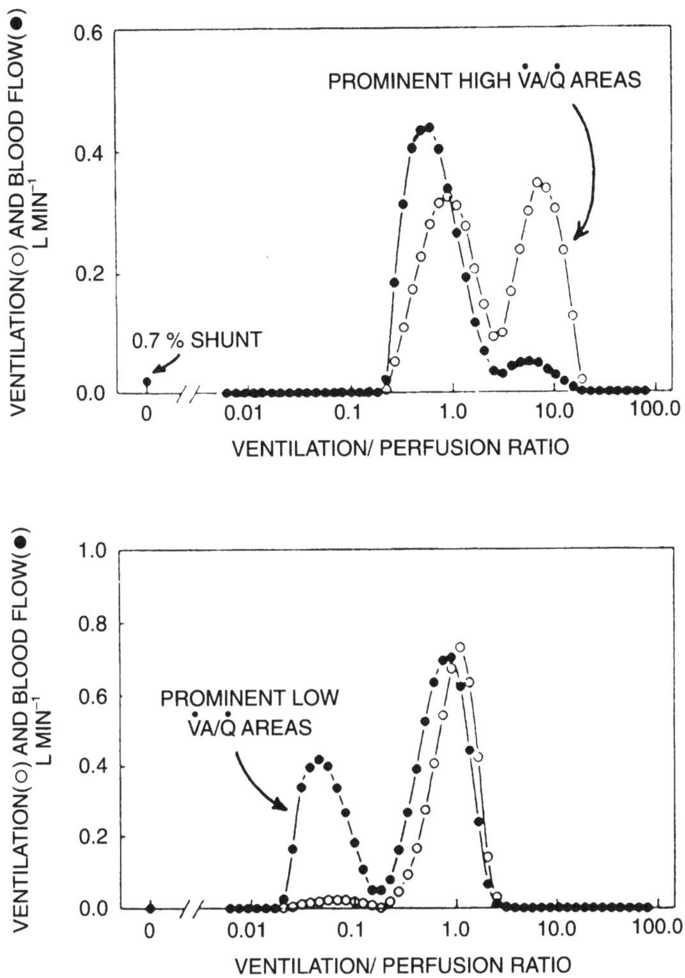

FIGURE 10.13. Characterization of the dispersion of ventilation-to-perfusion ratios in patients with predominantly Type A ("emphysematous," upper panel) and Type B ("bronchitic," lower panel) chronic obstructive pulmonary disease. Note that the Type A patient is characterized with large regions of high ventilation-to-perfusion ratios, but with little right-to-left shunting. The Type B patient does not manifest the regions of high ventilation-to-perfusion, but in this case there are appreciable areas of low ventilation-to-perfusion, again with no discernible right-to-lift shunt. (Taken from: West, J.B., and Wagner, P.D. (1991) Ventilation-perfusion relationships. In: *The Lung: Scientific Foundations*. Crystal, R.G., and West, J.B. (eds.), pp. 1289–1305. Raven Press, New York, NY.)

From such an analysis of the results of an adequate number of inert gases (usually six to eight) whose solubility or partition coefficients are sufficiently diverse, it has proved possible to construct a graphical representation of the $\dot{V}_A/\dot{Q}$ distribution in humans (including regions in which $\dot{V}_A/\dot{Q} = 0$; i.e., "shunts"), and the consequent effects on $(A\text{–}a)P_{O_2}$. For example, the distribution of $\dot{V}_A$ and $\dot{Q}$ are closely matched in a normal young subject (Fig. 10.12). In subjects with lung disease, however, the distributions can be shown to be poorly matched, with little or no right-to-left shunting in patients with chronic obstructive pulmonary disease (COPD) of either Type A or Type B. In patients with either diffuse interstitial fibrosis or pulmonary emboli, however, significant right-to-left shunts are demonstrable (Fig. 10.13).

The other significant feature of this technique is that, as it quantifies the $\dot{V}A/\dot{Q}$ and shunt effects on the Pa_{O_2}, the diffusion component can be determined by exclusion (i.e., if there is a residual component of Pa_{O_2} after the shunt and $\dot{V}A/\dot{Q}$ effects have been quantified, then what remains is considered to be the diffusion component).

The technique has also been employed to demonstrate that a component of the widening of the $(A\text{–}a)P_{O_2}$ seen during exercise even in normal young subjects is, surprisingly, attributable to an increased dispersion of $\dot{V}A/\dot{Q}$ (despite the improved topographical distributions of both $\dot{V}A$ and $\dot{Q}$). It has also shown that normal subjects at sea level can develop a diffusion impairment for O_2 exchange that accounts in part for the hypoxemia demonstrated in some highly "fit" subjects exercising at high metabolic rates. This effect is, of course, more prominent in subjects at high altitude.

Summary

Defining the gas-exchange function of the lung to be that of "alveolarizing" the mixed venous blood both obviates the tautology of using "arterializing" in the definition and provides a frame of reference for quantifying the effectiveness of the exchange process (i.e., the alveolar-to-arterial partial pressure difference for the gas of interest). Thus, while alveolar hypoventilation leads to arterial hypoxemia and hypercapnia, it does not widen the alveolar-to-arterial partial pressure differences. The other mechanisms of arterial hypoxemia do: (a) diffusion impairment, (b) right-to-left shunt, and (c) maldistribution of alveolar ventilation with respect to diffusion ($\dot{V}A/\dot{Q}$). However, establishing a single representative value for alveolar PO_2, for example, when there are large regional variations in gas-exchange efficiency and whose contributions to the expirate are flow-dependent, is naturally a challenging task. The issue is ordinarily resolved by using the "ideal" alveolar PO_2 as the frame of reference. The usefully factitious index is the alveolar PO_2 that would obtain were there to be no influence of mechanisms (a), (b), and (c) listed above, and is practically computed by assuming that the alveolar PO_2 exactly equals the arterial value. This assumption also forms the basis for the practical means of computing the physiological dead space fraction of the breath (VD/VT) which is high when alveolar (and arterial) PCO_2 differ greatly from the mixed expired PCO_2 as a result of total ventilation greatly exceeding alveolar ventilation. However, VD/VT is highly dependent upon breathing pattern, and so a high value may not, *per se*, indicate abnormal gas exchange. A further important factor relates to the time available for diffusional gas exchange in the pulmonary capillaries. The depends both upon the volume of the pulmonary capillaries (Qc) and their blood flow ($\dot{Q}c$), whose ratio (i.e., $Qc/\dot{Q}c$) provides the mean transit time—of what, especially in subjects with pulmonary vascular disease, could be a wide distribution of regional values. The extent of any disequilibrium between the alveolar and end-capillary gas partial pressures will depend on what has been termed the "equilibrium coefficient," $D/\dot{Q}c,\beta$, where D is the diffusion coefficient and β is the local slope of the dissociation curve for the particular gas: the combination of a small D and a large $\dot{Q}c$ and/or β predisposes to diffusion limitation of gas transfer, whereas the reverse predisposes to perfusion limitation. If an appropriate number (e.g., 6–8) of inert gases, of sufficiently diverse partition coefficients, are infused to a steady state into the mixed venous blood, then their retentions into the "arterialized" blood (or excretions into the alveolar gas) allow a functionally continuous representation of the lung's $\dot{V}A/\dot{Q}$ profile. As this also quantifies the regions for which $\dot{V}A/\dot{Q} = 0$ (i.e., shunt), the expected arterial PO_2 can be established, together with the contributions to any hypoxemia from $\dot{V}A/\dot{Q}$ and shunt; any remaining residual hypoxemia must be attributable to diffusion impairment. This important advance allows the mechanism(s) of arterial hypoxemia to be quantitatively apportioned, hence providing a focus for amelioration.

References

Dempsey, J.A., and Wagner, P.D. (1999) Exercise-induced arterial hypoxemia. J. Appl. Physiol. 87, 1997–2006.

Farhi, L.E. (1967) Elimination of inert gas by the lung. Respir. Physiol. 3, 1–11.

Hammond, M.D., Gale, G.E., Kapitan, K.S., Ries, A., and Wagner, P.D. (1986) Pulmonary gas exchange in humans during exercise at sea level. J. Appl. Physiol. 60, 1590–1598.

Hammond, M.D., and Hempleman, S.C. (1987) Oxygen diffusing capacity estimates derived from measured $\dot{V}_A/\dot{Q}$ distributions in man. Respir. Physiol. 69, 129–142.

Hughes, J.M.B. (1991) Diffusive gas exchange. In: Exercise, pulmonary physiology and pathophysiology. Whipp, B.J., and Wasserman, K. (eds.), pp. 143–171. Dekker, New York.

Johnson, R.L., Jr., Heigenhauser, G.J.F., Hsia, C.C.W., Jones, N.L., and Wagner, P.D. (1996) Determinants of gas exchange and acid-base balance during exercise. In: Handbook of physiology, Section 12: Exercise: Regulation and integration of multiple systems. Rowell, L.B., and Shepherd, J.T. (eds.), pp. 541–584. Oxford University Press, New York.

Nunn, J.F. (1987) Applied respiratory physiology. Third ed. p. 171. Butterworths, London.

Piiper, J. (1996) Pulmonary gas exchange. In: Comprehensive human physiology. Greger, R., and Windhorst, U. (eds.), pp. 2037–2049. Springer Verlag, Heidelberg.

Rahn, H., and Fenn, W.O. (1955) Graphical Analysis of the Respiratory Gas Exchange, pp. 1–380. American Physiological Society, Washington, D.C.

Riley, R.L., and Cournand, A. (1949) "Ideal" alveolar air and the analysis of ventilation-perfusion relationships in the lungs. J. Appl. Physiol. 1, 825–847.

Staub, N.C. (1991) Basic respiratory physiology, pp. 131–148. Churchill Livingstone, New York.

Wagner, P.D., and Gale, G.E. (1991) Ventilation-perfusion relationships. In: Exercise, pulmonary physiology and pathophysiology. Whipp, B.J., and Wasserman, K. (eds.), pp. 121–142. Dekker, New York.

Wagner, P.D., Laravuso, R.B., Uhl, R.R., and West, J.B. (1974) Continuous distributions of ventilation-perfusion ratios in normal subjects breathing air and 100% O_2. J. Clin. Invest. 54, 53–68.

Weibel, E.R., Taylor, C.R., and Hoppeler, H. (1992) Variations in function and design: testing symmorphosis in the respiratory system. Respir. Physiol. 87, 325–348.

West, J.B. (1987) Assessing pulmonary gas exchange. New Engl. J. Med. 316, 1336–1338.

Whipp, B.J. (1994) The bioenergetic and gas exchange basis of exercise testing. Clin. Chest Med. 15, 173–192.

Whipp, B.J., Wagner, P.D., and Agusti, A. (1997) Factors determining the response to exercise in healthy subjects. In: Clinical exercise testing. Roca, J., and Whipp, B.J. (eds.). European Respiratory Monograph Vol. 2, No. 6. Sheffield: European Respiratory Journals, pp. 3–31.

Recommended Readings

Altose, M.D., and Kawakami, Y. (eds.) (1999) Control of breathing in health and disease. Vol. 135 in Lung biology in health and disease series, Marcell Dekker, New York.

Arena, R.T., and Humphrey, R. (2001) Characteristic ventilatory expired gas values in patients with heart failure during exercise testing. Clin. Exerc. Physiol. 3, 17–26.

Bard, R.L., and Nicklas, J.M. (2000) New graphical method for evaluating gas exchange in congestive heart failure. Med. Scie. Sport. Exerc. 32, 870–876.

Lundgren, C.E.G., and Miller, J.N. (eds.) (1999) Lung biology in health and disease. Vol. 132, Marcel Dekker, New York.

Prefaut, C., Durand, F., Mucci, P., and Caillaud, C. (2000) Exercise-induced arterial hypoxaemia in athletes: a review. Sports Med. 30, 47–61.

West, J.B., and Mathiew-Costello, O. (1999) Structure, strength, failure and remodeling of the pulmonary blood-gas barrier. Ann. Rev. Physiol. 61, 543–572.

11
The Pulmonary Circulation in Health and Disease

MARTIN TRISTANI-FIROUZI, STEPHEN L. ARCHER, AND E. KENNETH WEIR

Control of blood flow through the pulmonary circulation is unique for several reasons. Unlike other organs, the lungs must accept the entire cardiac output. Despite receiving the entire cardiac output, pulmonary artery pressure must remain low to allow the exchange of oxygen and carbon dioxide across the thin layer of cells separating the capillaries and the alveoli. The maintenance of this low pressure circulation is dependent upon the metabolically active pulmonary vascular endothelium and the production of vasoactive mediators. Endothelial dysfunction can result in alterations in pulmonary vascular tone, *pulmonary hypertension*, or alterations in vascular permeability, *pulmonary edema*. The critical role of the endothelium in the maintenance of vascular tone and integrity will be discussed in this chapter.

Development of the Pulmonary Circulation

Fetal Pulmonary Circulation

Fetal circulation is unique in that gas exchange (i.e., removal of CO_2 and addition of O_2) occurs in the placenta rather than in the lungs. As such, pulmonary blood flow is low in the fetus, but is adequate to meet the metabolic needs of the developing lung. The fetus relies on several mechanisms to minimize pulmonary blood flow *in utero*. As

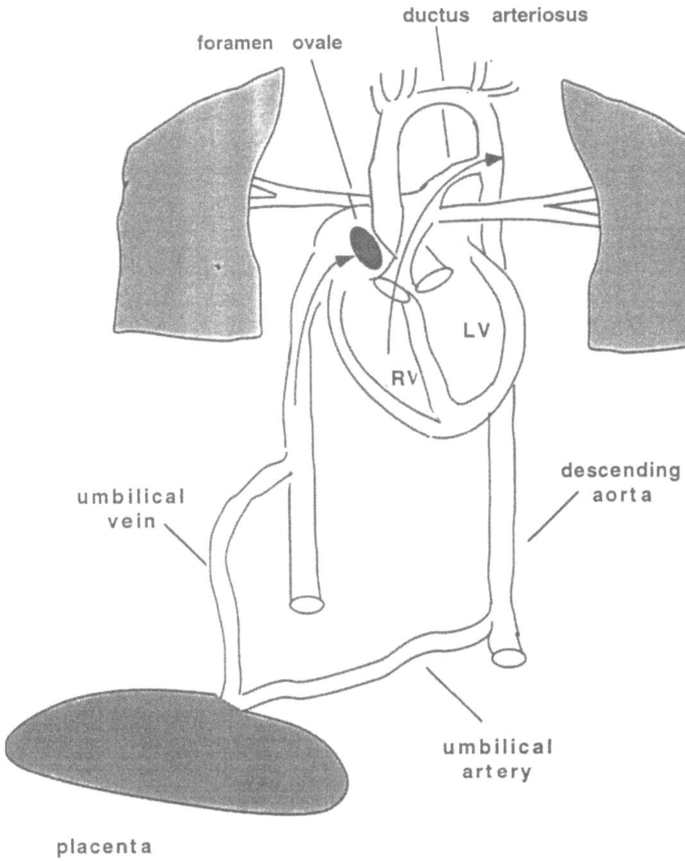

FIGURE 11.1. The fetal circulation. The foramen ovale and ductus arteriosus are two important fetal structures that serve to divert blood away from the pulmonary circulation. Oxygenated blood returning from the placenta courses through the umbilical vein to the right atrium, where it is diverted across the foramen ovale and into the left side of the heart. Most of the blood ejected from the right ventricle (RV) flows through the ductus arteriosus and into the descending aorta, effectively bypassing the lungs. LV = left ventricle.

illustrated in Fig. 11.1, the majority of oxygenated blood returning from the placenta is shunted across the foramen ovale, effectively bypassing the right ventricle. Blood ejected from the right ventricle primarily flows through the ductus arteriosus and into the descending aorta, thereby bypassing the pulmonary circulation. These two shunts minimize blood flow to the lungs, which receive only 5–10% of biventricular output (Rudolph, 1974). The major factor that contributes to low fetal pulmonary blood flow is the fact that pulmonary vascular resistance is higher than systemic resistance *in utero*. The large cross-sectional surface area of the placental microcirculation maintains low systemic vascular resistance, thus regulating in preferential flow through the ductus arteriosus and away from the pulmonary circulation.

Fetal pulmonary vascular resistance is maintained by a combination of anatomical and mechanical factors, the state of oxygenation, and the elaboration of vasoactive substances. The muscular layer (media) that surrounds fetal pulmonary arteries is thicker than the corresponding newborn and adult vessels (Rendas et al., 1978), in part, accounting for increased vasomotor tone *in utero*, as demonstrated in Fig. 11.2. Precapillary arteries comprise broad endothelial cells with numerous surface projections that decrease luminal diameter compared with their postnatal counterparts (Haworth, 1988). Likewise, fluid-filled alveoli compress adjacent vessels, thereby decreasing luminal diameter and increasing resistance to blood flow. Finally, the number of small pulmonary arteries in the fetus is less than it is in the newborn or adult (Rendas et al., 1978); consequently, the total cross-sectional vascular area is reduced.

The net effect of decreased cross-sectional area is an increase in resistance to blood flow, relative to postnatal life. More attention will be paid to this later. In addition to these anatomical and mechanical factors, elevated fetal pulmonary resistance is maintained by vasoactive mediators. The low fetal Po_2 (approximately 20 mmHg, compared with the normal adult Po_2, about 100 mmHg) contributes to increased pulmonary vascular resistance by acting through the mechanism of hypoxic pulmonary vasoconstriction. Further decrements in fetal Po_2 cause additional pulmonary vasoconstriction, and, conversely, increasing fetal Po_2 causes pulmonary vasodilation (Parker and Purves, 1967; Morin et al., 1988). Although the cellular mechanism of fetal hypoxic pulmonary vasoconstriction is not certain, oxygen-induced fetal pulmonary vasodilation is mediated through elaboration of endothelium-derived nitric oxide (EDNO) (Tiktinsky and Morin, 1993). EDNO which is produced under basal conditions acts to oppose the normally elevated pulmonary vascular resistance (Abman et al., 1990). Products of the cyclooxygenase and lipooxygenase pathways, thromboxane A_2, and leukotrienes are potent pulmonary vasoconstrictors, but they do not appear to modulate pulmonary artery tone *in utero* (Clozel et al., 1985; Cassin, 1993). In summary, elevated fetal pulmonary resistance is maintained by a combination of anatomical and mechanical factors as well as a balance between vasoconstrictor and vasodilator agents.

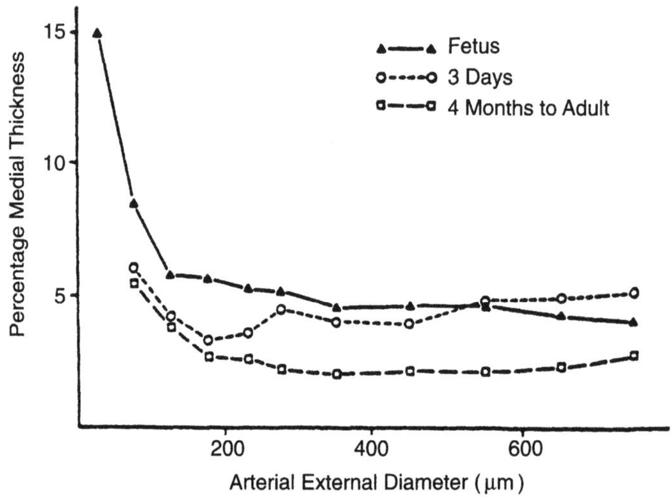

FIGURE 11.2. Comparison of the arteriolar medial wall thickness between fetal, newborn, and adult pulmonary arterioles. Fetal pulmonary arterioles possess thicker media compared with their newborn and adult counterparts, particularly the smallest vessels. This increased medial thickness may contribute to the increased vascular tone of the fetal pulmonary circulation. (*Source*: Reprinted from Rendas et al., 1978, with permission.)

Transition of the Pulmonary Circulation at Birth

Dramatic changes occur within the pulmonary circulation at birth and in the following weeks (Fig. 11.3). At birth, pulmonary artery blood flow increases immediately by 8–10-fold (Dawes et al., 1953; Cassin et al., 1964) and pulmonary artery pressure decreases by 50% within the first 24 hours of life (Emmanouilides et al., 1964). The increase in pulmonary blood flow elevates left atrial pressure, resulting in closure of the foramen ovale. In response to the rise in oxygen tension, the ductus arteriosus constricts at birth (Tristani-Firouzi et al., 1996), directing blood ejected from the right ventricle into the pulmonary circulation. Even though pulmonary blood flow comprises a fraction of the cardiac output *in utero*, the pulmonary circulation must accomodate the entire cardiac output at birth. This is accomplished by a rapid and dramatic decrease in pulmonary vascular resistance that is dependent upon a complex interaction between mechanical factors, morphological changes, endothelium-derived vasoactive substances, and the direct action of oxygen on pulmonary vascular smooth muscle cells. Ventilation of the fetal lung with a gas mixture designed to maintain Po_2 and Pco_2 at normal fetal levels results in pulmonary vasodilation (Cassin et al., 1964; Enhorning et al., 1966). Thus, physical expansion of the lungs alone stimulates pulmonary vasodilation. Lung expan-

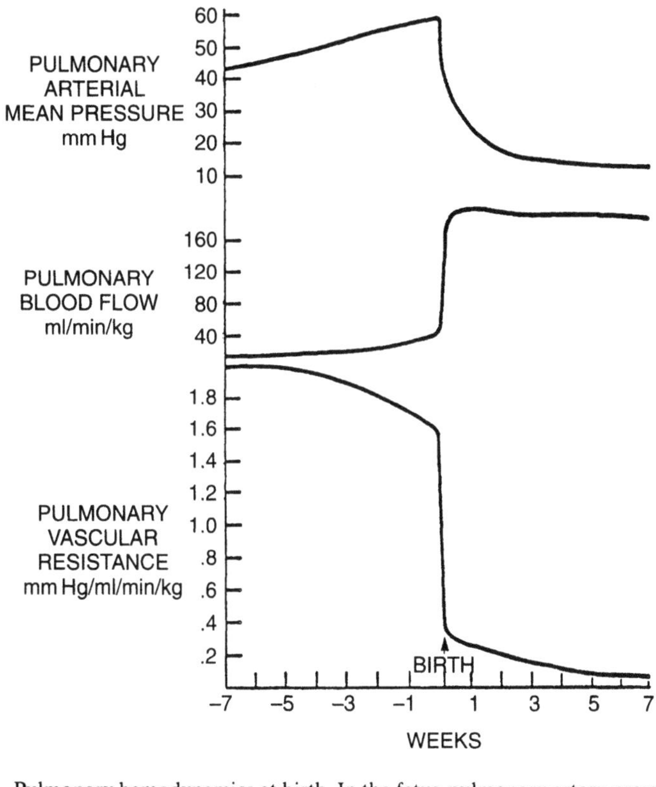

FIGURE 11.3. Pulmonary hemodynamics at birth. In the fetus, pulmonary artery pressure and vascular resistance are elevated and pulmonary blood flow is low. At birth, there is an immediate fall in pulmonary artery blood pressure and resistance, accompanied by an increase in pulmonary artery blood flow. Note how resistance continues to decline over the first several weeks of life. (*Source*: Reprinted from Rudolph, 1974, with permission.)

sion causes distension of extraalveolar blood vessels, which are tethered to the surrounding lung parenchyma, thereby increasing the diameter of the vessel lumen. As we shall soon note, resistance is inversely proportional to the fourth power of the radius; thus, small changes in diameter have significant effects on resistance. In addition, decreasing CO_2 and increasing O_2 content in the alveoli or blood independently cause pulmonary vasodilation (Cassin et al., 1964); however, the underlying mechanism remains uncertain. Finally, increased blood flow exerts shear stress on the surrounding endothelial cells, which may, in turn, stimulate release of EDNO and cause further vasodilation (Cooke et al., 1990; Cornfield et al., 1992).

Morphological changes accompany the fall in pulmonary vascular resistance at birth. Within minutes of birth, precapillary endothelial cells elongate and lose their intraluminal projections (Haworth, 1988), thereby decreasing resistance to blood flow by increasing luminal diameter. In addition, medial thickness decreases progressively over several weeks by rapid proliferation of new vessels and medial atrophy (Rendas et al., 1978). Growth of new vessels and decreased medial thickness allow resistance to fall to adult levels by approximately 3 months of age. The factors that control vascular remodeling in the perinatal period are poorly understood.

Vasoactive substances produced by the endothelium modulate the hemodynamic changes that occur at birth. Production of prostacyclin and EDNO by endothelial cells increases throughout late gestation and peaks around the time of birth (Shaul et al., 1993; Brannon et al., 1994). Inhibition of prostacyclin or EDNO synthesis attenuates, but does not abolish, the pulmonary vasodilation associated with birth (Velvis et al., 1991; Cornfield et al., 1992). Thus, these vasodilators produced by the endothelium contribute to the postnatal adaptation of the pulmonary circulation, but they are not the sole agents responsible for these changes.

In summary, therefore, the fall in pulmonary vascular resistance at birth is a complex process related to the interaction of numerous factors including gaseous expansion of the lung, alterations in CO_2 and O_2 content, increases in shear stress exerted by increased flow, changes in cell morphology, and elaboration of vasoactive mediators. Failure of the pulmonary circulation to make this transition from the fetal to the neonatal circulation results in persistent pulmonary hypertension of the newborn, a cause of significant neonatal morbidity and mortality. Elucidation of the mechanisms underlying normal regulation of the perinatal pulmonary circulation may improve treatment and perhaps prevent persistent pulmonary hypertension of the newborn.

Regulation of Pulmonary Vascular Tone

The Mechanics of Vascular Resistance

Resistance to flow through a vascular bed can be estimated based upon the Poiseuille-Hagen equation (i.e., the hydraulic equivalent of Ohm's law), which describes the mechanics of liquid flowing through a rigid, cylindrical tube (Roos, 1962):

$$R = P/Q = 8\,\eta l/r^4$$

Resistance to flow (R) is equal to the decrease in pressure between the inlet and outlet of the tube (P) divided by the rate of flow (Q). In the pulmonary circulation, vascular resistance (PVR) is equal to mean pulmonary artery pressure (PAP) minus mean left atrial pressure (LAP) divided by pulmonary blood flow (Q):

$$\text{PVR} = [\text{PAP} - \text{LAP}]/\text{Q}$$

Thus, an increase in pulmonary artery pressure or a decrease in flow can elevate resistance. Resistance is also dependent upon η (fluid viscosity), l (the length of the tube), and r (the radius to the fourth power). Resistance, r, includes the net ratio of all blood vessels combined, or the entire cross-sectional area of the vascular bed. As we will discuss, vasoconstrictors increase and, similarly, vasodilators decrease resistance by altering the radius of small blood vessels. Although the effect of changing the radius of one blood vessel is negligible, the net result of changing the radius of numerous vessels on the total cross-sectional area is dramatic.

As is usually the case in biological systems, concepts such as the Poiseuille-Hagan equation incompletely describe blood flow because blood is not a Newtonian fluid and blood vessels are not rigid cylindrical tubes. Most pulmonary blood vessels are thin-walled and their tone is influenced by transmission of alveolar pressure and/or the degree of lung expansion (West, 1985). Pressure within the alveoli is transmitted directly to surrounding capillaries (*intraalveolar blood vessels*). In some instances, alveolar pressure may exceed capillary pressure (see Zone 1, upcoming), causing the collapse of capillaries and cessation of blood flow. The tone of *extraalveolar vessels* is not dependent upon alveolar pressure; however, as these vessels are tethered to the surrounding lung parenchyma, they are influenced by the degree of lung expansion. Thus, at low lung volumes (e.g., what occurs in the fetal state or following atelectasis (i.e., airway collapse), extraalveolar vessels may be collapsed because of the absence of an outward distending force. Lung expansion exerts an outward radial force causing distention, an increase in luminal radius and a decrease in resistance to blood flow.

Pulmonary vascular resistance is also influenced by two additional factors; *recruitment* of new blood vessels and *distention* of existing vessels. Under normal circumstances (and to a greater degree in the fetus), some vessels are collapsed and do not conduct blood. If pressure or blood flow rises, these vessels conduct blood and, thereby decrease vascular resistance by increasing the total cross-sectional area of the vascular bed. At higher pressures, distention of blood vessels occurs, which again decreases resistance by increasing luminal radius. Recruitment and distension are compensatory mechanisms to accomodate increases in pressure or flow. Once these mechanisms are maximized, further increases in pressure or flow cause elevated resistance, or pulmonary hypertension.

The distribution of blood flow varies between the apex of the lung and the base. This differential distribution is secondary to differences in hydrostatic pressure differences exerted by blood in different regions of the lung (West, 1985). The distance between the apex and base in the adult is around 30 cm; therefore, if one assumes the pulmonary arterial system to be a continuous column of blood, the pressure difference from apex to base is equal to 30 cm H_2O, or 23 mmHg. This difference is significant, given the low pressure of the pulmonary arterial system. Figure 11.4 describes the effect of hydrostatic pressure and alveolar pressure on regional distribution of pulmonary blood flow.

The Role of the Endothelium in Regulation of Vascular Tone

Although passive or mechanical factors regulate pulmonary vascular tone, the endothelium produces constrictor and dilator agents which actively modulate tone. Vascular tone is determined by a delicate balance between the production of vasodilators (e.g., EDNO and prostacyclin), and vasoconstrictors (e.g., endothelin-1 and thromboxane A_2) (Fig. 11.5). Endothelial dysfunction may alter this balance and cause abnormalities in vasomotor tone. For example, in the presence of an intact endothelium, several receptor-operated agonists, such as bradykinin and acetylcholine, stimulate vasodilation, through elaboration of EDNO, endothelium-derived hyperpolarizing factor

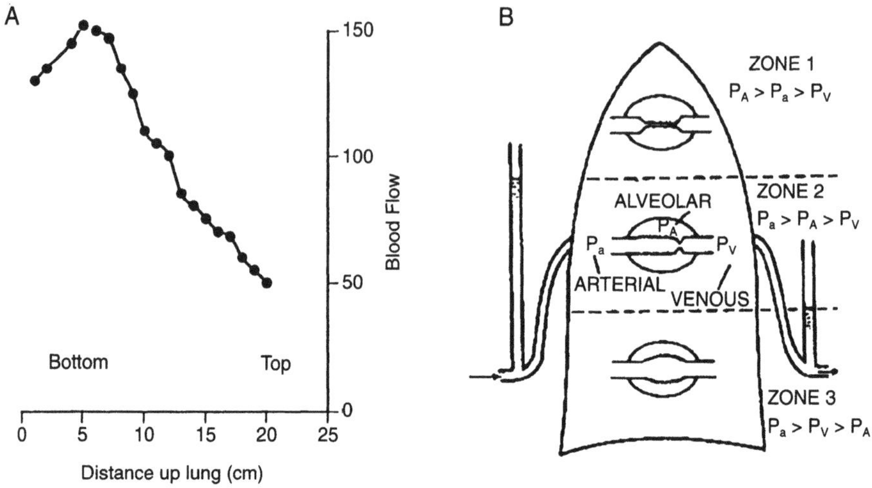

FIGURE 11.4. Distribution of pulmonary blood flow. (A) Blood flow throughout the lungs is unevenly distributed, with greater blood flow at the bottom compared with the top of the lung. (B) The uneven distribution of blood flow can be explained by assuming that blood perfusing the lung is a continuous column. Based on gravity, the pressure exerted by the column of blood is greatest at the base and least at the apex. At the top of the lung, zone 1, alveolar pressure (P_A) exceeds both pulmonary arteriolar pressure (P_a) and venous pressure (P_v), causing compression of blood vessels and decreased blood flow. Within zone 2, P_A is less than P_a, but greater than P_v. Within zone 3, the majority of blood flow occurs because both P_a and P_v exceed P_A. (*Source*: Modified with permission from West, 1985).

(EDHF), and prostacyclin. In the presence of endothelial dysfunction, however, these same agonists cause vasoconstriction. It therefore seems important to ascertain whether an imbalance is closely associated with certain abnormalities and whether the term pulmonary endothelial dysfunction needs to be restricted to specific situations such as hypoxia-induced pulmonary hypertension in which there is an imbalance between endothelium-derived constricting and relaxing factors. Expressed in another way, the observed alteration in vascular tone might well be characteristic of the pulmonary disease. This is the case in a number of pulmonary diseases such as chronic obstructive lung disease and cardiopulmonary bypass in addition to primary and secondary pulmonary hypertension (Chen and Oparil, 2000).

All vasoactive mediators produced by the endothelium exert their ultimate effect within vascular smooth muscle cells by altering the availability of intracellular calcium to bind to contractile proteins and generate the force of contraction. These mediators differ in the mechanism by which intracellular Ca^{2+} levels are altered, which include modulating ion channel function and influencing intracellular Ca^{2+} storage. In addition, these agents may alter the Ca^{2+} sensitivity of the contractile apparatus. These mechanisms will be described in the following.

Endothelium-Derived Nitric Oxide (EDNO)

By studying isolated blood vessels, Furchgott and Zawadzki first described a dilator substance whose production was dependent upon the presence of an intact endothelium (Furchgott and Zawadzki, 1980). The effluent from intact vessels caused endothelium-denuded vessels to dilate. This short-lived substance was termed endothelium-derived relaxing factor (EDRF), later changed to endothelium-derived nitric oxide (EDNO) after further investigation revealed the compound to be indistinguishable from nitric oxide (NO) (Ignarro et al., 1987; Palmer et al., 1987). The pioneering work of Furchgott, Ignarro, and Murad on the role of EDNO as a messenger molecule in the cardiovascular system ultimately led to the award to them of the 1998 Nobel Prize in Medicine (Gibbs et al., 1999). EDNO is an unstable radical generated by oxidation of the terminal guanidino nitrogen atom of L-arginine by the enzyme NO synthase (Palmer et al., 1988). EDNO rapidly diffuses from the endothelial cell into adjacent vascular smooth muscle cells, where it stimulates guanylate cyclase to produce cyclic guanosine mononucleotide phosphate (cGMP) (Gruetter et al., 1979; Murad et al., 1979). cGMP is an important second messenger molecule capable of activating several cellular processes, all of which contribute to subsequent vasodilation (Fig. 11.6). cGMP activates protein kinases that are postulated to phosphorylate and activate potassium (K^+) channels (Archer et al., 1994). K^+ channel activation promotes outward flux of K^+ ions, leaving behind negatively charged macromolecules, and thereby hyperpolarizing the cell membrane (i.e., the interior of the cell becomes more electronegative). Membrane hyperpolarization decreases the activity of voltage-dependent calcium channels and, thus, decreases Ca^{2+} flux into the cell interior (Nelson and Quayle, 1995). Vasorelaxation occurs because less intracellular Ca^{2+} is available to bind to the contractile apparatus. cGMP may also directly interact with ion channels (Miller, 1983).

EDNO production is stimulated by several mechanisms. As already mentioned, receptor-operated agonists (e.g., bradykinin and acetylcholine) stimulate EDNO production by stimulating NO synthase. Physical stimuli also promote NO synthase activ-

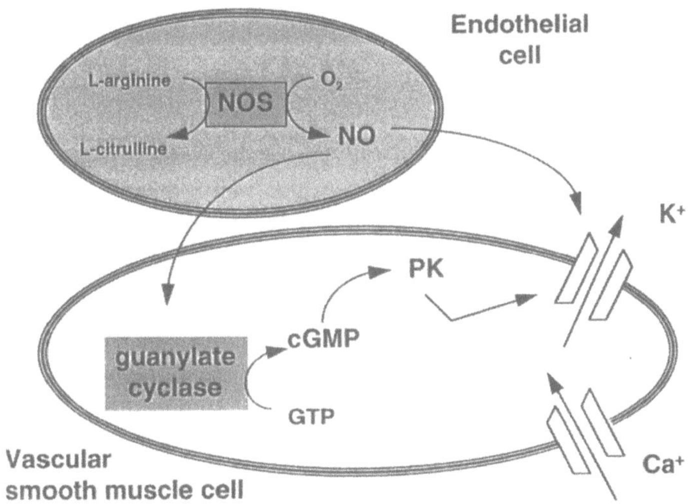

FIGURE 11.5. Diagrammatic representation of thromboxane A_2, endothelin-1, and prostacyclin receptors in a vascular smooth muscle cell. Vasoactive agents cause vasodilation or vasoconstriction by altering intracellular calcium concentrations. Both thromboxane A_2 (TXA) and endothelin-1 (ET-1) are coupled to phospholipase C (PLC) through receptor-bound G-proteins. PLC converts a membrane phospholipid to inositol triphosphate (IP_3), which stimulates release of calcium (Ca^{2+}) from the sarcoplasmic reticulum (SR). In addition, endothelin-1 stimulates influx of calcium through voltage-dependent calcium channels. Prostacyclin (PGI_2) receptor is coupled through G-proteins to adenylylate cyclase (AC), which converts adenosine triphosphate (ATP) to cyclic-adenosine monophosphate (cAMP). cAMP-dependent protein kinases (PK) activate potassium (K^+) channels, causing membrane hyperpolarization and subsequent closure of voltage-dependent calcium channels. cAMP-PK also stimulates uptake of calcium into the sarcoplasmic reticulum.

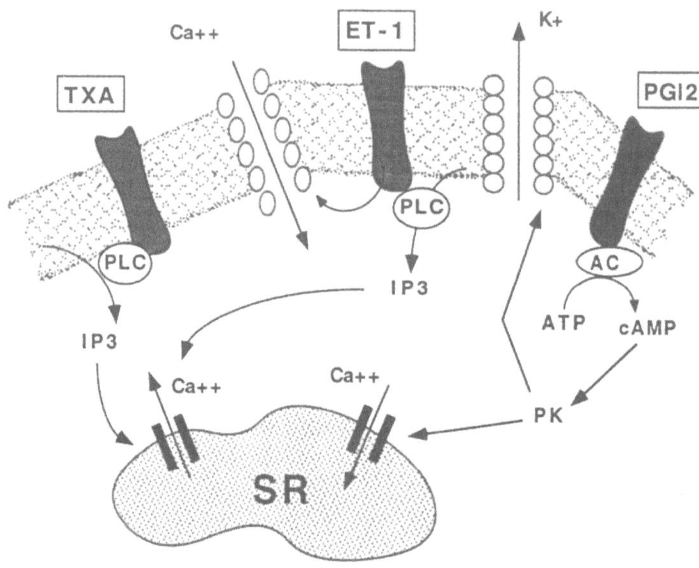

FIGURE 11.6. Mechanism of action of endothelium-derived nitric oxide. Nitric oxide synthase (NOS) combines L-arginine and oxygen (O_2) to form nitric oxide (NO) within endothelial cells. NO diffuses to adjacent vascular smooth muscle and stimulates production of cyclic guanosine monophosphate (cGMP) by guanylate cyclase. cGMP activates protein kinases (PK) which activate potassium (K^+) channels, causing membrane hyperpolarization, closure of voltage-activated calcium (Ca^{2+}) channels, a decrease in cytosolic Ca^{2+} levels and vasorelaxation. NO may directly stimulate K^+ channels as well. GTP, guanosine triphosphate.

ity. The physical force of blood flowing across the endothelial cell surface, shear stress, is capable of enhancing EDNO synthesis (Pohl et al., 1986; Rubanyi et al., 1986). In addition, pulsatile stretching of the vascular wall induces release of EDNO (Hutcheson and Griffith, 1991; Busse et al., 1993b). These mechanisms may play an important role in the increase in pulmonary blood flow occurring at birth. In addition, these mechanisms regulate a negative feedback loop that modulates the effect of vasoconstrictors. For example, vasoconstriction increases vascular shear stress (by decreasing luminal radius), which stimulates EDNO production (Busse et al., 1993a).

Basal production of EDNO plays an integral role in the regulation of vascular tone in systemic arteries. Inhibition of NO synthase causes systemic hypertension in most species, including humans (Calver et al., 1992; Ikeda et al., 1992; Isaacson et al., 1994). Even though it could be speculated that the pulmonary vascular resistance is markedly lower than its systemic counterpart because of greater EDNO production, this concept is not universally accepted. Inhibition of EDNO synthesis does not invariably cause pulmonary hypertension (Nishiwaki et al., 1992; Hampl et al., 1993). Furthermore, in species where EDNO inhibition causes pulmonary hypertension, the severity tends to be less than the magnitude of systemic hypertension (Stamler et al., 1994). Thus, EDNO does not appear to modulate resting vascular tone in the normal adult pulmonary circulation, at least not to the degree found in the systemic vasculature.

In contrast, under conditions where pulmonary vascular resistance is elevated, EDNO appears to play an important role in regulation of vascular tone. EDNO regulates basal pulmonary tone in the fetus (Abman et al., 1990) and modulates the transition from the fetal to the neonatal pulmonary circulation (Cornfield et al., 1992). In addition, basal EDNO production may be elevated in pulmonary hypertension as a compensatory mechanism to moderate the elevated resistance (Isaacson et al., 1994).

In summary, EDNO may play a smaller role in basal regulation of the low tone adult pulmonary circulation compared with the systemic vascular bed. Under conditions of elevated pulmonary vascular resistance, however, such as in the fetus and in pulmonary hypertension, the importance of EDNO cannot be ignored (see Pitt and St. Croix, 2002).

Prostacyclin

Like EDNO, the vasodilator prostacyclin is released from the endothelium in response to both receptor-operated stimulation as well as increases in shear stress (Busse et al., 1993a). These factors stimulate liberation of arachidonic acid from membrane phospholipid by phospholipase A_2, a membrane-bound enzyme. Prostacyclin synthase converts arachidonic acid to prostacyclin, which diffuses to adjacent vascular smooth muscle cells. Prostacyclin receptors on vascular smooth muscle cells are coupled to the enzyme adenylate cyclase through guanine nucleotide regulatory proteins (G-proteins) within the cell membrane. Activation of adenylate cyclase stimulates production of cyclic adenosine monophosphate (cAMP). cAMP, like cGMP, acts through protein kinases to activate K^+ channels and cause vasodilation (Walsh, 1991; Taguchi et al., 1995). In addition, cAMP-dependent protein kinases may stimulate uptake of Ca^{2+} into intracellular storage pools.

Thromboxane A_2

Thromboxane A_2 is a potent platelet aggregator and vasoconstrictor produced primarily by platelets and, to a lesser degree, by the endothelium (Ingerman-Wojenski et al., 1981). Thromboxane A_2 is coupled to phospholipase C through G-proteins in a manner similar to prostacyclin receptors. Activation of phospholipase C results in production of the second messenger, inositol 1,4,5-triphosphate (IP_3), which activates release of calcium from the sarcoplasmic reticulum. Evidence suggests that thromboxane also increases the calcium sensitivity of the contractile proteins (Schror, 1993). Some investigators have postulated that an imbalance in the production of thromboxane A_2 and prostacyclin (increased thromboxane relative to prostacyclin) within the lung contributes to development of pulmonary hypertension (Christman et al., 1992; Adatia et al., 1993).

Endothelin-1

Endothelin-1 is an endothelium-derived peptide with potent vasoconstrictor properties (Miyauchi and Masaki, 1999). Endothelin-1 production is stimulated by various agonists, including epinephrine, angiotensin II, and thrombin, which enhance the conversion of endothelin-1 to its active form by endothelin-converting enzyme (Lushcer, 1992). So far, two G-protein-coupled endothelin receptor subtypes have been cloned: One is ET_A present in vascular smooth muscle that mediates vasoconstriction (Arai et al., 1990); the other, ET_B, is present on endothelial cells and mediates vasodilation through activation of EDNO and prostacyclin (Sakurai et al., 1990). Thus, endothelin released from the endothelium can produce vasoconstriction as well as vasodilation, allowing for tight regulation of vascular tone through a negative-feedback mechanism. ET_A receptor activation increases intracellular Ca^{2+} by increasing flux through voltage-dependent Ca^{2+} channels (Goto et al., 1989), and by release of Ca^{2+} from intracellular storage pools (Pollock et al., 1995).

Current evidence supports the view that ET-1 plays a key role in priming the effector pathway in the small distal pulmonary arteries (Sylvester, 2001) and that the priming mechanism in all likelihood includes myofilament calcium sensitivity. Whether partial membrane depolarization is necessary is not yet certain.

It is unclear to what degree, if any, endothelin-1 participates in control of basal pulmonary vascular tone. The role of endothelin-1 in modulating tone under conditions of elevated pulmonary vascular tone is equally unclear. Numerous studies have demonstrated elevated circulating levels of endothelin-1 in pulmonary hypertension (Allen et al., 1993; Giaid et al., 1993; Hosoda, 1994; Yamamoto et al., 1994). It remains uncertain, however, whether increased endothelin-1 is a marker or a mediator of disease. Evidence that endothelin-1 may participate in or directly cause elevated pulmonary vascular resistance comes from studies demonstrating that treatment with an ET_A receptor antagonist attenuates chemically induced or hypoxia-induced pulmonary hypertension (Miyauchi et al., 1993; Bonvallet et al., 1994; Eddahibi et al., 1995; Oparil et al., 1995). It seems likely that such antagonists may also be effective in the treatment of primary pulmonary hypertension.

In summary, the endothelium produces a balance of constrictor and dilator agents that regulate the low vascular tone of the adult pulmonary circulation. Alterations in this balance may be responsible for the development or maintenance of pulmonary hypertension. Further investigation is necessary to elucidate the mechanisms underlying control of vascular tone under normal and pathological conditions.

Hypoxic Pulmonary Vasoconstriction

The pulmonary circulation is unique in that hypoxia causes pulmonary vasoconstriction. In contrast, hypoxia induces vasodilation of systemic arteries. As mentioned earlier, hypoxic pulmonary vasoconstriction is responsible for maintaining elevated resistance in the fetal pulmonary circulation, which is essential for fetal survival. In postnatal life, hypoxic pulmonary vasoconstriction serves the purpose of shunting deoxygenated pulmonary arterial blood away from atelectatic, hypoxic regions of the lung. By matching ventilation to perfusion, overall gas exchange is improved.

Vascular smooth muscle cells isolated from small pulmonary arteries constrict when placed in a low O_2 medium, which indicates that the endothelium is not an integral component for sensing changes in O_2 (Madden et al., 1992). Hypoxia inhibits K^+ channels on the cell membrane of pulmonary artery smooth muscle cells, causing membrane depolarization, activation of voltage-dependent Ca^{2+} channels, an increase in cytosolic Ca^{2+}, and vasoconstriction (Post et al., 1992; Weir and Archer, 1995). The mechanism whereby K^+ channels "sense" changes in partial pressure of oxygen (Po_2) is not known, but it may be related to the reduction-oxidation (redox) status of the cell. Cellular respiration depends on the flow of electrons from NAD(P)H through the electron transport chain, and ultimately to molecular O_2. Under conditions of decreased Po_2, electron flux through the electron transport chain decreases and NAD(P)H levels increase. The increased NAD(P)H may reduce other molecules (e.g., glutathione), contributing to an overall "reduced" cellular state. These reduced compounds inhibit K^+ currents in pulmonary vascular smooth muscle cells, mimicking the effect of hypoxia (Archer et al., 1993). Oxidizing agents exert the opposite effect (i.e., activation of K^+ channels) and act in a manner similar to normoxia.

In summary, hypoxic vasoconstriction is an important physiological response unique to the pulmonary circulation. This mechanism is essential for the maintenance of the elevated pulmonary vascular resistance in the fetus and in matching ventilation to perfusion in underventilated regions of the adult lung. The effector pathway leading to vasoconstriction lies in the small distal pulmonary arteries where oxygen is rate limiting to a system which is closely coupled to the contractile apparatus of smooth muscle (Sylvester, 2001). Further studies are needed to elucidate the exact mechanism by which the pulmonary vasculature senses changes in Po_2 and constricts in response to hypoxia.

Pathophysiology of the Pulmonary Circulation

Pulmonary Hypertension

Pulmonary hypertension is a general term that applies to any condition resulting in an acute or chronic increase in pulmonary artery pressure. In the adult, *pulmonary hypertension* is defined as a mean pulmonary artery pressure that exceeds 25 mmHg. Elevated pulmonary vascular resistance imposes an increased workload on the right heart, resulting initially in right-ventricular hypertrophy, and subsequently in right ventricular dilation and depressed cardiac function. Hypertension may occur secondary to an increase in left-atrial pressure because, as left-atrial pressure rises, pulmonary artery pressure concomitantly increases upstream. The Poiseuille-Hagen equation,

$$PVR = (PAP - LAP)/Q = 8\,\eta l/r^4$$

stipulates that increases in blood flow result in decreased resistance; however, chronically increased pulmonary blood flow causes pulmonary hypertension by altering the structure of small pulmonary arteries, with the net effect of decreasing the cross-sectional surface area of the pulmonary vasculature. In fact, most forms of chronic pulmonary hypertension are associated with alterations in the morphology and/or luminal radius of small pulmonary arteries.

Small pulmonary arteries (*intraacinar arteries*) are described pathologically as muscular, partially muscular, and nonmuscular (Meyrick, 1987). Chronic pulmonary hypertension is associated with distal extension of muscle into partially muscular and nonmuscular arteries, medial hypertrophy (an increase in the thickness of the muscular coat surrounding arteries), intimal proliferation (an increase in cellular content of the innermost layer of small pulmonary arteries), and obliteration of vessel lumen. It appears that distal extension of muscle and medial hypertrophy occur in response to increased pressure or stretch of blood vessels (Meyrick, 1991). The mechanisms underlying lumen obliteration and a consequent decrease in peripheral arterial volume remain to be determined.

Pulmonary hypertension may be classified in several ways (Table 11.1). The following is a modification of the classification originally proposed by Paul Wood in the 1950s (Archer and Weir, 1994):

- Increased downstream pressure
- Increased pulmonary blood flow
- Obstruction of proximal pulmonary arteries
- Diseases of small pulmonary arteries
- Hypoxic pulmonary vasoconstriction

Increased Downstream Pressure

Increased downstream pressure occurs in conditions associated with increased left-atrial pressure. Increases in downstream pressure are in turn transmitted back through the microcirculation to the pulmonary arteries, raising pulmonary artery pressure. Congestive heart failure is a common cause of increased left-ventricular and atrial pressure in older adults. It occurs as a result of coronary artery disease or systemic hypertension. Valvular disease (e.g., mitral regurgitation or stenosis), elevates left atrial pressure. Increases in left-atrial pressure are frequently associated with pulmonary vasoconstriction that further raises pulmonary artery pressure.

TABLE 11.1. Etiologies of pulmonary hypertension.*

Increased downstream pressure
Congestive heart failure
Mitral stenosis or regurgitation
Increased pulmonary blood flow
Left to right shunting (ventricular septal defect, atrial septal defect, patent ductus arteriosus, complex congenital heart disease)
Obstruction of proximal pulmonary arteries
Pulmonary thromboembolism
Extrinsic compression (tumor, mediastinal fibrosis)
Diseases of small pulmonary arteries
Microembolism (thrombus, tumor, *Schistosoma* ovae)
Collagen vascular disease (scleroderma, systemic lupus erythematosis)
Chronic obstructive pulmonary disease
Primary pulmonary hypertension
Hypoxic pulmonary vasoconstriction
High altitude residence
Primary hypoventilation (obstructive sleep apnea)
Neuromuscular disorders (Myasthenia gravis, poliomyelitis)
Chest wall deformities (kyphoscoliosis)

* This is not a complete list of etiologies of pulmonary hypertension.

Increased Pulmonary Artery Blood Flow

Chronically increased pulmonary artery blood flow causes pulmonary hypertension. Left-to-right shunts (i.e., the flow of fully oxygenated blood back through the pulmonary circulation) occur in conditions such as ventricular septal defect, atrial septal defect, patent ductus arteriosus, and other more complex congenital heart defects. Most of these conditions are diagnosed in childhood. When surgical repair is undertaken in early childhood, the associated pulmonary hypertension is often reversible (Rabinovitch et al., 1984). In some circumstances, a delay in surgery leads to progressive and irreversible pulmonary hypertension.

Obstruction of Proximal Pulmonary Arteries

Embolism of a peripheral venous blood clot to the proximal pulmonary arteries (pulmonary thromboembolism) may obstruct blood ejected from the right ventricle and elevate proximal pulmonary artery pressure. In a small number of patients, the thrombus fails to lyse and is incorporated into the wall of the proximal arteries. This can result in chronic thromboembolic pulmonary hypertension. Extrinsic compression of the proximal pulmonary arteries by a tumor or by fibrosis of adjacent mediastinal structures (fibrosing mediastinitis) may also cause pulmonary hypertension. In these forms of pulmonary hypertension, the distal pulmonary vasculature usually does not undergo the morphological changes described earlier. Pulmonary hypertension secondary to chronic proximal pulmonary artery obstruction may be correctable by surgical intervention.

Diseases of Small Pulmonary Arteries

Diseases of small pulmonary arteries may cause obstruction and/or obliteration of distal pulmonary arteries, thereby decreasing the cross-sectional area of the pulmonary vasculature. Microembolism of tumor, embolization of crushed tablet matrix injected intravenously by drug abusers, or, in the case of the parasitic disease, schistosomiasis, embolization of ova with associated cellular proliferation of the distal pulmonary arteries may cause pulmonary hypertension. Collagen-vascular diseases (e.g., scleroderma, systemic lupus erythematosis, and to a lesser extent, rheumatoid arthritis) may stimulate intimal proliferation and fibrosis of distal pulmonary arteries.

Chronic obstructive pulmonary disease (COPD) is associated with extension of muscle into nonmuscularized small pulmonary arteries and occurs as a result of a combination of obliteration, vasoconstriction, and obstruction of small pulmonary vessels (Archer and Weir, 1994). Pulmonary hypertension in COPD is usually mild at rest, but pulmonary artery pressure may increase significantly with exercise.

Primary pulmonary hypertension is a category reserved for those patients in whom identifiable causes of pulmonary hypertension have been excluded. Primary pulmonary hypertension is primarily a disease of young adults, particularly young women, although all ages may be affected (D'Alonso et al., 1991). The overall prognosis is poor with approximately one half of the patients succumbing to this illness within 3 years of diagnosis, in the absence of prostacyclin treatment of lung transplantation (D'Alonso et al., 1991). As one might expect, survival correlates with the severity of pulmonary vascular disease, as assessed by pulmonary artery pressure, right-atrial pressure, and cardiac output. Although the etiology of primary pulmonary hypertension remains unclear, several pathogenic mechanisms are listed in Fig. 11.7. These mechanisms cause thrombosis *in situ* and stimulate intimal proliferation.

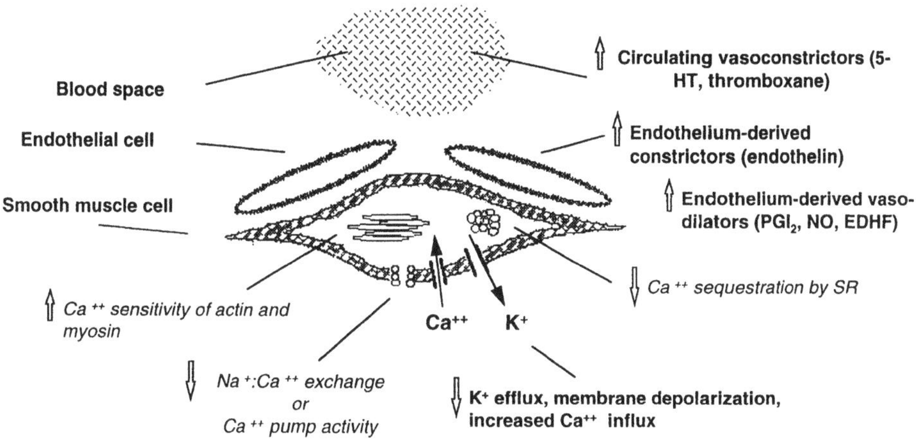

FIGURE 11.7. Mechanisms of vasoconstriction in primary pulmonary hypertension. Increased circulating vasoconstrictors [e.g., serotonin (5-HT) and thromboxane] as well as vasoconstrictors produced by the endothelium (e.g., endothelin-1) contribute to the vasoconstriction associated with pulmonary hypertension. The endothelium-derived vasodilators, prostacyclin (PGI_2), nitric oxide (NO), and endothelium-derived hyperpolarizing factor (EDHF) may be upregulated in primary pulmonary hypertension: however, they are unable to fully compensate for the increased vascular tone. Decreased potassium (K^+) channel activity causes membrane depolarization, activation of voltage-dependent calcium (Ca^{2+}) channels and increased cytosolic Ca^{2+} levels. Other theoretical considerations accounting for increased vasoconstriction include decreased Ca^{2+} sequestration by the sarcoplasmic reticulum (SR), decreased Na^+:Ca^{2+} exchanger or Ca^{2+} pump activity, and increased Ca^{2+} sensitivity of the contractile apparatus.

Hypoxic Pulmonary Vasoconstriction

Chronic alveolar hypoxia is a cause of pulmonary hypertension that acts by inducing vasoconstriction (Weir and Archer, 1995) as well as by stimulating the remodeling of small pulmonary arteries, as described earlier. Alveolar hypoxia may occur as the result of living at high altitude or of hypoventilation. Chronic hypoventilation occurs in individuals with obstructive sleep apnea, neuromuscular disorders of ventilation (myasthenia gravis, poliomyelitis), or chest wall deformities (kyphoscoliosis).

Treatment of Pulmonary Hypertension

Advances in our understanding of the role of the endothelium in vascular regulation have led to improvements in the treatment of pulmonary hypertension (Barst, 1999; Higenbottam et al., 1999). NO, prostacyclin, and calcium channel antagonists are important treatment modalities and prolong survival in selected patients with primary pulmonary hypertension.

Inhaled Nitric Oxide

Nitric oxide (NO) is a potent vasodilator that selectively dilates the pulmonary vasculature when inhaled. The pulmonary selectivity of inhaled NO is due to its avid binding to hemoglobin, forming methemoglobin. Inhaled NO is thus inactivated prior to reaching the systemic circulation and systemic vasodilation does not occur. NO improves ventilation-perfusion mismatch by dilating blood vessels that perfuse ventilated regions of the lung. Inhaled NO cannot diffuse into unventilated regions of the lung and, therefore, NO does not alter blood flow to these regions. Inhaled NO has proved useful in the treatment of pulmonary hypertension associated with the newborn period, congenital heart disease, adult respiratory distress syndrome, and primary pulmonary hypertension (Pepke-Zaba et al., 1991; Roberts et al., 1992; Frostell et al., 1993; Roberts et al., 1993; Rossaint et al., 1993; Adatia et al., 1994; Lonnqvist et al., 1994). The major disadvantage to inhaled NO therapy is that it must be continuously delivered due to its short half-life. In addition, NO is a toxic gas at high concentrations, whose delivery must be carefully monitored. Thus, inhaled NO is limited to treatment of pulmonary hypertension in the acute setting (e.g., in an intensive care unit), and is not well suited for chronic treatment of pulmonary hypertension.

Prostacyclin

Prostacyclin is a potent, short-acting vasodilator that is the most potent endogenous inhibitor of platelet aggregation discovered to date (Vane and Botting, 1995). The combination of vasodilator and antiplatelet properties make prostacyclin an attractive therapeutic choice for the treatment of pulmonary hypertension. Because of its short half-life, prostacyclin must be administered by continuous intravenous infusion. In clinical trials, long-term continuous intravenous prostacyclin, infused over several months, improves exercise tolerance and prolongs survival in patients with primary pulmonary hypertension (Jones et al., 1987; Barst et al., 1994). It is noteworthy that failure to respond to acute administration of prostacyclin does not preclude a favorable long-term response. Prostacyclin and calcium channel blockers may act by inhibiting cellular proliferation as well as reducing vasoconstriction. Thus, the dual action may account for the clinical improvement seen with chronic therapy.

Disadvantages to chronic intravenous prostacyclin infusion are the associated decrease in systemic arterial pressure and the problems associated with pump failure and line infection. In addition, tachyphylaxis may develop over time, requiring a progressive increase in dosage. Alternative means of delivery (e.g., as nebulized prostacyclin or prostacyclin analogs) appear promising (Bindl et al., 1994; Mikhail et al., 1995), although long-term clinical trials are necessary.

Calcium Channel Antagonists

The calcium channel antagonists nifedipine and diltiazem cause vasodilation by blocking influx of calcium into vascular smooth muscle cells through voltage-dependent calcium channels. Chronic oral administration of these agents improves exercise tolerance, causes regression of right-ventricular hypertrophy, and prolongs survival in selected patients with primary pulmonary hypertension (Rich et al., 1992). Only about 25% of patients respond with a greater than 20% reduction in pulmonary artery pressure and resistance (i.e., these patients are termed *responders*). The 5-year survival rate in responders is 94% versus 38% for nonresponders. It is still somewhat controversial whether calcium channel antagonists simply identify a cohort of patients who do well regardless of treatment, or whether treatment directly improves survival. The improvement in exercise tolerance and regression of right-ventricular hypertrophy suggests that calcium channel antagonists are in fact beneficial in selected patients. The major limitation to the use of oral calcium channel antagonists is the lack of pulmonary selectivity (i.e., systemic hypotension often occurs).

Anticoagulation

In patients with primary pulmonary hypertension, thrombosis of the small pulmonary arteries has been demonstrated pathologically and increased thrombin activity has also been reported (Archer and Weir, 1994). Based on these observations, oral anticoagulation has been recommended by some investigators. Two studies have reported improved survival in patients taking oral anticoagulants, independent of the use of other medications (Fuster et al., 1984; Rich et al., 1992); however, neither study was randomized in nature.

In summary, pulmonary hypertension occurs as a result of diverse etiologies. The prognosis is to some degree dependent upon the underlying cause. An understanding of the role of the endothelium as a modulator of vascular tone has led to the development of new treatment strategies for pulmonary hypertension and the possibility of improving quality of life and longevity.

Pulmonary Edema

Pulmonary edema occurs as a result of an imbalance in forces working to expel fluid from the capillaries, forces promoting movement of fluid into capillaries, and forces removing fluid from the *interstitial* space (i.e., the space surrounding blood vessels) (e.g., lymphatic drainage). Under normal circumstances, a small amount of fluid and protein cross the endothelium to enter into the interstitial space. This fluid is drained by the lymphatics that course in the perivascular spaces toward the hilum, and eventually empty into mediastinal lymph nodes or the thoracic duct. Transcapillary fluid movement was first described by Starling (1896), who stated that the rate of fluid movement across a blood vessel is directly proportional to the pressure gradient across the vessel membrane and the permeability of the membrane. The pressure gradient includes both hydrostatic and osmotic forces (Fig. 11.8). Starling's equation expressed algebraically is:

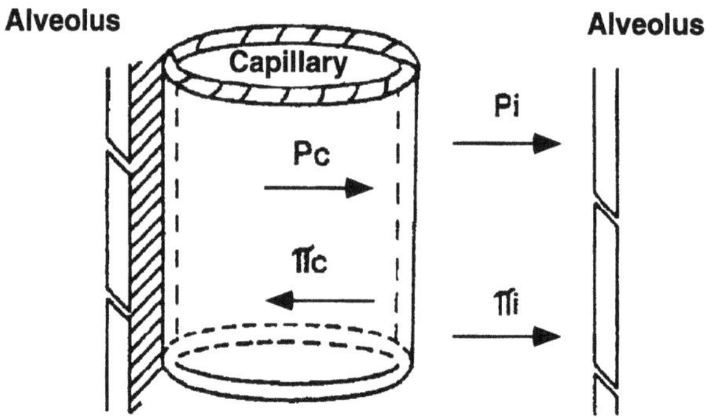

FIGURE 11.8. Starling forces and transcapillary fluid movement. Diagrammatic representation of a capillary surrounded by two alveoli. Transcapillary fluid movement is governed by a balance between hydrostatic forces acting to extrude fluid into the interstitium and osmotic forces drawing fluid into the capillary. The perfusion pressure within the capillary P_c, exceeds the interstitial pressure, P_i, favoring fluid movement into the interstitium; however, the osmotic force exerted by proteins within the capillary, π_c, is greater than its interstitial counterpart, π_i, which tends to draw fluid into the capillary. The net balance of hydrostatic and osmotic forces causes fluid movement into the interstitial space at a rate of about 20ml/hour, which is drained into the lymphatic system. (*Source*: Reprinted from Allison, 1991, with permission).

$$\text{net fluid out} = K_f[(P_c - P_i) - \sigma(\pi_c - \pi_i)]$$

where K_f = the capillary filtration coefficient, which reflects the permeability of the capillary membrane to fluid

P_c and P_i = the hydrostatic pressures exerted by fluid within the capillary and interstitial space, respectively

σ = the protein reflection coefficient and is related to the ability of proteins to cross the capillary membrane

If $\sigma = 1$, the membrane is impermeable to proteins; reciprocally, if $\sigma = 0$, the membrane is freely permeable. σ is estimated to be around 0.75 for plasma proteins in lung microvessels (Taylor and Ballard, 1992). π_c and π_i reflect the osmotic pressure exerted by proteins within the capillary and interstitial space, respectively. Blood cells and plasma proteins within blood vessels increase the capillary osmotic pressure, relative to the low osmotic pressure (low protein content) of the interstitial fluid. Thus, capillary osmotic pressure (π_c) is the major force opposing transcapillary fluid movement. Under normal circumstances, net fluid movement across the lung microvessels is around 20ml/hour, which is easily absorbed by the lymphatic system.

Several factors resist the formation of pulmonary edema. Lymphatics are capable of increasing flow severalfold under conditions of increased transcapillary fluid movement (e.g., as occurs with exercise or left-ventricular failure). Because fluid build-up occurs within the interstitial space, the pressure within the interstitium (P_i) increases, tending to decrease the hydrostatic pressure gradient across the capillary membrane. In addition, fluid entering the interstitium results in dilution of interstitial protein content, decreases π_i, and tends to decrease the osmotic gradient across the capillary membrane. These factors oppose transcapillary fluid movement, but when overwhelmed, fluid builds up within the alveoli and pulmonary edema develops. Pulmonary edema may develop as a result of increased hydrostatic pressure or increased vascular membrane permeability.

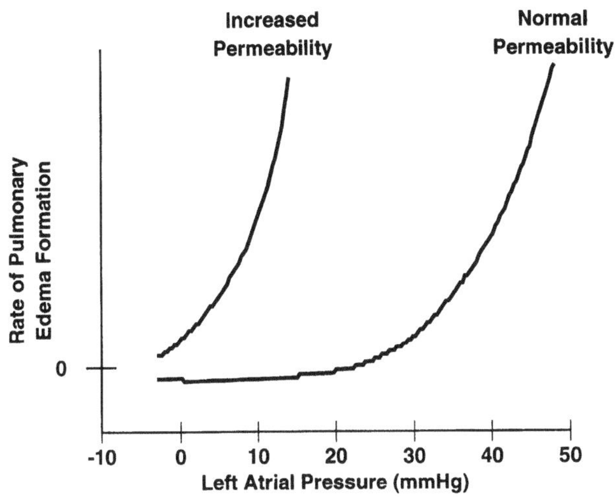

FIGURE 11.9. The relationship between left-atrial pressure and development of pulmonary edema. Increasing left-atrial pressure increases transcapillary fluid movement, but does not cause pulmonary edema because lymphatic drainage increases concomitantly. Left atrial pressure in excess of 28 mmHg, however, exceeds the ability of the lymphatics to drain fluid and pulmonary edema develops. Under conditions of increased capillary membrane permeability, pulmonary edema develops at much lower left atrial pressures. Decreasing left atrial pressure with a diuretic may decrease pulmonary edema formation in both circumstances. (*Source*: Modified with permission from Taylor and Ballard, 1992.)

Increased Hydrostatic Pressure

Increased hydrostatic pressure, as occurs in conditions associated with elevated left-atrial pressure, increases the transcapillary pressure gradient (by increasing P_c) and drives fluid into the interstitial space. A rise in lymphatic flow compensates initially for increased transcapillary fluid movement, but left-atrial pressures in excess of 28 mmHg are usually associated with the development of pulmonary edema. Figure 11.9 illustrates the effect of left-atrial pressure on pulmonary edema formation. Edema fluid resulting from increased hydrostatic pressure is low in protein content and is called *a transudate.*

Increased Permeability

Damage to the integrity of the microvascular endothelium results in an increase in permeability to both fluid and protein, thereby increasing K_f and decreasing σ. As the interstitial protein content increases, the osmotic gradient across the vascular membrane decreases as a result of the increase in π_i. The net effect of endothelial damage is that the factors opposing edema formation are greatly overwhelmed by the increased membrane permeability; consequently, alveoli flooding occurs. In contrast to the low protein content of interstitial fluid associated with increased hydrostatic pressure, increased permeability causes a high protein content fluid called an *exudate*. Endothelial damage and the resultant altered permeability are associated with sepsis, adult respiratory distress syndrome, radiation, drugs or toxin exposure (Block, 1992).

Treatment of Pulmonary Edema

The major clinical sequelae of alveolar flooding are decreased oxygenation and increased respiratory effort, thus leading to acute respiratory failure in some instances. Treatment is directed at reducing pulmonary edema formation and improving oxygenation. Decreasing the hydrostatic pressure gradient may be beneficial regardless of the etiology of pulmonary edema. As shown in Fig. 11.8, increased left-atrial pressure promotes transcapillary fluid movement under conditions of increased permeability as well as increased hydrostatic pressure. Thus, initial treatment of pulmonary edema should include decreasing mean left-atrial pressure toward normal (<12 mmHg) by administering a diuretic or by reducing left ventricular afterload. Administration of oxygen may improve hypoxemia associated with pulmonary edema. Inhaled NO may play a beneficial role in the treatment of one form of pulmonary edema, acute high-altitude pulmonary edema (Anand et al., 1998). In the case of respiratory failure, mechanical ventilation with positive end expiratory pressure (PEEP) is warranted. Unfortunately, no specific therapy is currently available to repair endothelial damage and reduce permeability.

Summary

Pulmonary circulation is unique in that the lungs accept the entire cardiac output, yet maintain a blood pressure markedly lower than the systemic circulation. The pulmonary vascular endothelium plays an integral role in the regulation of vascular tone throughout the developing pulmonary circulation and into adulthood. Dysfunction of the endothelium may contribute to pulmonary hypertension or pulmonary edema. Advances in our understanding of endothelial function has directly led to improved therapies for the treatment of pulmonary hypertension.

References

Abman, S.H., Chatfield, B.A., Hall, S.L., and McMurtry, I.F. (1990) Role of endothelium-derived relaxing factor during transition of pulmonary circulation at birth. Am. J. Physiol. 259, H1921–H1927.

Adatia, I., Barrow, S.E., Stratton, P.D., Miall-Allen, V.M., Ritter, J.M., and Haworth, S.G. (1993) Thromboxane A_2 and prostacyclin biosynthesis in children and adolescents with pulmonary vascular disease. Circulation 88, 2117–2122.

Adatia, I., Lillihei, C., Arnold, J.H., Thompson, J.E., Palazzo, R., Fackler, J.C., and Wessel, D.L. (1994) Inhaled nitric oxide in the treatment of postoperative graft dysfunction after lung transplantation. Ann. Thorac. Surg. 57, 1311–1318.

Allen, S.W., Chatfield, B.A., Koppenhafer, S.A., Schaffer, M.S., Wolfe, R.R., and Abman, S.H. (1993) Circulating immunoreactive endothelin-1 in children with pulmonary hypertension. Association with acute hypoxic pulmonary vasoreactivity. Am. Rev. Respir. Dis. 148, 519–522.

Allison, R.C. (1991) Initial treatment of pulmonary edema: a physiological approach. Am. J. Med. Sci. 302, 385–391.

Anand, I.S., Prasad, B.A., Chugh, S.S., Rao, K.R., Cornfield, D.N., Milla, C.E., et al. (1998) Effects of inhaled nitric oxide and oxygen in high-altitude pulmonary edema. Circulation 98, 2441–2445.

Arai, H., Hori, S., Aramori, I., Ohkubo, H., and Nakanishi, S. (1990) Cloning and expression of a cDNA encoding an endothelin receptor. Nature 348, 730–732.

Archer, S.L., Huang, J., Hampl, V., Nelson, D.P., Shultz, P.J., and Weir, E.K. (1994) Nitric oxide and cGMP cause vasorelaxation by activation of a charybdotoxin-sensitive K channel by cGMP-dependent protein kinase. Proc. Natl. Acad. Sci. USA 91, 7583–7587.

Archer, S.L., Huang, J., Henry, T., Peterson, D., and Weir, E.K. (1993) A redox-based O_2 sensor in rat pulmonary vasculature. Circ. Res. 73, 1100–1112.

Archer, S.L., and Weir, E.K. (1994) In: Clinical cardiology in the elderly, Chesler, E. (ed.) pp. 447–501. Futura Publishing Inc., Armonk, NY.

Barst, R.J. (1999) Recent advances in the treatment of pediatric pulmonary artery hypertension. Pediatr. Clin. North Am. 46, 331–345.

Barst, R.J., Rubin, L.J., McGoon, M.D., Caldwell, E.J., Long, W.A., and Levy, P.S. (1994) Survival in primary pulmonary hypertension with long-term continuous intravenous prostacyclin. Ann. Intern. Med. 121, 409–415.

Bindl, L., Fahnenstich, H., and Peukert, U. (1994) Aerosolised prostacyclin for pulmonary hypertension in neonates. Arch. Dis. Child. 71, F214–F216.

Block, E.R. (1992) Pulmonary endothelial cell pathobiology: implications for acute lung injury. Am. J. Med. Sci. 304, 136–144.

Bonvallet, S.T., Zamore, M.R., Hasunuma, K., Sato, K., Hanasato, N., Anderson, D., et al. (1994) BQ123, an ET_a-receptor antagonist, attenuates hypoxic pulmonary hypertension in rats. Am. J. Physiol. 266, H1327–H1331.

Brannon, T.S., North, A.J., Wells, L.B., and Shaul, P.W. (1994) Prostacyclin synthesis in ovine pulmonary artery is developmentally regulated by changes in cyclooxygenase-1 gene expression. J. Clin. Invest. 93, 2230–2235.

Busse, R., Fleming, I., and Hecker, M. (1993a) Signal transduction in endothelium-dependent vasodilation. Eur. Heart J. 14, 2–9.

Busse, R., Mülsch, A., Fleming, I., and Hecker, M. (1993b) Mechanisms of nitric oxide release from the vascular endothelium. Circulation 87, V-18–V-25.

Calver, A., Collier, J., Moncada, S., and Vallance, P. (1992) Effect of local intra-arterial NG-monomethyl-L-arginine in patients with hypertension: the nitric oxide dilator mechanism appears abnormal. J. Hypertens. 10, 1025–1031.

Cassin, S. (1993) The role of eicosanoids and endothelium-dependent factors in regulation of the fetal pulmonary circulation. J. Lipid Mediators 6, 477–485.

Cassin, S., Dawes, G.S., Mott, J.C., Ross, B.B., and Strang, L.B. (1964) The vascular resistance of the fetal and newly ventilated lung of the lamb. J. Physiol. 171, 61–79.

Chen, Y.F., and Oparil, S. (2000) Endothelial dysfunction in the pulmonary vascular bed. Am. J. Med. Scie. 320, 2123–2132.

Christman, B.W., McPherson, C.D., Newman, J.H., King, G.A., Bernard, G.R., Groves, B.M., and Loyd, J.E. (1992) An imbalance between the excretion of thromboxane and prostacyclin metabolites in pulmonary hypertension. N. Engl. J. Med. 327, 70–75.

Clozel, M., Clyman, R.I., Soifer, S.J., and Heymann, M.A. (1985) Thromboxane is not responsible for the high pulmonary vascular resistance in fetal lambs. Pediatr. Res. 19, 1254–1257.

Cooke, J.P., Stamler, J., Andon, N., Davies, P.F., McKinley, G., and Loscalzo, J. (1990) Flow stimulates endothelial cells to release a nitrovasodilator that is potentiated by reduced thiol. Am. J. Physiol. 259, H804–H812.

Cornfield, D., Chatfield, B., McQueston, J., McMurtry, I., and Abman, S. (1992) Effects of birth-related stimuli on L-arginine-dependent pulmonary vasodilation in ovine fetus. Am. J. Physiol. 262, H1474–H1481.

D'Alonso, G.E., Barst, R.J., Ayres, S.M., Bergofsky, E.H., Brundage, B.H., Detre, K.M., et al. (1991) Survival in patients with primary pulmonary hypertension: results of a national prospective study. Ann. Intern. Med. 115, 343–349.

Dawes, G.S., Mott, J.C., Widdicombe, J.G., and Wyatt, D.G. (1953) Changes in the lungs of the newborn lamb. J. Physiol. (Lond) 121, 141–162.

Eddahibi, S., Raffestin, B., Clozel, M., Levame, M., and Adnot, S. (1995) Protection from pulmonary hypertension with an orally active endothelin receptor antagonist in hypoxic rats. Am. J. Physiol. 268, H828–H835.

Emmanouilides, G., Moss, A., Duffie, E., and Adams, F. (1964) Pulmonary arterial pressure changes in human infants from birth to 3 days of age. J. Pediatr. 65, 327–333.

Enhorning, G., Adams, F., and Norman, A. (1966) Effects of lung expansion on the fetal lamb circulation. Acta Paediatr. Scand. 55, 441–451.

Frostell, C.G., Blomqvist, H., Hedenstierna, G., Lundberg, J., and Zapol, W.M. (1993) Inhaled nitric oxide selectively reverses human hypoxic pulmonary vasoconstriction without causing systemic vasodilation. Anesthesiology 78, 427–435.

Furchgott, R., and Zawadzki, J. (1980) The obligatory role of endothelial cells in the relaxation of arterial smooth muscle by acetylcholine. Nature 288, 373–376.

Fuster, V., Steele, P.M., Edwards, W.D., Gersh, B.J., McGoon, M.D., and Frye, R.L. (1984) Primary pulmonary hypertension: natural history and the importance of thrombosis. Circulation 70, 580–587.

Giaid, A., Yanagisawa, M., Langleben, D., Michel, R.P., Levy, R., Shennib, H., Kimura, S., et al. (1993) Expression of endothelin-1 in the lungs of patients with pulmonary hypertension. N. Engl. J. Med. 328, 1732–1739.

Gibbs, W.W., Nemecek, S., and Stix, G. (1999) The 1998 Nobel Prizes in Science. Sci. Am. 280, 16–19.

Goto, K., Kasuya, Y., Matsuki, N., Takuwa, Y., Kurihara, H., Ishikawa, T., et al. (1989) Endothelin activates the dihydropyridine-sensitive voltage-dependent Ca^{2+} channel in vascular smooth muscle. Proc. Natl. Acad. Sci. USA 86, 3915–3918.

Gruetter, C.A., Barry, B.K., McNamara, D.B., Gruetter, D.Y., Kadowitz, P.J., and Ignarro, L.J. (1979) Relaxation of bovine coronary artery and activation of coronary arterial guanylate cyclase by nitric oxide, nitroprusside and a carcinogenic nitrosoamine. Adv. Cyclic Nucleot. Res. 5, 211–224.

Hampl, V., Archer, S.L., Nelson, D.P., and Weir, E.K. (1993) Chronic EDRF inhibition and hypoxia: effects on pulmonary circulation and systemic blood pressure. J. Appl. Physiol. 75, 1748–1757.

Haworth, S.G. (1988) Pulmonary vascular remodeling in neonatal pulmonary hypertension. State of the art. Chest 93 (Suppl.), 133S–138S.

Higenbottam, T., Stenmark, K., and Simonneau, G. (1999) Treatments for severe pulmonary hypertension. Lancet 353, 338–340.

Hosoda, Y. (1994) Pathology of pulmonary hypertension: a human and experimental study. Pathol. Int. 44, 241–267.

Hutcheson, I.R., and Griffith, T.M. (1991) Release of endothelium-derived relaxing factor is modulated by both frequency and amplitude of pulsatile flow. Am. J. Physiol. 261, H257–H262.

Ignarro, L.J., Buga, G.M., Wood, K.S., Byrns, R.E., and Chaudhuri, G. (1987) Endothelium-derived relaxing factor produced and released from artery and vein is nitric oxide. Proc. Natl. Acad. Sci. USA 84, 9265–9269.

Ikeda, K., Gutierrez, O.G., and Yamori, Y. (1992) Dietary N^G-nitro-L-arginine induces sustained hypertension in normotensive Wistar-Kyoto rats. Clin. Exp. Pharmacol. Physiol. 19, 583–586.

Ingerman-Wojenski, C., Silver, M.J., Smith, J.B., and Macarak, E. (1981) Bovine endothelial cells in culture produce thromboxane as well as prostacyclin. J. Clin. Invest. 67, 1292–1296.

Isaacson, T.C., Hampl, V., Weir, E.K., Nelson, D.P., and Archer, S.L. (1994) Increased endothelium-derived nitric oxide in hypertensive pulmonary circulation of chronically hypoxic rats. J. Appl. Physiol. 76, 933–940.

Jones, D.K., Higginbaum, T.W., and Wallwork, J. (1987) Treatment of primary pulmonary hypertension with intravenous epoprostenol (prostacyclin). Br. Heart J. 57, 270–278.

Lonnqvist, P.A., Winberg, P., Lundell, B., Sellden, H., and Olsson, G.L. (1994) Inhaled nitric oxide in neonates and children with pulmonary hypertension. Acta Scand. Paediatr. 83, 1132–1136.

Lushcer, T.F. (1992) Endothelin: Systemic arterial and pulmonary effects of a new peptide with potent biological properties. Am. Rev. Respir. Dis. 146, S56–S60.

Madden, J.A., Vadula, M.S., and Kurup, V.P. (1992) Effects of hypoxia and other vasoactive agents on pulmonary and cerebral artery smooth muscle cells. Am. J. Physiol. 263, L384–L393.

Meyrick, B. (1987) In: Pulmonary circulation in health and disease, Will, J.A., Dawson, C.A., Weir, E.K., and Buckman, C.K. (eds.) pp. 27–39. Academic Press, Inc., BocaRaton.

Meyrick, B. (1991) Structure function correlates in the pulmonary vasculature during acute lung injury and chronic pulmonary hypertension. Toxicol. Pathol. 4, 447–457.

Mikhail, G.W., Gibbs, J.S., Richardson, M., Chester, A.D., Rogers, P., Wright, G., et al. (1995) The superiority of nebulized prostacyclin in the treatment of patients with primary and secondary pulmonary hypertension. Circulation (Suppl., Abstracts From the 68th Scientific Sessions) 92, I–242.

Miller, W.H. (1983) Physiological effects of cyclic GMP in the vertebrate retinal rod outer segment. Adv. Cyclic Nucleotide Res. 15, 495–511.

Miyauchi, T., and Masaki, T. (1999) Pathophysiology of endothelin in the cardiovascular system. Ann. Rev. Physiol. 61, 391–415.

Miyauchi, T., Yorikane, R., Sakai, S., Sakurai, T., Okada, M., Nishikibe, M., et al. (1993) Contribution of endogenous endothelin-1 to the progression of cardiopulmonary alterations in rats with monocrotaline-induced pulmonary hypertension. Circ. Res. 73, 887–897.

Morin, F., Eagan, E., and Norfleet, W. (1988) Development of pulmonary vascular response to oxygen. Am. J. Physiol. 254, H542–H546.

Murad, F., Arnold, W., Mittal, C.K., and Braughler, J.M. (1979) Properties and regulation of guanylate cyclase and some proposed functions for cyclic GMP. Adv. Cyclic Nucleot. Res. 11, 175–204.

Nelson, M.T., and Quayle, J.M. (1995) Physiological roles and properties of potassium channels in arterial smooth muscle. Am. J. Physiol. 268, C799–C822.

Nishiwaki, K., Nyhan, D.P., Rock, P., Desai, P.M., Peterson, W.P., Pribble, C.G., and Murray, P.A. (1992) N^{ω}-nitro-L-arginine and pulmonary vascular pressure-flow relationship in conscious dogs. Am. J. Physiol. 262, H1331–H1337.

Oparil, S., Chen, S.J., Meng, Q.C., Elton, T.S., Yano, M., and Chen, Y.F. (1995) Endothelin-A receptor antagonist prevents acute hypoxia-induced pulmonary hypertension in the rat. Am. J. Physiol. 12, L95–L100.

Palmer, R.M.J., Ferrige, A.G., and Moncada, S. (1987) Nitric oxide release accounts for the biological activity of endothelium-derived relaxing factor. Nature 327, 524–526.

Palmer, R.M.J., Rees, D.D., Ashton, D.S., and Moncada, S. (1988) L-arginine is the physiological precursor for the formation of nitric oxide in endothelium-dependent relaxation. Biochem. Biophys. Res. Commun. 153, 1251–1256.

Parker, H.R., and Purves, M.J. (1967) Some effects of maternal hyperoxia and hypoxia on the blood gas tensions and vascular pressures in the foetal sheep. Q. J. Exp. Physiol. 52, 205–221.

Pepke-Zaba, J., Higenbottam, T., Dinh-Xuan, A.T., Stone, D., and Wallwork, J. (1991) Inhaled nitric oxide as a cause of selective pulmonary vasodilation in pulmonary hypertension. Lancet 338, 1173–1174.

Pitt, B.R., and St. Croix, C. (2002) Complex regulation of iNOS in lung. Am. J. Respir. Cell Mol. Biol. 26, 6–9.

Pohl, U., Holtz, J., Busse, R., and Bassenge, E. (1986) Crucial role of endothelium in the vasodilator response to increased flow in vivo. Hypertension 8, 37–44.

Pollock, D.M., Keith, T.L., and Highsmith, R.F. (1995) Endothelin receptors and calcium signaling. FASEB 9, 1195–1204.

Post, J.M., Hume, J.R., Archer, S.L., and Weir, E.K. (1992) Direct role for potassium channel inhibition in hypoxic pulmonary vasoconstriction. Am. J. Physiol. 262, C882–C890.

Rabinovitch, M., Keane, J.F., Norwood, W.I., Castaneda, A.R., and Reid, L. (1984) Vascular structure in lung tissue obtained at biopsy correlated with pulmonary hemodynamic findings after repair of congenital heart defects. Circulation 69, 655–667.

Rendas, A., Branthwaite, M., and Reid, L. (1978) Growth of pulmonary circulation in normal pig-structural analysis and cardiopulmonary function. J. Appl. Physiol. 45, 806–817.

Rich, S., Kaufmann, E., and Levy, S. (1992) The effect of high doses of calcium-channel blockers on survival in primary pulmonary hypertension. N. Engl. J. Med. 327, 76–81.

Roberts, J.D., Lang, P., Bigatello, L., Vlahakes, G.J., and Zapol, W.M. (1993) Inhaled nitric oxide in congenital heart disease. Circulation 87, 447–453.

Roberts, J.D., Polaner, D.M., Lang, P., and Zapol, W.M. (1992) Inhaled nitric oxide in persistent pulmonary hypertension of the newborn. Lancet 340, 818–819.

Roos, A. (1962) Poiseuille's law and its limitations in vascular systems. Med. Thorac. 19, 224–238.

Rossaint, R., Falke, K.J., Lopez, F., Slama, K., Pison, U., and Zapol, W.M. (1993) Inhaled nitric oxide for the adult respiratory distress syndrome. N. Engl. J. Med. 328, 399–405.

Rubanyi, G., Romero, J.C., and Vanhoutte, P.M. (1986) Flow-induced release of endothelium-derived relaxing factor. Am. J. Physiol. 250, H1145–H1149.

Rudolph, A.M. (1974) In: Congenital diseases of the heart: clinical-physiological considerations in diagnosis and management, Rudolph, A.M. (ed.) pp. 29–48. Year Book Publishers Inc., Chicago.

Sakurai, T., Yanagisawa, M., and Takuwa, Y. (1990) Cloning of a cDNA encoding a non-isopeptide-selective subtype of the endothelin receptor. Nature 348, 732–735.
Schror, K. (1993) The effect of prostaglandins and thromboxane A_2 on coronary vascular tone-mechanisms of action and therapeutic implications. Eur. Heart J. 14, 34–41.
Shaul, P.W., Farrar, M.A., and Magness, R.R. (1993) Pulmonary endothelial nitric oxide production is developmentally regulated in the fetus and newborn. Am. J. Physiol. 265, H1056–H1063.
Stamler, J.S., Loh, E., Roddy, M.-A., Currie, K.E., and Creager, M.A. (1994) Nitric oxide regulates basal systemic and pulmonary vascular resistance in healthy humans. Circulation 89, 2035–2040.
Sylvester, J.T. (2001) Hypoxic pulmonary vasoconstriction. Circ. Res. 88, 1228–1230.
Taguchi, H., Heistad, D.D., Kitazono, T., and Faraci, F.M. (1995) Dilation of cerebral arterioles in response to activation of adenylate cyclase is dependent on activation of calcium-dependent K^+ channels. Circ. Res. 76, 1057–1062.
Taylor, A.E., and Ballard, S.T. (1992) Microvascular function: transvascular exchange of fluid in the airways. Am. Rev. Respir. Dis. 146, S24–S27.
Tiktinsky, M.H., and Morin, F.C. (1993) Increasing oxygen tension dilates fetal pulmonary circulation via endothelium-derived relaxing factor. Am. J. Physiol. 265, H376–H380.
Tristani-Firouzi, M., Reeve, H.L., Tolarova, S., Weir, E.K., and Archer, S.L. (1996) Oxygen-induced constriction of rabbit ductus arteriosus occurs via inhibition of a 4-aminopyridine-, voltage-sensitive potassium channel [see comments]. J. Clin. Invest. 98, 1959–1965.
Vane, J.R., and Botting, R.M. (1995) Pharmacodynamic profile of prostacyclin. Am. J. Cardiol. 75, 3A–10A.
Velvis, H., Moore, P., and Heymann, M.A. (1991) Prostaglandin inhibition prevents the fall in pulmonary vascular resistance as a result of rhythmic distension of the lungs in fetal lambs. Pediatr. Res. 30, 62–68.
Walsh, M.P. (1991) Calcium-dependent mechanisms of regulation of smooth muscle contraction. Biochem. Cell Biol. 69, 771–780.
Weir, E.K., and Archer, S.L. (1995) The mechanism of acute hypoxic pulmonary vasoconstriction: the tale of two channels. FASEB J. 9, 183–189.
West, J.B. (1985) In: Respiratory physiology—the essentials, West, J.B. (ed.) pp. 31–48. Williams and Wilkins, Baltimore.
Yamamoto, K., Ikeda, U., Mito, H., Fujikawa, H., Sekiguchi, H., and Shimada, K. (1994) Endothelin production in pulmonary circulation of patients with mitral stenosis. Circulation 89, 2093–2098.

Recommended Readings

Dudek, S.M., and Garcia, J.G. (2001) Cytoskeletal regulation of pulmonary vascular permeability. J. Appl. Physiol. 91, 1487–1500.
Fineman, J.R., Soifer, S.J., and Heymann, M.A. (1995) Regulation of pulmonary vascular tone in the perinatal period. Ann. Rev. Physiol. 57, 115–134.
Higenbottam, T., Stenmark, K., and Simonneau, G. (1999) Treatments for severe pulmonary hypertension. Lancet 353(9150), 338–340.
Michelakis, E.D., Archer, S.L., and Weir, E.K. (1995) Acute hypoxic pulmonary vasoconstriction: a model of oxygen sensing. Physiol. Res. 44(6), 361–367.
Pitt, B.R., and St. Croix, C. (2002) Complex regulation of iNOS in lung. Am. J. Respir. Cell Mol. Biol. 26, 6–9.
Stenmark, K.R., and Mecham, R.P. (1997) Cellular and molecular mechanisms of pulmonary vascular remodeling. Ann. Rev. Physiol. 59, 89–144.
West, J.B., and Mathieu-Costello, O. (1999) Structure, strength, failure, and remodeling of the pulmonary blood-gas barrier. Ann. Rev. Physiol. 61, 543–572.

12
Regulation of Lung Water: Role of the Bronchial Circulation

NIRMAL B. CHARAN AND PAULA CARVALHO

The main function of the lung is to allow transfer of oxygen and carbon dioxide molecules across the alveolar–capillary membrane. In order to achieve this task most efficiently, the lung must free itself of excess liquid, proteins, and other debris. Under normal conditions, there is continuous leakage of water from the alveolar capillaries into the interstitial space, which is eventually reabsorbed and restored to the circulation.

The normal extravascular water content of the lungs is estimated to be about 300–400 ml. About 100 ml of this volume is located in the immediate surroundings of the alveolar capillaries and like the intravascular capillary blood volume, it is spread as a thin film over a large surface area. The amount of extravascular water is regulated by the balance between the fluid that leaks from the pulmonary microvasculature and the fluid that is effectively removed from the lung. Thus, the term *pulmonary edema* is used to denote the condition in which there is an excess of water accumulation in the lung. Accumulation may occur when there is an increase in fluid filtration from the pulmonary capillaries and/or a decrease in reabsorption of fluid from the interstitial space.

Physiological Principles that Govern Lung Fluid Balance

According to Starling's equation, the movement of fluid and protein across the capillary wall is governed by the permeability of the capillary membrane and differences in hydrostatic as well as oncotic pressures across the capillary wall. Thus, the force that governs fluid filtration and fluid reabsorption across the capillary walls is given by the equation:

$$Jv = Lp \cdot S(\Delta P - \sigma \Delta \pi)$$

where Jv = the filtration rate
Lp = the microvascular hydraulic conductivity
S = the surface area available for filtration
ΔP and $\Delta\pi$ = the transvascular hydrostatic and colloid osmotic pressure differences
σ = the osmotic reflection coefficient

The ΔP equals (Pc-Pi) and $\Delta\pi$ equals ($\pi c - \pi i$), where Pc and Pi are the intracapillary and interstitial hydrostatic pressures, respectively; πc and πi are the corresponding colloid osmotic pressures. Both plasma (πc) and interstitial (πi) colloid osmotic pressures are determined by the protein concentration alone, and osmotic pressure exerted by small solutes (<10 kD) plays no role in fluid filtration.

There are two coefficients in this equation: one is $Lp \cdot S$, which is the same as K*f* for most physiologists; the other is σ. The filtration coefficient (K*f*) equals the product of the hydraulic conductance of the membrane (Lp) and the surface area (S) available for filtration. Here Lp is only a membrane parameter that is independent of exchange surface area, whereas K*f* includes available surface area for filtration. The coefficient σ is a reflection coefficient that varies from 0 to 1. It defines molecular sieving through the capillary membrane. In other words, it describes the effectiveness of the membrane in preventing the flow of solute compared with the flow of water across the membrane. In the pulmonary circulation, the only substances that do not equilibrate across the microvascular membrane are large plasma protein molecules. For these molecules, the reflection coefficient is 1.0. This implies that the microvascular membrane does not allow proteins to pass through it and, hence, full osmotic pressure exerted by proteins in the plasma is effective across the microvascular barrier. Thus, $\Delta\pi$ that governs osmotic flow is itself dependent on the intrinsic properties of the membrane. In most microvascular beds σ is slightly less than 1, and, if the microvascular barrier becomes more permeable to proteins, σ decreases. For example, when σ is 0, the capillary membrane is completely permeable to proteins; therefore, there is no osmotic pressure difference across the capillary wall. In this situation, there is no flow due to osmotic forces alone. On the other hand, flow due to $\Delta\pi$ is greater when σ is closer to 1.

The transmicrovascular convective flux (Jv) depends on the differences in both hydrostatic and oncotic pressures rather than on a single entity. Although πc and πi are written as if they are independent variables in the preceding equation, they are in fact coupled. This is because the interstitial protein concentration (πi) is dependent on the plasma protein concentration. For example, when there is a decrease in serum protein concentration, the interstitial protein concentration also decreases within a few hours; however, if πc changes rapidly, there is then a transient change in $\Delta\pi$.

Kedem and Katchalsky further defined the transport of proteins across the capillary membrane as follows:

$$Jp = PS\Delta C + (1-\sigma)\overline{C} \cdot Jv$$

where Jp = the transmicrovascular protein flux
P = the microvascular permeability coefficient
$\overline{C}$ = the mean protein concentration in the transendothelial transport pathway
ΔC = the transmembrane difference in protein concentration

In the Kedem-Katchalsky equation, convective flux (Jv) implies that transmicrovascular protein transport is filtration-dependent; therefore, intravascular pressure determines the filtration rate as well as protein transport. The transmembrane gradient of protein concentration may also govern diffusive protein transport, but to a much lesser extent.

The pathway taken by liquid and aqueous molecules as they flow across the microvascular barrier by diffusion and convection has been of great interest to physiologists. This concept of the "sieving" property of a membrane can be defined as $NCr^4/\Delta Xn$

where N = the number of pores/cm^2
C = a constant
r = the radius of these pores
ΔX = the thickness of the capillary wall
n = the viscosity of the filtering fluid

The diameter of these pores determines the size of the molecules that are allowed to sieve through the capillary membrane. Small pores, therefore, allow only the osmotic flow, whereas large pores permit sieving of the larger molecules (e.g., proteins). If a membrane has only small pores, only water flows through these pores. When this water enters the interstitium, it dilutes the interstitial protein concentration (decreases πi), resulting in an increase in osmotic gradient across the capillary wall. As described earlier, this favors fluid reabsorption and, therefore, buffers the net filtration force, preventing accumulation of water in the interstitium. In contrast, flow through the large pores allow proteins to filter which increases interstitial protein content (increases πi). In this situation, the osmotic gradient across the capillary membrane decreases and, therefore, the effectiveness of fluid reabsorption also decreases.

Under normal circumstances, forces tending to move fluid out of and into the circulation are almost balanced; therefore, only a small amount of fluid continuously leaks from the circulation. This fluid is eventually removed by the lymphatic channels. Fluid normally filters at the arterial end of the capillaries and is reabsorbed at the venular end, which keeps the interstitium relatively dry. If there is imbalance of forces or changes in the capillary permeability, however, excessive fluid accumulation in the interstitial space may occur, resulting in pulmonary edema. Thus, in terms of microvascular physiology, pulmonary edema may be due to either an increase in the "hydrostatic" pressure across the capillary membrane or an "increase in capillary permeability." Hydrostatic edema is caused by increased filtration due to an increase in transvascular pressure differences (ΔP and $\Delta \pi$), as long as the membrane barrier Lp and σ remain unchanged. Hydrostatic pulmonary edema occurs in any condition that causes an increase in pulmonary microvascular pressure, of which a typical example is an increase in left-atrial pressure due to any cause. The term *capillary permeability* edema is applied when there is damage to the capillary endothelium, resulting in increased fluid and protein filtration through the capillary membrane. Because injury to the endothelium results in loss of membrane barrier, the capillary filtrate is relatively more protein-rich than it is in hydrostatic edema, because in the latter case the capillary membrane is functionally intact. A typical clinical example of increased capillary permeability edema is adult respiratory distress syndrome (ARDS), in which injury to endothelial cells is known to occur in response to a number of etiological factors (e.g., sepsis, shock, trauma, and multiple transfusions).

The Movement of Liquid and Protein Across the Pulmonary Capillaries

Pulmonary capillaries are the predominant site for fluid filtration; however, leakage from the extraalveolar vessels has also been demonstrated. The excessive fluid that leaks from the pulmonary microvasculature first enters the interstitium. Extravascular ground substance of the interstitium consists of glycosaminoglycans, which are long-chained molecules of random configuration and varying charge. It also contains a large number of elastin, reticulin, and collagen fibers, interstitial cells of several varieties, and an extensive basal laminal system for the alveolar epithelial cells and the capillary endothelium. These macromolecules provide a meshlike network through which fluid percolates to reach the lymphatics. The swelling characteristic of tissue is dependent on these macromolecules. The most-studied molecule is hyaluronan. The intrinsic matrix property of the lung is important because swelling of the matrix develops when there is an increase in fluid filtration and the interstitium starts to fill with fluid that blunts increases in interstitial pressure. This contributes toward the maintenance of the pulmonary blood flow even in the presence of significant pulmonary edema.

The lung interstitium carries a net negative charge that is likely to influence the flow of ions through the interstitium. For example, interstitial transit of positively charged molecules is less impeded than the transit of negatively charged molecules. Presence of hyaluronic acid is thought to provide the negative charges that repel anions because treatment with hyaluronidase improves the flow of anions.

Once the capacity of the pulmonary microvasculature to absorb fluid reaches its maximum limits, the fluid makes its way along interstitial fibers toward the respiratory bronchioles. There are no lymphatic vessels within the alveolar walls; rather, they are confined to the loose connective tissue spaces around the bronchi, disappearing altogether at the level of the respiratory bronchiole. The movement of fluid from the filtration site to the initial lymphatics is passive and takes place along a hydraulic pressure gradient. Tissue movements and pressure swings that take place during breathing, facilitate the fluid movement. The total hydraulic driving pressure from the alveolar wall interstitium to the lymphatic capillaries is estimated to be about 4–5 cm H_2O. Lymph flow increases in the presence of increasing filtration rates, thus leading to the removal of excess fluid from the interstitium. Once lymphatic clearance is at its maximum, fluid starts to accumulate in the interstitium and around the peribronchial connective tissue space. As a consequence, the first site of fluid accumulation in pulmonary edema is the extraalveolar connective tissue space, followed by entry into the peribronchovascular loose connective tissue space, which has a relatively large holding capacity. This results in swelling of hyaluronan and other matrix glycosaminoglycans, which are classically recognized as the peribronchial cuffs of edema fluid.

Interstitial pulmonary edema can be detected in the lung when there is a 30–50% increase in the extravascular lung water content. The maximum capacity of the interstitial space is approximately 7 ml/kg, or approximately 500 ml of fluid in a 70 kg human. When the maximum capacity of the pulmonary interstitial space is exceeded, there is flooding of the airspaces with edema fluid. Both fluid and protein are normally not filtered into the alveoli because the alveolar epithelium has a very low permeability to small molecules. Moreover, there is evidence supporting the view that the alveolar flooding stage of edema does not occur unless total lung water has increased by about 50%. This could occur either with hydrostatic or increased-permeability edema. In both of these circumstances the protein concentration in alveolar edema fluid is equal to the protein concentration of the interstitial edema fluid. It is also noteworthy that in case of increased permeability edema, fluid enters the alveoli through the alveolar epithelium. In hydrostatic pulmonary edema, however, fluid enters alveolar space because of overflow of fluid from the peribronchial space. At the onset of alveolar flooding, edema fluid just covers the corners of the alveoli but, with more flooding, the alveoli get completely filled with fluid in an "all-or-none" manner, meaning that some alveoli are devoid of any fluid, whereas others are filled with fluid. This allows some gas exchange to occur in alveoli that are relatively normal. Once alveolar flooding starts, fluid rapidly enters the air spaces. It remains unclear why the alveolar epithelial barrier suddenly changes. It has been suggested that, with increasing interstitial edema, interstitial fluid pressure may increase sufficiently to force open the epithelial intercellular junctions. This is a moot point because the site of alveolar flooding in high-pressure edema is not the alveolar-epithelial barrier but rather the terminal airway epithelium.

In addition, there is controversy over the effect of increased alveolar pressure on fluid filtration from the pulmonary capillaries. One view holds that positive end-expiratory pressure (PEEP) can reduce extravascular lung fluid by decreasing transmural vascular pressures. The critic argues that if the alveolar pressure is increased by using high-peak inspiratory pressures, it can cause an increase in microvascular membrane permeability. The critic goes further by stating that the use of high tidal volumes during mechanical ventilation is known to cause marked changes in the permeability of the alveolar capillary membrane, resulting in pulmonary edema (see discussion—Amat. et al., 2000). This is important because most patients with ARDS are treated with mechanical ventilation. Other evidence shows that use of low tidal volumes improves survival of patients with ARDS. It is a matter of interest that gadolinium, which is a trivalent lanthanide element that effectively blocks nonselective, stretch-activated cation channels, blocks this phenomenon. This suggests that calcium channels might be involved in causing increased permeability edema.

Reabsorption of Interstitial Liquid and Protein

When fluid filtration from pulmonary capillaries increases, fluid starts to accumulate in the lung. This fluid is rapidly cleared from the interstitium so that gas exchange in alveolar units can be maintained as efficiently as possible. This view borne out in several animal studies, among which is the demonstration that instilling a crytalloid solution into the airspaces moves rapidly into the interstitium and then cleared from the lung with a half-time of approximately 3 hours. The question now is, How does the lung remove excess fluid in order to keep itself relatively dry? The answer lies in the fact that there are five adjacent anatomical structures that could play a role in the reabsorption of excess alveolar and interstitial fluid. These are listed in Table 12.1.

For example, take edema fluid that has a low protein concentration. It can be reabsorbed into the adjacent vascular plexuses, as seen in the case of fluid that transudates into the interstitium, as long as the high hydrostatic pressure within the capillaries does not exceed the oncotic pressure that normally forces the fluid back into the capillaries. Once fluid accumulates in the peribronchial and perivascular spaces, however, it is unlikely that it is absorbed into the large pulmonary vessels because these vessels have less surface area and their walls are thicker. It is significant that complete elimination of pulmonary blood flow fails to abolish interstitial fluid clearance, which suggests that, at least in these animal models, the pulmonary circulation is not the main route for fluid removal in hydrostatic or in increased capillary permeability pulmonary edema.

For many years, the lymphatic channels were thought to be the major pathway for lung fluid clearance because of their favorable anatomical location where excess fluid

TABLE 12.1. Pathways for fluid and protein clearance in the lung.

Pulmonary circulation	Pleural space
Lung lymphatics	Mediastinum
Airways	Bronchial circulation

and protein normally accumulate. In fact, this view led to the development of an animal model involving a lung lymph fistula that was used extensively to study the lung fluid balance. In this model, an increase in lung lymph flow implies an increase in net fluid flux, whereas changes in the ratio of plasma/lymph protein reflect permeability changes in the capillary membrane. Findings show that the lymphatics remove only 10–30% of interstitial edema fluid. On the other hand, it could be argued that protein-rich interstitial fluid is probably cleared primarily by the lymphatics because there is no osmotic or hydrostatic gradient favoring fluid reabsorption into the pulmonary microvasculature when this occurs. The protein concentration in the interstitial edema fluid also determines the rate at which a given volume of fluid is absorbed: The higher the protein concentration, the slower the rate of fluid clearance.

Once alveolar flooding occurs, some of the fluid is cleared by mucociliary transport to larger airways and is eventually expectorated. The movement of fluid across the alveolar epithelium is more complex and is inversely proportional to the concentration of the protein in the alveolar fluid. Excess alveolar fluid is removed from the distal air spaces by active sodium transport and this may be mediated by the β-adrenergic system. It has been found in an animal study that, following smoke inhalation, edema occurred in the lungs where the pulmonary artery has been occluded, but that it failed to occur where only the bronchial artery had been ligated. Alveolar fluid clearance is markedly reduced by inhibition of apical sodium channel uptake or Na^+-K^+-ATPase activity, suggesting that active metabolic processes play a role in the removal of fluid across the alveolar barrier.

Pleural effusions are not unusual in patients with pulmonary edema. The mechanism responsible for the formation of pleural effusions in pulmonary edema is not yet well understood. They do not occur in the early stages of pulmonary edema and, therefore, are unlikely to play an important role in fluid clearance. Nevertheless, with increasing rates of fluid filtration, the flow of interstitial liquid into the pleural space can be protective against the development of life-threatening alveolar edema. Transfer of excess fluid from lung interstitium to the pleural space, where the effects of fluid on lung function are relatively minor, helps to preserve gas exchange functions of the lung. It is also unlikely that the mediastinum plays an important role in the removal of edema fluid because vascular plexuses are not present in this location.

Role of the Bronchial Circulation in Lung Fluid Balance

Anatomical Considerations

The airways receive their blood supply from the bronchial circulation, which belongs to the systemic circulation. The bronchial arteries originate directly or indirectly from the aorta and enter the mediastinum. There a bronchomediastinal plexus is formed that supplies a number of mediastinal structures, including the esophagus, hilar lymph nodes, *vasa vasorum* of the thoracic aorta, pericardium, mediastinal pleura, and vagus nerve. Upon reaching the main bronchus on each side, small vessels originate that form an annulus surrounding the bronchus. The bronchial arteries then enter the lungs on each side through the hilum and, upon reaching the mainstem bronchus, they divide in the peribronchial connective tissue sheath. These arteries then follow the branches of the bronchial tree. In general, two or three branches of the bronchial artery wind around and supply the wall of the bronchus. They follow the airways as far as the terminal bronchioles. At the hilum the bronchial arteries have a diameter of approximately 1.5 mm, which decreases to approximately 0.5–0.75 mm at the point of entry into a bronchopulmonary segment. There appears to be no relationship between the diameter of the bronchial lumen and the size of the bronchial arteries.

The branches of bronchial arteries anastomose freely with each other, forming a plexus in the peribronchial space. They also give rise to small branches that penetrate the muscular layer to reach the bronchial mucosa, where they form the submucosal plexus. Thus, there is a submucosal as well as a peribronchial vascular plexus along the entire length of the bronchial tree down to the terminal bronchioles. The bronchial arteries also supply the walls of the pulmonary artery, the pulmonary vein, and the visceral pleura. The medial surface of the visceral pleura receives vessels from the annulus formed by the branches of the bronchial arteries at the hilum. The anterior, lateral, and interlobar visceral pleurae receive their blood supply from the distal branches of the bronchial arteries, which supply the airways and penetrate the lung parenchyma to reach the pleura.

The bronchial venous blood has two pathways for drainage. The venous blood from the proximal tracheobronchial tree (first two or three subdivisions, mainly extrapulmonary) drains into the bronchial veins, which become continuous with mediastinal vessels. These proceed to the azygous system and drain through the superior vena cava into the right heart. The bronchial microvascular plexus that follows the airways anastomoses with the pulmonary microvasculature along the airway and beyond the terminal bronchioles. Thus, the intrapulmonary bronchial venous flow drains into the pulmonary circulation and, hence, into the left atrium. Most of the bronchial venous blood drains into the pulmonary circulation through the bronchopulmonary anastomoses. A relatively smaller portion drains into the systemic venous system.

Role of the Bronchial Circulation in the Regulation of Lung Water

The question now arising is whether the bronchial circulation has a role in lung fluid balance. Studies that have looked into this question are unfortunately sparse. Nevertheless, there is a rationale for implicating the bronchial circulation in playing an important role in lung fluid balance. This is partly because the largest volume of interstitial pulmonary edema fluid accumulates in the bronchovascular connective tissue space and partly because bronchial vascular plexuses are abundant in this location. Only two reasons can be given why pulmonary edema fluid always accumulates in the peribronchial space: Either the fluid leaks from the bronchial microvasculature or the fluid leaks from some other vessels at a completely different site, but enters the peribronchial space, where it gets absorbed into the bronchial microvasculature.

Why should fluid leak from the bronchial microvasculature in hydrostatic pulmonary edema when the pressures are increased in the pulmonary microvasculature? One answer is the fact that there are numerous bronchovascular communications between the pulmonary veins and the bronchial vascular plexuses. A rise, therefore, in pulmonary venous pressure would be transmitted into the bronchial venous plexus. An increase in pulmonary vascular pressure may initially result in diversion of blood through the bronchopulmonary anastomoses into the systemic venous system. This may be a mechanism responsible for preventing increases in pulmonary microvascular pressures in response to modest elevations in left atrial pressure. What happens, then, if systemic venous pressure is elevated? It would impede bronchial venous drainage into the right side of the heart and result in increases in the bronchial venous pressure. This would be expected to divert bronchial venous blood into the low pressure pulmonary circulation. As a result, hydrostatic pressure in the bronchial microvasculature would not change significantly.

This prediction has been confirmed in one study where bronchial venous pressure was increased by infusing saline into the bronchial vein. In this type of experiment, neither lung lymph flow nor the lymph–plasma protein ratio changed, which suggests that increasing bronchial venous pressure alone does not result in an increase in fluid filtration from the bronchial vascular plexus. If, however, both the pulmonary vascular and the systemic venous pressures are increased, then fluid transudation from the bronchial vascular plexuses may well occur. In this instance, fluid flux from the bronchial vasculature is further facilitated because peribronchial interstitial pressure is more negative than that in the alveolar wall. Thus, an increase in pulmonary venous pressure, together with an elevated systemic venous pressure may result in higher transvascular hydrostatic pressure gradients in the bronchial microvasculature. Were this the case, fluid transudation from the bronchial venous plexuses can occur. This reasoning is supported by the fact that systemic venous hypertension in combination with fluid overload has been shown to cause a significant increase in lung water together with airway edema. This implies that, in biventricular heart failure, fluid filtration can occur from the bronchial microvasculature.

Fluid leakage from the bronchial circulation may also occur in response to increased capillary permeability. For example, administration of endotoxin, histamine, and bradykinin has been shown to make the bronchial venules highly permeable in both submucosal and peribronchial plexuses, resulting in peribronchial interstitial edema. This suggests that in certain pathological conditions (e.g., asthma and bronchitis) fluid leakage into the bronchial mucosa may occur from the bronchial circulation. It has also been shown that there is increased permeability in the bronchial vascular bed after synthetic smoke inhalation. In the absence of the bronchial circulation, there is approximately a 30% reduction in lung edema formation. In another animal study, smoke inhalation elicited lung edema when the pulmonary artery had been occluded, but there was no edema when only the bronchial arteries had been ligated. Together, these findings support the general view that leakage from the bronchial circulation contributes to the formation of pulmonary edema following smoke inhalation.

Last, the bronchial circulation may play a role in fluid reabsorption from the peribronchial, as well as from the submucosal plexus. It was demonstrated in one animal study that sodium chloride and water can be absorbed from the mucosal surface of the segmental bronchi, suggesting that the submucosal bronchial vascular plexus is capable of fluid absorption. In an animal model of hydrostatic pulmonary edema, we found that occlusion of the bronchial vein, which drains into the systemic venous system, resulted in a 25% increase in lung water, thus indicating that in the presence of pulmonary edema the bronchial circulation plays a role in fluid absorption. This finding has been confirmed in another study where it was found that 14% of interstitial pulmonary fluid is reabsorbed into the bronchial circulation during recovery from hydrostatic edema.

Summary

Regulation of lung water is important for the maintenance of normal gas exchange functions of the lung. Pulmonary edema may occur if there is an increase in fluid filtration due to an increase in hydrostatic pressure or an increase in vascular permeability. The lungs appear very efficient in removing excessive fluid that is filtered into the pulmonary interstitium. The lymphatics, pulmonary circulation, and bronchial circulation have an equally important role in fluid reabsorption. Pleural effusions occur rather late as they are indicative of severe pulmonary edema. The cellular and biochemical mechanisms responsible for fluid filtration and fluid reabsorption are currently being investigated, and these studies may eventually lead to new modalities for the treatment of pulmonary edema.

References

Amato, M.B.P., Barbas, C.S.V., Medeiros, D.M. et al. (2002) Acute respiratory distress syndrome network. Ventilation with lower tidal volumes as compared with traditional tidal volumes for acute lung injury and the acute respiratory distress syndrome. N. Engl. J. Med. 342, 1301–1308.

Bhattacharya, J. (1998) Physiological basis of pulmonary edema. In: Pulmonary edema. Matthay, M.A., and Ingbar, D.H. (eds.) pp. 1–36. Lung biology in health and disease series, vol. 116. Marcel Dekker, Inc., New York.

Charan, N.B. (1998) Regulation of lung water. Cardiologia 43, 1305–1314.

Charan, N.B., and Carvalho, P. (1992) The anatomy of the normal bronchial circulatory system in man and animals. In: Butler, J. (ed.) pp. 45–78. The bronchial circulation. [Lenfant, C. (executive ed.) Lung biology in health and disease series.] Marcel Dekker, Inc., New York.

Charan, N.B., Turk, G.M., and Dhand, R. (1984) Gross and subgross anatomy of bronchial circulation in sheep. J. Appl. Physiol. 57, 658–664.

Charan, N.B., Turk, G.M., and Hey, D.H. (1985) Effect of increased bronchial venous pressure on lung lymph flow. J. Appl. Physiol. 59, 1249–1253.

Drefuss, D., and Saumon, G. (1998) Ventilator-induced lung injury: lesson from experimental studies. Am. J. Respir. Crit. Care Med. 157, 294–323.

Effros, R.M., Mason, G.R., Hukkanen, J., and Silverman, P. (1998) New evidence for active sodium transport from fluid-filled rat lungs. J. Appl. Physiol. 66, 906–919.

Efimova, O., Volokhow, A.B., Iliafar, S., and Hales, C.A. (2000) Ligation of the bronchial artery in sheep attenuates early pulmonary changes following exposure to smoke. J. Appl. Physiol. 88, 888–893.

Folkesson, H.G., Norlin, A., Wang, Y., Abedinpour, P., and Matthay, M.A. (2000) Dexamethasone and thyroid hormone pretreatment upregulate alveolar epithelial fluid clearance in adult rats. J. Appl. Physiol. 88, 416–424.

Fukue, M., Serikov, V.B., and Jerome, E.H. (1996) Bronchial vascular reabsorption of low protein interstitial edema liquid in perfused sheep lungs. J. Appl. Physiol. 81, 810–815.

Kedem, O., and Katchalsky, A. (1958) Thermodynamic analysis of the permeability of biological membranes to non-electrolytes. Biochem. Biophys. Acta. 27, 229–246.

Pietra, G.G., and Magno, M. (1978) Pharmacological factors influencing permeability of the bronchial microcirculation. Fed. Proc. 37, 2466–2479.

Pietra, G.G., Szidon, J.P., Carpenter, H.A., and Fishman, A.P. (1974) Bronchial venular leakage during endotoxin shock. Am. J. Pathol. 77, 387–406.

Sakurai, H., Johnigan, R., Kiruchi, Y., Harada, M., Traber, L.D., and Traber, D.L. (1998) Effect of reduced bronchial circulation on lung fluid flux after smoke inhalation in sheep. J. Appl. Physiol. 84, 980–986.

Staub, N.C. (1974) Pulmonary edema. Physiol. Rev. 54, 678–811.

Staub, N.C. (1983) Alveolar flooding and clearance. Am. Rev. Respir. Dis. 127, S44–S51.

Taylor, A.E. (1981) Capillary fluid filtration. Circ. Res. 49, 557–575.

Recommended Readings

Bhattacharya, J. (1998) Physiological basis of pulmonary edema. In: Pulmonary Edema. Matthay, M.A., and Ingbar, D.H. (eds.) Lung biology in health and disease series. Vol. 116. pp. 1–36. Marcel Dekker, Inc., New York.

Charan, N.B. (1998) Regulation of lung water. Cardiologia 43, 1305–1314.

13
Airway Wall Liquid in Health and Disease

DEBORAH YAGER

Introduction

Airway Wall Liquid Is Essential for Proper Airway Function

In normal airways, liquid transport promotes several physiologically significant processes. Cellular hydration, which may change in response to alteration of cell size during extreme changes in lung volume or due to osmotic driving forces, is maintained by a continuous supply of water. The flux of nutrients and cells from the bronchial vasculature into the airway wall is accompanied by water, and vascularly derived water and cellular wastes are subsequently cleared by liquid flux via the lymphatics. The ability of the lung to clear inhaled particles that land on airway luminal surfaces strongly depends on a sufficiently hydrated mucociliary transport system. The humidity and temperature of inhaled air is adjusted as it flows through the airway tree, requiring evaporation or condensation and thus, regulation of bronchial blood flow and mucosal liquid flux. These observations suggest that a circulatory pathway for airway liquid is important for the maintenance of homeostasis in the airway wall. The exact pathways and mechanisms for control of normal airway liquid transport are not completely known, but their apparent complexities present many opportunities for malfunction to occur in disease states. In particular, excess accumulation of airway liquid within or on the surface of the airway wall can lead to airway obstruction, enhancement of the effects of smooth muscle shortening, and interference with mucociliary transport along the tracheobronchial tree.

Sources of Airway Wall Liquid

The majority of airway wall liquid is composed of plasma transudate or exudate derived from the bronchial microvasculature. The relevant functional anatomy of the bronchial microvasculature in humans includes an adventitial plexus, a densely interconnected network of vessels that begins near the hilum and follows the airway tree within the peribronchial connective tissue, and the submucosal plexus, which is a similar network that is found within the connective tissue of the submucosa (Paré et al., 1996). The two plexuses are joined by vessels that penetrate through the smooth muscle layer. These vessels continue to the level of the terminal bronchiole, where they form anastomoses with the pulmonary microcirculation. They are strategically located for rapid exchange of plasma filtrate from the bronchial microvasculature into airway wall interstitial compartments. Drainage of the airway circulation into the systemic venous circulation occurs via two routes: tracheal vessels into tracheal veins and bronchial vessels into the azygous and hemizygous veins or via anastomoses with the pulmonary venous system.

Plasma filtrate originating from the pulmonary microvasculature can travel from the alveolar region up the airway tree along luminal surfaces, thereby supplying the airway surface liquid layer covering the airway epithelium. Because the total cross-sectional area of the airways decreases as one moves proximally, the airways must reabsorb fluid to prevent luminal obstruction. The extent to which this occurs and the proportion of the total volume of airway wall liquid that might be supplied in this case is not known.

Measurement of Airway Wall Liquid

Two groups of measurement techniques will be considered: (1) techniques for measuring airway wall liquid content, and (2) techniques for measuring airway wall liquid flux, including measurement of vascular and epithelial permeability.

Airway Wall Liquid Content

Measurement of airway wall liquid content by the inert soluble gas technique and by gravimetric methods have been reviewed previously (Paré et al., 1996). In the inert soluble gas technique, a mixture of two gases with differing solubility coefficients is injected into the airways and the concentration of these gases is measured downstream. The theoretical thickness of the airway wall is computed, assuming that the insoluble gas remains in the lumen and a fraction of the soluble gas enters the airway wall. Gravimetric methods involve measurements of dry and wet tissue weights before and after provocation to determine net liquid gain or loss. A significant limitation of these methods is the inability to assess liquid accumulation quantitatively in specific airway wall compartments. This can be accomplished by morphometric analysis of airways in cross-section that have been preserved by fast-freezing techniques, where airway water appears as ice. Differences in compartment size in similar sized airways from treated and untreated groups represent net liquid loss or accumulation due to treatment, assuming that insufficient time exists for significant tissue remodeling from time of treatment to time of tissue freezing. Light microscopy of frozen dog airways revealed that liquid accumulated first in peribronchial and perivascular spaces prior to alveolar flooding during pulmonary edema (Staub et al., 1967). Peripheral guinea pig airways were frozen before and 30–60 seconds after treatment with intravenous histamine at a dose sufficient to induce maximal smooth muscle contraction and imaged in a frozen-hydrated state by low-temperature scanning electron microscopy (LTSEM) (Yager et al., 1996). Measurements of cross-sectional areas of the epithelial, lamina propria, submucosal, and adventitial compartments before and after histamine treatment indicated that liquid from a leaky bronchial microvasculature accumulated very rapidly (within 60 seconds of infusion) in all but the submucosal compartment in response to histamine, more than doubling total airway cross-sectional area with resultant loss of airway lumen, and that liquid shifts between compartments occurred as a result of histamine-induced muscular contraction. Although useful in their ability to localize airway liquid effects of bronchoprovocation, results of studies using these techniques must be carefully interpreted due to the potential for freezing artifacts. Liquid flux between cells and extracellular matrix can occur in response to osmotic gradients established by exuded solutes around ice crystals during freezing (Yager et al., 1994, 1996). Airway tissue must be frozen at a sufficiently fast rate to prevent the resulting shrinking or swelling of airway wall compartments.

Pathway of Airway Wall Liquid Flux

The pathway of airway wall liquid flux from the bronchial microvasculature into airway wall interstitial spaces and the airway lumen can be ascertained with the use of intravascular tracer macromolecules, as previously reviewed (Yager et al., 1995; Paré et al., 1996). The exact site of microvascular leak can be marked with particulate tracers such as Monastral blue or colloidal carbon. Monastral blue crosses the vascular endothelium during vascular permeability changes but remains trapped behind the endothelial cell basement membrane. Under normal conditions, plasma transudation occurs continuously along the bronchial microvasculature in response to hydrostatic or osmotic pressure gradients. During inflammation, plasma exudation has been localized to 7- to 80-μm diameter postcapillary venules, where contractile myoendothelial cells in the vascular lining react to mediators derived from humoral, neural, and cellular sources to form gaps as wide as 1.5 μm (Gabbiani et al., 1970; McDonald, 1987). Colloidal gold particles have successfully been used for tracking plasma exudation from the vasculature into the mucosa and from there onto the luminal surface (Erjefält et al., 1995). Changes in vascular permeability are often assessed by vascular injection of Evan's Blue, which is a vital dye that binds to serum albumin and remains in the bloodstream for several hours. The amount of albumin-bound dye present in the airway wall can be measured spectrophotometrically or by flourescence microscopy. Radiolabeled macromolecules are also useful for tracking the pathway of plasma exudation from vasculature to lumen, as well as the mucosal absorption of solutes from the airway lumen. The transport of larger radiolabeled bulk plasma proteins (albumin and fibrinogen, 70 and 340 kD, respectively) and dextrans (fluorescein isothiocyanate (FITC)-dextran, 70–156 kD) as well as smaller radiolabeled tracers [diethylenetriamine pentaacetic acid (DTPA), 492 D] and plasma derived peptides (bradykinin) has been assessed across the tracheobronchial mucosa of guinea pigs (O'Donnell et al., 1990; Erjefält and Persson, 1991a–c; Gustafsson and Persson, 1991; Erjefält et al., 1993a,b; Erjefält et al., 1994) in response to a wide variety of cell mediators. Amounts of albumin, fibrinogen, or bradykinin in nasal (Greiff et al., 1994; Svensson et al., 1994) and tracheobronchial (Salomonsson et al., 1992) lavage fluids have also been assessed in normal and asthmatic humans. In these instances, epithelial as well as vascular permeability is measured.

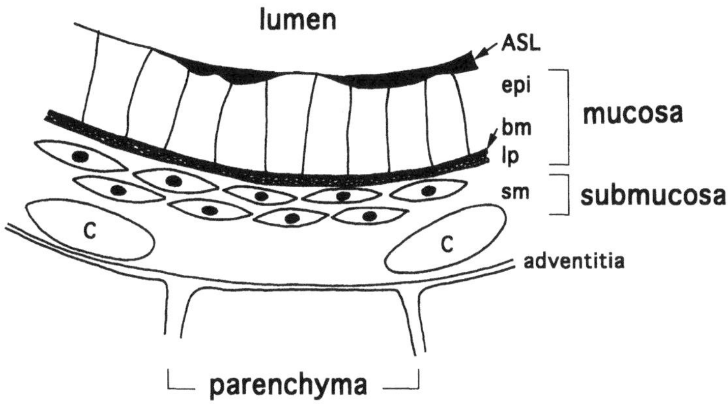

FIGURE 13.1. Schematic diagram of the wall of an intraparenchymal airway. ASL = airway surface liquid; c = cartilage; epi = epithelium; bm = basement membrane; lp = lamina propria; sm = smooth muscle. (*Source*: From Yager et al., 1995.)

Mechanical Consequences of Airway Liquid Accumulation

The increase in airway wall thickness resulting from accumulation of liquid could cause airway narrowing by simply displacing airway mass toward the lumen and decreasing the luminal cross-sectional area available for airflow. It has the more profound potential to exaggerate the changes induced by airway smooth muscle contraction significantly, leading to severe luminal obstruction and even airway closure. The physical mechanisms by which excess liquid can amplify airway narrowing after muscular contraction depend on the site of liquid accumulation within the airway wall. Possible mechanisms for the effects of airway liquid on airway caliber are considered later for each of four airway compartments, using airway wall terminology defined by Bai et al. (1994) (Fig. 13.1).

Luminal Accumulation of Liquid

Under relaxed conditions, the airway epithelium rests on an unfolded basement membrane and underlying lamina propria, forming the mucosal layer (Fig. 13.1). Muscular contraction forces the airway mucosa into folds that protrude into the luminal space. Figure 13.2 shows micrographs of three levels of induced bronchoconstriction in guinea pig airways (Yager et al., 1989). The mucosal folds form channels for liquid collection, which are shown schematically in the center column. Luminal cross-sectional area will be reduced from that before the presence of liquid if the interstices between epithelial projections fill with liquid, and this is shown as the light grey on the right. It can be seen that the larger the degree of smooth muscle contraction and the larger the epithelial projections, the more pronounced the luminal narrowing due to liquid filling of interstices, shown as the dark grey on the right. In addition, if the surface tension (γ) at the air–liquid interface is significantly greater than zero, the pressure drop across the interface will produce an additional force that will further narrow the airway lumen. As γ increases, the radius of curvature of the interfaces joining tips of epithelial projections increases, thereby maintaining the same pressure drop (Yager et al., 1991). Once the curvature of these interfaces matches the curvature of the airway lumen,

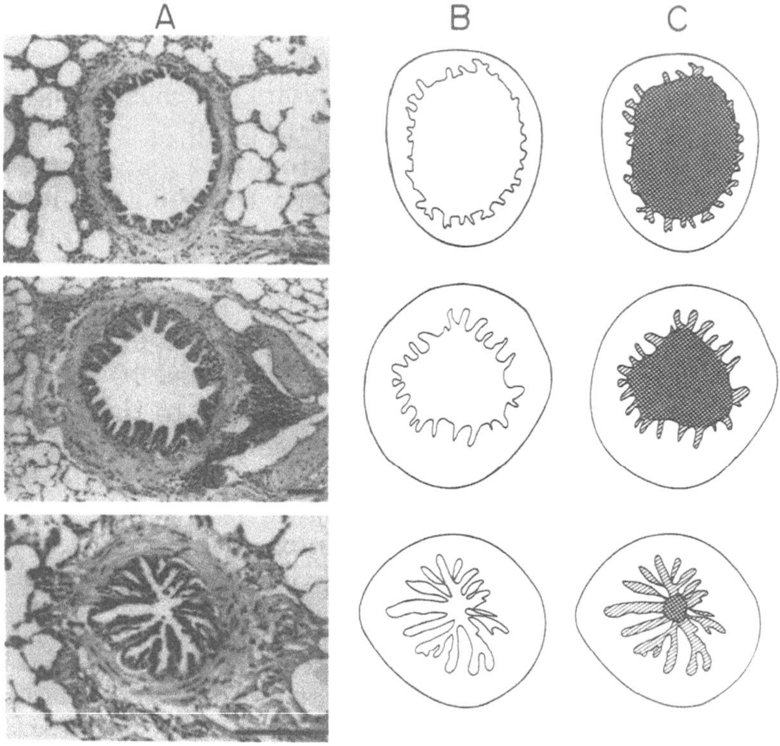

FIGURE 13.2. (A) photomicrographs of constricted airways from sensitized guinea pig lungs challenged with allergen, taken from histological sections, showing three ranges of responses. (B) Schematic diagrams of airway walls shown in (A), showing outer smooth muscle boundary and contours of epithelial projections. (C) Airway luminal cross-sectional areas before (hatched area) and after (cross-hatched area) liquid filling of interstices. bar = 100 μm. (*Source*: From Yager et al., 1989. Amplification of airway constriction due to liquid filling of airway interstices. J. Appl. Physiol. 66, 2873–2884.)

a point of instability is reached. The airway lumen will fill to closure if sufficient liquid is available.

Other forms of fluid instability in small constricted airways have been reported. Calculations by Hill et al. (1997) indicate that if the volume of airway surface liquid (ASL) is less than 2% of the luminal volume, then the ASL is thin and surface tension has no effect on airway mechanics. If fluid volume exceeds 2%, fluid pools form in the folds and interactions between surface tension of luminal liquid and bending stiffness of the mucosa inside the folds changes fold geometry as the airway constricts; the area of the air–liquid interface decreases as airway cross-sectional area decreases and surface tension contributes to airway compression. As the airway cross-sectional area decreases further, a point of fluid instability is reached; the interface of the fluid pool at the base of a fold suddenly jumps to the neck, bypassing the unstable half-filled position. Such fluid instabilities can induce airway instability in the form of abrupt airway closure and abrupt reopening at a higher opening pressure, particularly in small airways (Hill et al., 1997). For ASL volumes greater than 6% of luminal volume, fluid simply floods the space between the folds following airway constriction and subsequent mucosal buckling (Hill et al., 1997).

The pulmonary surfactant that normally lines the luminal surface of airways promotes airway stability by lowering surface tension; however, plasma exudate contains a variety of proteins (Phang and Keough, 1986) and phospholipid-derived mediators [e.g., lyso-platelet activating factor (lyso-PAF)] (Yager et al., 1989) which, like its metabolic precursor PAF, is structurally very similar to dipalmitoyl phosphatidylcholine, the primary surface-active component of pulmonary surfactant. Once incorporated into the surfactant monolayer, these compounds can interfere with surfactant function and raise surface tension, thereby promoting airway narrowing and possible closure.

Mucosal Accumulation of Liquid

The airway mucosa is composed of epithelium, basement membrane, and lamina propria; the lamina propria is composed of loose connective fibers (primarily collagen) interspersed with proteoglycans (Fig. 13.1) that form an ideal site for liquid accumulation. Any increase in airway wall mass interior to the smooth muscle can enhance the degree of luminal obstruction during muscular contraction simply by increasing the physical encroachment of luminal space. Thus, the obstructive effects of smooth muscle contraction can be amplified by liquid accumulation within the airway mucosa. In addition to causing physical encroachment of luminal space, collection of liquid in spaces interior to the smooth muscle can increase the degree of luminal narrowing following muscular contraction by adversely affecting the buckling pattern of the mucosa.

Lambert (1991) was the first to identify the importance of mucosal buckling as a means of resisting smooth muscle contraction. Treating the collagen-rich basement membrane as a linearly elastic thin-walled tube, he modeled how this tube resists a uniformly applied external pressure until a critical pressure is reached, at which point the membrane buckles reversibly into two or more folds. The critical buckling pressure was found to be proportional to the square of the number of folds; therefore, the greater the number of folds, the larger the external pressure required to initiate buckling. Figure 13.3(A) shows a plot of tube cross-sections for the case of 20 folds at several increasing transmural pressures. Just prior to reaching critical buckling pressure, the tube cross-section remains circular. As transmural pressure reaches critical buckling pressure, the

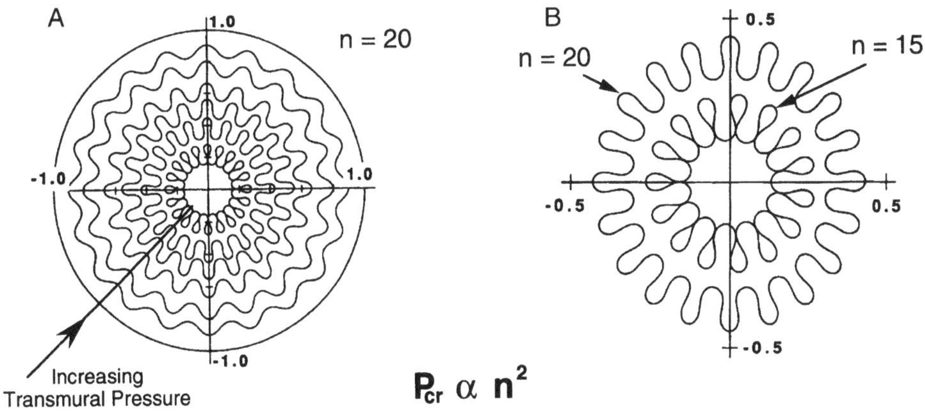

FIGURE 13.3. (A) plot of tube cross sections for the case of 20 folds at several increasing transmural pressures. (B) plot of tube cross sections for the case of 15 and 20 folds at the same transmural pressure difference. (*Source*: From Lambert, 1991.)

tube buckles in this instance into 20 folds, and as transmural pressure continues to increase the folds grow until the tips touch. Increasing the number of folds is a powerful way of resisting airway collapse [see Fig. 13.3(B) with cross-sections of tubes with 15 and 20 folds at the same transmural pressure difference]. For the same pressure difference, the case of fewer folds results in significantly more luminal narrowing.

Wiggs et al. (1997) extended this work by considering two mechanically distinct compartments inside the smooth muscle layer, each with its own Young's modulus and thickness. The innermost stiff and thin compartment represents the subepithelial collagen layer (basement membrane and/or lamina propria) of thickness t_i and Young's modulus E_i; the soft and thick outer compartment of the nonspecific connective tissue of the submucosa of thickness t_o and Young's modulus E_o. A preferred buckling mode for an airway of internal radius R was determined by minimizing the strain energy attributed to a specified buckling pattern for given values of t_i/R, t_o/R, and E_i/E_o, as computed by finite element analysis. They found that as the outer compartment becomes thicker, little change is seen in the buckling mode [Fig. 13.4(A)]. If the inner compartment is thickened, however, the number of folds in the preferred buckling mode dramatically decreases [Fig. 13.4(B)]. Stiffening of the inner layer relative to the outer layer (increasing E_i/E_o) has an intermediate effect on the buckling mode [Fig. 13.4(C)]. The mechanical effects of thickening and/or stiffening the inner layer are pronounced. An airway with a thin inner layer will resist narrowing by buckling into many folds [Fig. 13.5(A)]. A thickened or stiffened inner layer will significantly increase the critical buckling pressure required to induce folding, acting to protect the airway from collapse (Lambert et al., 1994). Once the critical buckling pressure is reached, however, the model predicts that the airway will buckle into a fewer number of folds, leading to profoundly increased luminal obstruction [Fig. 13.5(B)].

Wiggs' model also predicts that pressures vary from very low to very high at the subepithelial boundary; blood vessels located in a low-pressure region would be open and favor exudation into adjacent interstitium. Local plasma exudate rich in cytokines and growth factors could stimulate stress-induced remodeling, resulting in a thickened subepithelial layer of collagen, elastin, and proteoglycans. Liquid accumulation would add to the thickness and perhaps stiffness of this layer. Although at first protective by reducing local stress levels and raising critical buckling pressure, these events would reduce the number of lobes into which the inner wall folds during bronchoconstriction, thereby significantly increasing luminal obstruction.

Submucosal Accumulation of Liquid

Liquid accumulation in the submucosa (smooth muscle and nonspecific connective tissue between smooth muscle and lamina propria) can also affect the degree of luminal obstruction due to muscular contraction. Plasma leakage from the rich blood supply to this region could collect in the nonspecific connective tissue, a hydrophilic matrix composed of collagen, elastin, and proteoglycans. The resulting increase of tissue mass interior to the smooth muscle could amplify luminal narrowing during muscular contraction by physically increasing the amount of airway wall to be pushed into the lumen. Liquid collection in this space can also have a protective effect, as determined by the two-compartment model of Wiggs et al. (1997) described earlier. If the thickness of the outer layer t_o is increased, representing a thickened nonspecific connective tissue layer, the buckling mode is unaffected [Fig. 13.4(A)]. The critical buckling pressure is increased, however. In this manner, thickening this layer serves only to protect the airway wall during smooth muscle constriction, first by delaying the onset of buckling and then by not forcing the wall into fewer folds.

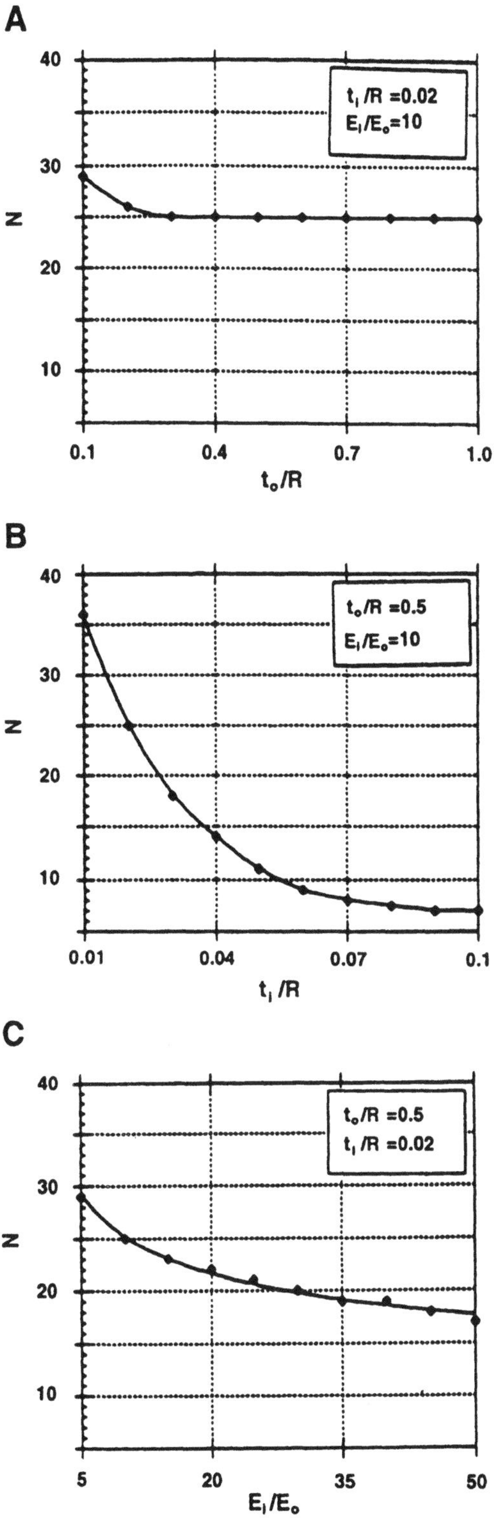

FIGURE 13.4. Results of finite-element linearized buckling analyses of a two-compartment model of airway wall, showing how the expected number of folds (N) varies as each of three model parameters is perturbed from an arbitrary reference case, holding the other two constant. (A) effect of varying outer thickness ratio (t_o/R). (B) effect of varying inner thickness ratio (t_i/R). (C) effect of varying stiffness ratio (E_i/E_o). (*Source*: From Wiggs et al., 1997.)

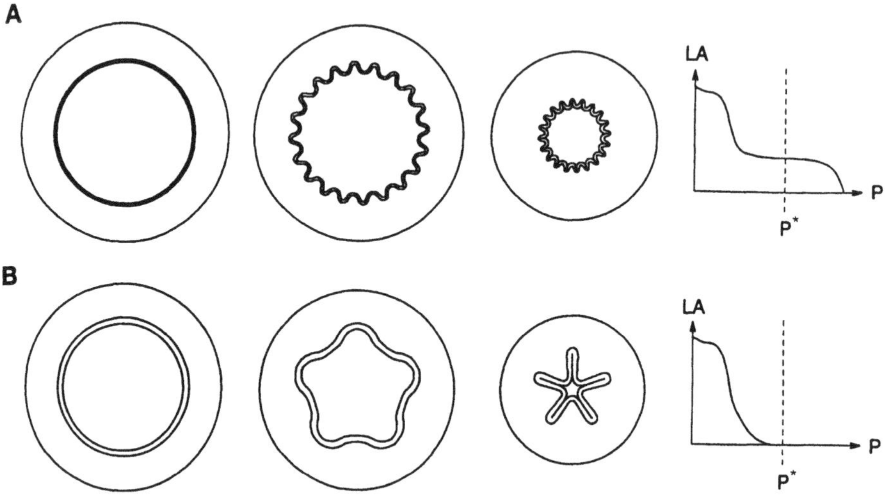

FIGURE 13.5. Schematic showing buckling of a two-compartment tube with a thin (A) and a thick (B) inner layer. Because of the lesser-fold pattern in B, tube can narrow to a greater extent than in (A) before folds push against one another, causing an increase in airway stiffness. In the extreme case shown, tube in B narrows to zero luminal area (LA) at a load corresponding to maximum effective pressure exerted by smooth muscle (P*). P, smooth muscle pressure. (*Source*: From Wiggs et al., 1997.)

Muscular tone would likely prevent accumulation of liquid within smooth muscle fibers, forcing it into the interstitium between fibers or between muscle bundles. If trapped between muscle fibers or bundles, excess liquid would resist smooth muscle contraction by presenting a hydrostatic load to the muscle that would increase with increased muscular shortening. In airways without nonspecific connective tissue (e.g., the peripheral airways of the guinea pig) liquid from submucosal microvascular beds could be displaced into the mucosa or adventitia during muscular contraction.

Adventitial Accumulation of Liquid

The loose peribronchial connective tissue of the adventitia is a preferential site of liquid accumulation after increases in pulmonary or bronchial microvascular pressure, blood flow, and/or permeability. Tissue compliance in this region is high and hydrostatic pressure is normally less than that in adjacent airway wall compartments, making it an ideal site for liquid collection. A gradient in pressure exists between alveolar wall interstitium (about 1 cm H_2O), adventitia around 50 μm pulmonary venules (about 0 cm H_2O), and in the hilum (about –2 cm H_2O) (Bhattacharya et al., 1984), where pressures are expressed relative to pleural pressure. As peribronchial pressure becomes more negative during airway constriction, liquid from microvessels in the adventitial plexus or from the underlying submucosa might preferentially accumulate in this space. The resulting adventitial thickening leads to loss of resistive loading to the smooth muscle, in terms of both muscle preload and afterload (Lambert et al., 1993; Macklem, 1996; Lambert and Paré, 1997). When increased adventitial thickness encroaches into the surrounding parenchyma and relaxes alveolar attachments, the local reduction in peribronchial stress around the airway reduces the muscle preload and moves the airway to a more compliant part of its area–pressure curve. Disassociation of smooth muscle from parenchymal attachments also reduces muscle afterload; less muscle tension is required for a given amount of shortening. The obstructive effects of this disassociation are most significant in small airways, where the larger percentage of wall cross-sectional area occupied by smooth muscle can more easily generate forces to overcome the resistance of a proportionally smaller mucosa.

Airway Wall Liquid in Disease States

Two disease states involving an imbalance of airway liquid flux will be discussed: asthma and cystic fibrosis. Although both diseases are characterized by a state of chronic inflammation, asthmatic airways experience excess accumulation of plasma exudate, whereas the surfaces of airways in cystic fibrosis are depleted of liquid.

Asthma

The asthmatic airway, even in mild cases, is chronically inflamed. The ability of several chemically distinct inflammatory cell mediators to induce both smooth muscle contraction and plasma exudation represents a powerful and interactive mechanism, whereby inflammation can alter airway responsiveness. The potential obstructive effect of plasma exudation depends on the site of accumulation and the relative timing between airway wall compartment filling and muscular contraction.

Many of the putative mediators of asthma are known to increase microvascular and epithelial permeability, thereby allowing liquid, macromolecules, and cells to enter airway wall interstitial spaces and lumen. Intravenously administered histamine (Evans et al., 1989; Yager et al., 1996), bradykinin (Ichinose and Barnes, 1990), the cysteinyl leukotrienes LTC_4, LTD_4, and LTE_4 (Hua et al., 1985; Evans et al., 1989), platelet-activating factor (O'Donnell and Barnett, 1987; Evans et al., 1989), and sensory neuropeptides substance P, neurokinins A and B, and calcitonin gene-related peptide (Rogers et al., 1988) have been shown to cause plasma exudation throughout the respiratory tract of guinea pigs, although their effects are maximal in different airway generations and wall compartments. Histamine and leukotrienes have been found to mediate the antigen-induced increase in vascular permeability in sensitized guinea pigs (Evans et al., 1988). When applied to the mucosal surface of guinea pigs, platelet-activating factor (O'Donnell et al., 1990) and histamine and bradykinin (Erjefält et al., 1993a) resulted in dose-dependent plasma exudation from the mucosal microvasculature into the airway lumen. Mucosal challenges with ovalbumin caused both an immediate and a late exudative response in sensitized guinea pigs (Erjefält et al., 1993b). The immediate response appears to be caused by the secondary release of histamine by activated mast cells, whereas the late response involved inflammatory mediators other than histamine. It is interesting that endogenously released nitric oxide from epithelial cells appears to suppress the macromolecular permeability of the mucosal microcirculation tonically, acting as a defense against inflammatory provocation (Erjefält et al., 1994).

Based on this work, Persson et al. (1991) proposed that even at threshold levels of inflammatory insults, plasma exudate does not remain in the airway wall, but that it is significantly exuded into the airway lumen in large tracheobronchial airways. Solute absorption from lumen to mucosa is conversely unaffected (Gustafsson and Persson, 1991). Activated plasma protein systems within the exudate increase interstitial protein oncotic pressure because of the higher molarity of the cleavage products, and local hydrostatic pressure is increased as water is recruited in response (Yager et al., 1995). Increases in osmotic and/or hydrostatic pressures after mucosal plasma exudation can separate epithelial cells, followed by luminal entry of plasma exudate (Persson et al., 1990). This represents a first line of defense that does not compromise the integrity of the epithelial barrier and quantitatively reflects the exudative response of the airways. In the airways of asthmatics, however, the airway epithelium becomes disrupted by continuous inflammatory insults, either directly or by cytotoxic proteins derived from migrating inflammatory cells. Local shear stress levels are also very high in subepithelial regions of constricted airways, favoring epithelial shedding (Wiggs et al., 1997).

Plasma exudation into the airway wall and onto the luminal surface can have direct mechanical consequences. In a model of airway narrowing based on detailed anatomic measurements of airway wall thickness in patients with asthma or COPD, Wiggs et al. (1992) found that small increases in mucosal thickness did not affect baseline resistance, but that it markedly enhanced the effect of smooth muscle shortening on airways resistance. These effects were most pronounced in peripheral airways and were exaggerated if the thickening was allowed to encroach on the lumen in the baseline state. The model further predicts that even though normal subjects exhibit a dose–response curve to bronchoactive agonists that will reach a response plateau, patients with mucosal thickening, as is commonly found in asthma, will have steep dose–response curves to bronchoactive agonists that will not reach a plateau within the range of increased airways resistance found *in vivo*. The mucosal thickening observed in the asthmatic airways of this study may have been caused by tissue remodeling and/or plasma exudation. Although important in identifying mucosal thickness as a predeterminant of luminal obstruction, this study did not account for the possibility of mucosal buckling or luminal entry of liquid.

As described earlier, muscular contraction forces everything inside the smooth muscle into folds, the size and number of which determine the magnitude of resistance to compression and amount of luminal compromise. The number of folds is likely the most critical determinant of the degree of luminal narrowing and is linked to bronchial hyperresponsiveness (Lambert, 1991). Modeling predicts that this number is most sensitive to the relative thicknesses of the lamina propria/basement membrane layer of the mucosa and the connective tissue layer of the submucosa (Wiggs et al., 1997). Mucosal challenge of sensitized guinea pigs with ovalbumin caused immediate entry of plasma exudate into the lamina propria, followed by flux of plasma between and around epithelial cells and into the airway lumen 3 and 6 minutes later, with the latter time point corresponding with peak rate of plasma entry (Erjefält et al., 1995). The delayed entry of luminal exudate is presumably due to the time required for mucosal liquid accumulation and build up of local hydrostatic pressure, followed by separation of epithelial cells. Allergen aerosol challenge of sensitized mice increased airway venular permeability to cause transient extravasation and lamina propria distribution of plasma in the large airways without crossing the epithelium into the airway lumen (Erjefält et al., 1998). If muscular contraction occurs immediately after challenge, simultaneous with liquid accumulation and subsequent thickening of the lamina propria, the mucosa would likely buckle into fewer folds and thereby pose less resistance to muscular contraction and penetrate deeper into the airway lumen. Plasma exudate that crosses the epithelium minutes later could fill the interstices between these folds and significantly amplify the degree of luminal narrowing. In addition to collecting in the lamina propria, plasma exudate also accumulated in the adventitia 60 seconds after intravenous histamine treatment of guinea pig airways (Yager et al., 1996). Adventitial liquid can also enhance luminal narrowing by decoupling the parenchyma from the airway wall, as previously discussed.

The amount of liquid normally lining the surface of peripheral airways was measured and found to have a strong dependence on lung volume (Yager et al., 1994), where the average thickness of the airway surface liquid (ASL) layer at total lung capacity (TLC) was twice that at functional residual capacity. Excess ASL, recruited during TLC excursions, could explain the observation that severely obstructed asthmatic patients with chronic airway obstruction who perform TLC maneuvers add to their bronchoconstriction rather than alleviate it, unlike asthmatics with acutely induced obstruction

(Lim et al., 1987). A consequence of the asthma diathesis in these patients may be greater apical water permeability in which water easily crosses the apical barrier into the ASL but cannot return.

Several therapeutic approaches are possible to reduce microvascular leakage in asthmatics by using drugs that either block the release of inflammatory mediators responsible for the leakage or block the effects of these mediators. Because of the large numbers of candidate inflammatory cells and cell mediators, it is perhaps more efficacious to pursue drug therapies that act at the microvascular site of leakage, irrespective of the cause of leak. Finding suitable drugs for human use has been slowed due to distinct differences in airway pharmacological responses between humans and animals, particularly guinea pigs. The mucosal exudation caused by tachykinin neuropeptides in guinea pigs is not found in human airways (Persson, 1991). Although a single topical treatment with glucocorticoids produced a prompt vascular antipermeability effect in guinea pigs, similar treatment in humans did not cause a direct vascular effect (Greiff et al., 1994). Common bronchodilators (e.g., xanthines and β 2-adrenoceptor) similarly appear to act directly on microvascular endothelial cells in guinea pigs (Erjefält and Persson, 1991), but they do not affect microvascular and epithelial permeabilities in humans (Svensson et al., 1994). These results suggest that the antiexudative effects of topical glucocorticoids or intravenously administered xanthines and β 2-adrenoceptor agonists in humans are indirectly caused by inhibition of cellular inflammatory processes rather than by direct effects on the bronchial microvasculature.

Cystic Fibrosis

Cystic fibrosis is a genetic disease in which there are defects of epithelial electrolyte transport. In the airways of these patients, the epithelium is impermeable to Cl^- and may have an increased rate of Na^+ absorption (Welsh, 1987). Because water transport across the epithelium into the ASL layer is closely associated with Cl^- transport across the apical epithelial cell membrane, and, conversely, water absorbtion with Na^+ transport in the reverse direction, ASL volume is depleted in these individuals, leading to malfunction of the tracheobronchial mucociliary transport system. This system consists of a viscous gel layer situated on top of a cilia bearing aqueous ASL layer, whereby the cilia beat freely in the ASL layer and the tips engage the gel and propel it toward the mouth. Inhaled particulates and sloughed cells trapped in the gel are in this manner cleared from the airway tree. Optimal coupling of cilia and gel layer requires close regulation of ASL thickness; normal airway epithelia are thought to regulate ASL volume (thickness) by isotonic ion and water transport (Boucher, 1994). In cystic fibrosis, the rate of isotonic ion and water absorption is postulated (Jiang et al., 1993; Boucher, 1994) and later measured in CF cell cultures (Matsui et al., 1998) to be abnormally high. The depleted ASL layer causes grossly inefficient ciliary beating and concentrates mucus, massively disrupting mucus transport and retaining mucus plaques that can act as initiating sites of airway infection.

Ionic composition and corresponding osmolarity of the ASL has also been postulated to affect antimicrobial factors on airway surfaces. A lowered [NaCl] ($\leq$50 mM) in ASL may activate defensins and create antimicrobial protection of normal airway surfaces (Quinton, 1994; Smith et al., 1996; Goldman et al., 1997; Travis et al., 1999). ASL hypotonicity is thought to be due to selective absorption of salt, but not water, across normal airway epithelium (Quinton, 1994; Widdicombe, 1997; Zabner et al., 1998). In cystic fibrosis, however, an iso- or hypertonic ([NaCl] > 100 mM) ASL caused by an impaired ability to absorb NaCl (Joris et al., 1993; Smith et al., 1996; Goldman et al., 1997; Travis et al., 1999) may interfere with defensin activity and exacerbate bacterial infections.

Verkman's group introduced a new technique for the measurement of ASL osmolality using fluorophore encapsulated liposomes (Jayarman et al., 2001) in combination with fluorescence microscopy. Having already shown that ASL Na, Cl, and pH in CF knockout mice and controls are the same, evidence has been obtained that rules out the isosmotic volume hypothesis formulated by Matsui et al. (2000). The Na, K, and Cl values seem sufficient to account for ASL osmolality. Such results have brought back into focus the importance of the viscosity of gland secretions in CF lung disease. This remains to be measured *in vivo*.

Summary

Airway liquid is an important component of airway wall homeostasis. The primary source of airway liquid, the bronchial microvasculature, is anatomically well suited for rapid exchange of plasma filtrate from the vasculature into airway wall interstitial compartments and the airway lumen. Disruption of the homeostatic balance of airway liquid can lead to disease states, depending on the site and composition of the liquid. In asthma, plasma exudation is an important part of the airway inflammatory response. It occurs very rapidly, seconds to minutes after bronchoprovocation, and might be protective against muscular contraction by increasing the load to the smooth muscle. Once the force generated by the smooth muscle surpasses mucosal critical buckling pressure, however, plasma exudation can be detrimental by increasing the tissue mass forced into the airway lumen, decreasing the number of folds into which the mucosa buckles, filling epithelial interstices or entire airway lumen, and/or reducing the parenchymal tethering forces at the outer airway wall boundary. In cystic fibrosis, abnormalities in ASL volume result in inefficient ciliary clearance of mucus. Understanding the source, pathways, and effects of airway liquid may lead to effective treatment modalities for these diseases.

References

Bai, A., Eidelman, D.H., Hogg, J.C., James, A.L., Lambert, R.K., Ludwig, M.S., et al. (1994) Proposed nomenclature for quantifying subdivisions of the bronchial wall. J. Appl. Physiol. 77(2), 1011–1014.

Bhattacharya, J., Gropper, M.A., and Staub, N.C. (1984) Interstitial fluid pressure gradient measured by micropuncture in excised dog lung. J. Appl. Physiol. 56(2), 271–277.

Boucher, R.C. (1994) Human airway ion transport. Part one. Am. J. Resp. Crit. Care Med. 150, 271–281.

Boucher, R.C. (1994) Human airway ion transport. Part two. Am. J. Respir. Crit. Care Med. 150, 581–593.

Erjefält, I., and Persson, C.G. (1991a) Allergen, bradykinin, and capsaicin increase outward but not inward macromolecular permeability of guinea-pig tracheobronchial mucosa. Clin. Exp. Allergy 21(2), 217–224.

Erjefält, I., and Persson, C.G. (1991b) Long duration and high potency of antiexudative effects of formoterol in guinea-pig tracheobronchial airways. Am. Rev. Respir. Dis. 144(4), 788–791.

Erjefält, I., and Persson, C.G. (1991c) Pharmacologic control of plasma exudation into tracheobronchial airways. Am. Rev. Respir. Dis. 143, 1008–1014.

Erjefält, I., Greiff, L., Alkner, U., and Persson, C.G. (1993a) Allergen-induced biphasic plasma exudation responses in guinea pig large airways. Am. Rev. Respir. Dis. 148(3), 695–701.

Erjefält, I., Luts, A., and Persson, C.G. (1993b) Appearance of airway absorption and exudation tracers in guinea pig tracheobronchial lymph nodes. J. Appl. Physiol. 74(2), 817–824.

Erjefält, J.S., Erjefält, I., Sundler, F., and Persson, C.G. (1994) Mucosal nitric oxide may tonically suppress airways plasma exudation. Am. J. Respir. Crit. Care Med. 150, 227–232.

Erjefält, J.S., Erjefält, I., Sundler, F., and Persson, C.G. (1995) Epithelial pathways for luminal entry of bulk plasma. Clin. Exp. Allergy 25, 187–195.

Erjefält, J.S., Andersson, P., Gustafsson, B., Korsgren, M., Sonmark, B., and Persson, C.G. (1998) Allergen challenge-induced extravasation of plasma in mouse airways. Clin. Exp. Allergy 28, 1013–1020.

Evans, T.W., Rogers, D.F., Aursudkij, B., Chung, K.F., and Barnes, P.J. (1988) Inflammatory mediators involved in antigen-induced airway microvascular leakage in guinea pigs. Am. Rev. Respir. Dis. 138(2), 395–399.

Evans, T.W., Rogers, D.F., Aursudkij, B., Chung, K.F., and Barnes, P.J. (1989) Regional and time-dependent effects of inflammatory mediators on airway microvascular permeability in the guinea pig. Clin. Sci. 76, 479–485.

Gabbiani, G., Badonnel, M.C., and Majno, G. (1970) Intra-arterial injections of histamine, serotonin, or bradykinin: a topographic study of vascular leakage. Proc. Soc. Exp. Bio. Med. 135, 447–452.

Goldman, M.J., Anderson, G.M., Stolzenberg, E.D., Kari, U.P., Zasloff, M., and Wilson, J.M. (1997) Human beta-defensin-1 is a salt-sensitive antibiotic in lung that is inactivated in cystic fibrosis. Cell 88, 553–560.

Greiff, L., Andersson, M., Svensson, C., Alkner, U., and Persson, C.G.A. (1994) Glucocorticoids may not inhibit plasma exudation by direct vascular antipermeability effects in human airways. Eur. Respir. J. 7(6), 1120–1124.

Gustafsson, B.G., and Persson, C.G.A. (1991) Asymmetrical effects of increases in hydrostatic pressure on macromolecular movement across the airway mucosa—a study in guinea-pig tracheal tube preparations. Clin. Exp. Allergy 21(1), 121–126.

Hill, M.J., Wilson, T.A., and Lambert, R.K. (1997) Effects of surface tension and intraluminal fluid on mechanics of small airways. J. Appl. Physiol. 82, 233–239.

Hua, X.Y., Dahlén, S.E., Lundberg, J.M., Hammarström, S., and Hedqvist, P. (1985) Leukotrienes C4, D4 and E4 cause widespread and extensive plasma extravasation in the guinea pig. Naunyn-Schmiedebergs Arch. Pharmacol. 330, 136–141.

Ichinose, M., and Barnes, P.J. (1990) Bradykinin-induced airway microvascular leakage and bronchoconstriction are mediated via a bradykinin-B2 receptor. Am. Rev. Respir. Dis. 142, 1104–1107.

Jayarman, S.Y., Song, Y., Vetrivel, I., Shankar, I., and Verkman, A.S. (2001) Noninvasive *in vivo* fluorescence measurement of airway surface liquid depth, salt, concentration and pH. J. Clin. Invest. 107, 317–324.

Jiang, C., Finkbeiner, W.E., Widdicombe, J.H., McCray, P.B., Jr., and Miller, S.S. (1993) Altered fluid transport across airway epithelium in cystic fibrosis. Science 262, 424–427.

Joris, L., Dab, I., and Quinton, P.M. (1993) Elemental composition of human airway surface fluid in healthy and diseased airways. Am. Rev. Respir. Dis. 148, 1633–1637.

Lambert, R.K. (1991) Role of bronchial basement membrane in airway collapse. J. Appl. Physiol. 71(2), 666–673.

Lambert, R.K., Wiggs, B.R., Kuwano, K., Hogg, J.C., and Paré, P.D. (1993) Functional significance of increased airway smooth muscle in asthma and COPD. J. Appl. Physiol. 74, 2771–2781.

Lambert, R.K., Codd, S.L., Alley, M.R., and Pack, R.J. (1994) Physical determinants of bronchial mucosal folding. J. Appl. Physiol. 77, 1206–1216.

Lambert, R.K., and Paré, P.D. (1997) Lung parenchymal shear modulus, airway wall remodeling, and bronchial hyperresponsiveness. J. Appl. Physiol. 83, 140–147.

Lim, T.K., Pride, N.B., and Ingram, R.H., Jr. (1987) Effects of volume history during spontaneous and acutely induced air-flow obstruction in asthma. Am. Rev. Respir. Dis. 135(3), 591–596.

Macklem, P.T. (1996) A theoretical analysis of the effect of airway smooth muscle load on airway narrowing. Am. J. Respir. Crit. Care Med. 153, 83–89.

Matsui, H., Davis, C.W., Tarrran, R., and Boucher, R.C. (2000) Osmotic water permeabilities of cultured, well-differentiated normal and cystic fibrosis airway epithelia. J. Clin. Invest. 105, 1419–1427.

Matsui, H., Randell, S.H., Peretti, S.W., Davis, C.W., and Boucher, R.C. (1998) Coordinated clearance of periciliary liquid and mucus from airway surfaces. J. Clin. Invest. 102, 1125–1131.

McDonald, D.M. (1987) Neurogenic inflammation in the respiratory tract: actions of sensory nerve mediators on blood vessels and epithelium of the airway mucosa. Am. Rev. Respir. Dis. 136, S65–72.

O'Donnell, S.R., and Barnett, C.J. (1987) Microvascular leakage to platelet activating factor in guinea-pig trachea and bronchi. Eur. J. Pharmacol. 138, 385–396.

O'Donnell, S.R., Erjefalt, I., and Persson, C.G.A. (1990) Early and late tracheobronchial plasma exudation by platelet-activating factor administered to the airway mucosal surface in guinea pigs—effects of Web-2086 and enprofylline. J. Pharmacol. Exp. Therap. 254(1), 65–70.
Paré, P.D., Yager, D., and Godden, D.J. (1996) Airway edema. In: Barnes, P., West, J., and Crystal, R. (eds.) The lung: scientific foundations, Raven Press, New York.
Persson, C.G., Erjefält, I., Gustafsson, B., and Luts, A. (1990) Subepithelial hydrostatic pressure may regulate plasma exudation across the mucosa. Int. Arch. Allergy Appl. Immunol. 92(2), 148–153.
Persson, C.G.A. (1991) Mucosal exudation in respiratory defence—neural or non-neural control. Int. Arch. Allergy Appl. Immunol. 94(1–4), 222–226.
Persson, C.G.A., Erjefalt, I., Alkner, U., Baumgarten, C., Greiff, L., Gustafsson, B., et al. (1991) Plasma exudation as a 1st line respiratory mucosal defence. Clin. Exper. Allergy 21(1), 17–24.
Phang, P.T., and Keough, K.M. (1986) Inhibition of pulmonary surfactant by plasma from normal adults and from patients having cardiopulmonary bypass. J. Thorac. Cardiovasc. Surg. 91, 248–251.
Quinton, P.M. (1994) Viscosity versus composition in airway pathology [editorial; comment]. Am. J. Respir. Crit. Care Med. 149, 6–7.
Rogers, D.F., Belvisi, M.G., Aursudkij, B., Evans, T.W., and Barnes, P.J. (1988) Effects and interactions of sensory neuropeptides on airway microvascular leakage in guinea-pigs. Br. J. Pharmacol. 95, 1109–1116.
Salomonsson, P., Gronneberg, R., Gilljam, H., Andersson, O., Billing, B., Enander, I., et al. (1992) Bronchial exudation of bulk plasma at allergen challenge in allergic asthma. Am. Rev. Respir. Dis. 146(6), 1535–1542.
Smith, J.J., Travis, S.M., Greenberg, E.P., and Welsh, M.J. (1996) Cystic fibrosis airway epithelia fail to kill bacteria because of abnormal airway surface fluid [published erratum appears in Cell 1996 Oct 18;87(2):following 355]. Cell 85, 229–236.
Staub, N.C., Nagano, H., and Pearce, M.L. (1967) Pulmonary edema in dogs, especially the sequence of fluid accumulation in lungs. J. Appl. Physiol. 22, 227–240.
Svensson, C., Alkner, U., Pipkorn, U., and Persson, C.G.A. (1994) Histamine-induced airway mucosal exudation of bulk plasma and plasma-derived mediators is not inhibited by intravenous bronchodilators. Eur. J. Clin. Pharmacol. 46(1), 59–65.
Travis, S.M., Conway, B.A., Zabner, J., Smith, J.J., Anderson, N.N., Singh, P.K., et al. (1999) Activity of abundant antimicrobials of the human airway. Am. J. Respir. Cell Mol. Biol. 20, 872–879.
Welsh, M.J. (1987) Electrolyte transport by airway epithelia. Physiol. Rev. 67(4), 1143–1184.
Widdicombe, J.G. (1997) Airway surface liquid: concepts and measurements. In: Rogers, D.F., and Lethem, M.I. (eds.) pp. 1–17, Airway mucus: basic mechanisms and clinical perspectives, Birkhauser, Basel.
Wiggs, B.R., Bosken, C., Paré, P.D., James, A., and Hogg, J.C. (1992) A model of airway narrowing in asthma and in chronic obstructive pulmonary disease. Am. Rev. Respir. Dis. 145, 1251–1258.
Wiggs, B.R., Hrousis, C.A., Drazen, J.M., and Kamm, R.D. (1997) On the mechanism of mucosal folding in normal and asthmatic airways. J. Appl. Physiol. 83, 1814–1821.
Yager, D., Butler, J.P., Bastacky, J., Israel, E., Smith, G., and Drazen, J.M. (1989) Amplification of airway constriction due to liquid filling of airway interstices. J. Appl. Physiol. 66, 2873–2884.
Yager, D., Cloutier, T., Feldman, H., Bastacky, J., Drazen, J.M., and Kamm, R.D. (1994) Airway surface liquid thickness as a function of lung volume in small airways of the guinea pig. J. Appl. Physiol. 77(5), 2333–2340.
Yager, D., Kamm, R.D., and Drazen, J.M. (1995) Airway wall liquid—sources and role as an amplifier of bronchoconstriction. Chest 107(3), S105–S110.
Yager, D., Martins, M.A., Feldman, H., Kamm, R.D., and Drazen, J.M. (1996) Acute histamine-induced flux of airway liquid—role of neuropeptides. J. Appl. Physiol. 80, 1285–1295.
Yager, D., Shore, S., and Drazen, J.M. (1991) Airway luminal liquid—sources and role as an amplifier of bronchoconstriction. Am. Rev. Respir. Dis. 143, S52–S54.
Zabner, J., Smith, J.J., Karp, P.H., Widdicombe, J.H., and Welsh, M.J. (1998) Loss of CFTR chloride channels alters salt absorption by cystic fibrosis airway epithelia in vitro. Mol. Cell 2, 397–403.

14
Correlations Between the Pulmonary Circulation and Gas Exchange in Health and Disease

BRYAN E. MARSHALL, H. FREDERICK FRASCH, AND CAROL MARSHALL

The development of blood gas electrodes in the 1950s enabled laboratory analyses of respiratory gas exchange to be extended readily to patients. Extension of the three-compartment model of gas exchange (Riley and Cournand, 1949) ultimately led to the analysis of gas exchange (Hlastala, 1984) as a distribution of ventilation–perfusion ratios (V/Q model). This latter currently provides the most detailed practical conceptual framework, and has been limited in clinical practice only by the technical difficulty of the multiple inert gas excretion method that is used to derive the distributions. These difficulties have largely been solved by advances in micropore membrane inlet mass spectrometry (Baumgardner et al., 1997) and rapid, simple, real-time bedside derivations should soon expand the clinical and research applications of V/Q ratio distributions. There has been intuitive recognition from the beginning that blood flow was as important for gas exchange as was ventilation, but, because of the relative inaccessibility of the pulmonary circulation, knowledge of the role of blood flow remained limited and the pulmonary vascular bed was regarded simply as a passive conduit that permitted high flow at low pressures. Introduction of the balloon-tipped or flow-directed catheter has rapidly changed this view and has revealed the fundamental responsivity of the pulmonary circuit to exogenous and endogenous mediators, particularly to hypoxia. At the same time detailed information about the biodynamic properties of the pulmonary vascular bed have enabled the development of improved computer models (P/Q model) that represent the behavior of the pulmonary circulation more faithfully. With the availability of a method (Marshall et al., 1994) to combine V/Q and P/Q Models (V/Q–P/Q model) it is finally possible to analyze and understand quantitatively the complex interactions that blood flow regulation brings to gas exchange, and *vice versa*. Using this approach the present discussion addresses questions, such as: Under what circumstances of health and disease does a vasoconstrictor improve oxygenation? What are the mechanisms underlying improvement of oxygenation when nitric oxide is inhaled or mixed venous oxygen tension is increased? What factors limit the influence on gas exchange and pulmonary blood flow distribution of changes in left-atrial pressure, mechanical ventilation, thoracotomy, or vasoactive drugs?

Ventilation–Perfusion Ratios: The V/Q Model of Gas Exchange

The pulmonary retention and elimination of an inert gas, infused in a steady state, can be represented by the arterial/mixed venous and the expired/mixed venous gas tension ratios. These ratios are simple functions of the total ventilation, total blood flow, solubility of the gas, and ventilation–perfusion ratio. If a mixture of six inert gases with widely different solubilities is administered, the pulmonary shunt, dead space, and distribution of V/Q ratios, expressed as means and log standard deviations (log SD) for ventilation and perfusion, can be derived using iterative computer routines (Wagner, Saltzman, and West, 1974). From the V/Q Model the values for normal lungs for the mean is approximately 0.8, and the upper limits of normal for the log SD is 0.6; for the physiological shunt, it is about 10%, and for anatomic dead space it is about 30% of the minute volume. With severe lung disease values for log SD exceeding 2.5 are observed, and the mean V/Q values vary widely, but shunt and dead space may exceed 80% of the cardiac output or ventilation, respectively. Exchange of the respiratory gases are accurately represented when calculated on the basis of these values for the V/Q Model and the impairment of oxygenation is correlated, particularly with shunt and broadening of the V/Q Ratio distribution for blood flow (West and Wagner, 1997).

Pressure–Flow Curves: The P/Q Model

When flow in the pulmonary vascular bed is increased from zero to above normal values, the pressure increase is related by a curve because the effective conductance (reciprocal resistance) of the bed progressively increases to some maximum that is thereafter constant. Although recruitment of vessels may account for some of the curvature at the lowest flows, measurement of the number, dimensions, and elastic properties of the entire pulmonary vascular tree have demonstrated a pattern of properties that support a generalized model (P/Q model) where the curvilinearity is entirely accounted for by the distension of vessels within specific transmural pressure limits (Fung, 1984). This model therefore accounts for the influence of pleural and alveolar gas pressures, left-atrial outflow pressures, and active and passive vasoconstriction on the position and shape of the pressure–flow curve.

Hypoxic pulmonary vasoconstriction (HPV) is the most unique property of pulmonary arteries to constrict in response to hypoxia. It is also the most obvious direct link between blood flow distribution and oxygen exchange. The stimulus (P_{SO_2}) for HPV (Marshall et al., 1994) is the oxygen tension in the smooth muscle of small pulmonary arteries (with diameters <500 μm) and is determined by both alveolar (P_{AO_2}) and mixed venous (P_{VO_2}) oxygen tensions ($P_{SO_2} = P_{AO_2}^{0.6} \times P_{VO_2}^{0.4}$). By stimulating HPV in hypoxic regions of the lung, blood flow is diverted to less hypoxic regions and oxygenation improved; however, severe systemic arterial hypoxemia (P_{AO_2} <50 mmHg) may be associated with inhibition of HPV (Brimioulle et al., 1994). This is probably because the vasa vasorum of the pulmonary arteries are from the systemic bronchial circulation and release dilator compounds when hypoxic (Marshall et al., 1991).

Interaction of Ventilation–Perfusion and Pressure–Flow: The V/Q–P/Q Model

The derivation of the V/Q–PQ model, which combines the two preceding models (Marshall et al., 1994), is illustrated in the diagram (Fig. 14.1), which shows the influence of HPV on the pulmonary artery pressure and gas exchange properties of the lungs in a patient with severe adult respiratory distress syndrome (ARDS). Apart from demonstrating the principles of the derivation, this figure shows the influence of HPV in reducing shunt and particularly in narrowing the distribution of the V/Q ratio distribution for the perfusion. The dual effect of HPV is to improve gas exchange and increase pulmonary artery pressure, but whether the desirable outcome of reduced hypoxemia or the undesirable one of pulmonary hypertension predominates depends primarily on the amount of the lung that is affected. Thus, in localized atelectasis the stimulus for HPV is the Pvo_2 and the diversion of blood flow away from the atelectatic region is very effective so that oxygenation is preserved with little overall increase in pulmonary vascular resistance. On the other hand, if the alveolar and/or mixed venous oxygen tensions are decreased (e.g., with severe lung disease, breathing hypoxic gas mixtures, or cardiac impairment), generalized HPV throughout the lung induces pulmonary hypertension and increases right heart work without any useful flow diversion.

The progressive influence of HPV on the interaction between gas exchange and pulmonary blood flow is illustrated in Fig. 14.2 by the changes in Pao_2, pulmonary shunt and pulmonary artery pressure (the latter corresponds to pulmonary vascular resistance because cardiac output, left atrial pressure, and all other relevant parameters are maintained constant) when the volume of the lung that is atelectatic is increased in the absence and presence of an abnormal V/Q ratio distribution.

Nonhypoxic Pulmonary Vasoconstriction and Nitric Oxide

In pathological states it is common for vessels to be narrowed and/or the pulmonary vascular bed to be diminished as a result of edema, inflammation, thrombosis, embolism or fibrosis. These changes diminish the reserve, but the considerable redundancy of the pulmonary circuit at rest permits loss of approximately 40% of the vascular bed before pulmonary hypertension is evident. By itself the effect on gas exchange is mainly to increase dead space, but to do so with little effect on oxygen exchange; however, the role of active pulmonary vasoconstriction as a result of endogenous mediators or exogenous drug administration has been revealed to be an important influence, particularly when combined with inhaled nitric oxide. Nitric oxide is a potent vasodilator, but it is so rapidly eliminated by binding to hemoglobin that it acts only on pulmonary arteries in the ventilated regions of the lungs when it is inhaled.

There are four considerations in the interpretation of the efficacy of nitric oxide (No). First, pulmonary vascular resistance declines steadily with NO dose, but Pao_2 increases with NO at 1–10 ppm doses when NO is administered in concentrations from 1 to 50 ppm to patients with impaired gas exchange, breathing air or slightly oxygen-enriched gas mixtures. It then declines again at concentrations greater than 10 ppm (Gerlach et al., 1993). One explanation for this is because the NO concentrations achieved in individual alveoli depend on the V/Q ratio; therefore, at low concentrations inhibition of constriction only occurs in the high V/Q regions, resulting in greater perfusion of the better oxygenated regions and improved Pao_2. At higher concentra-

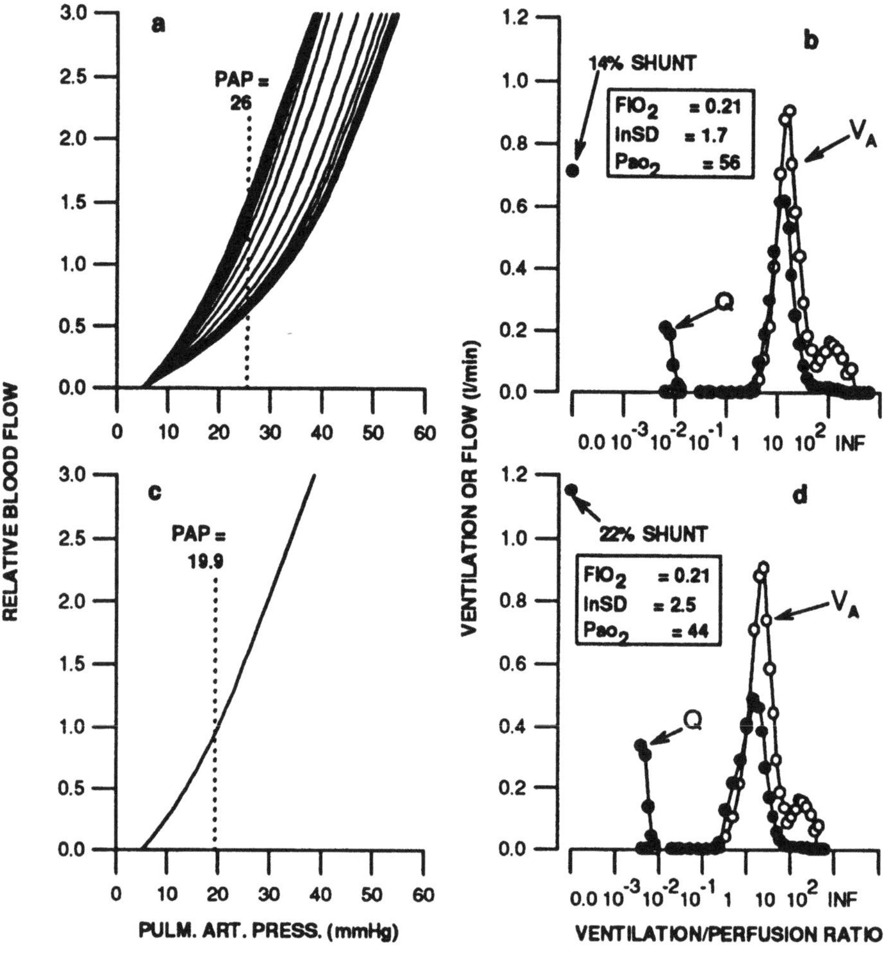

FIGURE 14.1. Method for relating V/Q ratio distributions and pressure–flow curves in a patient with Adult Respiratory Distress Syndrome. (C) The general conditions (cardiac output, left-atrial pressure, alveolar pressure, pleural pressure, vascular bed patency, and vascular narrowing) define a single pressure flow line in the absence of HPV. (B) The V/Q ratio distributions for ventilation and perfusion for 37 lung compartments are shown as derived from the multiple inert gas elimination technique in a patient with ARDS (inset summarizes state) and assuming the presence of HPV. For each compartment the alveolar and mixed venous oxygen tensions are known, and the stimulus (Pso_2) for HPV is calculated. (A) The compartmental HPV, from (B), is applied to the pressure flow curve of (C) to obtain a family curves shifted to the right by increasing HPV. There is only one pulmonary artery pressure that satisfies both the individual compartmental and total flow requirements for these curves. (A) and (B) therefore define the existing conditions in this patient. (D) Because HPV alters both the flow and the gas exchange to each compartment, the effect of removing HPV is calculated iteratively until the solution converges so that the arterial oxygen tension changes less than 0.1% with the final iteration. It is apparent that the increased shunt, and log SD together with the reduced Pao_2 that occur with ARDS in (B), are considerably worsened when HPV is absent in (D), whereas the pulmonary artery pressure is improved.

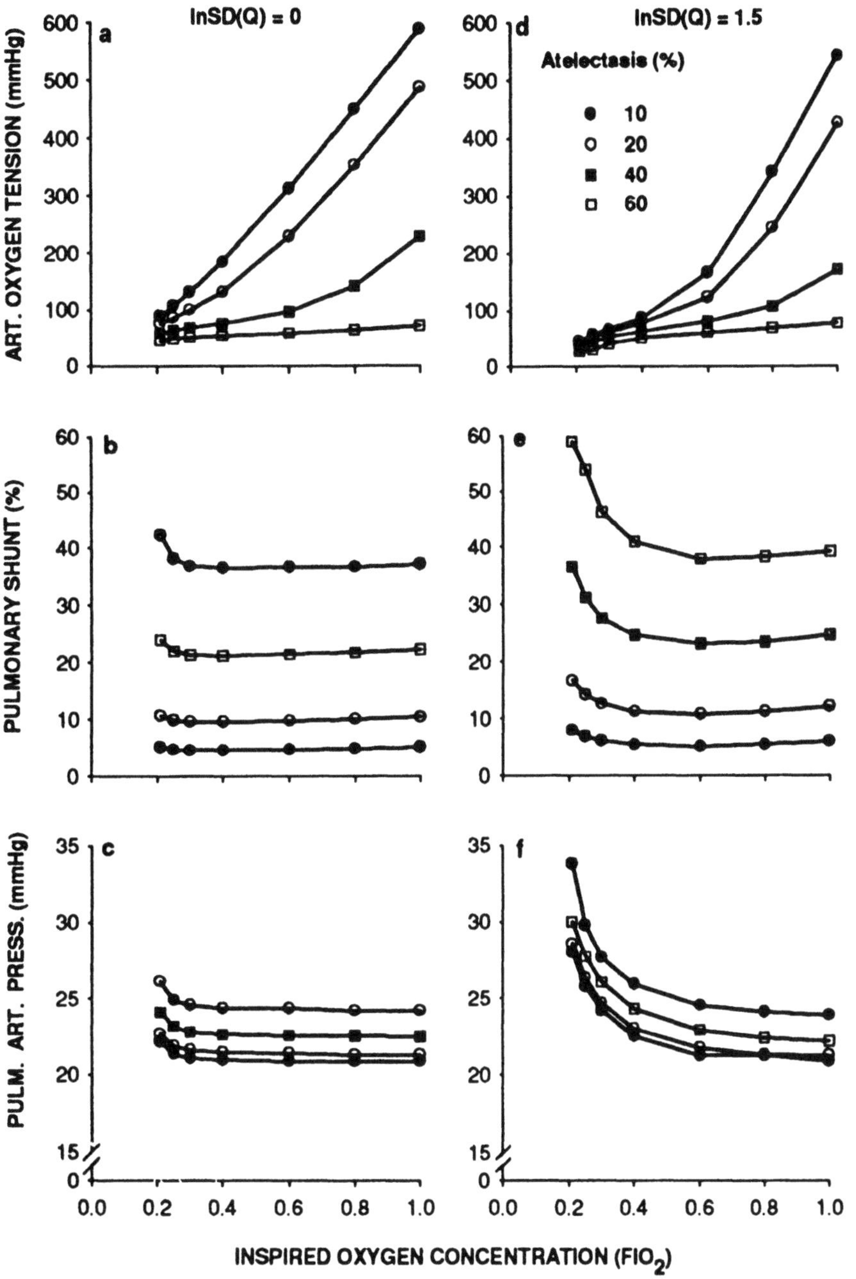

FIGURE 14.2. The influence of increasing atelectasis on arterial oxygen tension, pulmonary shunt and pulmonary artery pressure as the inspired oxygen is increased. The left panels (A–C) are for lungs with no variation in V/Q ratio distribution (log SD = 0), whereas the right panels (D–F) correspond to an abnormal distribution (log SD = 1.5). Arterial oxygen decreases as atelectasis increases, moreso in the presence of V/Q maldistribution. Note that when atelectasis is 40% or more, PaO_2 is relatively insensitive to FIO_2. The activity of HPV in reducing shunt below the atelectasis value is a function of PAO_2, PvO_2 and pulmonary vascular pressures, and is most effective in the middle range of FIO_2, especially when V/Q maldistribution is present.

tions, however, vasoconstriction is inhibited throughout the lung and oxygenation becomes impaired as HPV is reduced in low V/Q regions.

Second, most of the improvement of oxygenation with NO is not due to narrowing the V/Q ratio distribution, but rather to reduction of pulmonary shunt. For NO to improve oxygenation, by reducing shunting when the lung is ventilated with 60–100% oxygen, there must be nonhypoxic constriction in the ventilated lung regions (Fig. 14.3). The constriction can be part of the pathophysiologic process or deliberately introduced for this purpose by infusing short-acting prostaglandin $F_{2\alpha}$, the pulmonary specific vasoconstrictor almitrine, or any general vasoconstrictor that acts on small pulmonary arteries (Puybasset et al., 1995). If arterial constriction is due to passive causes, active constriction is lost, inhibited, or confined to vessels other than small pulmonary arteries, then NO has no therapeutic effect on oxygenation or pulmonary vascular resistance.

A third concern is the assumption that NO acts only on the vessels associated with ventilated alveoli. In many disease states, affected alveoli are heterogeneously distributed throughout the lung so that NO may diffuse from ventilated regions to reach adjacent, nonventilated regions, at reduced concentrations. In addition, there is some evidence that NO is oxygen dependent with a greater vasodilator effect when oxygen is reduced. These two effects can account for the dissociation between changes in pulmonary vascular resistance and oxygenation that is often observed when NO is administered clinically.

A fourth consideration is that for all of the preceding comments, improvement of oxygenation is maximum when HPV is fully active. Inhaled NO will still reduce resistance and improve oxygenation when pulmonary shunt is present with nonhypoxic vasoconstriction even when HPV is absent (Fig. 14.3); however, the effect is enhanced by the presence of HPV (Fig. 14.3). HPV is reduced or abolished by drugs, trauma, sepsis, and changes in redox state. Changes of intracellular redox state are also believed to be the basis for the initiation of hypoxic constriction in pulmonary artery smooth muscle (Marshall et al., 1996; Yuan et al., 1994).

Mixed Venous Oxygen Tension

The mixed venous oxygen tension is an important stimulus for HPV, and oxygenators are increasingly being used to supplement gas exchange in critically ill patients (Calderon et al., 1999). It is therefore essential to examine the interactions between HPV, Pvo_2, and NO in the absence and presence of nonhypoxic constriction of small arteries (Fig. 14.3).

Summary

The dynamic interactions between the oxygen-exchanging and blood-flow distribution properties of the lung have been discussed. These interactions are part of the normal physiology, but are particularly evident in disease states where they contribute both to the homeostatic responses and to the pathophysiology. Computer models are permitting quantitation of effects that hitherto have only been intuitively appreciated because of the number of variables and the nonlinearity and complexity of these relationships. It is becoming evident that these considerations are fundamental to the understanding of advances in therapy in critically ill patients.

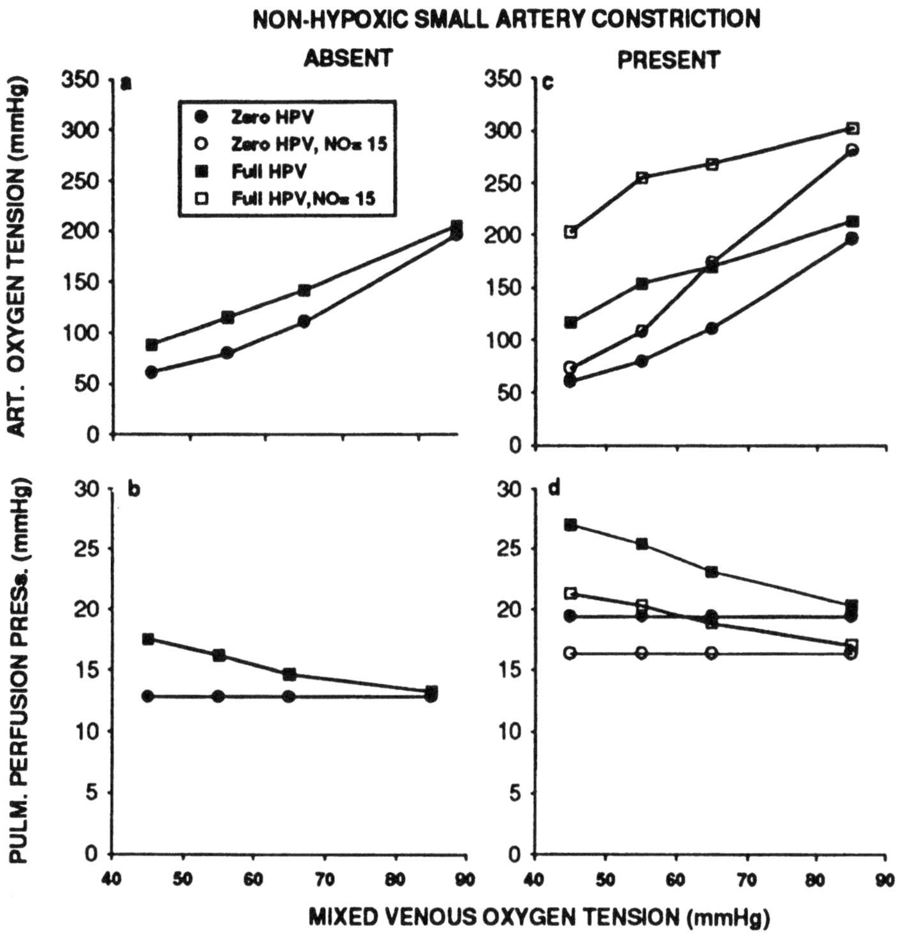

FIGURE 14.3. The importance of nonhypoxic small artery constriction and the interaction with mixed venous oxygen tension. All data are calculated assuming a patient with respiratory failure (shunt = 60%, ln SD = 1.3, and cardiac output 80% of normal), with FIO_2 of 1.0 and with ventilation adjusted to maintain $PVCO_2$ at 46 mmHg. HPV is assumed to be fully active. (A) and (B) Absence of nonhypoxic small artery constriction demonstrates that even in the absence of HPV, increasing the PVO_2 causes an impressive increase in PaO_2, whereas pulmonary perfusion pressure remains constant. The influence of HPV is to increase both the PaO_2 and perfusion pressure at low oxygen tension. Note, however, that in the absence of nonhypoxic, small artery constriction NO has no influence and the data with and without NO are superimposed. (C) and (D) Moderate nonhypoxic small artery constriction is applied, resulting in increased perfusion pressure, but now NO is associated with reduction of perfusion pressure and enhancement of PaO_2 changes. It is evident that the influence of HPV is also enhanced by the presence of the additional small artery constriction.

References

Baumgardner, J.J., Choi, I.-C., v Noordegraaf, A., Frasch, H.F., Neufeld, G.R., and Marshall, B.E. (2000) Sequential VA/Q distributions in the normal rabbit by micropore membrane inlet mass spectrometry. J. Appl. Physiol. 89, 1699–1708.

Brimioulle, S., Lejeune, P., Vachiery, J.-G., Delcroix, M., Hallemans, R., Leeman, M., and Naeije, R. (1994) Stimulus-response curve of hypoxic pulmonary vasoconstriction in intact dogs: effects of ASA. J. Appl. Physiol. 77, 476–480.

Calderon, M., Reyes, P., Tovar, A., Nunez, E., Lagunas, J., Soberanes, A., et al. (1999) Low flow veno-venous ECMO via subclavian catheter for severe respiratory failure. Heart Surg. For. 2, 38–40.

Fung, Y.C. (1984) Biodynamics: circulation. pp. 290–364. Springer-Verlag, New York.

Gerlach, H., Rossaint, R., Pappert, D., and Falke, K.J. (1993) Time-course and dose-response of nitric oxide inhalation for systemic oxygenation and pulmonary hypertension in patients with adult respiratory distress syndrome. Eur. J. Clin. Invest. 23, 499–502.

Hlastala, M.P. (1984) Multiple inert gas elimination technique. J. Appl. Physiol.: Respirat. Environ. Exercise Physiol. 56, 1–7.

Marshall, B.E., Clarke, W.R., Costarino, A.T., Chen, L., Miller, F., and Marshall, C. (1994) The dose response relationship for hypoxic pulmonary vasoconstriction. Resp. Physiol. 96, 231–247.

Marshall, C., Mamary, A.J., Verhoven, A.J., and Marshall, B.E. (1996) Pulmonary artery NADPH-oxidase is activated in hypoxic pulmonary vasoconstriction. Am. J. Resp. Cell. Mol. Biol. 15, 633–644.

Marshall, B.E., Marshall, C., Magno, M., Lilagen, P., and Pietra, G.G. (1991) Influence of bronchial arterial Po_2 on pulmonary vascular resistance. J. Appl. Physiol. 70, 405–415.

Riley, R.L., and Cournand, A. (1949) Ideal alveolar air and the analysis of ventilation-perfusion relationships in the lung. J. Appl. Physiol. 1, 825–847.

Wagner, P.D., Saltzman, H.A., and West, J.B. (1974) Measurement of continuous distributions of ventilation/perfusion ratios: theory. J. Appl. Physiol. 36, 588–599.

West, J.B., and Wanger, P.D. (1997) Ventilation-perfusion relationships. In: Crytal, R.G., West, J.B., Weibel, E.R., and Barnes, P.J. (eds.) The lung: scientific foundations. Vol. 2. pp. 1693–1710. Lippincott-Raven Press, Philadelphia.

Yuan, X.L., Todd, M.L., Rubin, L.J., and Blaustein, M.P. (1994) Deoxyglucose and reduced glutathione mimic effects of hypoxia on K^+ and Ca^{2+} conductances of pulmonary artery cells. Am. J. Physiol. 267, L52–63.

Recommended Readings

Marshall, B.E., Hanson, C.W., Frasch, H.F., and Marshall, C. (1994) Role of HPV in pulmonary gas exchange and blood flow distribution: I. Physiological concepts. Intens. Care Med. 20, 291–297.

Puybasset, L., Rouby, J.J., Murgeon, E., Cluzel, P., Souhil, Z., and Law-Koune, J.D. (1995) Factors influencing cardiopulmonary effect of inhaled nitric oxide in acute respiratory failure. Am. J. Respir. Crit. Care Med. 152, 318–328.

15
Respiratory Regulation of Acid–Base Balance in Health and Disease

EUGENE E. NATTIE

Introduction

Homeothermic mammals, including humans, normally maintain both intracellular and extracellular pH within a rather narrow range of values, with the normal values being approximately 6.9 and 7.4, respectively. Because intracellular pH is difficult to measure, most physiological and clinical evaluations focus on extracellular pH, which is measured in samples of arterial or venous blood. Among the different body compartments, including vascular and extravascular, as well as intra- and extracellular, blood is easily accessible and the pH values in blood are presumed to reflect whole body pH, including the intracellular compartment. Values of blood extracellular fluid pH above 7.7 and below 6.9 in humans can result in morbidity and mortality. A clear understanding of the importance of pH and how to interpret clinically available acid–base data is of utmost importance to the physician.

Using both chemical and evolutionary perspectives, I shall explain in this chapter why mammalian extracellular pH is normally 7.4 and how small changes in pH can determine regulatory responses, and result in pathophysiology. I shall then outline an approach to examine acid–base data including clinical examples.

Water, pH, Neutrality, and Temperature

The physiological solvent, water, is a complex and unique substance. Its concentration is quite high, at approximately 55 M at mammalian body temperature, especially when compared with the concentration of physiological solutes, which are in the mM range at their highest values. Water dissociates into H^+ and OH^- ions, which are at quite low concentrations at equilibrium. These ions exist in complex molecular formations due to attraction between their charge and the charge distribution of undissociated water molecules that exist as a dipole. At 25°C, with pH of 7.0, the H^+ and OH^- concentrations are 10^{-7} M, and the solution is said to be neutral. Addition of H^+ ions, with accompanying anions to neutral water, decreases the pH. The water is then said to be acidic. Similarly, an increase in pH relative to this neutral value defines an alkaline solution.

The dissociation of water increases with an increase in temperature, resulting in greater amounts of *both* H^+ and OH^-. Neutrality still prevails, but the absolute pH value is lower. The coefficient for the relationship of neutral pH and temperature is –0.017 pH units / °C, which means that the pH of neutrality at mammalian body temperature, 37°C, is approximately 6.8, a value below the normal mammalian extracellular fluid pH, which is 7.4, but close to the normal mammalian intracellular pH of 6.9 (Rahn, Reeves, and Howell, 1975).

Given the narrow range of values for a tolerable extracellular pH in humans, a look at the pH values in the animal kingdom, including ectotherms, yields some surprising findings. In general, the extracellular pH of ectothermic vertebrates and invertebrates varies inversely with temperature. pH decreases as temperature increases, and the coefficient is approximately –0.017 pH units / °C.

In the few cases examined, the intracellular pH also varies with temperature with a similar coefficient. At the prevailing ambient temperature, "normal" pH then might be 8.0 in a frog. This is a pH incompatible with life in a homeotherm at 37°C, yet the frog seems quite happy with it in physiological terms (Erasmus, Howell, and Rahn, 1970/71).

These data present an intriguing problem. Either ectotherms are insensitive to pH in comparison to the mammal, or pH *per se* does not universally define the variable that causes physiological insult. The following set of observations *suggest* a clue to this problem. In respect to temperature and pH, the blood of humans and that of the ectotherm *in vitro* both behave as does that of the ectotherm tested *in vivo*. An increase in temperature results in a decrease in pH with a coefficient of –0.017 pH units / °C. (see Fig. 15.1). This temperature–pH relationship appears to be ubiquitous in its presence, which suggests that it is fundamentally biochemical in nature. This problem (i.e., the effect of temperature on pH) has direct clinical relevance in the situation of hypothermia. When body temperature is lowered in order to decrease the metabolic rate of body and brain during surgery, blood pH increases in humans as predicted from the observations in the ectotherm. Understanding the significance of the temperature–pH relationship will help us to understand this clinical situation, as well as to provide us with some general principles that explain how pH changes affect physiological function.

Neutrality and Relative Alkalinity

We note here that the variation of pH with temperature in ectotherms and homeotherms essentially has the same coefficient as does the change in the neutral pH of water with temperature. In the case of extracellular pH, a constant relative alkalinity is maintained (i.e., the difference in extracellular pH vs. the neutral pH of water is

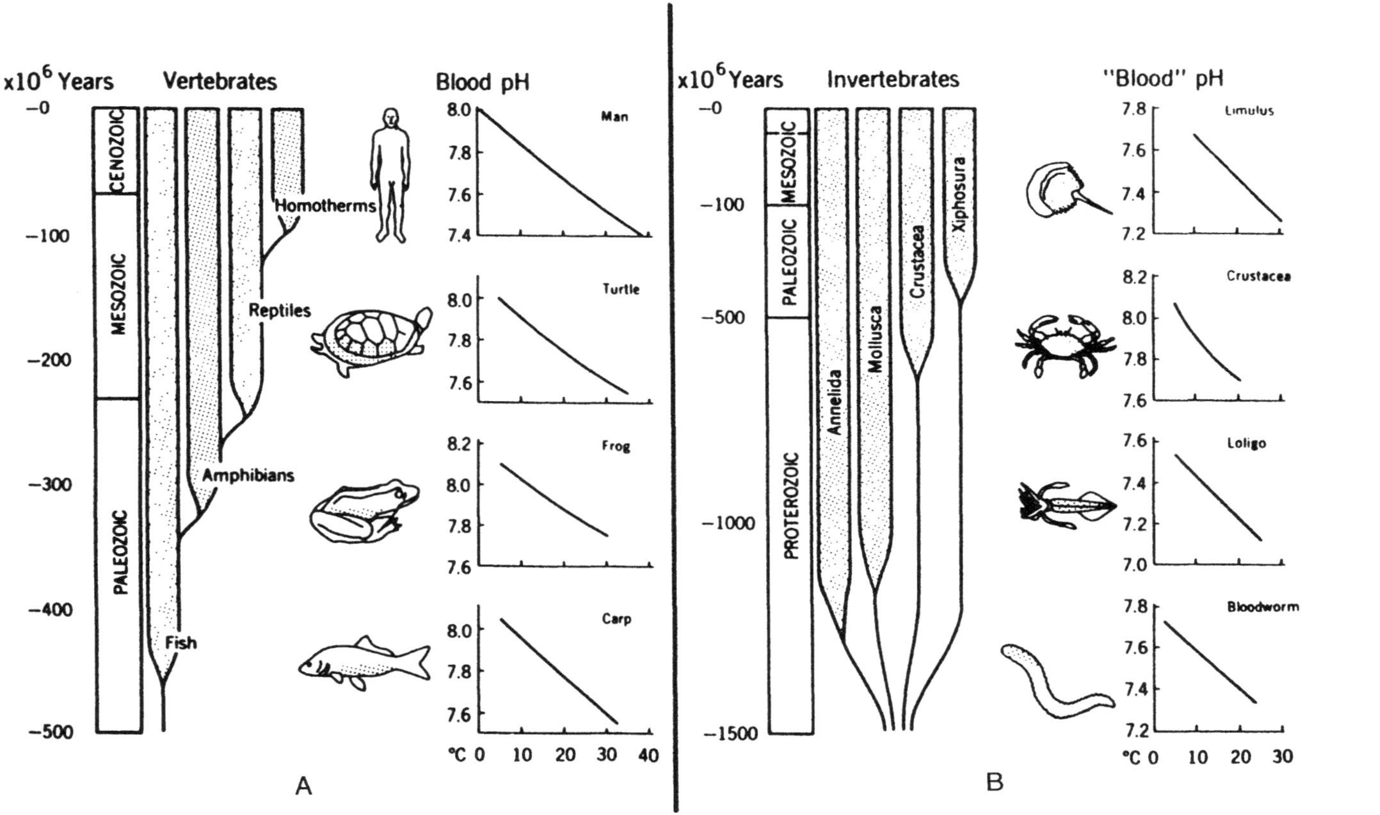

FIGURE 15.1. The widespread presence of the blood pH–temperature relationship among vertebrates (A) and invertebrates (B) indicates its fundamental nature. (*Source*: From Rahn et al., 1975, Am. Rev. Rsp. Dis. 112, 165–172. with permission of the publishers.)

constant as temperature changes). In the intracellular compartment, the few existing observations suggest that with temperature changes intracellular pH remains at the value for that of neutral water. Thus, intracellular neutrality is maintained in all animals as temperature changes. (This last statement still requires more evidence to be universally accepted among acid–base investigators, so it is stated here as a working hypothesis). This constancy of intracellular neutrality and extracellular relative alkalinity with varying temperature raises two questions: How does it occur? What is its significance?

Imidazole–Histidine

Body water contains a variety of solutes, including inorganic and organic charged and uncharged substances. Many of these substances can react with H^+ ions binding them and removing them from solution, or releasing them and adding them to solution depending on the circumstances. Such substances are called *buffers*, and their effectiveness as buffers is greatest when they are half in the undissociated and half in the dissociated forms. The pK′ of a substance that can gain or lose H^+ ions is a measure of its tendency to do so. When the pK′ and the prevailing pH are equal, the substance is half dissociated (i.e., it is an excellent buffer).

Even though there are many buffer substances both inside and outside of the cell, the imidazole ring of the amino acid histidine has a pK′ of 7.0 at 37°C, making it an ideal buffer (i.e., at a physiological pH of 7.0, histidine is 50% dissociated). It is also present on many proteins (most especially hemoglobin), and is quantitatively a very important physiological buffer.

To ask the question, what accounts for the change in body fluid pH as temperature is changed (with a coefficient of approximately –0.017 pH units / °C), we can reconstitute the various constituents of body fluids and measure the response of pH to changes in temperature in a stepwise fashion. In the absence of imidazole–histidine, the effect of temperature on the dissociation of water and other acidic substances present in body fluids is such that the pH change is much less than what is observed in the blood of ectotherm and endotherm or in water alone. In the presence of imidazole–histidine, the pH of the solution changes with temperature, as does that of water alone. This means that the effect of temperature on pH as measured in blood and cell is dependent on the presence of imidazole–histidine. In its absence, the decrease in pH with increasing temperature is much less.

Moreover, the pK′ of imidazole–histidine varies with temperature almost exactly as does the pH of ectotherm and endotherm blood and the best estimates of intracellular pH. This means that imidazole–histidine remains half in the dissociated and half in the undissociated forms at any temperature-defined pH. Its contribution to the H^+ ion concentration of the solution is constant (i.e., it is independent of temperature). This constant fractional dissociation of imidazole–histidine as temperature changes has far-reaching implications.

Protein Charge State

The significance of the maintenance of a constant neutral cell pH and a constant relative alkalinity of the extracellular fluid when temperature changes lies in the relationship between pH, temperature, and proteins. Proteins are formed by covalent bonds that connect the basic structural carbon atoms of the amino acids; however, the tertiary protein structure as its amino acid chain folds within itself gives life to the protein. This tertiary structure is determined by hydrogen bonds formed between reactive groups on the amino acids. Any protein has a distribution of surface charge that reflects its tertiary structure and the subsequent topographical organization of charged groups. The interaction of adjacent charged groups defines the actions of the protein. These charge interactions and the hydrogen bonds are sensitive to changes in pH.

To the extent that this protein charge state is determined by imidazole–histidine (and similar groups), protein charge state, tertiary structure, and function will remain unchanged when temperature changes. This is due to the fact that the dissociation of imidazole–histidine, and hence its charge on the protein, remains constant as the temperature changes. As temperature and cell pH change, the fractional dissociation of imidazole–histidine groups remains constant. To the extent that imidazole–histidine groups define the charge state of a protein, then such proteins will remain in their designed tertiary structure, and their physiological function will thus remain unaltered. This mechanism allows the maintenance of normal protein function in the face of temperature changes that result in significant changes in pH. It has been called the *α-stat hypothesis*, a term that means that the fractional dissociation of imidazole, called α imidazole, remains constant (Nattie, 1990).

If in the homeotherm pH changes at constant temperature, then the pH-sensitive imidazole–histidine groups change their charge, resulting in a new protein conformation or structure. This has two implications. First, imidazole–histidine can provide one way for proteins to "sense" pH changes. There are many proteins that are pH sensitive that are involved in the regulation of pH when pH changes occur under constant temperature conditions. Second, under greater pH stress, such pH sensitive groups can alter protein structure and function in a pathophysiological sense. This can result in effects that are detrimental to normal physiological function. In addition, severe pH changes can alter the hydrogen bonds in a more general way, resulting in a completely dysfunctional protein.

Conclusions

In the homeotherm, humans, the importance of maintaining pH within the physiological range is to maintain proteins in their normal, functional, tertiary structure. Small changes in pH can alter protein function via effects on groups like imidazole–histidine that are pH sensitive and determine the function of the protein by their location on the amino acid chains. Earlier life forms, which are ectothermic, used these same, or similarly designed, proteins. The choice by nature of imidazole–histidine as a key amino acid conferred two special properties on physiological solutions. First, it results in the maintenance of a cell pH that is equivalent to the neutral pH of water as temperature changes. Second, it allows the maintenance of a constant protein charge state and unchanged protein function as temperature changes. With the development of homeothermy, we are left with the legacy of proteins that are pH sensitive under constant temperature conditions that are part of our normal physiology.

When temperature does change in humans (e.g., in hypothermia for surgery), pH does increase as predicted by our ectothermic ancestry. What should the physician do? Trained to guard carefully the normal pH of 7.4 in blood when changes occur at normal temperature, should the physician treat this temperature-induced alteration in pH? The jury is still out on this question. The analysis would suggest that with temperature-induced changes in pH, protein charge state and protein function are maintained in their normal state. Thus, the blood pH in hypothermia should be maintained at the alkaline value predicted from the temperature–pH analysis.

An Acid–Base System for Humans

Increased Pco_2 in Air Breathers; Volatile, Respiratory Acid

Our analysis so far has ignored a key set of variables that are very important in human acid–base physiology. When lifeforms existed only in water, obtaining oxygen from the water required high flow rates through the gas exchange apparatus. This resulted in very low values remaining in the body fluids of the highly diffusible end-product of metabolism, carbon dioxide. With the evolution to air breathing, oxygen could be more easily obtained, resulting in less required air movement through the gas exchange apparatus. This lowering of the ventilatory requirement resulted in a rise in the level of carbon dioxide in the blood. Our measure of this level is the partial pressure of carbon dioxide, Pco_2. This value reflects the amount of carbon dioxide dissolved in the blood and is a measure of the reactivity of carbon dioxide in other chemical reactions and in diffusion from one site to another. In fish the Pco_2 is 2–5 mmHg; it is 40 mmHg in humans. This rise in the Pco_2 in the evolution from water to air breathing has important acid–base consequences because CO_2 is an acid.

$$CO_2 + H_2O = H_2CO_3 = H^+ + HCO_3^- \qquad (15.1)$$

The hydration of CO_2 forms carbonic acid that quickly dissociates to hydrogen and bicarbonate ions. CO_2 can be thought of as a volatile acid in the sense that the CO_2 dissolved in solution is in equilibrium with any gas phase into which it comes in contact, In humans this would be in the lung at the alveolar–capillary membrane. Thus, the level of alveolar ventilation is a prime determinant of the level of dissolved CO_2, measured by the Pco_2. The other major determinant of the level of CO_2 is, of course, the rate of CO_2 production by the tissues. In the steady state, the Pco_2 in arterial blood is directly proportional to body CO_2 production and inversely proportional to the level of alveolar ventilation.

$$\text{arterial } Pco_2 \approx CO_2 \text{ production/alveolar ventilation} \qquad (15.2)$$

Pco_2, HCO_3^-, pH, and Buffers

In normal humans at rest, the balance in Eq. (15.2) is such that the Pco_2 is 40 mmHg. The accompanying normal pH is 7.4 (or H^+ = 40 nmols/L) and the normal HCO_3^- is 24 mmols/L. Note the much greater concentration of HCO_3^- than H^+. If no other substances were present (i.e., blood was a simple bicarbonate solution), an increase in Pco_2 say from 40 to 80 mmHg (produced by a decrease in the level of alveolar ventilation) would drive the reaction to the right by mass action, resulting in an increase in the concentration of H^+ from 40 to 80 nmols/L (pH to 7.1) and an equimolar increase in the concentration of HCO_3^- from 24 to 24.00040 mmols/L. This change in bicarbonate is unmeasurably small and is explained by the large initial difference in the concentrations of hydrogen and bicarbonate ions.

In fact, blood is not simply a bicarbonate solution. There are many other substances present, including ions and buffers. We can summarize the effects of such buffers in:

$$CO_2 + H_2O = H_2CO_3 = H^+ + HCO_3^- + Buf^- = HBuf^- \qquad (15.3)$$

If we repeat in blood the scenario described earlier in which we increase the Pco_2 from 40 to 80 mmHg, the response is quite different. The reaction is driven to the right as before, but in this case the presence of the buffers (Buf^-) titrates the increase in hydrogen ions. More hydrogen ions are formed, as are more bicarbonate ions. The concentration of HCO_3^- will increase by the extent that hydrogen ions combine with the buffers. In blood of a normal human in this case, the HCO_3^- concentration will rise by 5 mmol/L, reflecting the binding of 5 mmol/L of H^+ ions. The actual H^+ ion concentration in the blood will be 66 nmol/L (pH = 7.18), which is a value that is less acidic than that observed in the simple bicarbonate solution in the absence of buffers (Eq. 15.1).

The buffers limit the acidosis produced by the increase in P_{CO_2}. They would similarly limit the alkalosis produced by a decrease in P_{CO_2}. Beyond this important physiological function, the fact that the bicarbonate concentration changes measurably during changes in P_{CO_2} can be used to develop a scheme for understanding acid–base variables.

The Acute CO_2 Titration Curve

If we repeat the experiment described for Eq. (15.3) for a number of levels of P_{CO_2}, we then can describe the relationship of the blood bicarbonate concentration to the P_{CO_2} at which it is in equilibrium. This is called an *acute CO_2 titration curve* and one is shown in Fig. 15.2 (Cohen and Kassirer, 1982).

The wide dark band for the bicarbonate values describes the responses of a normal population of subjects exposed acutely to the various P_{CO_2} levels. Remember that this change in bicarbonate concentration when the P_{CO_2} is changed reflects the presence of buffers for hydrogen ions. Thus, a steeper relationship in this plot of bicarbonate versus P_{CO_2} would indicate the presence of a greater buffer concentration. A flat curve would indicate the absence of any buffers. This CO_2 titration curve tells us what the bicarbonate concentration should be when the only stress imposed is an acute change in the P_{CO_2}.

This relationship between bicarbonate and the P_{CO_2} implies that there are accompanying changes in pH. These can easily be calculated. Given Eq. (15.1) it is possible, with knowledge of the dissociation constants, to define a quantitative relationship for pH, P_{CO_2}, and HCO_3^-. This is called the *Henderson–Hasselbalch relationship* (Cohen and Kassirer, 1982).

$$\mathrm{pH} = \mathrm{pK}' + \log\left(\left[\mathrm{HCO_3}^-\right]/\mathrm{S} \times P_{CO_2}\right) \tag{15.4}$$

where pK′ (a combined dissociation constant) = 6.1

S (the solubility of CO_2 = 0.03 mmol/L/mmHg under normal physiological conditions.

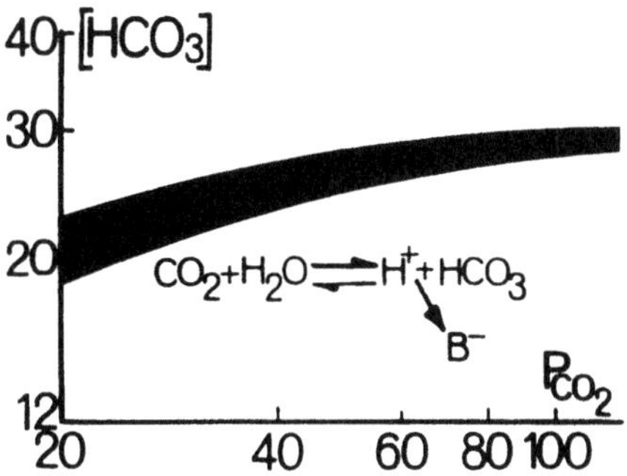

FIGURE 15.2. A CO_2 titration curve for man. Plasma HCO_3^- is plotted versus the arterial P_{CO_2}. B^- in the equation refers to the buffer anion concentration. The dark band represents the normal range of values for the HCO_3^- responses to an acute change in the P_{CO_2}. This change in HCO_3^- results from the presence of buffers for H^+ ions (see text). (*Source*: The figure is from a privately published monograph on acid–base balance by John C. Mithoefer.)

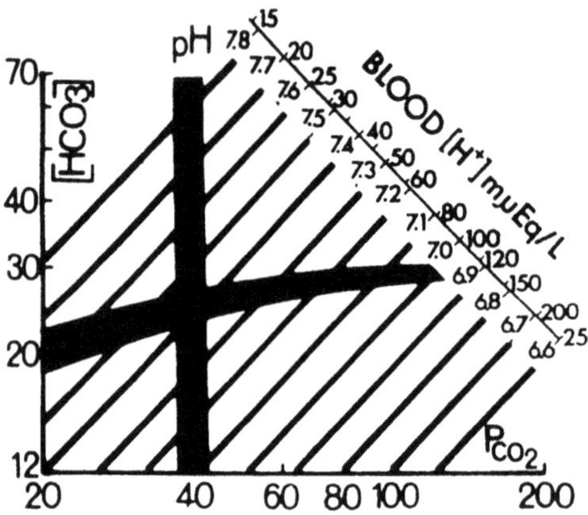

FIGURE 15.3. Same as Fig. 16.2 with the addition of pH isopleths and a band representing the normal range of values for arterial P_{CO_2} (see text). (*Source*: The figure is from a privately published monograph on acid–base balance by John C. Mithoefer.)

Under constant conditions, this relationship always holds and allows one to calculate one of the three acid–base variables, pH, P_{CO_2}, or HCO_3^-, if any two of the others are known. In modern clinical practice, the pH and P_{CO_2} are usually measured from an arterial blood sample obtained anaerobically and the HCO_3^- concentration calculated. The clinician can then use these three variables to discern the acid–base condition of the patient.

As we develop our scheme for evaluating these variables, we can use Eq. (15.4) to provide information useful to the interpretation of the acute CO_2 titration relationship between HCO_3^- and the P_{CO_2}. On our graphical representation of this relationship, we can calculate all the combinations of HCO_3^- and P_{CO_2} that define a given pH (e.g., the normal pH of 7.4). A line joining all of these calculated points will run diagonally from lower left to upper right. This line, a pH isopleth, defines all values of HCO_3^- and P_{CO_2} that have a value of 7.4 as the accompanying pH. We can then do this for a series of pH values and draw each of these isopleths on the CO_2 titration curve. This is shown in Fig. 15.3.

Figure 15.3 shows the acute CO_2 titration curve that we derived and showed in Fig. 15.2. It also shows the pH isopleths as described earlier. The broad vertical band above the P_{CO_2} of 40 mmHg shows the normal range of P_{CO_2} values. The pH isopleths are graphical representations of the Henderson–Hasselbalch relationship (Eq. 15.4) and they describe all combinations of P_{CO_2} and HCO_3^- that define any pH value. These relationships persist in any condition as long as temperature is unchanged (temperature affects the pK′ and S values used for the constants in the Henderson–Hasselbalch equation). The broad bands represent the ranges of normal physiological values for the P_{CO_2} and for the response of the HCO_3^- to an acute change in P_{CO_2}. Thus, for example, if a patient had an acute increase in P_{CO_2} from 40 to 80 mmHg (caused perhaps by respiratory depression secondary to drug ingestion) we can see, using Fig. 15.3, that the HCO_3^- would increase by approximately 5 mmol/L, to a value of approximately 29 mmol/L, and the pH would be 7.18. This is an example of an acute respiratory acidosis, whereby the initial stress is a failure of adequate ventilation producing an increase in P_{CO_2}. This increase results in a greater amount of H^+, much of which is buffered. The pH reflects this acidosis. Before we see how useful this graph and the relationships within it can be, we must add a few additional concepts.

Ionic Compensations; The Kidney

In the example presented of an acute respiratory acidosis the fall in pH is actually diminished because of the presence of buffers in the blood. Remember that their effectiveness is measured indirectly by the measured increase in HCO_3^- in response to the change in P_{CO_2}. The body has additional defenses for pH regulation that require more time to be effective. These involve changes in the pattern of excretion of ions by the kidney. It is beyond our scope to consider these in detail here, but, we can summarize their effects in respect to acid–base regulation (Valtin and Gennari, 1987).

Renal excretion of HCO_3^- is dependent importantly on the level of the P_{CO_2}. An increase in P_{CO_2} results in more retention of HCO_3^- (and excretion of H^+). By mass action, this raises pH back toward a normal value. In Fig. 15.3 we can see the effect of what would happen if our patient with an increase in P_{CO_2} to 80 mmHg had this change sustained for 1 or 2 days. This is known to produce renal retention of HCO_3^- that would increase the HCO_3^- from 29 mmol/L, the value observed with an acute increase in P_{CO_2}, to, say, 35 mmol/l. The new pH (Fig. 15.3) would be 7.3, a value less acidic than 7.18. These renal mechanisms are usually described as a secondary pH regulatory response in contrast to the initial pH regulatory effects of blood buffers. The renal response begins as soon as the P_{CO_2} is changed, but it takes hours to days to have its full effect. If the primary event is a decrease in P_{CO_2}, then the renal responses with respect to HCO_3^- (and H^+) occur in the opposite direction. It is important to note that as a general rule these regulatory responses, which serve to minimize pH changes, never correct the pH back to the normal value. This is a key point for subsequent interpretation of acid–base values.

Metabolic, Nonvolatile Acids

So far we have covered two general ways by which pH can change. First, a P_{CO_2} change directly influences pH immediately via hydration to carbonic acid and immediate dehydration to H^+ and HCO_3^-. Second, a P_{CO_2} change influences renal excretion of HCO_3^- (H^+), which helps to compensate for the initial pH change. Other processes can also alter the HCO_3^- concentration.

Metabolic events can produce H^+ (e.g., lactic acid), which is in part buffered by the normal HCO_3^-. This use of the P_{CO_2}—HCO_3^- system results in the production of CO_2 at the expense of HCO_3^-. The volatile CO_2 can be excreted by the lungs (ventilation is actually increased by the acid pH stimulation of chemoreceptors, which enhance breathing). The HCO_3^- concentration decreases. This pattern of events is called a metabolic acidosis. Here the primary event is a decrease in HCO_3^- and pH caused by the overproduction of a metabolic acid. The decrease in P_{CO_2} is a ventilatory compensation for this acidosis in that by increasing the rate of CO_2 excretion, pH is actually increased.

The opposite occurs if a metabolic process results in an increase in the normal HCO_3^- concentration. This event, which is a metabolic alkalosis, produces an increase in pH. In many cases this alkaline pH inhibits ventilation (via the pH sensitive chemoreceptors) raising the P_{CO_2}. The initial alkalosis is tempered by this respiratory compensation in that the rise in P_{CO_2} adds H^+ to the system.

A Schema

We now understand that the three acid–base variables used most frequently in the clinical setting are: pH, P_{CO_2}, and HCO_3^-. They are linked by many processes that can be summarized by Eq. (15.3). Any change in the amount of alveolar ventilation directly changes the Pco_2, which immediately changes pH, and, because of the presence of buffers, the HCO_3^- concentration. The change in HCO_3^- due to an acute change in Pco_2 is small [i.e., within a few millimoles per liter (mmol/L)], and this provides a cornerstone for our analysis scheme. Chronic changes in Pco_2 affect renal processes that add or remove HCO_3^- (or H^+). Metabolic abnormalities can also affect pH and HCO_3^-. Our goal is to understand these relationships in a way that will allow quick and accurate determination of the acid–base state given any combination of measured values for pH, Pco_2, and HCO_3^-.

Our approach is based on the CO_2 titration curve shown in Fig. 15.3 and the fixed relationship expressed via the Henderson–Hasselbalch relationship in Eq. (15.4) and shown as the diagonal pH isopleths in Fig. 15.3. The Pco_2–HCO_3^- relationship tells us an important fact: With acute changes in Pco_2 the resultant change in HCO_3^- is a few mmol/L. Larger changes in HCO_3^- must be accounted for by some other process (e.g., renal or metabolic). A final axiom is that compensatory mechanisms do not return the initial pH change entirely back to the normal value. Let us apply these ideas to examples of each major type of acid–base disorder to see how this approach works in practice.

Clinical Approach to Acid–Base Disorders

A Suggested Approach

When the three acid–base variables are reported, examine them in the following order.

The pH

Is the pH acidic (<7.4) or alkaline (>7.4)? This defines the disorder as an acidosis or an alkalosis.

The Pco_2

Is the Pco_2 greater than normal (>40 mmHg)? If it is and the pH is acidic, then we have a respiratory acidosis as one component of the disorder. Whether this acidosis is acute or chronic, and whether it is compensated by renal mechanisms, depends on the level of increase of HCO_3^-. An increase of a few mmol/L indicates an acute event. An increase of greater than a few mmol/L is a more chronic event. The degree of acidosis will be less in this compensated case. If the Pco_2 is below normal (<40 mmHg) and the pH is alkaline, then we have a respiratory alkalosis as one component of the disorder. Whether this alkalosis is acute or chronic depends on the level of the decrease in HCO_3^-. A decrease in HCO_3^- of only a few mmol/L is consistent with an acute respiratory alkalosis. If the HCO_3^- is decreased by a greater amount, then we have a chronic partially compensated respiratory alkalosis. If the pH is acidic and the Pco_2 is less than normal, then there must be another cause for the acidosis, a metabolic acidosis. If the pH is alkaline and the Pco_2 are above normal, then there must be another cause for the alkalosis, a metabolic alkalosis.

The HCO_3^-

The HCO_3^- can be increased in three ways: (1) an acute increase in Pco_2 due to the presence of buffers will increase it (but only by a few mmol/L) (see Eq. 15.3); (2) a chronic increase in Pco_2 will increase the HCO_3^- via renal processes; (3) metabolic disorders can primarily increase the HCO_3^-. The HCO_3^- be decreased in three ways: (1)

an acute decrease in P_{CO_2} due to the presence of buffers will decrease it (but only a few mmol/L) (see Eq. 16.3); (2) a chronic decrease in P_{CO_2} will decrease the HCO_3^- via renal processes; (3) metabolic disorders can primarily decrease the HCO_3^-.

We shall now apply this approach to six sets of acid–base data and show how this approach can be used.

Acute Respiratory Acidosis

Data: pH = 7.25, P_{CO_2} = 60 mmHg, and HCO_3^- = 26 mmol/L

Using our suggested approach, we see that the pH is acidic and the P_{CO_2} is greater than normal, which suggests a respiratory acidosis. The HCO_3^- is increased a few mmol/L greater than the normal value of 24 mmol/L, which is a change that is consistent with that seen in an acute increase in P_{CO_2}. These data are entirely consistent with a diagnosis of an acute respiratory acidosis. An example would be a patient with depressed ventilation secondary to a drug overdose.

Chronic Respiratory Acidosis

Data: pH = 7.35, P_{CO_2} = 60 mmHg, and HCO_3^- = 33 mmol/L

The pH is less than 7.3 and the P_{CO_2} is increased greater than normal, which suggests the presence of a respiratory acidosis. In this case, the HCO_3^- is increased by more than a few mmol/L. The cause could be renal processes reflecting a compensatory response, or it could be a metabolic source. Given the acid pH and the elevated P_{CO_2}, the most likely explanation is renal compensation. Hence, our diagnosis is a respiratory acidosis-compensated. An example would be a patient with chronic obstructive pulmonary disease (COPD) with chronic CO_2 retention as a result of abnormal pulmonary gas exchange. This results in chronic respiratory acidosis with renal compensation.

Acute Respiratory Alkalosis

Data: pH = 7.60, P_{CO_2} = 22 mmHg, and HCO_3^- = 20 mmol/L

The pH is alkaline and the P_{CO_2} is decreased to less than the normal value of 40 mmHg, which suggests the presence of a respiratory alkalosis. The HCO_3^- is decreased by a few mmol/L a change consistent with that seen with an acute decrease in P_{CO_2}. These data are consistent with a diagnosis of an acute respiratory alkalosis. A clinical example would be a patient with psychogenic hyperventilation caused by anxiety or stress.

Chronic Respiratory Alkalosis

Data: pH = 7.45, P_{CO_2} = 22 mmHg, and HCO_3^- = 14 mmol/L

The pH is alkaline and the P_{CO_2} is decreased the less than the normal value of 40 mmHg, which suggests the presence of a respiratory alkalosis. Here, the HCO_3^- concentration is decreased by more than a few mmol/L. Such a decrease can be the result of renal responses to a chronic decrease in P_{CO_2} or to a metabolic production of acid. Given the alkaline pH and the decrease in P_{CO_2} the most likely diagnosis is respiratory alkalosis-compensated. This pattern is clinically unusual, such as after prolonged exposure to a hypoxic environment (e.g., high altitude). Here the hypoxic stimulation of ventilation via the peripheral chemoreceptors produces a prolonged hyperventilation and decreased Pco_2. Another example would be prolonged hyperventilation in a patient on a mechanical ventilator.

Metabolic Acidosis

Data: pH = 7.20, Pco_2 = 22 mmHg, and HCO_3^- = 8 mmol/L

The pH is less than 7.3 indicating the presence of an acidosis. The Pco_2 is decreased, indicating that the acidosis is not respiratory in origin. The HCO_3^- is decreased and by more than the few mmol/L expected by a decrease in Pco_2 were that the only event taking place. A decrease in HCO_3^- can result from renal compensation to a respiratory alkalosis (unlikely here) or from a metabolic disorder, producing an acidosis. This pattern of acid–base variables is most likely the result of a metabolic acidosis-compensated. Here the primary event is the excess production of a metabolic acid that decreases HCO_3^- as it is buffered. The compensation is the decrease in Pco_2 caused by stimulation of ventilatory chemoreceptors by the acidic pH. This can be seen clinically in patients with lactic acidosis induced by tissue hypoxia usually as a result of poor perfusion by the cardiovascular system.

Compare these acid–base values to those for a respiratory alkalosis-compensated. In both cases the Pco_2 is decreased as is the HCO_3^-. The degree of the decrease in HCO_3^- and the related absolute pH value, allow one to make a clear distinction between them. In the respiratory alkalosis–compensated, the pH is alkaline and the decrease in HCO_3^- less pronounced in comparison to the metabolic acidosis–compensated.

Metabolic Alkalosis

Data: pH = 7.60, Pco_2 = 45 mmHg, and HCO_3^- = 43 mmol/L

The pH is alkaline; hence, we have an alkalosis. The Pco_2 is not decreased, however; in fact, it is increased to greater than the normal value. Thus, it is not a respiratory alkalosis. The HCO_3^- is increased, and by more than the few mmol/L, which would be the result of this increase in Pco_2 were that the only problem here. An increase in HCO_3^- such as this can be caused by renal compensation for a chronic elevation of the Pco_2 (an unlikely event here) or by a metabolic disorder. These data are consistent with a metabolic alkalosis, with the primary event being the increase in HCO_3^- and the accompanying increase in pH. This pH increase can inhibit the ventilatory chemoreceptors resulting in a decrease in ventilation and an increase in Pco_2. This latter sequence of responses is a respiratory compensation to the metabolic alkalosis in that the increase in CO_2 adds a source of H^+ to counterbalance the primary alkalosis. The degree and constancy of this ventilatory compensation to a metabolic alkalosis varies among patients and situations. Metabolic alkalosis can be produced by prolonged vomiting. In this case, the body sustains a loss of both hydrogen and K ions because the gastric juices are very acidic and contain K.

Mixed Disturbances; The Importance of the Clinical Setting

In each of the presentations of acid–base variables, we have ignored the clinical setting (i.e., the other clinical data in regard to history, physical examination, and laboratory values). The physician should never do this. The clinical setting is vital in the actual diagnosis of an acid–base disorder. The acid–base data may not be as clear-cut as it is in the preceding examples, and the clinical setting will be of great help in making some distinctions. To illustrate this we shall analyze the following mixed disturbances, acid–base disorders with more than one cause.

Data: pH = 7.45, Pco_2 = 60 mmHg, and HCO_3^- = 40 mmol/L

The pH is alkaline, so, by definition, the diagnosis is an alkalosis. The Pco_2 is increased to greater than the normal value of 40 mmHg to a value of 60 mmHg. This is due to a decrease in alveolar ventilation, which could reflect inability to maintain normal ventilation as it is in a patient with chronic lung disease. It could also reflect inhibition of ventilatory chemoreceptors by alkalosis in a patient with normal pulmonary function. The HCO_3^- is increased by more than the few mmol/L expected solely from the acute CO_2 titration curve. It can be explained by renal compensation for a chronic increase in Pco_2 or by a metabolic source of HCO_3^-. We have previously suggested that no physiological compensation is perfect, which means that compensatory processes do not correct the pH back to the normal value. It is possible, however, for a patient to have more than one acid–base disorder. This can result in confusing sets of acid–base variables. Here, these data are consistent with two possibilities: one is a metabolic alkalosis-compensated, the other a mixed disorder; a metabolic alkalosis together with a respiratory acidosis-compensated via renal processes. The increase in HCO_3^- in the latter case would be explained by both the renal response to the increase in Pco_2 and the metabolic source. One way to make this distinction is to put the acid–base data in the context of the patient. Here, if the patient has normal pulmonary function, then the diagnosis of a metabolic alkalosis-compensated is more likely. If the patient is known to have chronic lung disease and has had CO_2 retention (i.e., an increase in Pco_2) demonstrated on prior occasions, then a mixed disorder of respiratory acidosis-compensated and metabolic alkalosis is more likely.

Data: pH = 7.00, Pco_2 = 50 mmHg, and HCO_3^- = 12 mmol/L

By definition, this is, an acidosis. The Pco_2 is increased indicating the presence of a respiratory acidosis. Were this the only event we would expect the HCO_3^- to be increased; either by a few mmol/L, which reflects blood buffers, or by more than a few mmol/L, which reflects renal compensation. Here the HCO_3^- is decreased to less than the normal value of 24 mmol/L. With the observed increase in Pco_2, the only cause of a decrease in HCO_3^- is a source of metabolic acid. Thus, in this case, we have a mixed disturbance—a respiratory and metabolic acidosis.

Figure 15.4 is a reproduction of Fig. 15.3 with numbered areas delineated to show the values of pH, Pco_2, and HCO_3^-, which would accompany various acid–base diagnoses. It might be helpful to go over this figure now, making sure that you understand within the context of the approach presented earlier how each diagnosis is obtained. One caveat must be added here. Areas 4 and 5 represent the region that describes various combinations of increased Pco_2 and increased HCO_3^-. How to define precisely the boundary between Area 4 and Area 5 depends on how much respiratory compensation can occur to a metabolic alkalosis. In fact, in some situations the degree of ventilatory inhibition, and the accompanying increase in Pco_2, can be quite dramatic. In these cases, there would be a larger Area 4. The purpose of this figure is *not* to provide areas for memorization. It is to provide an exercise to use the principles and approach outlined earlier.

Intracellular pH Regulation

It is worth emphasizing here that the clinical approach to acid–base diagnosis and subsequent treatment is based on measurements of acid–base variables made in blood. Whole body intracellular pH and intracellular pH of specific organ tissues are of equal if not greater importance. The presumption is that regulation by physiological systems in normal conditions, and by the physician in disease, of blood pH or an estimate of

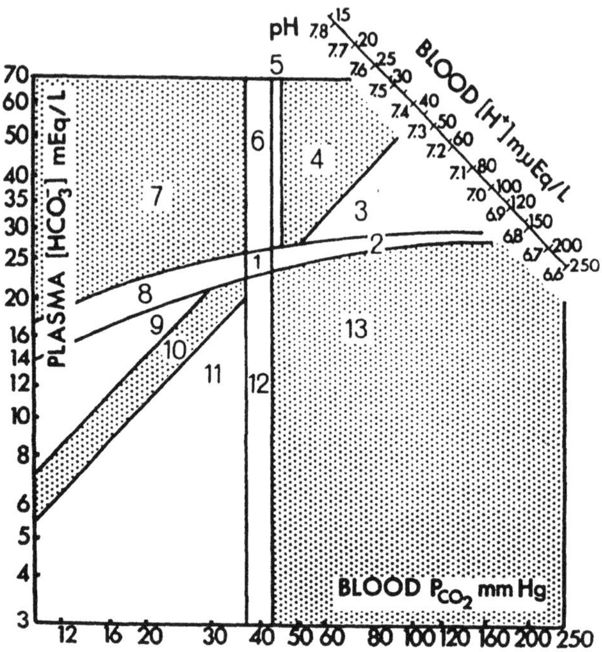

FIGURE 15.4. Like Fig. 16.3 with numbered areas on the plot showing different categories of acid–base disturbance as described later. Interpretation—(1) Normal. (2) Respiratory acidosis. (3) Respiratory acidosis—compensated. (4) Respiratory acidosis and Metabolic alkalosis (mixed). (5) Metabolic alkalosis—compensated. (6) Metabolic alkalosis. (7) Respiratory alkalosis and Metabolic alkalosis (mixed). (8) Respiratory alkalosis. (9) Respiratory alkalosis—compensated. (10) Metabolic acidosis and Respiratory alkalosis (mixed). (11) Metabolic acidosis—compensated. (12) Metabolic acidosis. (13) Metabolic acidosis and Respiratory acidosis (mixed). (*Source*: The figure is from a privately published monograph on acid–base balance by John C. Mithoefer.)

intracellular pH serves an important function. The maintenance of normal blood or extracellular fluid pH and P_{CO_2} should affect intracellular pH. In practical terms, it is easier to obtain reliable estimates of pH, P_{CO_2}, and HCO_3^- in blood, and the P_{CO_2} values give a direct index of the level of ventilation. This approach has a long and successful history; however, pH_i itself can be regulated. For example, raised P_{CO_2} (e.g., of the order of 50 or 60 mmHg) produces a fall in pH_i. This is reversed by physiochemical buffering and by the active extrusion of H^+ ions (H^+-ATPase). Apart from a proton pump, there is an Na/H exchanger, whereby H^+ efflux and Na influx are governed by the Na gradients that are generated and maintained by the Na/K pump. An additional system, the Cl/ HCO_3^- exchanger, plays a role in correcting alkalosis. Its level of activity is dependent on pH_i. Note that both the Na^+/H^+ and Cl^-/HCO_3^- exchangers are electrically neutral systems.

The Control of Ventilation by Chemoreceptors

In respect to CO_2, the central chemoreceptors rather than the peripheral chemoreceptors of the carotid body largely control ventilation (respiration). The contribution of the peripheral chemoreceptors to the resting ventilatory drive resulting from tonic excitation of the brain stem respiratory neurons is estimated to be about 25% (see Nattie, 1999).

The central chemoreceptors, like the peripheral chemoreceptors, respond both to small changes in arterial blood Pco_2 as well as to pH_e (more precisely, pH_i). These receptors are located in regions of the brainstem that are more widely distributed than was previously thought because apart from the ventrolateral medulla, their sites are identified in the nucleus of the solitary tract, locus ceruleus, ventral respiratory group, caudal medullary raphe, and fastigial nucleus of the cerebellum.

Although the sensitivity and threshold of each of those regions that contain chemoreceptors to CO_2 may vary, there is general agreement that the total response to Pco_2 follows a linear course over the 40–100 mmHg Pco_2 range. Plots of ventilation ($\dot{V}$) (ordinate) versus $Paco_2$ (abscissa) yield a straight line the slope of which is a measure of CO_2 sensitivity. During sleep this slope is reduced and apnea (from the Greek *apneusis*, meaning inspiratory breathholding) may intervene when the Pco_2 falls below a threshold value.

Both H^+ and HCO_3^- do cross the blood–brain barrier albeit more slowly than does CO_2 which crosses rapidly and stimulates ventilation as the result of reducing pH. Neverthless in chronic metabolic acídosis, ventilation is stimulated, whereas in acute metabolic ($\uparrow HCO_3^-$, $\downarrow H^+$) alkalosis, ventilation is inhibited.

The Effects of Hypoxia

Hypoxia acts as a general depressant. The carotid body chemoreceptors are the only sites in the body that respond to hypoxia in an excitatory manner. In other words, they act as O_2 sensors. That this is indeed the case is shown by the fact that surgical removal of the carotid body results in ventilatory depression if the animal is rendered hypoxic. Plots of $\dot{V}$ versus Po_2 yield a hyperbola, the shape of which is accentuated by a raised Pco_2 (hypercapnea). Put in another way, hypercapnia augments the ventilatory response to hypoxia and *vice versa*. The increase in ventilatory response to a raised Pco_2 during hypoxia occurs when pO_2 falls below 60 mmHg. The combination of a raised Pco_2 and hypoxia is the equivalent of asphyxia.

Acknowledgments. The first section of this chapter describes the original work of Hermann Rahn and Robert B. Reeves and their colleagues. The approach used to describe and "diagnose" acid–base disorders, via the CO_2 titration curve, is that of John C. Mithoefer. It is described in a privately published volume, "Acid–Base Balance," which I have used as a foundation for teaching this subject, as well as for writing this chapter. Figures 16.2, 16.3, and 16.4 are from this volume. The author is also indebted to his colleagues, S. Marsh Tenney, Heinz Valtin, James C. Leiter, and Donald Bartlett, for many acid–base discussions.

References

Cohen, J.J., and Kassirer, J.P. (1982) Acid–base. Little, Brown, Boston.

Erasmus, B. deW., Howell, B.J., and Rahn, H. (1970/71) Ontogeny of acid–base balance in the bullfrog and chicken. Respir. Physiol. 11, 46–53.

Nattie, E. (1990) The alphastat hypothesis in respiratory control and acid–base balance. J. Appl. Physiol. (brief review) 69, 1201–1207.

Nattie, E. (1999) CO_2 brainstem chemoreceptors and breathing. Prog. Neurobiology 59, 299–331.

Rahn, H., Reeves, R.B., and Howell, B.J. (1975) Hydrogen ion regulation, temperature, and evolution. Am. Rev. Resp. Dis. 112, 165–172.
Valtin, H., and Gennari, F.J. (1987) Acid–base disorders. Little, Brown, Boston.

Recommended Readings

Goss, G., and Grinstein, S. (1995) Mechanisms of intracellular pH regulation. In: Bittar, E.E., and Bittar, N. (eds.) Principles of medical biology. Chap. 7, Vol. 4. JAI Press Inc., Greenwich, CT.
Nattie, E. (1999) CO_2 brainstem chemoreceptors and breathing. Prog. Neurobiology 59, 299–331.
Rose, B.D., and Renike, H.G. (1996) Acid–base physiology and metabolic alkalosis. In: Rose, B.D., and Renike, H.G. (eds.) Renal pathophysiology—the essentials. Williams and Wilkins, Baltimore.
Seldin, D.W., and Giebisch, G. (eds.) (1990) The regulation of acid–base balance. Raven Press, New York.
Staub, N. (1998) Transport of oxygen and carbon dioxide. Tissue oxygenation. In: Berne, R.M., and Levy, M.N. (eds.) Physiology, fourth ed. Mosby, St Louis.

16
Exercise and Breathing in Health and Disease

M. YOUNES

One of the most devastating complications of respiratory disease is the limitation it imposes on the individual's capacity to engage in physical activity. As disease progresses in severity, the capacity to exercise progressively decreases. In very severe disease, the subject becomes dyspneic on minimal exertion, or even at rest. In this chapter, the normal respiratory responses to exercise will first be reviewed. This will be followed by a description of how different disorders affect these responses.

Exercise and Breathing in Health

Exercising muscles need more oxygen (O_2) and generate more carbon dioxide (CO_2) than do muscles at rest. The rates of O_2 consumption ($\dot{V}O_2$) and CO_2 production ($\dot{V}CO_2$) increase in direct proportion to power output (i.e., the rate of doing work). During very heavy exercise healthy young adults normally consume 40–50 ml/minute of O_2/kg ideal body weight. This is to be contrasted with the resting rate of 3–4 ml/min/kg. The maximum $\dot{V}O_2$ ($\dot{V}O_2$ max), thus maximum power output, decreases progressively with age, reaching 20–25 ml/min/kg in healthy 70 year olds. Even then, $\dot{V}O_2$max is greater than five times the resting rate.

Gas Exchange

The function of the respiratory system is to maintain adequate oxygenation in the arterial blood supplying the muscles (P_{AO_2}). Furthermore, it must maintain P_{ACO_2} at a level that ensures a reasonably normal pH. Muscle function, and indeed almost all body functions, are quite sensitive to blood pH. Because P_{AO_2} and P_{ACO_2} are functions of the alveolar levels (P_{AO_2} and P_{ACO_2}), the respiratory system must maintain "normal" values of P_{AO_2} and P_{ACO_2}, despite much greater O_2 consumption and CO_2 production. It should be intuitively obvious that the only way this can be done is by increasing the rate of air exchanged between alveoli and atmosphere (alveolar ventilation, $\dot{V}_A$). The relations between $\dot{V}_A$, metabolic rate, and alveolar partial pressures are expressed formally by the following equations:

$$P_{ACO_2} = \frac{\dot{V}CO_2}{\dot{V}_A} \cdot 0.863 \qquad (16.1)$$

and

$$P_{AO_2} = P_IO_2 - \frac{\dot{V}O_2}{\dot{V}_A} \cdot 0.863 \qquad (16.2)$$

Where P_IO_2 is the partial pressure of O_2 in inspired air, $\dot{V}O_2$ is in milliliters per minute (STPD), $\dot{V}_A$ is in liters per minute (BTPS), and 0.863 is a multipurpose constant. (See footnote for the basis and derivation of these equations).*

These equations indicate that in order for P_{AO_2} and P_{ACO_2} to remain at their normal resting levels during exercise, alveolar ventilation must increase in direct proportion to metabolic rate. The proportionality of $\dot{V}_A$ to $\dot{V}O_2$ or to $\dot{V}CO_2$ is about 20:1. In other words, in order to maintain P_{AO_2} and P_{ACO_2} at resting levels, $\dot{V}_A$ must increase about 20 L/minute for every 1 L/minute increase in $\dot{V}O_2$ and $\dot{V}CO_2$. Because not all fresh air breathed reaches the alveoli (because of dead space), actual ventilation at the mouth would have to increase to a greater extent.

For all practical purposes, one may assume that $\dot{V}_A$ [in] and $\dot{V}_A$ [out] are the same (they can be very slightly different when R is different from unity). Thus,

$$\dot{V}O_2 = \dot{V}_A(F_IO_2 - F_AO_2)$$

Solving for F_AO_2,

$$F_AO_2 = F_IO_2 - \dot{V}O_2/\dot{V}_A$$

To convert to partial pressure units, both sides are multiplied by barometric pressure less water vapor pressure (BP – 47). This yields,

$$F_AO_2(BP-47) = F_IO_2(BP-47) - \left[\left(\dot{V}O_2/\dot{V}_A\right)(BP-47)\right]$$
$$P_AO_2 = P_IO_2 - \left(\dot{V}O_2/\dot{V}_A\right)(BP-47)$$

For the preceding equation to be correct, $\dot{V}O_2$ and $\dot{V}_A$ must be expressed in the same units (e.g., L/minute, STPD, etc.); however, $\dot{V}O_2$ is expressed conventionally in STPD units, whereas $\dot{V}_A$ is expressed in BTPS units. Furthermore, the calculation can be rendered mentally easier to execute if $\dot{V}O_2$ is expressed in milliliters per minute, whereas $\dot{V}_A$ is expressed in liters per minute. When these conversions are combined, (BP – 47) is converted to a constant, 0.863. Thus, if P_IO_2 is 150 mmHg, $\dot{V}O_2$ is 240 ml/minute STPD, $\dot{V}_A$ is 4.0 L/minute BTPS, P_{AO_2} is,

---

* The rate of O_2 exchange between alveoli and atmosphere ($\dot{V}O_2$) is the difference between the amount of O_2 entering the alveoli per min ($\dot{V}O_2$ [in]) less the amount of O_2 leaving the alveoli per minute ($\dot{V}O_2$ [out]). $\dot{V}O_2$ [in] is the product of inspired alveolar ventilation ($\dot{V}_A$ [in]) and the concentration of O_2 in inspired air (F_IO_2). $\dot{V}O_2$ [out] is the product of expired alveolar ventilation ($\dot{V}_A$ [out]) and O_2 concentration in alveolar gas F_AO_2). It follows that

$$\dot{V}O_2 = \left[\dot{V}_A[\text{in}] \cdot F_IO_2\right] - \left[\dot{V}_A[\text{out}] \cdot F_AO_2\right]$$

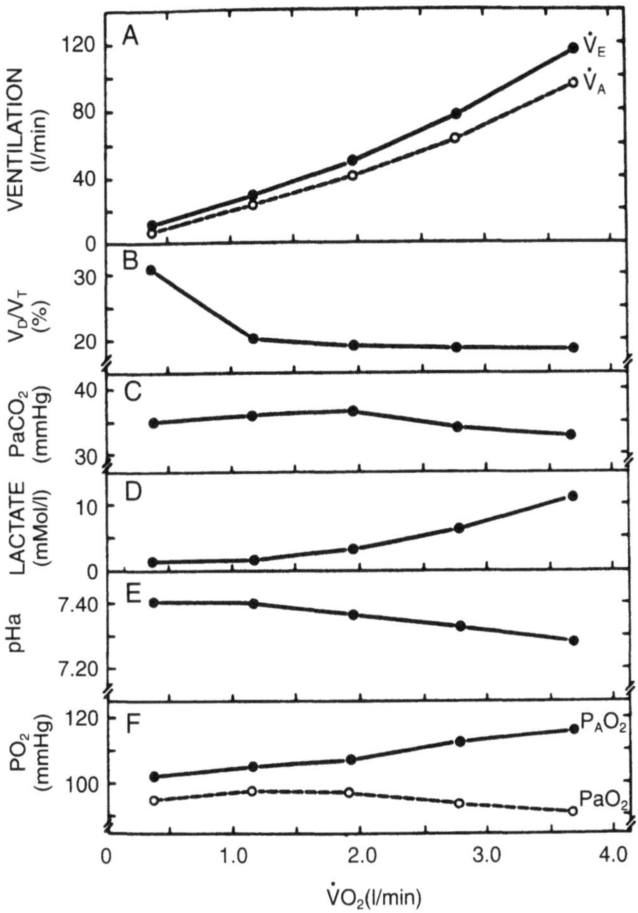

FIGURE 16.1. Changes in total ventilation ($\dot{V}_E$), alveolar ventilation ($\dot{V}_A$), deadspace fraction (V_D/V_T), arterial P_{CO_2} (P_{ACO_2}), lactate, arterial pH (pHa), alveolar P_{O_2} (P_{AO_2}), and arterial P_{O_2} (P_{AO_2}) at different levels of exercise. The latter is expressed in terms of O_2 consumption ($\dot{V}O_2$). (*Source*: Data from Wagner, P. D., Glen, G. E., Moon, R. E., et al: Pulmonary gas exchange in humans exercising at sea level and simulated altitude. J. Appl. Physiol. 61, 260, 1986.)

$$P_{AO_2} = 150 - [240/4.0] \cdot 863 = 98$$

The same applies to CO_2 exchange between alveoli and atmosphere. Because CO_2 is virtually absent in inspired air, however, the relation simplifies to the form given in Eq. (16.1) in the text.

Figure 16.1 shows the normal response to progressive exercise in healthy young adults of average size. Ventilation increases steadily as level of exercise ($\dot{V}O_2$) increases. In Fig. 16.1(A), note that the difference between total ventilation ($\dot{V}_E$) and alveolar ventilation ($\dot{V}_A$) is quite small, reflecting the low fraction of wasted ventilation (V_D/V_T). In fact, V_D/V_T decreases during exercise, primarily because of the larger tidal volume [Fig. 16.1(B)]. Anatomical dead space (i.e., volume of conducting airways) then becomes a smaller fraction of tidal volume. It should be evident that if V_D/V_T were larger, or if it did not decrease with exercise, a greater increase in $\dot{V}_E$ would be required to achieve the same increase in $\dot{V}_A$, which is a fact that is of considerable relevance in disease (see later).

The increase in $\dot{V}_A$ during mild and moderate exercise matches the increase in CO_2 production so that P_{ACO_2}, and hence P_{aCO_2}, changes little from resting levels [Fig. 16.1(C)]. With heavy exercise $\dot{V}_A$ increases relatively more than $\dot{V}CO_2$ with, according to Eq. (16.1), a consequent decline in P_{CO_2} [Fig. 16.1(C)].

There is always a certain degree of anaerobic glycolysis in exercising muscles. This results in lactate production. At low levels of exercise lactate is produced at a sufficiently small rate that there is little change in blood lactate level [Fig. 16.1(D)]; However, with heavier exercise, lactate production by exercising muscles exceeds the ability of nonexercising tissues to handle. Thus, blood lactate level increases. A level of 10 mmol/L is not unusual with very heavy exercise. The level of exercise at which blood lactate begins increasing is commonly referred to as the anaerobic threshold (AT). This occurs in normal subjects at about 50% of $\dot{V}O_2$max. Identification of AT is of some clinical usefulness because its occurrence at low levels of exercise has usually been observed in conditions associated with reduced O_2 delivery to the exercising muscles. In practice, AT is usually estimated from ventilatory data (change in the slope of the $\dot{V}_E/VCO_2$ and $\dot{V}_E/VO_2$ relations), obviating the need for numerous serum lactate determinations.

The metabolic acidosis, for which serum lactate is a marker, results in excess H^+ production. This, in turn, is buffered by HCO_3^-. As a result, HCO_3^- level decreases as lactate increases. The change in $[HCO_3^-]$ is nearly equimolar with the change in lactate.

During heavy exercise $\dot{V}_A$ increases to a greater extent relative to $\dot{V}CO_2$, resulting in a decrease in P_{ACO_2}. This, according to the Henderson–Hasselbalch equation, would tend to moderate the decrease in pH that would otherwise result from the metabolic acidosis. Nonetheless, pH does decrease in heavy exercise [Fig. 16.1(E)], although not to the same extent that would take place in the absence of relative hyperventilation.

As a result of the increase in $\dot{V}_A$, alveolar P_{O_2} remains high throughout exercise. In fact, the relative hyperventilation in moderate and heavy exercise results in alveolar P_{O_2} values that are higher than at rest Fig. 16.1(F). P_{AO_2}, however, does not reflect this increase and, in fact, tends to decrease slightly with heavy exercise. In well-trained athletes, who exercise to levels well above those of average normal individuals, marked reductions in P_{AO_2} are sometimes observed despite the high (relative to rest) P_{AO_2}. It is important to recognize that hypoxemia may develop during exercise in normal subjects because failure to do so may lead to a mistaken impression of the presence of disease and unnecessary investigations. In contrast to patients with lung disease, however, in whom hypoxemia may develop at modest levels of exercise, hypoxemia in normal subjects occurs only with very high levels of exercise, usually exceeding predicted $\dot{V}O_2$ max for the subject.

The mechanism of the widening of the alveolar-arterial PO_2 difference in normal subjects (A–a DO_2) is not fully established. In general, the difference between P_{AO_2} and P_{AO_2} can be due to the presence of shunting, ventilation-perfusion ($\dot{V}/\dot{Q}$) mismatching and/or to diffusion limitation. The amount of shunting and $\dot{V}/\dot{Q}$ mismatching that normally exist would have a greater effect on A–a DO_2 during exercise because of the marked reduction in mixed venous O_2 content (due to greater O_2 extraction by the tissues)*; however, evidence indicates that $\dot{V}/\dot{Q}$ mismatching may increase during heavy exercise in normal subjects. This may be related to the development of mild interstitial edema. In addition, as cardiac output increases, red cell transit time through pulmonary capillaries decreases. A minimum exposure time of red cells to alveolar gas is required for red cell P_{O_2} to reach alveolar P_{O_2}. As transit time decreases, the likelihood of red cells exiting pulmonary capillaries without equilibration with alveolar P_{O_2} increases, and the magnitude of the difference between P_{AO_2} and capillary P_{O_2} also increases (diffusion limitation). Whereas it was thought earlier that there is sufficient transit time reserve in normal subjects to preclude diffusion limitation during exercise, evidence suggets that this is a major mechanism of the normal widening of the A–a DO_2 during exercise.

Pattern of Breathing

Ventilation may be increased through a greater tidal volume, a higher respiratory rate or both. Figure 16.2 shows the pattern usually adopted by normal subjects during light, moderate, heavy, and very heavy exercise. The position of the tidal volume excursions relative to lung volume extremes (i.e., total lung capacity and residual volume) is also shown.

As the level of exercise increases, ventilation is augmented primarily by producing larger tidal volumes. Respiratory rate changes little. Once tidal volume reaches 50–60% of vital capacity, it tends not to increase any further. The increase in ventilation with higher levels of exercise is attained by increasing respiratory rate. In fact, tidal volume may decrease near-exhaustive exercise as respiratory rate rises above 40/minute.

The other important feature of the normal response is that the increase in tidal volume is accomplished partly by encroaching on expiratory reserve. In other words, end-expiratory volume decreases, less than passive functional residual capacity (see Fig. 16.2). This decrease occurs with even mild exercise. The subsequent change in end-expiratory volume at higher levels varies considerably among subjects. On average, in young subjects, it tends not to change beyond the value reached at mild exercise (about 300–500 ml below FRC). The ability to force lung volume to decrease, less than passive FRC is to a large extent a function of the maximum flow rate that can be generated during expiration (see later). Older normal persons have decreased maximum expiratory flows (analogous to mild obstructive disease). As a result, their ability to reduce end-expiratory lung volume during exercise is impaired.

The pattern adopted by normal subjects, as illustrated in Fig. 16.2, offers important advantages in terms of respiratory muscle function. First, and as indicated earlier, the increase in tidal volume results in a lesser wasted ventilation (VD/VT), improving ventilatory efficiency. The respiratory muscles will have to do less extra work to achieve the same increase in alveolar ventilation. The decrease in end-expiratory volume offers several additional advantages described below.

---

* According to the shunt equation

$$\dot{Q}S/\dot{Q}T = [C_cO_2 - CaO_2]/[C_cO_2 - C_{\bar{v}}O_2]$$

where $\dot{Q}S/\dot{Q}T$ is the shunt flow as a fraction of total flow, $C_{\bar{c}}O_2$ is O_2 content in pulmonary capillary blood that is related to P_{AO_2}, CaO_2 is O_2 content in arterial blood, and $C_{\bar{v}}O_2$ is O_2 content in mixed venous blood. It can also be appreciated that for the same $\dot{Q}S/\dot{Q}T$ and P_{AO_2} (hence $C_{\bar{c}}O_2$), a reduction in $C_{\bar{v}}O_2$ must result in a lower CaO_2, and hence P_{AO_2}. For similar reasons, a reduction in $C_{\bar{v}}O_2$ would aggravate the effect of a given degree of $\dot{V}/\dot{Q}$ mismatch on arterial P_{O_2}.

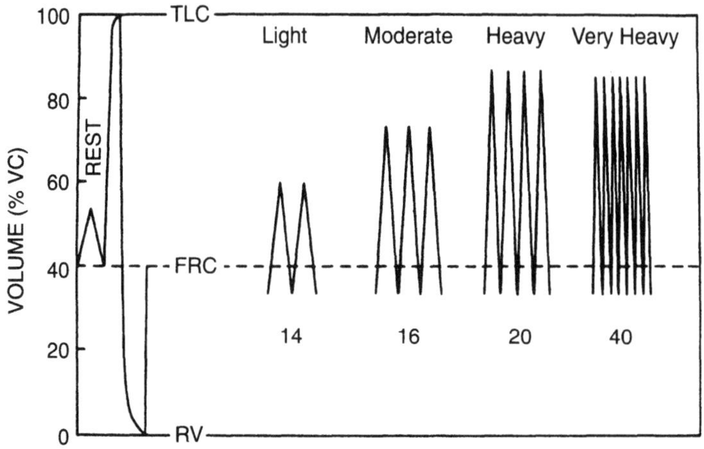

FIGURE 16.2. Breathing pattern at different levels of exercise. Respiratory rate is indicated at bottom of each section. TLC-total lung capacity; FRC-functional residual capacity; RV-residual volume; VC-vital capacity.

It may be recalled that the relation between lung volume and elastic recoil of the respiratory system is sigmoidal in shape and intersects the pressure axis at FRC, which is at about 40% of vital capacity (Fig. 16.3). At FRC the elastic recoil of lung and chest wall cancel out. This is the neutral position of the respiratory system to which the system will always tend to recoil. If volume is reduced less than FRC during expiration, by action of expiratory muscles, the system will tend to recoil back to FRC as soon as expiratory muscles relax at the beginning of inspiration, thereby assisting the inspiratory muscles in expanding the chest during inspiration. In this fashion, expiratory muscles share in doing the work of breathing, reducing the contribution of inspiratory muscles and helping to protect them from fatigue.

The second advantage of reducing end-expiratory volume (EEV) is that it permits tidal volume to increase substantially without encroaching on the flat upper portion

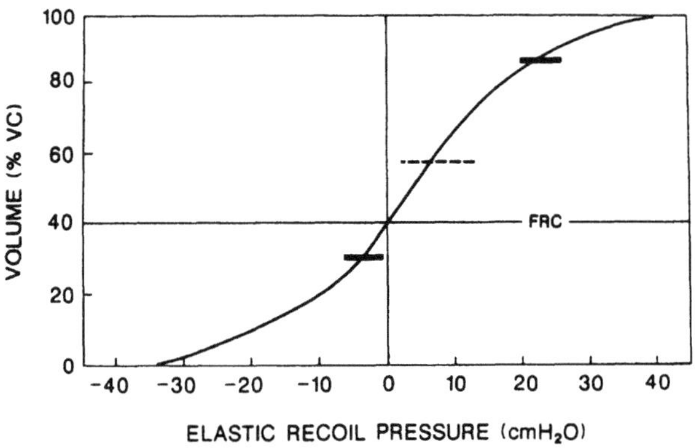

FIGURE 16.3. Elastic recoil pressure of the respiratory system at different volumes. The solid horizontal bars indicate the normal tidal volume extremes during heavy exercise. Note that the compliance of the system (slope of the line) is high in this range and decreases above it and below it. The dashed horizontal line is average lung volume during heavy exercise. FRC-passive functional residual capacity; VC-vital capacity.

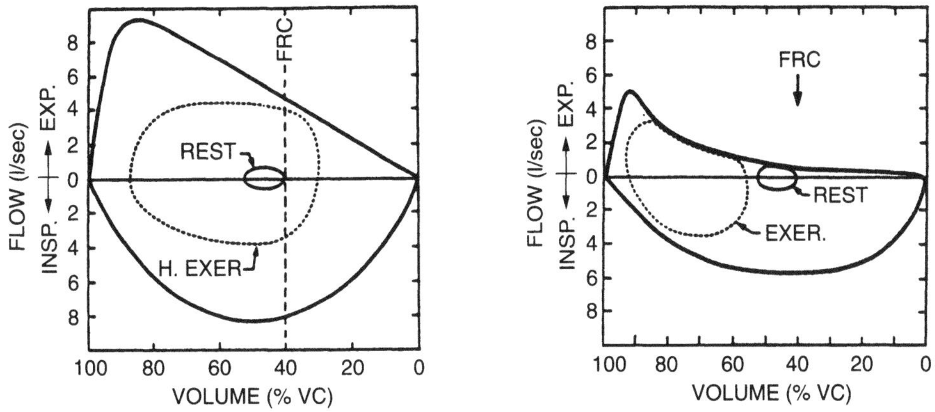

FIGURE 16.4. (Left) Maximal flow–volume curve for a healthy young adult of average size (solid outer loop). Note the tremendous reserve at rest (difference between small loop and maximal curve). Dotted circle is behavior during heavy exercise. FRC-functional residual capacity. (Right) Maximal inspiratory and expiratory flow–volume curve in a patient with COPD. Note the reduction in maximum expiratory flow rates relative to the normal subject. During heavy exercise patient must breathe at high lung volumes to take advantage of the greater expiratory flow rate.

of the elastic recoil curve. In this top portion, the respiratory system is quite stiff (Fig. 16.3) and increasing tidal volume by encroaching on this region, as opposed to reducing EEV, can be accomplished only at a very high cost in terms of inspiratory muscle pressure.

Third, the reduction in EEV helps the inspiratory muscles to operate near their optimal length. It may be recalled that the pressure that can be generated by inspiratory muscles decreases as their length decreases (i.e., as lung volume increases). The average length with the strategy normally adopted (dashed line, Fig. 16.3) is still close to FRC (where optimal length is).

Normal individuals can take advantage of these benefits of lowering EEV only because of the tremendous reserve in their maximum expiratory flow. Figure 16.4 illustrates why this is so. It may be recalled that expiratory flow can be increased with increasing effort only up to a point. Beyond this point ($\dot{V}$max) greater efforts fail to increase flow on account of choking in the airways. It may also be recalled that $\dot{V}$max decreases as a function of lung volume, resulting in the nearly triangular appearance of the familiar maximum expiratory flow–volume curve (Fig. 16.4, left). Nonetheless, $\dot{V}$max is still substantially near FRC in normal subjects. For example, in a healthy 5′ 10″, 20-year-old man, $\dot{V}$max near FRC is still about 5 L/second. In other words, subjects can lower their EEV without constraining their flow rates. As Fig. 16.4 (left) shows, expiratory flow just barely approaches the maximum flow curve only at maximum exercise, and even then it does so only at end-expiration. Patients who do not have this reserve in expiratory flow cannot take advantage of this strategy, and this contributes importantly to their disability (see later).

Mechanism of Ventilatory Adjustments

Figure 16.5 shows the pattern of ventilatory response following a step transition from rest to a constant level of moderate exercise. There is an immediate increase in ventilation (Phase I), followed by a gradual further increase (Phase II), leading to a final

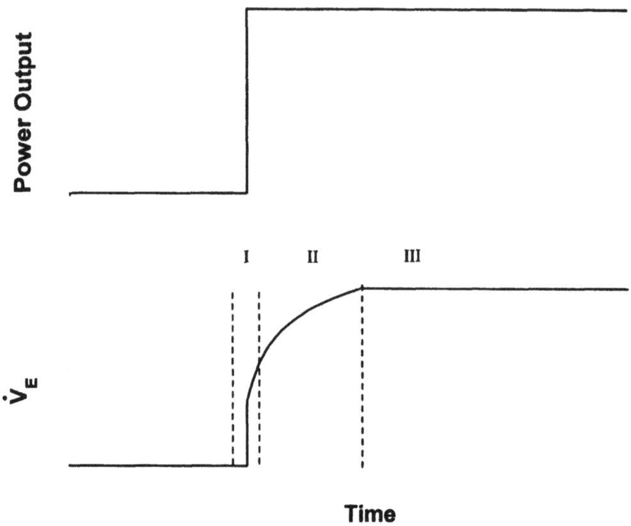

FIGURE 16.5. In response to a step change in power output ventilation increases in three phases, an immediate (Phase I), followed by a gradual increase (Phase II), leading to a more stable level of ventilation (Phase III).

steady level of ventilation (Phase III). Phase I is not as prominent in work to higher work transitions. Furthermore, with high-intensity work, Phase III may not be constant. Rather, ventilation (and heart rate) progressively increase even though power output may be constant.

The mechanism(s) of these ventilatory adjustments is (are) not known with certainty. A variety of mechanisms have been postulated. These include:

1. Neurogenic drive associated with neural signals that cause locomotion. Thus, brain centers that result in locomotion automatically activate the respiratory centers to increase ventilation.
2. Receptors in the exercising muscles are stimulated by metabolic by-products or changes in electrolyte concentrations (notably potassium). Activity from these muscle receptors (so-called metaboreceptors) stimulates the respiratory centers via somatic afferents.
3. Chemoreceptors' activity is altered as a result of small changes in mean P_{CO_2}, changes in amplitude of CO_2 oscillations around the same mean, development of lactic acidosis with a consequent increase in H^+, or because of increased concentrations of circulating humoral factors, notably catecholamines and K^+.
4. Cardiovascular receptors responding to the hemodynamic changes in exercise.
5. Intrapulmonary receptors whose activity is modulated by CO_2 or other humoral factors.
6. Behavioral responses.
7. Increase in body temperature.

There is evidence to support each of these mechanisms; however, there is no consensus as to their relative importance. It is likely that different mechanisms assume greater or lesser importance under different conditions. For example, behavioral responses must clearly be the dominant mechanism for the anticipatory increase in ventilation that frequently precedes the onset of exercise. Lactic acidosis must play a role, the importance of which increases with intensity and duration of exercise. Temperature may be important only with long-term exercise. The role of chemoreceptors may vary depending on the net effect of other mechanisms. Thus, when all other mechanisms combined are inadequate to match ventilation to metabolism, PCO_2 will rise and chemoreceptors would then provide an extra stimulus to ventilation. On the other hand, if other mechanisms overcompensate, with a consequent reduction in PCO_2 and H^+, chemoreceptor acitivity may actually decrease, thereby moderating the hypocapnia and alkalemia that would otherwise result.

Exercise and Breathing in Respiratory Diseases

Chronic Obstructive Pulmonary Disease

The principal and invariable abnormality in chronic obstructive pulmonary disease (COPD) is reduction in maximal expiratory flow; patients cannot exhale at the same rate as do normal people (Fig. 16.4, right). Other abnormalities that are not invariable but often present include:

- Loss of lung elastic recoil with secondary hyperinflation
- Increased resistance to airflow during inspiration
- Inspiratory muscles unable to generate as much pressure as normal, partly because they are operating at a less advantageous mechanical position (e.g., short, flat diaphragm due to hyperinflation) and, occasionally, because of malnutrition
- Abnormalities in ventilation-perfusion relationships resulting in widened alveolar–arterial O_2 gradient and hypoxemia
- Loss of gas-exchanging surface area because of destruction of lung parenchyma (emphysema)
- Pulmonary hypertension secondary to reduction in vascular bed (emphysema) and/or hypoxic pulmonary vasoconstriction
- An increase in ventilation during exercise. On average, the ventilation at a given level of exercise is higher in COPD patients than it is in normals. The cause of this is not known. This may contribute importantly to exercise limitation (see later).

The maximal ability of patients with COPD to ventilate is decreased principally because of the expiratory flow limititation, but also because of the increased inspiratory resistance and weaker muscles. This maximal possible ventilation, called *maximum voluntary ventilation* (MVV), can be approximated to FEV_1 (i.e., MVV = 35 FEV_1).* Because FEV_1 decreases in COPD, the ability of COPD patients to meet the ventilatory demands of exercise is impaired.

Figure 16.6 shows the relation between exercise ventilation and MVV in a typical normal subject and in a typical patient with severe COPD. The two subjects have the same age, sex, and height, so that the predicted (i.e., normal) maximal exercise capac-

* In reality, there is no single value that reflects maximum possible ventilation. As with other muscles, respiratory muscles will fatigue if forced to work hard. The point at which fatigue (respiratory endurance time) occurs is a function of the intensity of the work done by the muscles and the duration of this activity. Thus, the maximal level of ventilation that can be sustained for several minutes is much less than the maximal level that can be sustained for a few seconds. Other factors also affect the relation between ventilation and respiratory endurance time (e.g., respiratory muscle strength, breathing pattern, lung volume, PAO_2 and $PACO_2$, and so on). The relation commonly used to estimate MVV (i.e., MVV = 35 FEV_1) is empirical and should be viewed only as approximate.

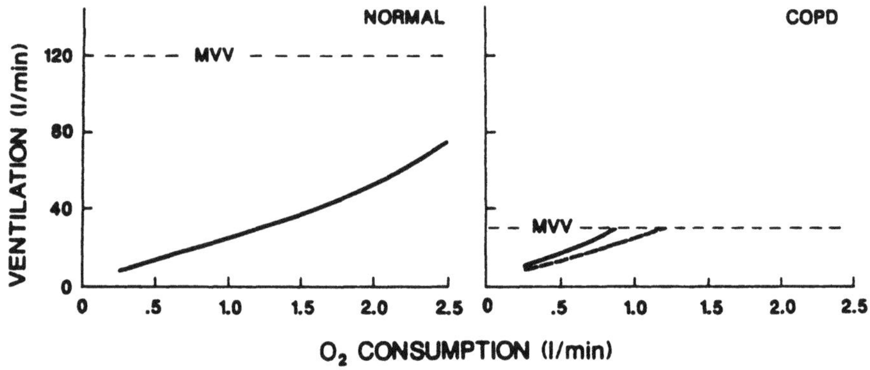

FIGURE 16.6. Relationship between exercise level, expressed as O_2 consumption, and ventilation in a normal subject and in a COPD patient with the same age, sex, and height. Maximal voluntary ventilation (MVV) is much reduced in the COPD patient. Note that ventilation at maximal exercise is well below the subject's ventilatory ceiling in the normal subject. In this patient, the ceiling is reached at low level of exercise (solid curve) mostly because MVV is low, but also because ventilation is somewhat higher than normal at any level of exercise (difference between solid and dashed curves).

ity ($\dot{V}O_2max$) and predicted ventilation are similar. The actual MVV, which is often referred to as the *ventilatory ceiling*, is shown by the horizontal dashed line. As may be expected, it is much reduced in the COPD patient. At the highest level of exercise, ventilation of the normal subject is still considerably less than the maximal ability to ventilate, which reflects considerable ventilatory reserve. In the COPD patient ventilation at maximum exercise reaches or exceeds the estimated ventilatory ceiling. This observation has led to the notion that exercise is ventilation-limited in patients with COPD.

Apart from lowering maximal possible ventilation, the reduction in maximal expiratory flow makes it very difficult for the patient to lower end-expiratory volume during exercise. Because of the very low expiratory flow rates near and below FRC in COPD (see Fig. 16.4, right), it would take considerable time to reduce EEV. For example, maximal flows of 0.4 L/minute are not unusual near FRC in COPD. If the patient decides to continue exhaling until lung volume decreases 300–500 ml below FRC, the patient will then have to spend a considerable time during each breath breathing at a low flow rate. This would reduce maximal ventilation. The only way for these patients to increase ventilation substantially is to raise the EEV in order to take advantage of the relatively high flow rates at higher lung volumes (Fig. 16.4, right). In fact, most patients with COPD do exactly that. End-expiratory volume progressively increases relative to resting FRC, as level of exercise increases. At the highest level of exercise, EEV may be 0.8–1.0 L above its resting level. This phenomenon has been termed *dynamic hyperinflation* to distinguish it from the static hyperinflation that may be present all the time because of reduced elastic recoil. Dynamic hyperinflation is a transient increase in lung volume, higher than passive FRC, because of inadequate time for emptying. The consequences of this are serious in terms of inspiratory muscle function. First, because expiratory muscles fail to reduce EEV below passive FRC, they cannot assist in doing the inspiratory work as they normally do (see earlier). The inspiratory muscles must, therefore, carry the full burden of expanding the respiratory system.

Second, because of the dynamic hyperinflation, the elastic recoil at end-expiration is positive (EEV above passive FRC). The inspiratory muscles must first generate enough pressure to neutralize this recoil (which is essentially an expiratory force) before any air can begin to move in. The situation is somewhat analogous to trying to reverse the direction of a moving object. This is clearly harder than getting it to move beginning from a stationary position (i.e., starting at passive FRC).

Third, in order to obtain a reasonable tidal volume starting from a high EEV the inspiratory muscles must expand the system to almost its full capacity (total lung capacity). Because of the stiffness in this range (see Fig. 16.3) the work involved is disproportionately higher.

Finally, the inspiratory muscles are forced to operate at an even more disadvantageous position because of the dynamic hyperinflation.

It is of interest to note that even though the primary mechanical problem affects expiration, the muscles that suffer most are the inspiratory muscles. These muscles both have to work harder, and their ability to generate pressure is already reduced (see Factor 3, earlier). To generate a given increase in ventilation, the inspiratory muscles must therefore use a much greater fraction of their own maximum power output.

The situation of the inspiratory muscles is further aggravated by the effect of the disease on the ventilatory requirement of exercise. Because inspiratory muscles, as with other muscles, must use more O_2 and generate more CO_2 if they work harder, the metabolic cost of a given level of external work (e.g., climbing two flights of stairs) is higher than it is in a normal subject (to account for the extra cost of breathing). Furthermore, as indicated earlier, COPD patients often have a higher ventilation than normal at any given level of exercise (Fig. 16.6, right). Given the reduced ventilatory ceiling, a higher ventilatory response to exercise translates into the patient reaching maximum ventilation sooner (Fig. 16.6, right). It has been shown that variability in this factor (i.e., $\dot{V}_E$ at a given exercise) is one of the most important determinants of exercise tolerance in COPD and is, to a considerable extent, responsible for differences in exercise tolerance between patients with the same mechanical abnormality (i.e., same reduction in FEV_1).

The adverse effect of COPD on inspiratory muscle function can be summarized as:

1. Increased ventilatory demand at a given external power output
   a. Greater metabolic requirements at a given power output due to increased metabolic cost of breathing
   b. The ventilatory response to exercise is usually increased
2. More work required from inspiratory muscles to attain a given increase in ventilation
   a. Inspiratory pressure is required to counteract the elastic recoil at end-expiration as a result of dynamic hyperinflation
   b. Loss of contribution of expiratory muscles to lung expansion
   c. Increased elastic work as the inspiratory muscles are forced to expand the system almost to TLC
   d. Increased inspiratory resistance
3. Decreased ability of inspiratory muscles to generate pressure
   a. Already hyperinflated state at rest
   b. Further hyperinflation during exercise
   c. Malnutrition

It is evident that inspiratory muscles are in a precarious situation in COPD. Despite these difficulties, the inspiratory muscles manage to defend alveolar ventilation during exercise, as evidenced by an unchanged P_{ACO_2}, in all but the most severe cases of

COPD. Even there, only some patients sustain an increase in P_{ACO_2} with exercise. CO_2 retention does not limit exercise in COPD. The two most common complaints cited by patients when asked why they stopped exercise are dyspnea and fatigue, particularly of the exercising muscles.

The mechanism of dyspnea during exercise in COPD has not been established. The most likely mechanisms are:

- There is a greater effort by inspiratory muscles to attain the ventilatory requirements of exercise (see earlier).
- There is a greater increase in pulmonary artery pressure during exercise due to increased pulmonary vascular resistance (see Factor 6). Pulmonary hypertension is usually associated with considerable dyspnea on exertion (see later).
- There is dynamic compression of the airways. Deformation of the airways results in an unpleasant sensation. If pleural pressure is increased during expiration beyond the level required to achieve maximum flow ($\dot{V}$max), the airways choke and the airways downstream from the choke point are compressed and deformed.
- Left-ventricular dysfunction may exist during exercise in some patients that results in pulmonary congestion. Most patients with COPD are in the older age group and may have coexistent coronary artery disease. Myocardial ischemia may be aggravated during exercise due to greater myocardial work, particularly if arterial hypoxemia exists due to COPD. The hypertrophied right ventricle (cor pulmonale secondary to COPD) may also interfere with left ventricular filling during diastole.

Although counterintuitive, many patients with severe COPD stop exercise because of leg fatigue while assigning dyspnea a low rating as bona fide the cause of terminating exercise. Possible reasons for this include:

- Impaired cardiac output response to exercise due to cardiac or pulmonary vascular disease. O_2 delivery to the exercising muscles, therefore, is suboptimal.
- Hypoxemia commonly exists in COPD and is, sometimes, aggravated by exercise. This would also cause a lower O_2 delivery to the exercising muscles.
- Deconditioning often occurs in these patients as a result of chronic inactivity.
- A degenerative muscle disorder is postulated to occur in some patients.

Patients with COPD vary greatly in their ability to tolerate exertion. Although exercise tolerance is directly correlated with FEV_1 there is considerable scatter in this relation, which indicates that factors other than expiratory flow limitation contribute to dyspnea. One patient may have an almost normal exercise tolerance for a given reduction in FEV_1, whereas another may be quite dyspneic on mild exertion. Different contributions from the preceding mechanisms may account for this variability. Psychological factors may also be important; individuals vary greatly in their tolerance of unpleasant sensation. It has been shown that variability in ventilatory response during exercise ($\dot{V}_E$ at a given level of exercise) is an important determinant of the interindividual differences in exercise tolerance among COPD patients. The reason COPD patients with the same mechanical severity display very differnt ventilation during exercise is not clear.

Patients with COPD are frequently hypoxemic at rest. This is mostly due to $\dot{V}/\dot{Q}$ mismatching, although hypoventilation (as evident by hypercapnia) contributes in some. Except in the most severe cases, P_{O_2} does not change during exercise in a way that would significantly affect O_2 saturation. Even in severe cases P_{AO_2} decreases with exercise only in some patients, whereas it stays the same or even improves in the others. When P_{AO_2} decreases during exercise this is often the result of relative hypoventilation evident as an increase in P_{ACO_2}. At times, P_{AO_2} decreases despite an adequate alveolar

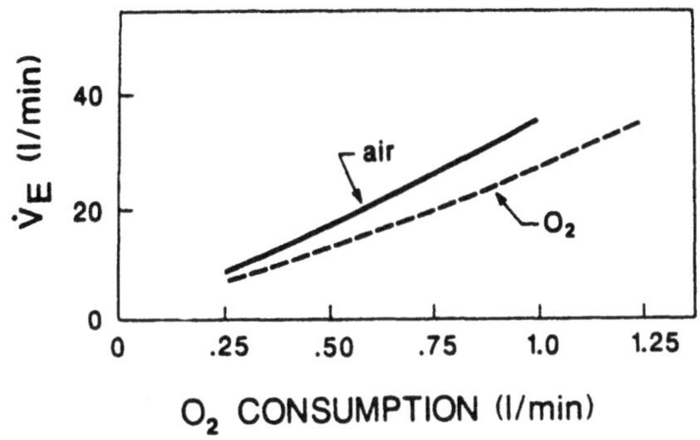

FIGURE 16.7. Effect of O_2 breathing on ventilatory response to exercise in COPD.

ventilation. In these cases greater $\dot{V}/\dot{Q}$ mismatching during exercise or excessive reduction in mixed venous O_2 content (see footnote, p. 5) accounts for the increased hypoxemia. Where hypoxemia exists during exercise, it may aggravate the respiratory distress due to activation of chemoreceptors and a consequent increase in respiratory drive (usually referred to as the *hypoxic drive*). When patients with COPD and hypoxemia are exercised while breathing O_2, the dyspnea level at a given power output is less and they can exercise longer and to higher levels. By removing the hypoxic drive, during O_2 breathing, patients can tolerate a higher P_{ACO_2}, and therefore ventilate less at a given power output. The decreased dyspnea can be attributed to the reduced work of inspiratory muscles. Increased endurance of respiratory muscles due to improved O_2 delivery, or a decrease in pulmonary artery pressure due to higher alveolar P_{O_2}, may also contribute. It is of interest that although ventilation is lower at a given power output, ventilation at maximum exercise is the same with or without O_2 (Fig. 16.7).

Restrictive Lung Disorders

By definition, restrictive lung disorders decrease lung compliance (i.e., for a given applied pressure, there is less lung expansion). This can occur following lung resection (less units to expand), but it is more commonly observed in diffuse lung disorders that result in an increase in lung water (e.g., pulmonary edema), lung cellularity (e.g., lung inflammation, diffuse malignancy) or fibrous tissue (pulmonary fibrosis). Several hundred specific disease entities can ultimately result in diffuse pulmonary fibrosis and lung stiffness. Because of the increase in lung stiffness, it is difficult to expand the chest. Total lung capacity, inspiratory capacity, and vital capacity are, therefore, reduced; however, because flow rates are normal, or even supernormal, the maximum ability to ventilate (MVV) is not as seriously affected as in obstructive disorders; the patient can compensate for the smaller tidal volume by increasing respiratory rate.

In addition to the increased stiffness, and as with most diffuse lung diseases, involvement of air spaces and blood vessels is not uniform or well matched, resulting in $\dot{V}/\dot{Q}$ mismatching. In addition, in some of these disorders the diffusion distance between air and blood is increased due to thickening of the alveolar–capillary membrane. The time required for blood to equilibrate with alveolar P_{O_2} is therefore lengthened.

An additional feature of relevance to exercise responses is the development of pulmonary hypertension in many of these disorders, particularly when severe. The heart is also occasionally involved in the same process that affects the lung (e.g., autoimmune diseases, sarcoidosis).

The ventilatory response to exercise in these patients differs from normal in several important respects. First, because of the excessive work of generating larger tidal volumes, tidal volume does not increase as much. Rather, ventilation increases primarily as a result of higher respiratory frequency. Respiratory rates in excess of 60/minute are not uncommon and rates in excess of 100/minute are occasionally observed. It is interesting, however, that the highest tidal volume observed in exercise is, as in normals, about 50–60% of the patient's own vital capacity (which is reduced).

Second, progressive hypoxemia is common during exercise, particularly in disorders that increase the thickness of the alveolar–capillary membrane. This occurs despite the normal or above normal alveolar Po_2 (see later). Although the widening of the A–a O_2 difference is due in part to the reduction in mixed venous O_2 content in the presence of $\dot{V}/\dot{Q}$ mismatching (see footnote on p. 5), there is convincing evidence that diffusion limitation contributes to this hypoxemia during exercise. As a result of the increase in blood flow during exercise, the transit time of red cells along the capillary is reduced, and less and less time is available for blood and alveolar Po_2 to equilibrate. When the diffusion distance is increased and equilibration is delayed, as in these disorders, the decrease in transit time would result in earlier appearance of diffusion of limitation. Unlike the case of normal subjects who may develop desaturation during extremely heavy exercise (see earlier), in these patients desaturation begins at relatively low levels and progresses rapidly as exercise increases in intensity.

Third, ventilation is usually higher at a given level of exercise, than in normal subjects. Particularly with diffuse interstitial processes (as opposed to lung resection or localized fibrosis), patients strive to maintain a lower P_{ACO_2} during exercise indicating an excessive "drive" or stimulus to breathe. The extra stimulus may be the result of exercise induced hypoxemia stimulating the peripheral chemoreceptors or to inappropriate metabolic acidosis secondary to limited increase in cardiac output (cardiac involvement or due to high pulmonary vascular resistance). In many cases, however, an extra chemical stimulus (low Po_2 or abnormally high H^+) cannot account for the alveolar hyperventilation and it is presumed that the interstitial process activates intrapulmonary receptors that provide an added stimulus to breathe.

The extra drive to breathe entails a greater alveolar ventilation (to lower the P_{ACO_2}), and the patient must also ventilate more to attain a given increase in $\dot{V}_A$. Because of the $\dot{V}/\dot{Q}$ mismatching, V_D/V_T is higher than it is normal. Failure of tidal volume to increase substantially during exercise further compounds this problem in that the normal decrease in V_D/V_T during exercise is not fully realized.

Dyspnea on exertion is very common in these disorders and is the usual reason for exercise limitation. Factors that contribute to this sensation include:

- Increased work of inspiratory muscles because of greater than normal ventilation and greater work per unit ventilation (due to the higher stiffness)
- Increased respiratory drive due to any of the factors discussed earlier
- Greater increase in pulmonary vascular pressure in the presence of pulmonary vascular involvement or cor pulmonale

Pulmonary Vascular Disorders

In pulmonary vascular disorders the resistance to blood flow in the pulmonary circulation is increased. This most commonly occurs in association with other types of lung disease (i.e., obstructive or restrictive) due to involvement of the pulmonary vessels in the disease process and/or as a result of vasoconstriction induced by hypoxemia (*cor pulmonale*). The underlying pathology is less frequently limited to the pulmonary vessels. Recurrent pulmonary embolization is the most common cause. Autoimmune diseases, most notably scleroderma, are sometimes associated with pulmonary hypertension due to spasm and hypertrophy of the smooth muscle in the pulmonary arteries. A similar condition occasionally occurs in the absence of a systemic autoimmune process (primary pulmonary hypertension). The pulmonary veins are very rarely constricted as a result of fibrotic reactions in the mediastinum. Finally, some pulmonary vessels may be congenitally absent or markedly narrowed.

Patients with isolated pulmonary vascular disease (i.e., not associated with other lung disease) usually have normal lung volumes and flow rates. Their ability to ventilate is therefore not impaired, yet, they are frequently limited, often to an extreme extent, by dyspnea on exertion. They also tend to develop progressive hypoxemia during exercise. Three basic abnormalities account for these problems.

First, because the vascular involvement is not uniform, the distribution of perfusion is also not uniform. The $\dot{V}/\dot{Q}$ mismatching results in hypoxemia, which can be aggravated during exercise. Some units receive very little or no perfusion. These act as "dead space." V_D/V_T is often very high in these patients, necessitating a greater increase in ventilation in order to maintain P_{ACO_2} (i.e., alveolar ventilation) at a given level.

Second, the constriction of pulmonary vessels makes it difficult to increase cardiac output appropriately during exercise. As a result, the arteriovenous O_2 difference must be wider at a given level of exercise because the same amount of O_2 must be extracted from a smaller amount of blood. This widened A–V O_2 difference, along with an already reduced arterial O_2 content, results in much lower mixed venous O_2 content at a given level of exercise. This aggravates the effect of $\dot{V}/\dot{Q}$ mismatching on arterial P_{O_2} (see footnote on p. 5), resulting in greater arterial hypoxemia during exercise. As with other conditions in which cardiac output response is limited, there is usually an abnormal increase in lactic acid level during exercise, resulting in inappropriate metabolic acidosis. The acidosis and hypoxemia provide added stimuli to breathing. The patient, therefore, strives to attain a greater alveolar ventilation which, as indicated earlier, is quite costly due to the higher V_D/V_T.

Third, in order to increase cardiac output despite the constricted vessels, the right ventricle must generate greater pressures. Right-ventricular and pulmonary-artery pressures often rise to dramatic levels (sometimes exceeding systemic values) during exercise. There is some evidence that increased pressure in these compartments results in hyperventilation and dyspnea. In patients with a patent foramen ovale, the increase is right-sided cardiac pressure during exercise may cause mixed venous blood to flow from right to left atrium, thereby aggravating the hypoxia.

Exercise and Breathing in Nonrespiratory Diseases

Several nonrespiratory disorders place a greater demand on the respiratory system during exercise. When the respiratory system is healthy, these extra demands may easily be accommodated. In the presence of respiratory disease, however, these added ventilatory requirements result in greater dyspnea and exercise limitation than would otherwise occur. These disorders can be grouped under the following headings.

Disorders That Impair O_2 Delivery to the Tissues

The amount of O_2 available to exercising muscles is a function of blood flow and amount of O_2 per unit of blood. Blood flow to exercising muscles may be abnormal because of local abnormalities in the blood vessels supplying the muscles (i.e., peripheral vascular disease) or as a result of inadequate increase in total blood flow (i.e., cardiac output). Valvular, myocardial, and pericardial disease can all result in an

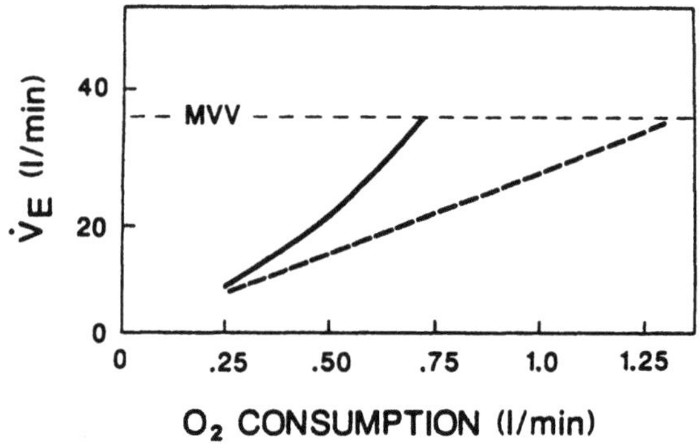

FIGURE 16.8. Relationship between O_2 consumption and ventilation ($\dot{V}_E$) with (solid line) and without an additional ventilatory stimulus (e.g., acidosis or hypoxemia) in a patient with limited reserve. Ventilatory ceiling (MVV) is reached at a lower level of exercise, where additional (to exercise) ventilatory stimuli exist.

inadequate cardiac output response to exercise. O_2 content per unit of blood may be low in the presence of anemia, when significant amounts of the hemoglobin are nonfunctional (e.g., carboxyhemoglobin, methemoglobin, and sulfhemoglobin) or in the presence of hemoglobins with abnormal O_2 affinity.

As discussed in relation to pulmonary vascular diseases, which also impair O_2 delivery during exercise, a lower O_2 delivery at a given level of exercise results in a lower mixed venous O_2 content. In the presence of lung disease (hence $\dot{V}/\dot{Q}$ mismatching) this can aggravate the hypoxemia. Furthermore, although a cause-and-effect relationship is not well established, disorders that impair O_2 delivery are often associated with inappropriate metabolic acidosis during exercise. This, along with greater arterial hypoxemia (if present), would increase "respiratory drive," resulting in greater ventilation and more dyspnea. This is particularly important in the patient with coexistent lung disease. For example, Fig. 16.8 shows the effect of hyperventilation in a patient whose MVV is reduced to 40 L/minute (instead of a predicted 100 L/minute) because of chronic lung disease. If the patient's ventilatory response to exercise is normal (dashed line) hat patient may be able to reach a VO_2 of 1.3 L/minute (intersection of dashed line and ventilatory ceiling). If the patient is forced to hyperventilate because of an extra ventilatory stimulus (e.g., acidosis or hypoxemia), hat patient will reach a ventilatory ceiling much sooner (intersection of solid line and horizontal line). If the cause of hyperventilation (which is often nonrespiratory) is corrected, the patient would be less dyspneic, and would be able to exercise more, despite having no change in lung disease. This interaction is important to remember because the causes of hyperventilation are commonly more amenable to correction (e.g., treating anemia or replacing a valve) than chronic lung disease.

In patients with valvular or myocardial disease of the left cardiac chambers there is the added complication of pulmonary congestion during exercise. This decreases pulmonary compliance, increases pulmonary vascular pressures, and may result in pulmonary edema. All these changes may result in more hyperventilation and dyspnea.

Disorders Associated with a Reduced Plasma Bicarbonate

(HCO_3^-) is reduced in metabolic acidosis (e.g., chronic renal failure). In an effort to normalize pH, ventilation is usually increased and P_{ACO_2} is lower than normal. (HCO_3^-)

can also be reduced as a consequence of chronic alveolar hyperventilation (compensated respiratory alkalosis). Although this is most often associated with respiratory disease, some patients have a low P_{ACO_2} and (HCO_3^-) for no apparent respiratory reason (? central origin). In order to maintain a reasonable pH in the presence of a reduced HCO_3^-, P_{ACO_2} must be maintained at a lower-than-normal level throughout exercise. According to Eq. 1, this entails a greater proportionate increase in $\dot{V}_A$. Thus, if P_{ACO_2} is to be maintained at 30 mmHg, instead of 40 mmHg, $\dot{V}_A$ must be 33% higher at all levels of exercise. For reasons similar to those discussed in the above category and in Fig. 17.8, this would result in greater dyspnea and exercise limitation in the patient with already abnormal pulmonary function.

Disorders Resulting in Greater Metabolic Cost for a Given Physical Task

In obesity, the patient must move a greater mass (i.e., do more work) to accomplish the same task (e.g., climbing one flight of stairs or walking a mile). A given task therefore requires more O_2 and generates more CO_2. The patient must accordingly ventilate more. For a given level of lung function, the obese person will be more dyspneic than a nonobese person while carrying out daily activities.

In articular and neuromuscular disorders, movements are awkward and muscle use is not as efficient (i.e., costs more energy for a given task) as in normal persons. For similar reasons, the ventilatory cost of carrying out daily activities will be higher even though the relation between metabolic rate (i.e., $\dot{V}O_2$ or $\dot{V}CO_2$) and ventilation is normal.

Recommended Readings

Bauerle, O., Chrusch, C.A., and Younes, M. (1998) Mechanisms by which COPD affects exercise tolerance. Am. J. Resp. Crit. Care Med. 157, 57–68.

Clinical exercise testing with reference to lung diseases; indications, standardization and interpretation strategies. ERS Task Force on Standardization of Clinical Exercise Testing. European Respiratory Society. Eur. Resp. J. (1997) Nov;10(11), 2662–2689.

Eldridge, F.L., and Waldrop, T.G. (1991) Neural control of breathing during exercise. In: Whipp, B.J., and Wasserman, K. (eds.) Exercise: pulmonary physiology and pathophysiology. Lung biology in health and disease, vol. 52, chap. 11, pp. 309–370. Dekker, New York.

Marciniuk, D.D., and Gallagher, C.G. (1996) Clinical exercise testing in chronic airflow limitation. Med. Clin. N. Am. 80(3), 565–587.

Wagner, P.D., and Gale, G.E. (1991) Ventilation-perfusion relationships. In: Whipp, B.J., and Wasserman, K. (eds.) Exercise: pulmonary physiology and pathophysiology. Lung biology in health and disease, vol. 52, chap. 4, pp. 121–142. Dekker, New York.

Whipp, B.J., and Ward, S.A. (1998) Determinants and control of breathing during muscular exercise. Br. J. Sports Med. 32(3), 199–211.

Younes, M. (1984) Interpretation of clinical exercise testing in respiratory disease. Clin. Chest Med. 5, 189–206.

17 High-Altitude Physiology and Pathophysiology

JOHN T. REEVES

Altitude exposure provides a natural laboratory for the study of the effects of low oxygen (hypoxia). Oxygen is the most necessary of substances in our environment, and without it we become unconscious within seconds. When O_2 is available, but in low concentrations, the body has elaborate defense systems and compensatory mechanisms that are progressively called into play as the duration of hypoxia increases. When these mechanisms fail or when they are stretched beyond their capacity, organ malfunction occurs. This chapter will focus primarily on the mechanisms of response to hypoxia and to a lesser extent on the disorders that result when the response is inadequate. For more information the reader is referred to other reviews (Penulosa and Sime, 1971; Reeves, 1973; Reeves and Grover, 1974; Dempsey and Forster, 1982; Winslow and Monge, 1987; Kawashima, 1989; Hackett and Roach, 1990; Weil, 1990; Reeves and Groves et al., 1991; Moore, 1992; Oelz et al., 1992; Honigman, 1993; Ward, 1995).

Barometric Pressure

Altitude

As one leaves sea level to go to high altitudes, barometric pressure (P_B) in the atmosphere decreases [Fig. 17.1(A)]. In this chapter, P_B versus altitude is shown as pressure actually measured at terrestrial elevations in the temperate zones. [P_B is often reported as the slightly lower standard curve established by the International Civil Aviation Organization (ICAO), which is a global average.] Air is compressible and the barometric pressure falls in an exponential fashion from approximately 760 mmHg at sea level to 253 mmHg as measured on the summit of Mt. Everest. Because of currents high in the atmosphere, the ambient barometric pressure falls from winter to summer, the magnitude of which is, for example, 8 mmHg in Colorado.

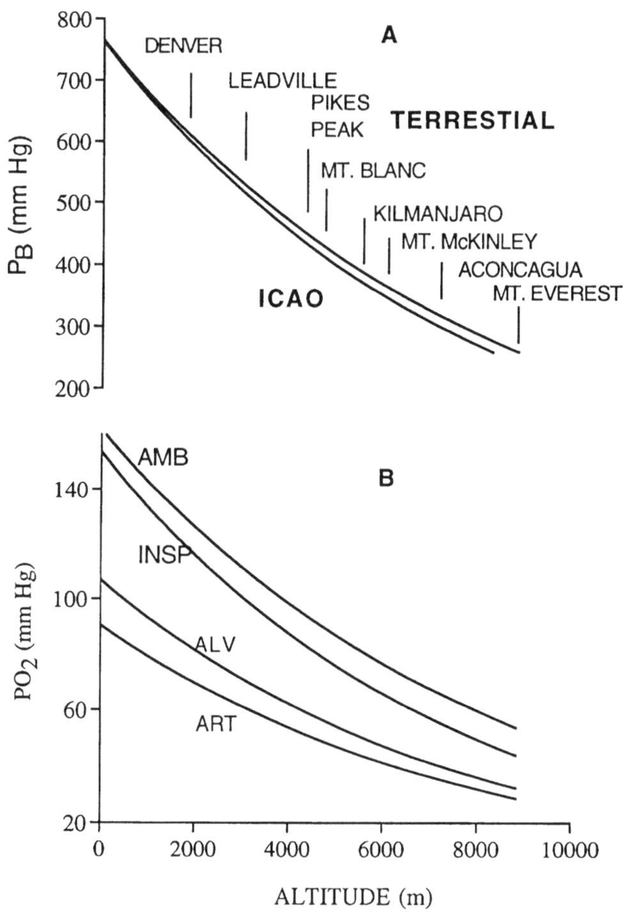

FIGURE 17.1. (A) Atmospheric pressure as related to altitudes measured at various terrestrial elevations (upper line) and as reported as a global average by the International Civil Aviation Organization (lower line). (B) Partial pressures of oxygen for increasing altitude, as measured in the ambient air (AMB), in the airways after the inspired air has been warmed and humidified (INSP), in the alveolus after CO_2 has been added (ALV), and in the arterial blood (ART).

Oxygen Pressure: Ambient Air to Alveolus

Because the concentration of oxygen in the air is constant at 20.93% throughout the atmosphere, atmospheric partial pressure of oxygen (Po_2, in millimeters of mercury) falls as the barometric pressure falls [Eq. (17.1), Fig. 17.1(B)]. The partial pressure of water (PH_2O) is 47 mmHg at the body temperature of 37°C and is independent of altitude. At all altitudes, when the inspired air reaches the bifurcation of the trachea, it is humidified at body temperature, resulting in an inspired oxygen pressure (P_Io_2) [Eq. (17.2), Fig. 17.1(B)] that is less than that in the ambient air. During air breathing, the body is not able to alter the P_Io_2, which therefore is the "iron clad" physical limitation of oxygen availability. Further, when the inspired air reaches the alveolus, it is diluted further by carbon dioxide; therefore, alveolar Po_2 is less than that inspired [Fig. 17.1(B)]. The alveolus contains essentially four gases: N_2, H_2O, O_2, and CO_2. The partial pressures of these must total the barometric pressure [Eq. (17.3)]. Because the P_Io_2 is approximately equal to the P_B minus the sum of alveolar PN_2 plus PH_2O, the alveolar oxygen pressure (Pao_2) is approximately equal to the P_Io_2–the alveolar carbon dioxide pressure ($Paco_2$), Eq. (17.3).

$$Po_2 = P_B \times 0.2093 \tag{17.1}$$

$$P_Io_2 = (P_B - 47) \times 0.2093 \tag{17.2}$$

$$Pao_2 = P_B - (PaN_2 + Paco_2 + PaH_2O); \approx P_Io_2 - Paco_2 \tag{17.3}$$

Oxygen Pressure: Alveolus to Arterial Blood

For every altitude or barometric pressure, the arterial oxygen tension (Pao_2) must be less than that in the alveolus (Pao_2) by the amount of the alveolar–arterial Po_2 gradient (A–a Po_2) [Fig. 17.1(B)]. At sea level, where the arterial Po_2 is on the flat part of the Hb-O_2 dissociation curve, any venous admixture causes a relatively large drop in Pao_2, and most of the A–a Po_2 gradient is made up of venous admixture (normally from the bronchial veins). At altitudes where the Pao_2 is on the steep part of the Hb-O_2 dissociation curve, however, venous admixture has a relatively small effect on arterial Po_2, and the A–a Po_2 gradient results primarily from limitation of pulmonary diffusion; therefore, as the oxygen molecule in ambient air is inhaled [Fig. 17.1(B)] the Po_2 falls as the inspired air is diluted with H_2O, the alveolar air is diluted with CO_2, and the molecule moves across the alveolar–capillary diffusion barrier from alveolus to capillary blood. To the extent that there is venous admixture, the arterial Po_2 is less than that in the capillary. The relation of arterial Po_2 to altitude is summarized in Fig. 17.1(B) for the well-acclimatized person.

Tissue Responses to Altitude Hypoxia

Although, oxygen may be sensed in the adrenal gland (Bülbring et al., 1934) and the central nervous system (Reis, personal communication), chemoreceptors for oxygen are currently known to exist in the carotid body, the pulmonary arterioles, and the juxtaglomerular apparatus of the kidney (Fig. 17.2). These chemoreceptors apparently continue to respond to low oxygen for many years, because altitude residents usually maintain increases in, respectively, pulmonary ventilation, pulmonary arterial pressure, and red cell volume.

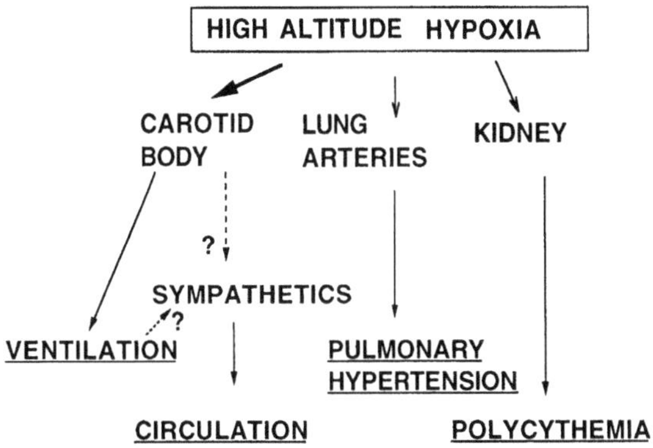

FIGURE 17.2. Schema reflecting stimulation of tissues considered to contain oxygen chemoreceptors and the responses of these tissues.

Carotid Body

Of all oxygen sensors, the best described is the carotid body, which is located on the internal carotid artery at its origin. Evidence has pointed toward the previously unknown mechanism by which the carotid body senses oxygen (Lopez-Barneo et al., 1988). Hypoxia appears to reversibly inhibit an outward K^+ current in the carotid body Type-1 cell, resulting in membrane depolarization the magnitude of which directly relates to the Po_2 in the environment. It seems likely, then, that the mechanism of oxygen sensing involves the closure of a voltage-sensitive K^+ channel that makes the interior of the cell less negative and therefore prone to depolarization. A reduction in Po_2 induces carotid body firing, which initiates an increase in ventilation within seconds. With continued stay at altitude, ventilation continues to increase over several days to reach a new and stable plateau, which is a process called *ventilatory acclimatization*. As discussed later, such acclimatization seems to be crucial for well being. Ventilation increases in a hyperbolic fashion as Po_2 falls, approaching a vertical asymptote (large increments in ventilation) as Po_2 falls toward approximately 30 mmHg, and a horizontal asymptote (little change in ventilation) for Po_2 values increasing to more than 100 mmHg. Although there is no "threshold" for the increase in ventilation, examination of Fig. 17.3(A) shows that the increase in ventilation becomes most marked when the arterial Po_2 falls to values of less than 70–80 mmHg (i.e., corresponding to altitudes greater than 1500–2000 m).

Pulmonary Arterial Smooth Muscle

Hypoxia probably acts directly on the pulmonary arteriolar smooth muscle to induce vasoconstriction and to raise resistance to flow in the hypoxic vessels. Evidence suggests that the mechanism involves hypoxia-induced closure of voltage-sensitive K^+ channels (Post et al., 1992) in a response that may be analogous to that of the carotid body. Hypoxic vasoconstriction is selective because lung arteries less than about 300 μ in diameter constrict, whereas larger arteries, capillaries, and veins constrict much less or not at all, and systemic arterioles dilate with hypoxia. The stimulus–response curve for pulmonary arterial pressure in residents of high altitude shows a hyperbolic function which is similar in shape to that described for ventilation [Fig. 17.3(A)]. Although hypoxic vasoconstriction begins within seconds, it persists, with residents of high altitude usually maintaining elevated pulmonary arterial pressures.

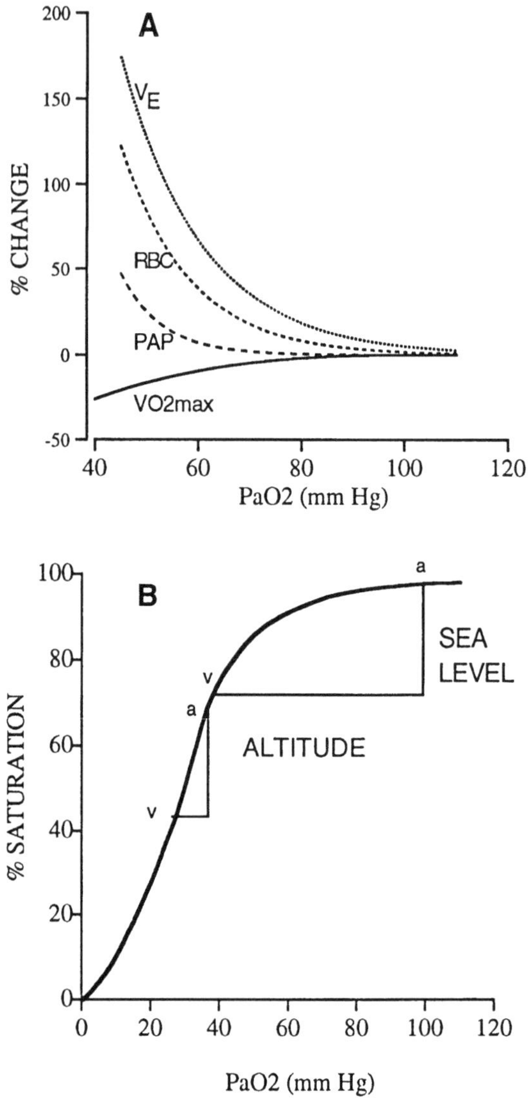

FIGURE 17.3. (A) Stimulus–response curves for ventilation (VE), red cell volume (RBC), and pulmonary arterial pressure (PAP) as a function of arterial oxygen tension at the altitude of residence for well adapted normal persons. The unbroken line indicates the decrease in maximal oxygen uptake (VO_{2max}) at lower arterial oxygen pressures. (B) Hb-O_2 dissociation curve at body temperature and pH of 7.40 illustrating a large decrease in Pao_2 from arterial (a) to mixed venous (v) blood during sea level rest, and the smaller decrease which occurs at altitude.

The question arises as to what function is served by hypoxia acting locally to constrict pulmonary arterioles. In the fetus, where the Pao_2 of 30 mmHg is normal, hypoxic vasoconstriction certainly contributes to the high vascular resistance in the collapsed, water-filled lung. At birth, the placenta is suddenly lost, and the lung must quickly become the organ of oxygenation. The first breath oxygenates the alveoli, relieves hypoxic vasoconstriction, and allows blood flow to the alveoli. The mechanism normally serves the newborn well because those lung segments that fail to inflate retain hypoxic vasoconstriction and are poorly perfused, whereas oxygenated segments are well perfused. Such a local mechanism to match perfusion to ventilation is well suited to defend arterial oxygenation and hence oxygen transport in the newborn period of life. Indeed, failure of pulmonary vasodilation with air breathing threatens life in the neonate.

It is not clear, however, that there is a useful function for hypoxic pulmonary vasoconstriction in the adult at altitude. There, the hypoxic inspired air causes a low Po_2 in all alveoli, resulting in diffuse pulmonary vasoconstriction and a global increase in resistance and in pulmonary hypertension. In normal adults at sea level, perfusion is well matched to ventilation at rest and during exercise. At altitude, increasing pressure simply loads the right ventricle, without substantially adding to the defense of arterial oxygenation. Hypoxic pulmonary vasoconstriction in adults may be a vestige of a mechanism essential for survival of the fetus and newborn because the constrictor response is much smaller in adults than it is in newborns, lethal pulmonary hypertension develops much more quickly in newborns than it does in adults, and, as discussed later, populations well adapted to life at altitude have relatively little pulmonary hypertension and relatively small hypoxic pulmonary pressor responses (Groves et al., 1993). If so, a vigorous pulmonary pressor response may be a detriment rather than a benefit at high altitude.

Juxtaglomerular Apparatus

Hypoxia stimulates release of erythropoietin from the kidney within hours, but the newly formed reticulocytes do not begin to enter the circulation for about 7 days. The rate of increase in red-cell volume shows considerable interindividual variation at altitude [Fig. 17.4(A)], but the stimulus (Po_2)–response (red-cell volume) curve for long-term residents at altitude is hyperbolic, as are the stimulus–response curves for ventilation and for pulmonary hypertension [Fig. 17.3(A)]. Using altitude as a measure of stimulus and using either hematocrit [Fig. 17.4(B)], or hemoglobin as the response, results in a curve similar to that of Fig. 17.3(A), which relates red-cell volume to Pao_2. Because each gram of hemoglobin carries 1.36 ml of oxygen, increasing hemoglobin increases the capacity of blood to carry O_2 [Eq. (17.4)], and, for a given saturation, increases the O_2 content of the blood [Eq. (17.5)].

$$\text{Blood } O_2 \text{ Capacity } (\text{ml } O_2/100\,\text{ml blood}) = 1.36 \times \text{Hb} \tag{17.4}$$

$$\text{Blood } O_2 \text{ Content } (\text{ml } O_2/100\,\text{ml blood}) = \\ \%\ \text{Saturation} \times \text{Blood } O_2 \text{ Capacity} \tag{17.5}$$

At high altitude, for a given arterial O_2 saturation (or Po_2), the greater the carrying capacity, the greater the arterial oxygen content. Because plasma volume becomes reduced within a few days at altitude [Fig. 17.4(A)], the total blood volume is reduced until the increase in red-cell volume is sufficient to restore blood volume to the sea level norm. As noted later, however, disadvantages of excess polycythemia seen in chronic mountain sickness include increased blood viscosity and microcirculatory thrombosis and/or hemorrhage.

Implications of the Responses to Hypoxia

The similar shape of the stimulus–response curves for the carotid body, the pulmonary arterioles, and the juxtaglomerulus apparatus, together with the new evidence for O_2-

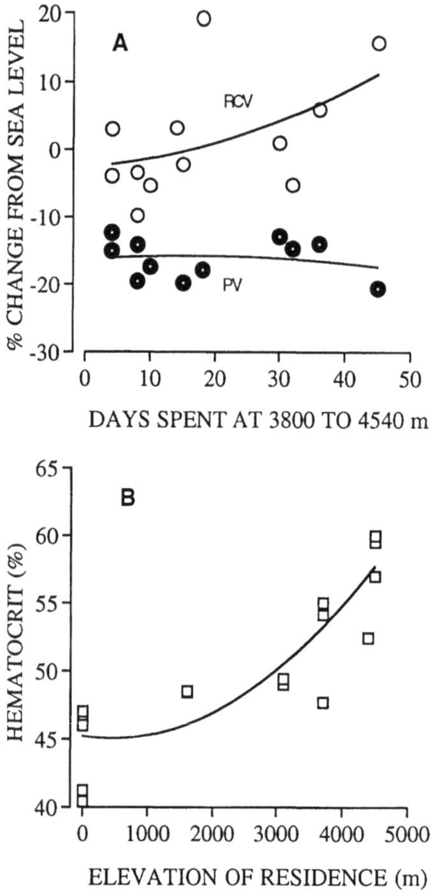

FIGURE 17.4. (A) Changes from sea level values over time at altitude for red cell volume (RCV, open circles) and for plasma volume (PV, filled circles). (B) Hematocrit as related to altitude of residence for healthy, well-adapted persons.

sensitive K^+ channels in two of the tissues, have supported the concept of a common chemoreceptor mechanism. Further, all three tissues appear to respond over the same Po_2 range [Fig. 17.3(A)]. The tissues show an increasing response when the arterial Po_2 falls to the steep part of the Hb-O_2 dissociation curve. It has not been explained why the stimulus response–curves of these 3B tissues resemble the Hb-O_2 dissociation curve rotated about its abscissa, but it is of interest.

Although O_2 sensors within the body may act over a similar O_2 pressure or saturation, the response times differ for the systems involved. Ventilation and pulmonary vascular resistance show an initial increase within seconds, with subsequent increments, respectively, over days and weeks. As indicated by the appearance of new reticulocytes, the bone marrow requires at least 4–5 days to generate new erythrocytes; thus, the increase in red-cell volume lags behind the increases in ventilation and pulmonary vascular resistance. It is clear, then, that within an individual, adaptations to hypoxia differ, depending whether the hypoxia lasts for seconds, days, weeks or months. When arterial oxygen concentrations are so low that the adaptations can no longer maintain organ function, the individual suffers illness, and even death.

The primary result of the increase in ventilation and in red-cell volume is defense of the arterial oxygen content against hypoxia. Because tissue oxygen requirements at rest and during exercise do not decrease at altitude, oxygen transport must be maintained despite the reduced environmental Po_2. In the acclimatized, resting subject, the increases in ventilation and hemoglobin successfully maintain the arterial oxygen content nearly at sea level values, for altitudes up to approximately 4000 m. When oxygen uptake is increased by exercise, however, arterial oxygen saturation and content fall, and maximal oxygen uptake becomes progressively limited with increasing altitude (decreasing arterial Po_2) [Fig. 17.3(A)]. On the summit of Mt. Everest, the reduction is to about one quarter that at sea level. Even at this extreme altitude, efficiency of skeletal muscle appears to be preserved. Thus, over the range from sea level to the summit of Mt. Everest, increasing altitude results in a parallel fall in work and maximal oxygen consumption, whereas the oxygen consumed is closely determined by the work done at all altitudes. With increasing altitude the adaptations cannot completely compensate for the hypoxia of altitude, and the greater the altitude, the greater the impairment of oxygen transport. The overall effects of hypoxia on the oxygen transport system perhaps can be seen most clearly by examining the links in the chain transporting oxygen from the environment to the tissue mitochondria, when the system is stressed to its maximum (i.e., during near maximal exercise at extreme altitude).

The Oxygen Transport Chain

Description

For convenience, the chain [Fig. 17.5(A)] may be considered to have four links, two of which are active and provide for the mass movement of air (ventilation) and blood

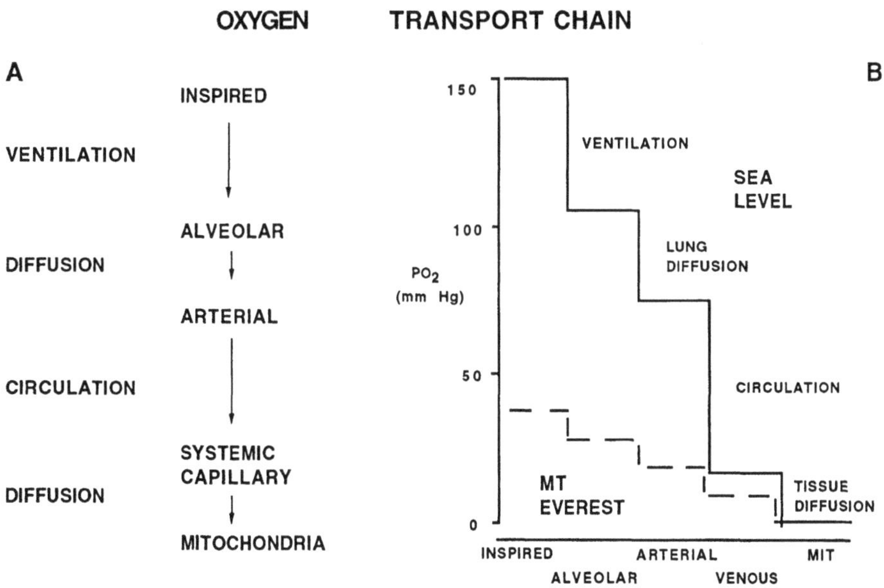

FIGURE 17.5. (A) Schema for the oxygen transport chain, emphasizing the main links of ventilation, lung diffusion, circulation, and tissue diffusion. (B) Schematic illustration of the Po_2 pressures during maximal exercise for each of the four links in the oxygen transport chain at sea level (unbroken line), and at the simulated summit of Mt. Everest, 8848 m (broken line). Mitochondria are shown by the letters MIT.

(circulation). The other two are passive and rely on diffusion of oxygen along a pressure gradient from the alveolus to the lung capillary (or arterial) blood, and from the systemic capillary to the tissue mitochondria. Thus, the bellows action of the respiratory system brings the inspired air to the alveolus. It then passively diffuses across the alveolar–capillary membrane into the blood. Cardiac action then delivers the oxygen within the blood to the systemic capillaries, where it moves by passive diffusion into the cells to be burned by the mitochondria. At all altitudes, either at rest or during exercise, the oxygen pressure (Po_2) must fall successively with each link in the chain. For example, at sea level the overall Po_2 fall from the inspired air to the mitochondria (near zero Po_2), is approximately 150 mmHg [Fig. 17.5(B)]. Because the inspired Po_2 is much less at the equivalent of the summit of Mt. Everest, the overall Po_2 fall to the mitochondria, and the fall at each successive link of the chain must be less [Fig. 17.5(B)].

A considerable feat is a climb to the highest point on earth, the summit of Mount Everest, without oxygen. Quite apart from coping with environmental dangers, the climbers must have performed maximal exercise in a very low oxygen environment. This section will focus on the physiology behind their achievement, which teaches us about normal responses to the twin stresses of exercise and hypoxia. The following discussion is based largely on a study of eight young men who were decompressed in an altitude chamber for 40 days to the simulated summit of Mount Everest, at an altitude of 8848 m (Groves et al., 1987; Houston et al., 1987; Reeves et al., 1987; Wagner et al., 1987; Sutton et al., 1988; Cymerman et al., 1989).

Ventilation

The first link, ventilation, is key because if oxygen does not get from atmosphere to alveoli, the other links become irrelevant. Indeed, increasing altitude markedly increases ventilation [Fig. 17.6(A)], where, for a given oxygen uptake, the volume of air respired (by convention, expressed as volume expired, V_E) is increased. For each of the various altitudes shown in Fig. 17.6(A), the highest oxygen uptake represents maximum effort, which decreases with increasing altitude, as indicated early. Despite the decrease in maximum oxygen uptake, however, the maximum ventilation does not decrease, but is relatively constant across all altitudes. Further, ventilation at maximum effort is close to that which the body can generate, which suggests that even at extreme altitude respiratory muscle function is sufficiently well maintained that maximum ventilation is altitude independent.

Whether or not ventilation is maximal at all altitudes for maximal exercise is an important question because it bears on the key limiting factors for exercise at all altitudes. The ventilation depicted in Fig. 17.6(A), measured at **B**ody **T**emperature, ambient **P**ressure, and **S**aturated with water vapor (designated $V_{E\,BTPS}$), represents the actual movement of the diaphragm and chest wall and, therefore, the bellows function of the respiratory system; however, if we consider oxygen molecules transported to the lungs, the number is related to ventilation measured under **S**tandard conditions of **T**emperature (0°C), **P**ressure (760 mmHg), and **D**ry (designated $V_{E\,STPD}$). At sea level, $V_{E\,STPD}$ is only about 10% less than $V_{E\,BTPS}$. At the summit of Mt. Everest, however, where the air is "thin," $V_{E\,STPD}$ is about one fourth of $V_{E\,BTPS}$. The remarkable finding, which has been known for many years (Christensen, 1937), is that ventilation when expressed under standard conditions is independent of altitude. Our own data from Operation Everest II supports this concept [Fig. 18.6(B)], suggesting that for any given oxygen uptake the same number of oxygen molecules is inspired at all altitudes. The respiratory system seems to function in a remarkably well-regulated manner, such that

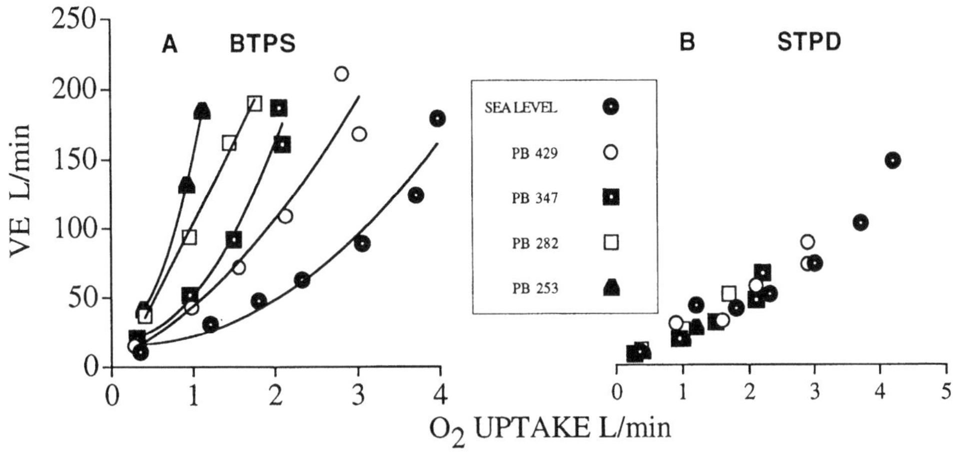

FIGURE 17.6. (A) Relationship of ventilation at body temperature, pressure, and saturated with water ($V_{E\,BTPS}$) for serial exercises to maximum effort at sea level (filled circles), and the following high altitudes listed by barometric pressure, P_B in millimeters of mercury 429 open circles, 347 filled square, 282 open square, and the equivalent of the summit of Mt. Everest 253 filled hexagon. (B) The data from (A) are shown for ventilation at standard conditions of temperature (0°C), pressure (760 mmHg), and dry.

it brings to the lung approximately 1.7×10^{23} molecules of O_2 for each liter of O_2 consumed, regardless of the location of the subject or the severity of the effort (Reeves et al., 1991).

If we make certain reasonable assumptions, then, there are potentially important conclusions to be drawn from these data. Assuming that the respiratory system is ventilating at near capacity at maximal exercise at high altitude, then the number of oxygen molecules brought to the alveolus will depend on air density (i.e., barometric pressure). If so, the close relationship of the number of O_2 molecules ventilated and those taken up [Fig. 17.6(B)] raise the possibility that work performed depends heavily on O_2 transport to the alveolus. Taken together, these two assumptions would lead to the speculation that ventilation is particularly important in limiting maximum exercise capacity at low barometric pressure.

Such a speculation is of interest because circulation is thought to be the dominant limiting factor at sea level (Wagner, 1988). Endurance athletes have larger hearts, larger blood volumes, and larger cardiac output values than do nonathletes. Although one cannot discount other links in the oxygen transport chain, exercise performance in health can usually be more closely linked to circulatory function than to other variables. Indeed, in our measurements at sea level, maximum oxygen uptake was variable between subjects, and was closely linked to maximum cardiac output (Reeves et al., 1990). The variability in maximum oxygen uptake was markedly reduced at altitude, particularly at the equivalent of the summit of Mt. Everest. Further, elite mountain climbers, including those who have scaled Mt. Everest without oxygen, have very ordinary exercise performance at sea level. Even a close examination of their skeletal muscle reveals no distinguishing characteristics (Oelz et al., 1986). The implication is that determinants of exercise performance at altitude that may be primarily respiratory differ from those at sea level, which may depend more heavily on circulation.

Because ventilation is key to alveolar oxygenation at altitude, and knowing that the alveolus contains the four gases mentioned earlier, we can ask how increased ventilation makes more oxygen molecules available to the alveolus. The inspired air is saturated or nearly so before it reaches the alveolus, so the number of water molecules must remain the same regardless of ventilation. The inert gases in the air, primarily N_2, are in equilibrium with the tissues, and therefore cannot be materially changed by increasing ventilation. Thus, for a given oxygen uptake, increased ventilation can effectively change only the balance between the two remaining gases, O_2 and CO_2. In fact, it is approximately true that with hyperventilation, CO_2 is exchanged, molecule for molecule, for O_2. For a constant metabolic rate, as P_{CO_2} falls, P_{O_2} rises.

The P_{CO_2} falls as alveolar ventilation (V_A, total ventilation minus dead space ventilation) rises for any given CO_2 production (VCO_2):

$$P_{CO_2} = (F \times VCO_{2STPD})/V_{ABTPS} \quad (17.6)$$

where F is a factor that converts VCO_2 from STPD to BTPS conditions and which also converts alveolar CO_2 from a fractional concentration to a partial pressure (Reeves, 1991). Because F is nearly constant from sea level to the summit of Mt. Everest, for a given CO_2 production, the P_{CO_2} depends almost entirely on the alveolar ventilation measured under BTPS conditions. Thus, we see that the elimination of CO_2 from the body depends upon BTPS ventilation, which measures the bellows action of the lungs, whereas the transport of oxygen from ambient air to alveolus depends on STPD ventilation, which relates to the number of oxygen molecules transported. These are the two sides of the same coin because for the same metabolic rate, lowering the alveolar P_{CO_2} raises the alveolar P_{O_2}.

By convention, ventilatory measurements are reported BTPS, and respiratory physiologists use the P_{CO_2} as a convenient measure of effective ventilation. Carbon dioxide, which is a water soluble molecule, also crosses the alveolar–capillary membrane with much greater facility than does oxygen. The result is that the arterial–alveolar P_{CO_2} gradient is so small that alveolar and arterial P_{CO_2} values are often used interchangeably.

At sea level, increasing metabolic rate from rest to maximal exercise simultaneously increases CO_2 production and alveolar ventilation such that the P_{CO_2} remains relatively constant at approximately 40 mmHg [Fig. 17.7(A)]. After acclimatization to high altitude, however, for a given metabolic rate, ventilation BTPS is higher than it is at sea level, and the P_{CO_2} is progressively lower as altitude increases. The broken line [Fig. 17.7(A)] connects P_{CO_2} values at maximal exercise for various altitudes, and illustrates the large ventilation-mediated reduction in P_{CO_2} with increasing altitude. Thus, even though the P_{CO_2} at sea level is near 40 mmHg for all metabolic rates, that at the summit of Mt. Everest both at rest (not shown) and at maximal exercise is approximately 8–10 mmHg [Fig. 17.7(A)].

This reduction in alveolar P_{CO_2} at altitude reduces the O_2 pressure gradient from the inspired air to the alveolus [Fig. 17.8(A)]. At sea level, even though the alveolar P_{O_2} is 40 mmHg less than that in the inspired air, P_{IO_2} of 150 mmHg is sufficiently high that the alveolar P_{O_2} is also high enough for good arterial oxygenation. The resulting Pa_{O_2} of 110 mmHg is adequate to maintain oxygen transport of several liters per minute. At the summit of Mt. Everest, however, the inspired P_{O_2} is only 43 mmHg. If the Pa_{CO_2} and hence, the P_{IO_2} to Pa_{O_2} gradient remains at 40 mmHg, (i.e., no increase in ventilation), the alveolar P_{O_2} would be 3 mmHg, and life would not be possible. It is because the alveolar P_{O_2} is only 8–10 mmHg less than that in the inspired air, the alveolar P_{O_2} on the summit of Mt. Everest is 33–35 mmHg, values that are compatible with life and which even allow modest exercise. Without an increase in ventilation at altitude, human beings breathing air would barely be able to climb Mt. Blanc, and certainly could not scale Mt. Everest. Thus, the ventilation-mediated reduction in P_{CO_2} reduces the P_{O_2}

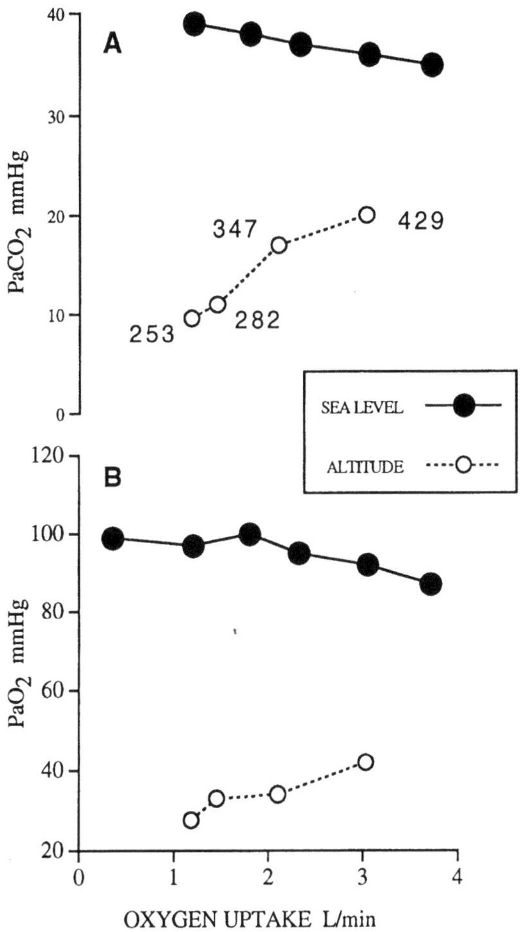

FIGURE 17.7. (A) Relation of arterial carbon dioxide pressures (Pa_{CO_2}) to oxygen uptake at sea level (filled circles) for exercises of increasing intensity to maximum effort. Pa_{CO_2} measurements at maximal effort at various high altitudes (open circles) listed by barometric pressures as in Fig. 17.6. (B) Relation of arterial oxygen pressure (Pa_{O_2}) at rest and during exercise to maximal effort at sea level (filled circles) for the same subjects shown in (A). The open circles indicate the values at the high altitudes shown in (A).

gradient from the inspired to the alveolar air, resulting in an alveolar oxygen concentration higher than would otherwise be possible.

Pulmonary Diffusion

Having reached the alveolus, the O_2 molecule must cross the alveolar–capillary membrane into the pulmonary capillary blood. The movement is by diffusion, which is a passive process, driven by the O_2 pressure gradient from alveolus to blood. Diffusion of a gas depends upon its mobility through aqueous tissue (determined largely by molecular size and solubility), and the properties (mainly thickness and surface area of the alveolar–capillary membrane) of the air–blood interface. When the pulmonary capillaries are fully recruited, lung diffusion is maximal, and the lung diffusing capacity (D_L)

is proportional to the maximal oxygen uptake divided by the O_2 pressure gradient from alveolus (Pao_2) to pulmonary capillary blood ($Pcapo_2$):

$$D_L \approx VO_{2max}/(P_AO_2 - P_{cap}O_2) \tag{17.7}$$

$$D_L \approx VO_2/(P_AO_2 - Pao_2) \tag{17.8}$$

Many stimuli (supine posture, mild exercise, increased ventilation, or hypoxia) appear to recruit fully lung capillaries. Therefore, it is approximately true that for exercise at sea level and for rest or exercise at altitude, the alveolar capillary membrane functions

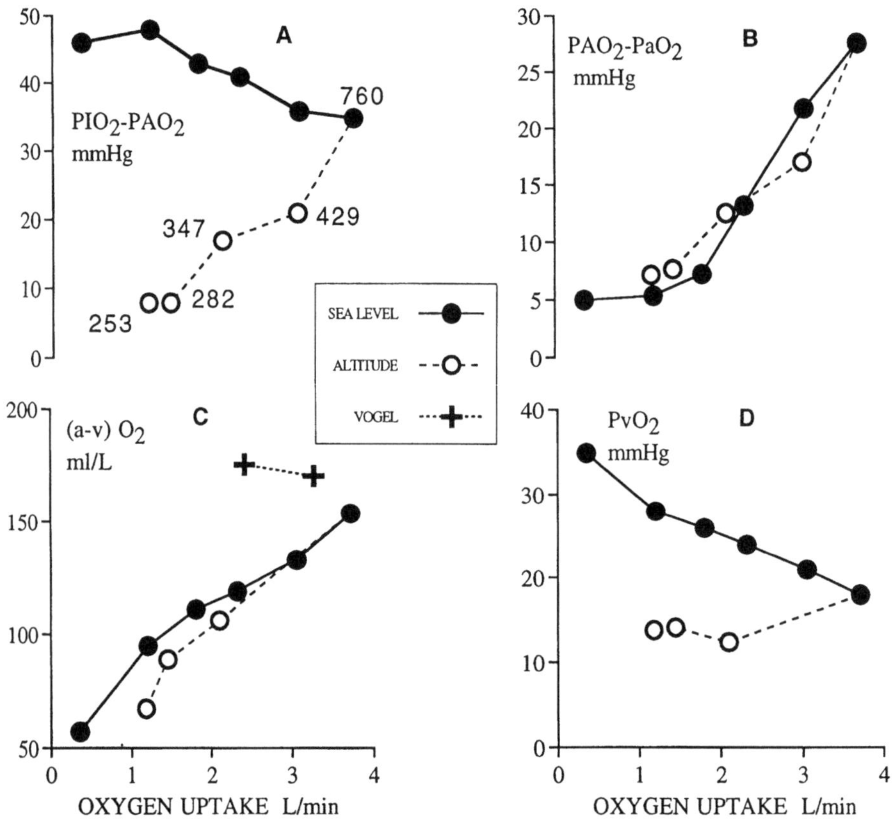

FIGURE 17.8. Relation of the pressure gradient for oxygen to the amount of oxygen transported at rest and during exercise at each link in the oxygen transport chain at sea level and high altitude. Shown as filled circles are the measurements at rest and during exercises to maximum effort at sea level. The open circles are the measurements at maximal effort at various altitudes indicated in Figs. 17.6 and 17.7. (*Source*: Data are taken from Sutton and Reeves, 1988; Groves and Reeves et al. 1987, Reeves and Groves et al., 1987.) (A) The O_2 pressure gradient for the ventilatory link as oxygen is transported from the inspired air (P_IO_2) to the alveolus (P_AO_2). (B) The O_2 pressure gradient for the pulmonary diffusion link as oxygen is transported from the alveolar air (P_AO_2) to the arterial blood (Pao_2). (C) The O_2 pressure gradient for the circulation link as oxygen is transported from the arterial blood (Pao_2) to the systemic capillary (represented by the mixed venous blood, Pvo_2). (*Source*: Data from Vogel, Hartley et al., 1974, are also shown.) (D) The O_2 pressure gradient for the tissue diffusion link as oxygen is transported from the systemic capillary blood (Pvo_2) to the mitochondria ($Po_2 = 0$).

at its maximal O_2 diffusivity (i.e., its diffusion capacity). Although the $Pcapo_2$ cannot be measured directly in humans, it may be approximated by the arterial oxygen pressure (Pao_2). At high altitude, therefore, and for exercise at sea level, Eq. (17.7) may be simplified to Eq. (17.8).

The implication is that the lung membrane diffusing capacity of an individual is a constraint that will not improve and within which he or she must live at altitude. If so, the pressure gradient from alveolus to arterial blood should be approximately proportional to oxygen uptake, where the line of proportionality passes through the origin. Such a relationship is seen (Fig. 17.8), where the high altitude values fall on those for sea level. This relationship is important because it implies (1) that the pulmonary vascular bed is fully recruited at altitude, (2) pulmonary O_2 diffusion capacity is not increased or decreased at altitude, and (3) at all altitudes, including sea level, the pressure gradient across the pulmonary capillary membrane is determined by the amount of oxygen to be transferred.

Except for well-trained athletes, an increasing O_2 pressure gradient from alveolus to blood is not usually a problem at sea level, where the alveolar Po_2 is high enough (>110 mmHg) to maintain arterial Po_2 of 80 or more, even at maximal exercise [Fig. 17.7(B)]. The result is that arterial oxygenation occurs on the flat portion of the Hb-O_2 dissociation curve. Only exceptional athletes may have a cardiovascular system capable of transporting such large amounts of oxygen that the membrane diffusion capacity of the lung may not be sufficient to maintain arterial oxygenation, even at sea level.

At altitude, however, arterial oxygen pressure decreases from rest to exercise because of the limitations inherent in lung membrane diffusing capacity. For example, at the summit of Mt. Everest, where the inspired Po_2 is only 43 mmHg, the resting alveolar Po_2 must be severely hypoxic, no matter what the ventilation. Exercise, by increasing the oxygen uptake, also increases the alveolar to arterial Po_2 gradient [Fig. 17.8(B)]. Because the $P_{A}O_2$ does rise, the Pao_2 must fall. Even at rest, the arterial oxygen saturation is on the steep part of the Hb-O_2 dissociation curve (resting SaO_2 = 58%), and increasing O_2 uptake required for maximal exercise causes the arterial oxygen saturation to fall to approximately 50%. It is clear that the low environmental oxygen present at altitude is associated with arterial hypoxemia at rest; with exercise, the diffusion limitation inherent in the normal lung further lowers arterial oxygen levels. In the extreme circumstance (i.e., the summit of Mt. Everest at maximal exercise) the arterial Po_2 in healthy individuals falls to average values of only 28 mmHg [Fig. 17.7(B)].

Circulation

Given that the arterial oxygenation at altitude is reduced at rest and falls even further with exercise, one wonders what adaptations are made by the circulation to maintain oxygen transport. It is perhaps not surprising that those adaptations vary depending on the altitude. At altitudes of 3000–4500 m, a given oxygen uptake occurs with an a–v O_2 difference, which is higher (i.e., a cardiac output that is lower) than is found at sea level. For example, maximal exercise at sea level and the summit of Pikes Peak (4300 m, PB 461 mmHg) found no change in a–v O_2 difference despite a lower oxygen uptake, [Fig. 17.8(C)] (Vogel et al., 1974). The findings therefore reflect a decreased cardiac output for a given oxygen uptake. Other investigators have similar findings for submaximal exercise so that there is agreement in the literature that cardiac output at rest or during exercise falls with residence for days or weeks at these altitudes.

Published reports, however, are also clear that a given oxygen uptake at more extreme altitudes has an a–v O_2 difference that is similar to that at sea level, as illustrated by Fig. 17.8(C). The implication for altitudes of approximately 6000 m and above is that cardiac output is not reduced at rest or during exercise; rather, it is appropriate for the amount of oxygen to be transported. It is not known why circulatory control is apparently different for the more moderate versus the more extreme altitudes. One suggestion, yet to be tested, is that the extreme hypoxemia at the higher altitudes overrides local arteriolar vasoconstriction, thereby acting to lower vascular resistance and to raise cardiac output. This possibility is reinforced by the finding that for a given oxygen uptake, cardiac output tends toward higher values and a–v O_2 differences toward lower values [Fig. 17.8(C)] at the summit of Mt. Everest.

Regardless of the normal behavior of the a–v O_2 difference or the cardiac output at altitude, it is primarily the shape of the Hb-O_2 dissociation curve causes the Po_2 gradient from artery to vein to become narrowed [Fig. 17.3(B)]. Thus, at sea level, where the arterial Po_2 is high and oxygenation is on the flat part of the curve, there is a large fall in Po_2 from artery to vein, and hence to systemic capillary. At altitude where the arterial Po_2 lies on the steep part of the curve, the Po_2 fall from artery to vein, or capillary, is much reduced. Thus, the shape of the Hb-O_2 dissociation curve, more than the cardiac output values, acts to preserve systemic capillary Po_2 and hence O_2 transport at high altitude.

Tissue Diffusion

The final link in the O_2 transport chain is the diffusion of O_2 from the systemic capillary to the mitochondria. Assuming that at maximal effort the Po_2 in the venous blood is close to that in the capillary and that the Po_2 in the mitochondria is near zero, then the gradient from capillary to mitochondria may be closely related to the Po_2 in the venous blood. Measurements of the mixed venous blood at maximal effort indicate that the Po_2 at sea level (19 mmHg) is only slightly higher than that on the summit of Mt. Everest (13 mmHg) [Fig. 17.8(D)]. Unlike the lung, skeletal muscle capillaries are not fully recruited with mild exercise, and it is not even established whether they are fully recruited with maximal exercise. It is likely, however, that the relatively well-maintained Po_2 pressure gradient from capillary to mitochondria is by virtue of the reduced amount of O_2 that is transported at high altitude.

Summary for Oxygen Transport

The O_2 transport during maximal exercise at altitude may then be summarized [Fig. 17.5(B)]. At sea level, where the O_2 uptake is approximately 4 L/minute, there are large Po_2 pressure drops at each link in the O_2 transport chain, particularly in the ventilatory and circulatory links, which are concerned with the mass movement of air and blood. At altitude, the pressure drop is less at each link. For the first link (ambient air to alveolus) ventilation primarily reduces the pressure drop by decreasing the alveolar Pco_2 thereby providing for a higher alveolar Po_2 than would otherwise be possible. For the second link (alveolus to arterial blood) the Po_2 pressure gradient falls primarily because less O_2 can be transported in the hypoxic environment. For the third link (arterial to systemic capillary blood) the fall in Po_2 gradient is primarily due to the shape of the Hb-O_2 dissociation curve. Finally, for the fourth link (systemic capillary blood to mitochondria) the Po_2 pressure gradient falls primarily because less O_2 can be transported in the hypoxic environment, as was the case for diffusion within the lung. The various factors that decrease the Po_2 gradient at each link in the O_2 transport chain therefore help the defend the Po_2 in the systemic capillaries (systemic veins), even at maximal exercise at extreme altitude on Mt. Everest.

The common thread running through oxygen transport for all altitudes is the defense of tissue oxygenation. At high altitude, that defense runs up against three hard facts: (1) the amount of oxygen required for a task does not decrease at altitude, (2) there is no adaptation by which an individual can overcome the low $P_{I}O_2$ at altitude, and (3) two of the four links in the transport chain are passive, depending on the laws of diffusion. Despite numerous alterations in the active links in the chain, tissue oxygenation can ultimately be defended by only one strategy (i.e., a limitation of performance).

Sympathoadrenal System

β Adrenergics

Although respiratory acclimatization is more important than cardiovascular and metabolic adaptations from the point of view of O_2 transport, the latter account for many of the observed changes at altitude. The cardiovascular changes appear to be mediated largely through the autonomic nervous system, which is composed of two limbs, the parasympathetics, and the sympathoadrenal system. This section will focus on the latter.

Hemodynamics

The heart rate rises on arrival at altitude (4300 m) as a result both of β adrenergic stimulation and parasympathetic withdrawal. Resting measurements of blood epinephrine levels are sufficiently variable that they do not show much change at altitude (Young et al., 1991), but those taken during steady-state submaximal exercise show an increase on arrival and after 3 weeks residence, which is a fall toward, but not to sea level values (Mazzeo et al., 1991). It seems likely that over several weeks both β adrenergic stimulation and heart rate return toward, but not completely to, the sea-level norm. The time course of changes in exercise epinephrine levels is consistent with the time course of the hypoxemia at altitude, which raises the possibility that hypoxia stimulates epinephrine release. Myocardial contractility is increased at high altitude. In goats, when the β-adrenergics are blocked, hypoxia becomes a myocardial depressant (Tucker et al., 1976). In individuals with intact adrenergics, high-altitude exposure is accompanied by an increase in the rate and magnitude of left-ventricular contraction, despite the fact that left-ventricular cavity diameter has decreased (Saurez et al., 1987). Even at extreme altitude, ventricular function curves for both the right and left ventricles remain normal (Reeves et al., 1987). Increased heart rate and contractility are consistent with known epinephrine actions, and they may be attributed, at least in part, to increased β-adrenergic tone. Thus, a reasonable hypothesis is that the effects of hypoxia on heart rate and myocardial contractility are mediated via the β-adrenergic sympathetics.

Metabolism

Epinephrine administration is known to increase metabolic rate. Metabolic rate increases at altitude and contributes to the weight loss observed at 4300 m (Butterfield, et al., 1992). Adrenergic stimulation seems to be responsible for the rise in basal metabolic rate at altitude because metabolic rate changes have the same time course as do heart rate and epinephrine levels, and the metabolic rate changes are prevented by administration of the β-adrenergic blocker, propranolol (Reeves, Mazzeo et al., 1992).

In addition, adrenergic stimulation appears to determine, in part, the choice of metabolic fuel at altitude. Thus, blood lactate rises on arrival at altitude. This rise is blunted, but not prevented by β-blockade (Young et al., 1991; Reeves, Wolfel et al., 1992; Wolfel, personal communication). Tracer studies indicate that both the rate of appearance and rate of disappearance of lactate are increased on arrival, which suggests that lactate is a source of metabolic fuel for muscle during exercise (Brooks et al., 1991; Brooks et al., 1992; Brooks 1998). Glucose turnover (in men) is also increased at altitude (Brooks et al., 1990), and the increase is augmented by β-adrenergic blockade (Roberts et al., 1996). Reliance on fat as fuel in men is decreased at altitude, and this decreases further with β-adrenergic blockade (Roberts, Butterfield et al., 1996). Thus, the metabolic fuel for rest and exercise is altered by altitue exposure and part of the mechanism is the activity of the β-adrenergic system.

α Sympathetics

α-sympathetic tone appears to follow a somewhat different time course from that of the β system in men (Mazzeo et al., 1991) and women (Mazzeo et al., 1998). For example, norepinephrine levels in urine and blood rise slowly at altitude and achieve maximal values between the first and second week. The increase in α-sympathetic activity with altitude acclimatization, as measured by norepinephrine release by the legs, occurs both at rest and during exercise (Mazzeo et al., 1995). This time course is coincident with the rise in arterial pressure [Fig. 17.9(A and B)]. From an examination of published literature (Wolfel et al., 1994), the rise is temporary both for pressure and urinary catecholamine excretion. Long-term residents of altitude have not been found to have elevated norepinephrine levels, nor do they usually have elevated systemic arterial pressures. For the first few weeks after arrival at altitude, however, arterial pressure and resistance are elevated, and these follow the same time course as does the rise in blood norepinephrine and urinary catecholamine excretion. This raises the possibility that activation of the α sympathetics is responsible for the changes in the systemic arterial circulation.

Increased α-sympathetic activity, constricts both veins and arteries, and the increased β-adrenergic stimulation, together provide some insight into the complex changes that occur in cardiac output over the first few days at altitudes of approximately 4000 m. Cardiac output at either rest or exercise is increased for the first 2 or 3 days at these altitudes, but it then falls to levels less than those observed at sea level and remains low for several weeks (Wolfel et al., 1991).

Even though the mechanisms of these changes are not clearly established, there are clues. Hypoxemia-induced adrenergic stimulation on arrival (and there is parasympathetic withdrawal) augments heart rate and contractility, both of which would contribute to an increased output soon (Hughson 1994). In addition, there is some increase in α-sympathetic neural stimulation over the first 2–3 days which could contribute to the observed venoconstriction, thereby increasing venous tone, venous pressure and cardiac filling. These factors probably combine to result in the initial increase in output. Support for a role of the β-adrenergic system is that men who have received propranolol to block the β receptors have a blunted increase in resting or exercising output at 4300 m (Wolfel, Selland et al., 1998).

After a few days with improvement in arterial oxygenation, however, the saturation increases and β-adrenergic stimulation becomes less than on arrival. In addition, sustained venoconstriction probably contributes to a reduction of about 20% in plasma volume (Grover et al., 1998), and the plasma volume decrease occurs even in the presence of dense β-adrenergic blockade. The resulting hemoconcentration increases the

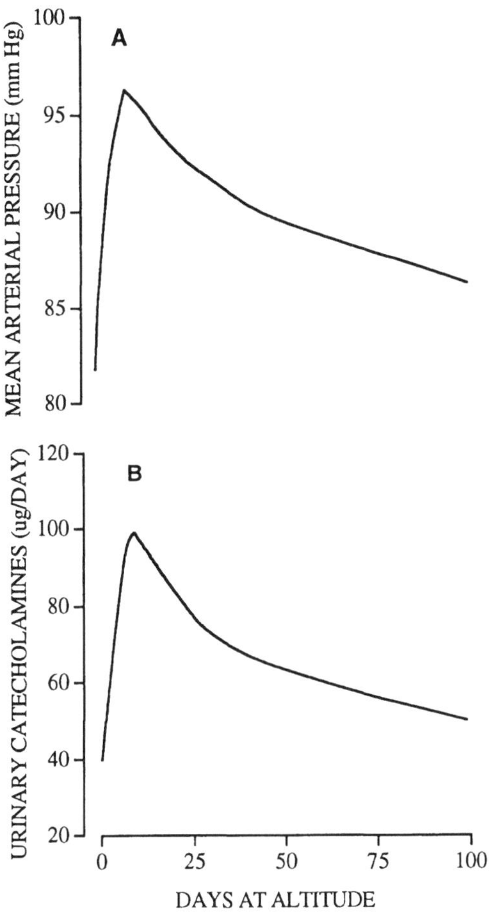

FIGURE 17.9. (A) Schematized data from that summarized by Wolfel et al. (1994) showing the changes in mean systemic arterial pressure over time at altitudes from 3500 to 4300 m. (B) Schematized data from that summarized by Wolfel et al. (1994) showing the changes in urinary excretion of total catecholamines over time at altitude.

hematocrit, which together with the increase in saturation causes a large increase in arterial oxygen content. Arterial constriction increases arterial resistance. The combination of these changes is accompanied by a decrease below sea level values in cardiac output by the 4th or 5th day at 4300 m altitude (Grover et al., 1986).

Summary for the Sympathoadrenal System

From the preceding, most of the cardiovascular effects of altitudes up to approximately 4300 m are attributable to the parasympathetic and the sympathoadrenal systems. The heart rate, metabolic rate, and myocardial contractility are largely mediated by the β-adrenergic system, where parasympathetic withdrawal contributes to the heart rate response. Arterial pressure and resistance and venous tone (Grover et al., 1986) are probably increased via the alpha sympathetic system. Plasma volume decrease and, hence, the initial hemoconcentration probably reflects the increased venous tone. The increased arterial resistance plus the decreased plasma volume contribute to the fall in cardiac output. All of these changes are not simultaneous, however, because the adrenal release seems to occur earlier than the increase in circulating norepinephrine; however, circulating norepinephrine levels may not reflect increased α sympathetic neural traffic early in the altitude exposure. More definitive experiments showing how the neural traffic is altered over time at altitude are needed.

Given the substantial number and magnitude of the cardiovascular adaptations that take place over the first days and weeks at altitude, one expects that the capacity for oxygen transport would be enhanced. It is surprising, however, that maximal oxygen uptake as measured on arrival at 4300m shows either no increase or only a slight increase after 3 weeks residence (Grover et al., 1986), i.e., adequate time for circulatory adjustments and ventilatory acclimatization. Why? Although the reason is not clear, one possibility is that for some reason with maximal effort at altitude, the basic limitation is the low inspired oxygen pressure. What then, is the net effect of the preceding alterations in the sympatho-adrenal system and even in ventilation? There are clues that submaximal exercise can be continued longer and with more comfort after, than before acclimatization. Further, improved oxygenation at rest is associated with relief of the symptoms of various acute altitude illnesses.

Acute High Altitude Illness

Acute high-altitude illness is a syndrome that is punctuated by several symptoms, notably dyprea at rest and on exertion, nausea, headache, cough, insomnia, and mental disturbances. Cold and exposure often aggravate this syndrome. It can be prevented by gradual adaptation or by pretreatment with the drug acetazolamide, a brown carbonic anhydrase inhibitor.

Symptoms which usually occur in some persons during the first 3 to 4 days of altitude ascent of over 1500m (4921ft) are most likely due primarily to hypoxia, because improvement rapidly follows descent or treatment with oxygen. Part of the problem is that several days are required for ventilation to increase to its new plateau (Dempsey and Forster, 1982; Weil, 1990), because the body cannot simultaneously defend arterial oxygenation and pH. Increased ventilation, which brings more O_2 to the alveoli, causes alkalosis by removing CO_2 from the blood. Alkalosis inhibits the ventilatory response to hypoxia. Over time, the body solves the dilemma of defending both oxygenation and pH, by reducing the intracellular bicarbonate concentration, probably somewhere within the central nervous system, thereby restoring the ratio of $[HCO_3^-]/[CO_2]$, and hence the intracellular pH, toward normal. As the pH falls, the inhibitory influence of alkalosis diminishes and ventilation increases. For example, when man goes from sea level to 4300m, ventilatory acclimatization (a falling P_{CO_2} and a rising P_{O_2}) proceeds rapidly at first and then more slowly until a stable plateau is reached at 10 days. Hypoxemia is more severe on arrival than after acclimatization and altitude illnesses occur during this acclimatization period. Whether the symptoms are causally linked to the acclimatization process or are simply coincidental is not clear.

Acute Mountain Sickness

The central nervous system, particularly the cerebral hemispheres are more sensitive to hypoxia than other tissues such as peripheral nerves, muscle or heart (Reeves, Houston et al., 1989). Therefore, cerebral symptoms, designated acute mountain sickness (AMS), which probably represent mild cerebral edema occur, and include headache, anorexia, nausea, fatigue, sleeplessness, (Hackett and Rennie, 1976; Hochstrasser et al., 1986; Honigman et al., 1993). Such symptoms occur in 25% of persons going quickly from sea level to 3000m and in 75% of those going to 4300m (Honigman et al., 1993). Symptoms can largely be prevented by maneuvers that allow for increased ventilation (e.g., slow ascent), or by using acetazolamide taken a day or two in advance of exposure. Acetazolamide increases breathing by causing acidosis via renal excretion of bicarbonate. Because both slow ascent and acetazolamide administration increases breathing, they may ameliorate symptoms by reducing hypoxemia, which is useful in practice, but does not provide insight into the mechanism of the disorder.

High Altitude Cerebral Edema

When the AMS victims present with objective, life-threatening cerebral signs, the term high altitude cerebral edema (HACE) has been applied, even though the disorder probably represents only a more severe form of AMS. The more severe cerebral signs (e.g., apathy, ataxia, mental confusion), probably result from brain swelling (vasogenic edema followed by cytotoxic edema). Increased intracranial pressure has been measured and is consistent with the frequent presence of vomiting and papilloedema. Cerebral edema in the absence of treatment (oxygen, descent) can result in coma and brain herniation. It is not known how hypoxia causes the brain cells to swell. Leakage of protein and water through the blood–brain barrier is a real possibility (vasogenic leak).

High Altitude Pulmonary Edema

The lung is less susceptible than the brain to hypoxia; therefore, high altitude pulmonary edema (HAPE) is less common than AMS. The presence of HAPE, however, is more ominous because mortality is high in the absence of treatment (Reeves and Grover, 1974; Hackett and Roach, 1990; Reeves and Schoene, 1991; Oelz et al., 1992). Hypoventilation at altitude is often but not always present (Selland et al., 1993). Although the mechanism of HAPE is unknown, the leading hypotheses are an increase in lung microcirculatory pressure (Hultgren et al., 1971) and a simultaneous increase in membrane permeability (Schoene et al., 1988). Pressure sufficient to rupture capillaries has been proposed by West and Mathieu-Costello (1992). Other causal mechanisms include a diminished pulmonary circulation capacity (Gibbs, 1999). Hypoxia is known to increase pulmonary arterial pressure, and increased pressure markedly potentiates permeability-related leak. Leakage may lead to an inflammatory process. Exercise, and possibly cold, appear to increase the incidence of HAPE, possibly because they increase pulmonary arterial pressure. By the same token, agents that lower pulmonary arterial pressure (calcium antagonists) appear to be useful in both treating and preventing HAPE (Oelz et al., 1992).

Peripheral Edema at High Altitude

Hypoxia is also associated with peripheral edema, particularly in warm weather. Possibly high skin blood flow facilitates accumulation of subcutaneous fluid, which need not be dependent and may occur unilaterally in the face or in the hands. Because cerebral, lung, and peripheral edema all occur early after rapid ascent in persons who are not well acclimatized, and because all three may occur simultaneously, the possibility must be considered that the hypoxia is associated with a bodywide problem in water handling; however, the nature of this problem and even whether the three disorders are due to a common mechanism remains to be elucidated.

Chronic Mountain Sickness

Some long-term residents of high altitude hypoventilate (Kryger et al., 1978), resulting in excessive hypoxemia, renal release of erythropoietin, and polycythemia. Polycythemia that is excessive (hematocrits greater than 60–65%), has come to be the hallmark of chronic altitude illness, and is referred to as chronic mountain sickness (CMS) or Monge's disease (Winslow and Monge, 1987; Penulosa and Sime, 1991; Monge et al., 1992; see review et seq., Reeves 1998). Those afflicted are usually males over 40 years of age, who have hemoglobin and hematocrit values (respectively greater than 18 g% and 60% at 3000 m and greater than 20 g% and 65% at 3600 m) that are higher than expected for the altitude of residence, where other causes of polycythemia can be ruled out. Symptoms include lassitude, mental confusion, and right-heart failure, which have been attributed to the hypoxemia and high blood viscosity. One cause of the hypoventilation is progressive loss of the drive to breathe, the hypoxic ventilatory response, over several decades at altitude. The longer the residence and the higher the altitude, the greater the loss of the ventilatory response to hypoxia. Another cause of the hypoxemia is hypoventilation at night, when arterial oxygen saturations may fall for several minutes to values as low as 50%. Improvement or even cure of the syndrome by descent, nocturnal oxygen, or ventilatory stimulation supports the view that an excessive hypoxic stimulus leads to excessive polycythemia.

Adaptations Within High Altitude Populations

Increased ventilation and increased hematocrit within individuals at altitude reflect increased pumping of air or viscous blood, which are both energy-intensive adaptations. Adaptations that are less energy intensive develop in persons born at altitude or evolve over generations in populations at altitude (Moore et al., 1992). For example, animals (Johnson et al., 1985) or persons who are born and live their developmental years at altitude have increased lung volumes and increased alveolar–capillary surface areas for O_2 diffusion. In addition, in certain populations (e.g., the Tibetans) hypoventilation during sleep and the decrease with time in the hypoxic ventilatory response at altitude is reduced or does not occur. In addition, Tibetans have mean pulmonary arterial pressures that are approximately those found in sea level residents and are lower than in other altitude residents (Moore et al., 1992; Groves et al., 1993). As a consequence of larger lungs and a better-sustained ventilation throughout life, and low pulmonary arterial pressure, acute and chronic mountain sickness is rare in this population. In addition, Tibetan women living in Lhasa give birth to infants that are of nearly normal birth weight and are heavier than infants born to other high altitude groups. This is consistent with the poorly understood phenomena of neonatal hypoxia tolerance (Singer, 1995). For example, the fetus lives and grows at a Po_2 that corresponds to an altitude of 8000 m.

In the Andes, the llama and alpaca have remarkably right-shifted Hb-O_2 dissociation curves to allow efficient loading of oxygen at low Po_2. High-flying birds (e.g., the bar-headed goose) can tolerate extreme alkalosis, which right shifts their dissociation curve to maintain arterial oxygenation during migration over the Himalayas. The human fetus and newborn, like diving mammals, have developed tissue tolerance to acidosis so that metabolic activity can be carried out for relatively long times by glycolysis in the absence of oxygen. Adaptation to high blood-lactate levels is also a cardinal feature seen in natives of the Andes, Tibet, and East Africa. Hochachka and co-workers (1999) have put forward the view that the similar hypoxia physiology among these natives may represent the ancestral condition for humans.

Altitude Maladaptation in Cattle (Brisket Disease)

In 1915 "brisket disease" was described in cattle brought from low altitude to reside at 3000 m in Colorado. Subsequent studies found that cattle have a remarkably vigorous pulmonary vasoconstrictor response to hypoxia, and calves with brisket disease at high altitude have hypoxia-induced pulmonary hypertension with right heart failure (Reeves et al., 1979). In the bovine species, heart failure fluid accumulates in the loose areolar tissue in the "brisket" area ventral to the sternum rather than in the legs. Pulmonary hypertension occurs more rapidly the higher the altitude of residence (greater hypoxic stimulus) and the younger the animal (greater response) (Fig. 17.10).

Such studies in cattle have contributed to the concept that hypoxia adversely affects the lung circulation more in the newborn than in the adult. Hypoxia in the newborn period acts to maintain high lung arterial tone as in the fetus. The presence of the muscular arterioles in the newborn allows for marked vasoconstriction and rapidly accelerating hypertension when hypoxia is maintained. In addition, the neonatal mammal has a remarkable propensity for an exuberant cellular response to stress. For the lung circulation, pulmonary hypertension maintained for only a few days leads to remarkable medial hypertrophy and proliferation of the adventitial fibroblast. Excessive amounts of collagen and elastin are laid down around the arterioles. The presence of

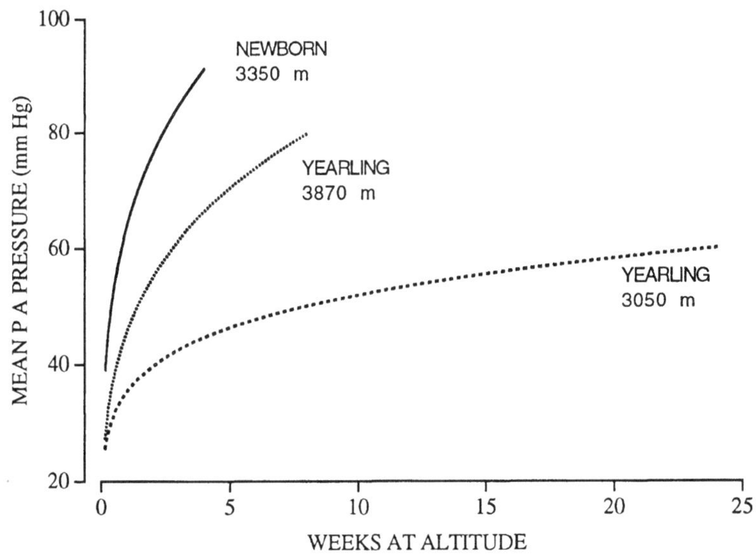

FIGURE 17.10. Mean pulmonary arterial pressure over time at altitude for newborn calves at 3350 m altitude (unbroken line) and for steers born and raised at low altitude but taken to 3870 m at one year of age. Even though the altitude stress is less for the newborns, they show a more brisk pulmonary pressor response.

both the medial thickening and the matrix act as a restrictive collar around the vascular lumen, inhibiting the capacity of the lung arterioles to relax. Such studies have emphasized that although the hypoxic pulmonary hypertension of altitude begins as vasoconstriction, the cellular response inhibits the ability of the vessels to relax. Although persons with chronic lung disease frequently develop right-heart failure, it is now clear that a syndrome not unlike brisket disease in cattle occurs in humans at altitude (Anand and Chandrashekhar, 1992).

Summary

Hypoxia is a common human condition. Compared with normoxia in the adult, we were all hypoxic in gestation. From birth onward many individuals have had to contend with episodes of hypoxia, whether it be local as in the newly expanded neonatal lung or more generalized hypoxia as from hypoventilation or anemia. Hypoxia may be prolonged, as it is in altitude residence or in lung disease. The body has several sensors to detect the presence of hypoxia and to respond to it. The responses may be cellular (metabolism) or related to organs (heart, vessels), or it may integrate whole systems (respiration, autonomics). When the stimulus is large and the response is inappropriately large or small, however, organ malfunction occurs. Populations evolving at altitude probably develop less energy-intensive strategies to live in hypoxic environments than do individuals who must cope acutely with altitude exposure. The low barometric pressure of altitude provides a natural laboratory to study the effects of hypoxia and their mechanisms.

References

Anand, I.S., and Chandrashekhar, Y. (1992) Subacute mountain sickness syndromes: role of pulmonary hypertension. In: Sutton, J.R., Coates, G., and Houston, C.S. (eds.), pp. 241–251. Hypoxia and mountain medicine. Queen City Printers, Burlington, VT.

Brooks, G.A., Butterfield, G.E., Wolfe, R.R., Groves, B.M., Mazzeo, R.S., Sutton J.R. et al. (1990) Increased dependence on blood glucose after acclimatization to 4300m. J. Appl. Physiol. 70, 919–927.

Brooks, G.A., Butterfield, G.E., Wolfe, R.R., Groves, B.M., Mazzeo, R.S., Sutton J.R. et al. (1991) Decreased reliance on lactate during exercise after acclimatization to 4300m. J. Appl. Physiol. 71, 333–341.

Brooks, G.A., Wolfel, E.E., Groves, B.M., Bender, P.R., Butterfield, G.E., Cymerman, A. et al. (1992) Muscle accounts for glucose disposal but not blood lactate appearance during exercise after acclimatization to 4300m. J. Appl. Physiol. 72, 2435–2445.

Brooks, G.A., Wolfel, E.E., Butterfield, G.E., Cymerman, A., Roberts, A.C., R., Mazzeo, R.S., and Reeves, J.T. (1998) Poor relationship between arterial [lactate] and leg net release during exercise at 4300m altitude. Am. J. Physiol. 275, R1192–R1201.

Bülbring, E., Burn, J.H., and de Elio, F.J. (1934) The secretion of adrenaline from the perfused suprarenal gland. J. Physiol. 107, 222–232.

Butterfield, G.E., Gates J., Fleming, S., Brooks, G.A., Sutton, J.R., and Reeves, J.T. (1992) Increased energy intake minimizes weight loss in men at high altitude. J. Appl. Physiol. 72, 1741–1748.

Christensen, E.H. (1937) Sauerstoff aufnahme and respiratorische Functionen in gross Hohen. Skand. Arch. Physiol. 76, 88–100.

Cymerman, A., Reeves, J.T., Sutton, J.R., Rock, P.B., Groves, B.M., Malkonian, M.K. et al. (1989) Operation Everest II: maximal oxygen uptake at extreme altitude. J. Appl. Physiol. 66, 2446–2453.

Dempsey, J.A., and Forster, H.V. (1982) Mediation of ventilatory adaptations. Physiol Rev. 62, 262–331.

Droma, T.S., McCullough, R.G., McCullough, R.E., Zhuang, J.G., Cymerman, A., Sun, S.F., et al. (1991). Increased vital and total lung capacities in Tibetan compared to Han residents of Lhasa (3658 m). Am. J. Phys. Anthro. 86, 341–351.

Grover, R.F., Selland, M.A., McCullough, R.G., Dahms, T.E., Wolfel, E.E., Butterfield, G.E., et al. (1998) β-Adrenergic blockade does not prevent polycythemia or decrease in plasma volume in men at 4300 m altitude. Eur. J. Appl. Physiol. 77, 264–270.

Grover, R.F., Weil, J.V., and Reeves, J.T. (1986) Cardiovascular adaptations at high altitude. In: Pandolf, K.B. (ed.) pp. 269–302, Exercise, sport, science reviews, vol. 14. Macmillan Publishing Co., New York.

Groves, B.M., Droma, T.S., Sutton, J.R., McCullough, R.G., McCullough, R.E., Zhuang, J.G., et al. (1993). Minimal hypoxic pulmonary hypertension in normal Tibetans at 3658 m. J. Appl. Physiol. 74, 312–318.

Groves, B.M., Reeves, J.T., Sutton, J.R., Wagner, P.D., Cymerman, A., et al. (1987) Operation Everest II: elevated high-altitude pulmonary resistance unresponsive to oxygen. J. Appl. Physiol. 63, 521–530.

Hackett, P.H., and Rennie, D. (1976) The incidence, importance and prophylaxis of acute mountain sickness. Lancet 2, 1149–1155.

Hackett, P.H., and Roach, R.C. (1990) High altitude pulmonary edema. J. Wilderness Med. 1, 3–26.

Hochackka, P.W., Rupert, J.L., and Monge, C. (1999) Adaptation and conservation of physiological systems in the evolution of human hypoxia tolerance. Comp. Biochem. Physiol. 124 (Part A), 1–17.

Hochstrasser, J., Nanzer, A., and Oelz, O. (1986) Altitude edema in the Swiss Alps. Observations on the incidence and clinical course of 50 patients. Schweiz. Med. Wochenschr. 28, 866–873.

Honigman, B., Theis, M.K., Koziol-McLain, J., Roach, R., Yip, R., Houston, C., and Moore, L.G. (1993) Acute mountain sickness in a tourist population at moderate altitudes. Ann. Int. Med. 118, 587–592.

Houston, C.S., Sutton, J.R., Cymerman, A., and Reeves, J.T. (1987) Operation Everest II: man at extreme altitude. J. Appl. Physiol. 63, 877–882.

Hughson, R.L., Yamamoto Y., McCullough, R.E., Sutton, J.R., and Reeves, J.T. (1994) Sympathetic and parasympathetic indicators of heart rate control at altitude studied by spectral analysis. J. Appl. Physiol. 77, 2537–2542.

Hultgren, H.N., Grover, R.F., and Hartley, L.H. (1971) Abnormal circulatory responses to high altitude in subjects with a history of high altitude pulmonary edema. Circulation 44, 759–770.

Johnson, R.L., Cassidy, S.S., Grover, R.F., Schutte, J.E., and Epstein, R.H. (1985) Functional capacities of lungs and thorax in beagles after prolonged residence at 3100 m. J. Appl. Physiol. 59, 1773–1782.

Kawashima, A., Kubo, K., Kobayashi, T., and Sekiguchi, M. (1989) Hemodynamic responses to acute hypoxia, hypobaria, and exercise in subjects susceptible to high altitude pulmonary edema. J. Appl. Physiol. 67, 1982–1989.

Kryger, M., Glas, R., Johnson, V.D., Scoggin, C.S., Grover, R.F., and Weil, J.V. (1978) Impaired oxygenation during sleep in excessive polycythemia of high altitude: improvement with respiratory stimulation. Sleep 1, 3–7.

Lopez-Barneo, J., Lopez-Lopez, J.R., Urena, J., and Gonzales, C. (1988) Chemotransduction in the carotid body: K+ modulated by Po_2 in type 1 chemoreceptor cells. Science 242, 580–582.

Mazzeo, R.S., Bender, P.B., Brooks, G.A., Butterfield, G.E., Groves, B.M., Sutton, J.R., et al. (1991) Arterial catecholamine responses during exercise with acute and chronic high altitude exposure. Am. J. Physiol. 261, E419–E424.

Mazzeo, R.S., Brooks, G.A., Butterfield, G.E., Podolin, D.A., Wolfel, E.E., and Reeves, J.T. Acclimatization to high altitude increases muscle sympathetic activity both at rest and during exercise. (1995) Am. J. Physiol. 269, R201–R207.

Monge, C., Arregui, A., and Leon-Velarde, F. (1992) Pathophysiology and epidemiology of chronic mountain sickness. Int. J. Sports Med. 13, S79–S81.

Moore, L.G., Curran-Everett, L., Droma, T.S., Groves, B.M., McCullough, R.G., McCullough R.E., et al. (1992) Are Tibetans better adapted? Int. J. Sports Med. 13, S86–S88.

Oelz, O., Howald, H., di Prampero, P.E., and Reeves, J.T. (1986) Physiological profile of world-class high-altitude climbers. J. Appl. Physiol. 60, 1734–1742.
Oelz, O., Maggiorini, M., Ritter, M., Noti, C., Waber, U., Vock, P., and Bartsch, P. (1992) Prevention and treatment of high altitude pulmonary edema by a calcium channel blocker. Int. J. Sports Med. 13, S65–S68.
Penulosa, D., and Sime, F. (1971) Chronic cor pulmonale due to loss of altitude acclimatization (chronic mountain sickness). Am. J. Med. 50, 728–743.
Post, J.M., Hume, J.R., Archer, S.L., and Weir, E.K. (1992) Direct role for potassium channel inhibition in hypoxic pulmonary vasoconstriction. Am. J. Physiol. 262, C882–C890.
Reeves J.T. (1973) Pulmonary vascular responses to high altitude. Cardiovasc. Clin. 5 Clinical-pathological correlations 2, 81–95.
Reeves, J.T., and Schoene, R.B. (1991) When lungs on mountains leak. N. Engl. J. Med. 325, 1306–1307.
Reeves, J.T., and Grover, R.F. (1974) High-altitude pulmonary hypertension and pulmonary edema. In: Yu, P.N., and Goodwin J.F. (eds.) pp. 99–118. Progress in cardiology IV Febiger, Philadelphia.
Reeves, J.T., Groves, B.M., Cymerman, A., Sutton, J.R., Wagner, P.D., Turkevich, D., and Houston, C.S. (1990) Cardiac filling pressures during cycle exercise at sea level. Resp. Physiol. 80, 147–154.
Reeves, J.T., Groves, B.M., Sutton, J.R., Wagner, P.D., Green H.J., Cymerman, A., and Houston, C.S. (1991) Adaptations to hypoxia: lessons from Operation Everest II. In: Simmons, D.H. (ed.) pp. 23–50, Current pulmonology, Mosby Year Book Publishers, St. Louis.
Reeves, J.T., Houston, C.S., and Sutton, J.R. (1989) Operation Everest II: resistance and susceptibility to chronic hypoxia in man. J. R. Soc. Med. 82, 513–514.
Reeves, J.T., Mazzeo, R.S. Wolfel, E.E., and Young, A.J. (1992) Increased arterial pressure after acclimatization to 4300 m: possible role of norepinephrine. Int. J. Sports Med. 13, S18–S21.
Reeves, J.T., Groves, B.M., Sutton, J.R., Wagner, P.D., Cymerman, A., et al. (1987) Operation Everest II: preservation of cardiac function at extreme altitude. J. Appl. Physiol. 63, 531–539.
Reeves, J.T., Houston, C.S., Sutton, J.R. (1989) Operation Everest II: resistance and susceptibility to chronic hypoxia in man. J. R. Soc. Med. 82, 513–514.
Reeves, J.T., Wagner, W.W., McMurtry, I.F., and Grover, R.F. (1979) Physiological effects of high altitude on the pulmonary circulation. In: Robertshaw, D. (ed.) pp. 289–310. Int. Rev. Physiol. III, vol. 20. University Park Press, Baltimore.
Reeves, J.T., Wolfel, E.E., Green, H.J., Mazzeo, R.S., Young, A.J., Sutton, J.R., and Brooks, G.A. (1992) Oxygen transport during exercise at altitude and the lactate paradox: Lessons from Operation Everest II. In: Holloszy, J.O. (ed.) pp. 275–296. Exercise & sports sciences review, vol. 20. Williams & Wilkins, Baltimore.
Reeves, J.T., Monge, C.C., Leon-Velarde, F., Moore, L.G., Asmus, I., Curran, L., et al. (1998) Symposium on chronic exposure to hypoxia and chronic mountain sickness (CMS). In: Ohno, H., Kobayashi, T., Masuyama, S., and Nakashima, M. (eds.) Press Committee pp. 105–166. Progress in mountain medicine and high altitude physiology. Third World Congress on Mountain Medicine and High Altitude Physiology, Matsumoto, Japan.
Roberts, A.C., Reeves, J.T., Butterfield, G.E., Mazzeo, R.S., Sutton, J.R., Wolfel, E.E., and Brooks, G.A. (1996) Acclimatization to 4300-m altitude decreases reliance on fat as a substrate. J. Appl. Physiol. 80, 605–615.
Roberts, A.C., Butterfield, G.E., Cymerman, A., Reeves, J.T., Wolfel, E.E., and Brooks, G.A. (1996) Acclimatization to 4300-m altitude decreases reliance on fat as a substrate. J. Appl. Physiol. 81, 1762–1771.
Schoene, R.B. Swenson, E.R., Pizzo, C.J., Hackett, P.H., Roach, R.C., Mills, W.J., et al. (1988) The lung at high altitude: bronchoalveolar lavage in acute mountain sickness and high altitude pulmonary edema. J. Appl. Physiol. 64, 2605–2613.
Selland, M.A., Stelzner, T.J., Stevens, T., Mazzeo, R.S., McCullough, R.S., and Reeves, J.T. (1993) Pulmonary function and hypoxic ventilatory response in subjects susceptible to high-altitude pulmonary edema. Chest 103, 111–116.
Singer, D. (1999) Neonatal tolerance to hypoxia: a comparative-physiological approach. Comp. Biochem. Physiol. (Part A), 123, 221–234.

Suarez, J., Alexander, J.K., and Houston, C.S. (1987) Enhanced left ventricular systolic performance at high altitude during Operation Everest II. Am. J. Cardiol. 60, 137–142.
Sutton, J.R., Reeves, J.T., Wagner, P.D., Groves, B.M., Cymerman, A. Malconian, M.K., et al. (1988) Operation Everest II: oxygen transport during exercise at extreme simulated altitude. J. Appl. Physiol. 64, 1309–1321.
Tenney, S.M. (1962) Physiological adaptations to life at high altitude. Mod. Concepts Cardiovasc. Dis. 31, 713–718.
Tenney, S.M. (1990) Avian Physiology and performance at high altitude. In: Sutton, J.R., Coates, G., Houston, C.S. (eds.) pp. 2–3. Hypoxia: the adaptations. BC Dekker, Philadelphia.
Tucker, C.E., James, W.E., Berry, M.A., Johnstone, C.J., and Grover, R.F. (1976) Depressed myocardial function in goats at high altitude. J. Appl. Physiol. 41, 356–361.
Vogel, J.A., Hartley, L.H., Cruz, J.C., and Hogan, R.P. (1974) Cardiac output during exercise in sea level residents at sea level and high altitude. J. Appl. Physiol. 36, 169–172.
Wagner, P.D., Sutton, J.R., Reeves, J.T., Cymerman, A., Groves, B.M., and Malkonian, M.K. (1987) Operation Everest II: pulmonary gas exchange during simulated ascent of Mt. Everest. J. Appl. Physiol. 63, 2348–2359.
Wagner, P.D. (1988) An integrated view of the determinants of maximum oxygen uptake. In: Oxygen transfer from atmosphere to tissues. pp. 245–256. Plenum Press, New York.
Ward, M.P., Milledge, J.S., and West, J.B. High (1995) Altitude medicine and physiology. Chapman & Hall Medical, New York.
Weil, J.V. (1990) Lesson from high altitude. Chest 97, 70S–76S.
Weil, J.V., Jamieson, G., Brown, D.W., and Grover, R.F. (1968) The red cell mass—arterial oxygen relationship in normal man. J. Clin. Invest. 47, 1627–1639.
West, J.B., and Mathieu-Costello, O. (1992) Stress failure in pulmonary capillaries: a mechanism for high altitude pulmonary edema. In: Sutton, J.R., Coates, G., and Houston, C.S., (eds.) pp. 229–240. Hypoxia and mountain medicine. Queen City Printers, Burlington, VT.
Winslow, R.M., and Monge, C. (1987) Hypoxia, polycythemia, and chronic mountain sickness. The Johns Hopkins University Press, Baltimore, MD.
Wolfel, E.E., Groves, B.M., Brooks, G.A., Butterfield, G.E., Mazzeo, R.S., Moore, L.G., et al. (1991) Oxygen transport during steady-state submaximal exercise in chronic hypoxia. J. Appl. Physiol. 70, 1129–1136.
Wolfel, E.E., Selland, M., Mazzeo, R.S., and Reeves, J.T. (1994) Systemic hypertension at 4300m is related to sympatho-adrenal activity. J. Appl. Physiol. 76, 1643–1651.
Wolfel, E.E., Selland, M.A., Cymerman, A., Brooks, G.A., Butterfield, G.E., Mazzeo, R.S., et al. (1998) O_2 extraction maintains O_2 uptake during submaximal exercise with beta-adrenergic blockade at 4300m. J. Appl. Physiol. 85, 1092–1102.
Young, A.J., Young, P.M., McCullough, R.E., Moore, L.G., Cymerman, A., and Reeves, J.T. (1991) Effect of beta-adrenergic blockade on plasma lactate concentration during exercise at high altitude. Eur. J. Appl. Physiol. 63, 315–322.

Recommended Readings

Gibbs, J.S. (1999) Pulmonary hemodynamics: implications for high altitude pulmonary edema (HAPE). Adv. Exptl. Med. Biol. 474, 81–91.
Hackett, P.H. (1999) High altitude cerebral edema and acute mountain sickness. A pathophysiology update. Adv. Exptl. Med. Biol. 474, 23–45.
Heath, D., and Williams, D.R. (1997) Man at high altitude. Churchill Livingstone, Edinburgh.
Ramirez, G., Bittle, P.A., Rosen, R., Ralde, H., and Pineda, D. (1999) High altitude living: genetic and environmental adaptation. Aviation Space Envtl. Med. 70, 73–81.
Ward, M.P., Milledge, J.S., and West, J.B. (1995) High altitude medicine and physiology. Chapman & Hall, New York.
Weibel, E.R. (1999) Understanding the limitation of O_2 supply through comparative physiology. Resp. Physiol. 118, 85–93.
West, J.B. (1998) High life. Oxford University Press, New York.

18
Lung Immunology and Host Defense

KEITH C. MEYER

The respiratory tract is potentially exposed to a diverse array of gases, particulates, and microorganisms via inhalation or aspiration. Because the respiratory tract is directly exposed to substances in the surrounding environment, numerous defense mechanisms protect this interface between the external environment and internal tissues (Table 18.1). Inhalation of organic or inorganic dusts, exposure to toxic gases, aspiration of gastric acid, smoke inhalation, and inhalation or aspiration of infectious agents can induce lung disease. Whether foreign substances cause lung disease is determined by (1) patterns of deposition in the respiratory tract; (2) the total burden of inhaled or aspirated substances; (3) the adequacy of clearance mechanisms; (4) the adequacy of mucosal protective mechanisms; and (5) the presence of properly functioning immune responses to foreign substances. A given agent may cause no disease in some individuals, disease limited to the conducting system (tracheobronchial tree), or disease involving the distal lung parenchyma (gas-exchange areas). Because defenses that do not require antigen-specific inflammatory responses represent an important first line of defense that is backed up by specific immune reactions, this chapter will review the complex and integrated system of mechanical barriers and phagocytic defenses that exist in the respiratory tract in addition to specific immune responses that protect the lung from injury mediated by exogenous agents.

Pulmonary Defense Mechanisms

Defenses of the Upper Airway

Microorganisms can reach the airways via inhalation of aerosols or aspiration of pharyngeal secretions (Fig. 18.1). Particulates or microbes are humidified as they pass into the upper respiratory tract, and aerodynamic and dimensional properties of these moisturized particles determine the depth of penetration and area of deposition in the

TABLE 18.1. Pulmonary defense mechanisms.

Mechanical defenses	Cough Impaction of particles in proximal airways Barrier function of mucus and epithelium Mucociliary clearance
Phagocytic defenses	Macrophages Polymorphonuclear leukocytes
Immune defenses	Surfactant Lymphocytes Immunoglobulin Dendritic cells Phagocytic cells Cytokines Adhesion molecules Antimicrobial peptides
Injury limitation	Antioxidants Antiproteases

lower respiratory tract. Reflex mechanisms and mechanical barriers are the basic protective mechanisms in the upper tract. Reflex mechanisms include coughing and sneezing, and the impaction of inhaled particles on nasal hairs or upon upper airway mucosa in areas of turbulent airflow represent one type of mechanical barrier. Mucus overlying the epithelial mucosa prevents penetration by deposited substances, and inhaled or aspirated substances are cleared by the ciliary action of mucosal epithelial cells. In addition, the normal naso-oropharyngeal bacterial flora help prevent colonization by bacterial or fungal pathogens that are not part of the normal flora, and secretory immunoglobulin A antibodies or surfactant-associated proteins may prevent viruses or pathogenic bacteria from adhering to epithelial surfaces.

The Conducting Airways

Dichotomous branching points of larger bronchi may disrupt laminar airflow and facilitate deflection and deposition of particulates suspended in the airstream onto the mucosa. Very small particles (i.e., generally less than 0.01 μm in diameter) are relatively unlikely to reach distal alveolar surfaces due to diffusional deposition in the nasopharynx or proximal tracheobronchial mucosa, or due to a complete lack of retention in the lung. Large particulates (≥10 μm) are predominantly deposited in areas of turbulent airflow in the upper airway, but particulates of 5–8 μm in diameter can be deposited on the tracheobronchial mucosa and must be cleared by mucociliary action. Particulates of 2–5 μm in diameter may be deposited in bronchoalveolar regions, where clearance may take up to 100 days or longer.

A pseudostratified, ciliated columnar epithelium lines the mucosa of the tracheobronchial tree (Fig. 18.2). Interspersed goblet cells and submosal glands produce the mucus that overlies the mucosal surface that stretches from the trachea to the respiratory bronchioles. Beating cilia constantly propel mucus proximally in the tracheobronchial tree. This mucociliary "escalator" is an important mechanism by which particulates deposited in the conducting airways are cleared by being transported to the proximal trachea, where mucus is expectorated. Individuals with abnormal cilia (e.g., Kartagener's syndrome) or mucus (e.g., cystic fibrosis) are prone to recurrent lower respiratory tract bacterial infection, which leads to bronchiectasis and

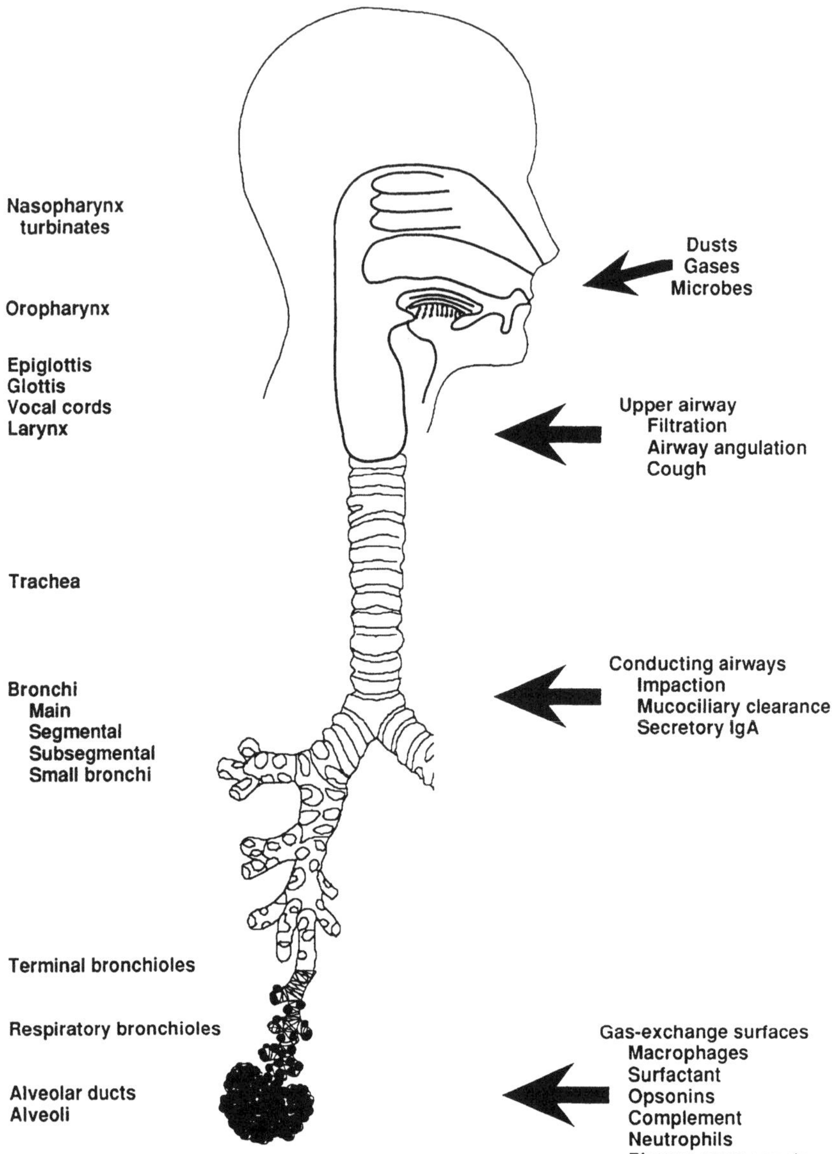

FIGURE 18.1. Host defenses of the upper and lower respiratory tract. Inhaled ambient air is filtered in the nasopharynx and continually branching, conducting airways. Microorganisms and large particulates impact on the mucosa and can be removed by mucociliary clearance and cough. Airspace macrophages are predominantly responsible for clearing small particulates or bacteria that reach alveolar surfaces where mucociliary clearance does not extend, and specific immune responses that involve components of systemic immunity can be mounted if foreign substances cause alveolar epithelial inflammation and increased mucosal permeability.

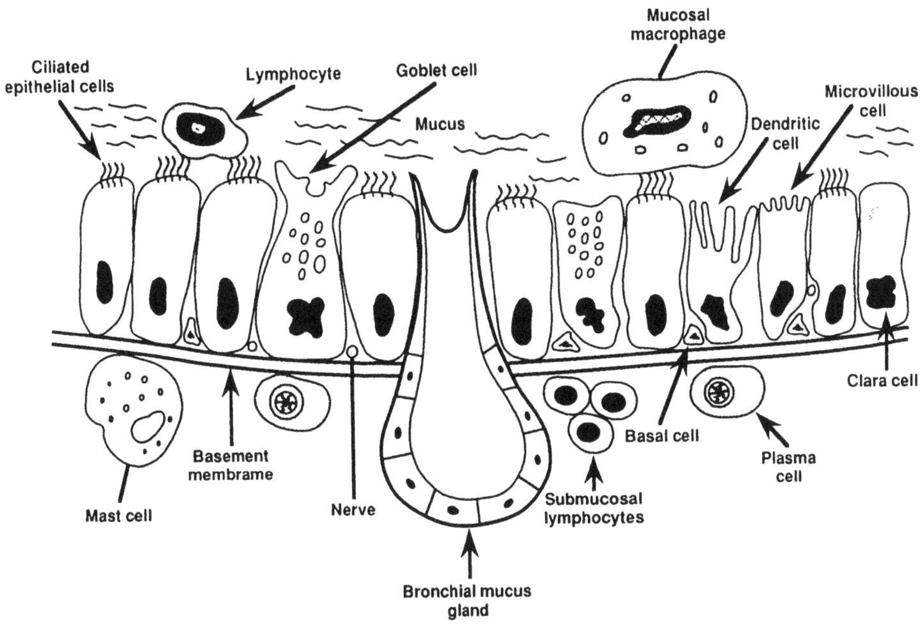

FIGURE 18.2. Mucosal components of conducting airways. The surface of the mucosa with its pseudostratified, ciliated epithelium and overlying mucus layer is shown. A mucus-producing submucosal gland and goblet cells are present. The overlying mucus is propelled toward the larynx by the cilia (the mucociliary "escalator"). Submucosal lymphocytes, plasma cells, and mast cells are present. Plasma cells can secrete immunoglobulins, and mast cells can release mediators (e.g., histamine). Macrophages and smaller numbers of lymphocytes can be found on the lumenal airway surface. Dendritic cells, which are potent accessory cells that present antigen to lymphocytes, are found in the mucosa. A plexus of adrenergic and cholinergic nerve fibers extends throughout the airway, and the mucosa is supplied by a rich vascular network from the bronchial circulation.

progressive pulmonary dysfunction. In addition to mechanical clearance mechanisms, immunoglobulins, iron-binding proteins, and other antimicrobial peptides confer additional protection to the mucosa of the conducting airways.

The Gas-Exchange Surface

The air-exchange surface, comprised of alveolar ducts and alveoli, has been estimated to have a surface area in aggregate that averages approximately 75 m^2 and ranges up to 150 m^2 in the adult human. The conducting bronchial structures repetitively divide as air flows more distally in the lung, and cross-sectional surface area increases as airways approach distal gas-exchange surfaces. Because the large increase in cross-sectional area at the level of the respiratory bronchioles and alveoli causes virtual absence of flow velocity of inhaled air, forceful impaction of particulates on the surface of alveolar units tends not to occur. Dynamic expansion and closure of alveoli with respiration, however, may force particulates against alveolar walls and cause their deposition in alveolar lining fluid. Because the mucociliary escalator and other defenses of the conducting airways are absent in alveolar units, host defenses depend upon such factors as phagocytosis by alveolar macrophages and access to lymphatic drainage. Some particles reaching the alveolar surfaces are cleared in days to weeks, but others may not be cleared for up to 1 year.

The alveolar Type-II epithelial cell-derived, surfactant-rich film that lines the epithelial surfaces of alveolar units can interact with substances that reach these areas and facilitate their neutralization and clearance (Fig. 18.3). Immunoglobulins, complement components, fibronectin, and surfactant apoproteins can adhere to the surface of microorganisms via specific receptors (opsonization), thereby facilitating phagocytosis by alveolar macrophages or polymorphonuclear leukocytes that possess cell membrane receptors for these glycoproteins. If irritation or inflammation increases epithelial permeability, plasma components can traverse the endothelial–epithelial barrier and gain access to alveolar spaces. These plasma components or other locally produced substances may have protective effects, but they can also disrupt the function of pulmonary surfactant. Other substances in epithelial lining fluid protect alveolar units from injury;

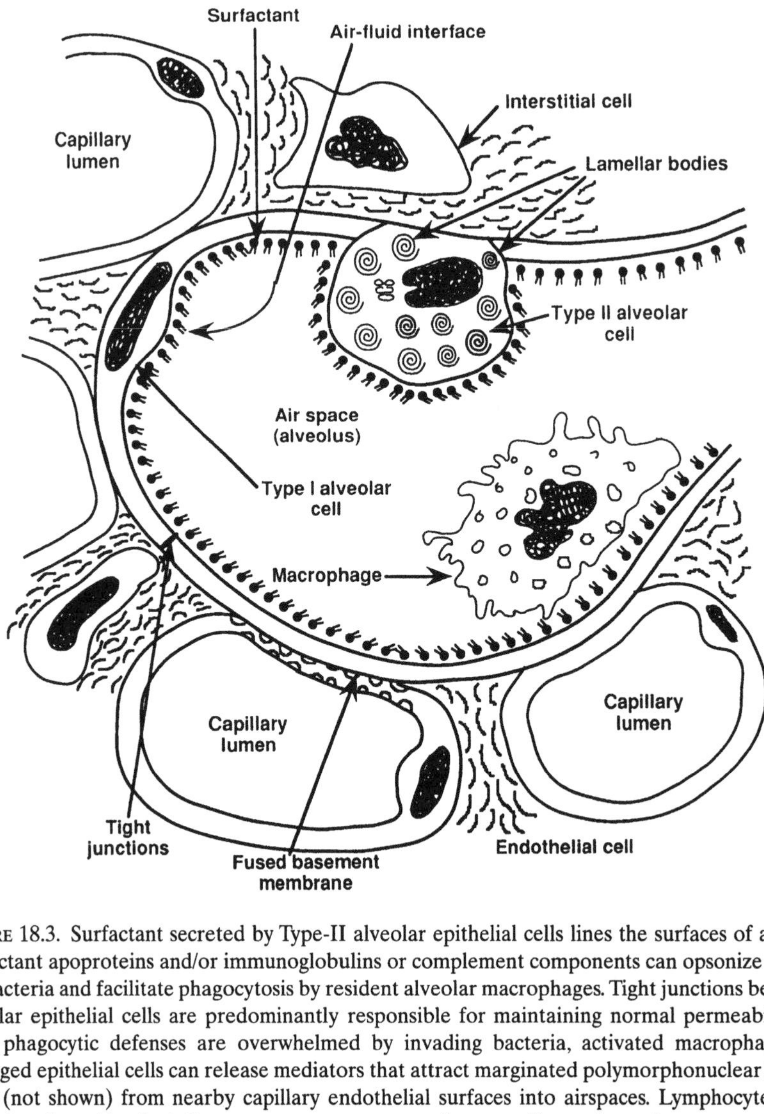

FIGURE 18.3. Surfactant secreted by Type-II alveolar epithelial cells lines the surfaces of alveoli. Surfactant apoproteins and/or immunoglobulins or complement components can opsonize invading bacteria and facilitate phagocytosis by resident alveolar macrophages. Tight junctions between alveolar epithelial cells are predominantly responsible for maintaining normal permeability. If local phagocytic defenses are overwhelmed by invading bacteria, activated macrophages or damaged epithelial cells can release mediators that attract marginated polymorphonuclear leukocytes (not shown) from nearby capillary endothelial surfaces into airspaces. Lymphocytes (not shown) activated by the inflammatory response may release mediators that enhance macrophage phagocytosis and bacterial killing.

TABLE 18.2. Pulmonary antiprotease and antioxidant defenses.

Antiproteases	α_1-Protease inhibitor (α_1-antitrypsin)
	Secretory leukoprotease inhibitor
	α_1-Antichymotrypsin
	Tissue inhibitor of metalloproteases (TIMP)
	Elafin
	α_2-Macroglobulin
Extracellular antioxidant defenses	Glutathione
	Albumin
	Superoxide dismutase
	Lactoferrin
	Transferrin
	Ceruloplasmin
	Bilirubin
Intracellular antioxidant systems	Superoxide dismutase
	Catalase
	Glutathione peroxidase
Nonenzymatic antioxidants	Vitamin E (tocopherol)
	β-Carotene
	Vitamin C (ascorbate)
	Taurine
	Uric acid
	Albumin

for example, glutathione, which is the most abundant extracellular antioxidant in epithelial surface fluid, can neutralize tissue-damaging oxidants, and antiproteases neutralize proteolytic enzymes such as neutrophil elastase (Table 18.2).

In the normal lung, free cells on alveolar surfaces are predominantly macrophages. When alveolar units are rinsed with normal saline (bronchoalveolar lavage), approximately 85–90% of the retrieved cells are macrophages, 10% are lymphocytes, and relatively few are neutrophils (1–3%) or eosinophils (<1%). The relative number of cells on alveolar surfaces is generally increased when inflammation is present. Lymphocytes, neutrophils, eosinophils, or monocytes can infiltrate the lung interstitium and reach epithelial surfaces. These inflammatory cells adhere to endothelial surfaces via specific receptors (adhesion molecules and their complementary ligands) and migrate into the lung interstitium or across epithelial surfaces in response to chemoattractant glycoproteins (e.g., interleukin-8 and monocyte chemoattractant protein-1) or other chemoattractants that are released at sites of inflammation. The identification and characterization of cell–cell and cell–substrate adhesion molecules, as well as of chemokines, is a rapidly expanding area of research.

Clearance of Inhaled Particulates

Once deposited upon the mucus film that lines the surface of conducting airways, particulates are generally propelled via the mucociliary escalator to the large bronchi and trachea. Macrophages adhering to the mucosa or migrating in the mucus film may ingest such particulates. However, particulates that penetrate deeply into the lung may be deposited on the mucosa of the respiratory bronchioles, alveolar ducts or alveoli where the mucociliary escalator does not reach. Although a small proportion of such particulates may be translocated to ciliated airways via surfactant, most particles (1) directly penetrate the surfactant-rich film lining the epithelium and the mucosa to be cleared via transcellular pathways, (2) gain access to the lymphatic system, or (3) are ingested by alveolar macrophages and cleared via intracellular degradation. Clearance may be ineffective if an overwhelming burden of particulate matter is deposited or if phagocytosis is ineffective (e.g., asbestos fibers causing macrophage cell death and lysis).

Pulmonary Immune Responses

Adaptive immunity, also termed antigen-specific immunity, is mediated by T and B lymphocytes derived from pluripotent stem cells in fetal liver and bone marrow. It is characterized by the generation of antigen-specific immunoglobulins by B cells and antigen-specific surface receptors by T cells. Although it is estimated that approximately 10^4 specific immunoglobulin and 10^{18} T cell receptors are generated via somatic gene rearrangement, relatively small numbers of memory T cells are retained for any specific antigen. Although T cells must passage through the thymus to complete their development, adaptive responses are generated in secondary lymphoid tissues that include the spleen, lymph nodes, and mucosa-associated lymphoid tissue that is present throughout the lung.

The innate immune system, which does not rely on antigen-specific lymphocyte responses, is a constantly active system. Although innate responses do not rely on immunologic memory, such responses are immediate and involve pathogen-specific, pattern-recognition receptors (e.g. mannan-binding lectins that bind carbohydrate moieties on gram-positive and gram-negative bacteria, fungi, and some parasites and viruses), phagocytes, signalling receptors (e.g. toll-like receptors that induce expression of cytokines and costimulatory molecules), the alternative complement pathway, and antimicrobial peptides. The highly conserved, pathogen-associated molecular structures that are recognized and bound by receptors of the innate immune system are shared by entire classes of microbes and are not found on host tissues. These microbial structures include mannans, lipopolysaccharide, peptidoglycan, lipoteichoic acids, glucans, bacterial DNA, and double-stranded RNA. Innate immunity, particularly the production of antimicrobial peptides and other antimicrobial molecules by epithelial cells, plays a critically important role in protection against infection in the lung with its extensive epithelial surfaces. Innate immune mechanisms also produce signals which trigger adaptive immunity or act in an adjuvant fashion to augment and regulate acquired immune responses via expression of co-stimulatory molecules.

Immune Cells of the Normal Lung

Antigen-Presenting Cells

Macrophages and dendritic cells are the predominant cell populations in the lower respiratory tract that can recognize, ingest, and process antigenic substances. Once antigen is ingested and partially degraded, antigenic determinants are displayed on the surface of these cells and recognized by antigen-reactive lymphocytes. Major histocompatibility complex (MHC) gene products on the antigen-presenting cell surface, particularly class II molecules, play a major role in presenting antigen to lymphocytes.

Lung macrophages are often referred to as alveolar macrophages; however, macrophages both reside on alveolar epithelial surfaces, and they are found in the mucus layer lining airway epithelium, the lung interstitium, the pulmonary microvasculature, or the pleural spaces. Other reticuloendothelial system counterparts of the lung macrophage are Kuppfer cells in the liver, peritoneal macrophages, bone osteoclasts, or central nervous system microglial cells. Lung macrophage populations are derived both from the influx and differentiation of peripheral blood monocytes and from *in situ* replication of resident macrophages. Macrophages both ingest, process, and present antigen, and are capable of secreting a plethora of monokines and growth factors that can have autocrine, paracrine, or endocrine effects on cell behavior. Macrophage-derived cytokines can regulate the behavior of lymphocytes, neutrophils, endothelial cells, or fibroblasts. Macrophages also scavenge particulates, ingest and remove macromolecular debris, ingest and kill microorganisms, and maintain and repair the lung parenchyma by secreting factors that influence the behavior of endothelial cells and fibroblasts. Nonmobile cells (e.g., fibroblasts, endothelial cells, or epithelial cells), however, are not merely passive cells. They initiate or modulate immune responses via responses to various immunostimulatory signals or via production and secretion of cytokines, antimicrobial peptides, and other immunoregulatory molecules.

Although pulmonary macrophages can serve as accessory cells for antigen-induced lymphocyte activation and proliferation, they do so relatively poorly as compared with peripheral blood monocytes or dendritic cells. They can even inhibit lymphocyte activation. Dendritic cells, which are also referred to as interdigitating cells or Langerhans' cells, are derived from bone marrow precursors and are of lymphocytic lineage. These morphologically distinct cells are present in many tissues and strongly express class II MHC surface molecules. Dendritic cells are very potent accessory cells and can be up to 100 times more potent than monocyte/macrophages in inducing T lymphocyte proliferation. They are widely distributed throughout the normal lung in the epithelium of airways, in the alveolar interstitium, and along visceral pleurae. Because dendritic cells are well situated to interact with antigens that reach and penetrate bronchial or alveolar walls, and because they are potent activators of T lymphocytes, they are thought to play a pivotal role in a wide spectrum of pulmonary immune responses. Dendritic cells rapidly translocate to draining lymph nodes upon encountering stimulatory antigens.

Lymphocytes

Lymphoid tissue is directly adjacent to the air–tissue interface throughout the conducting and gas exchange surfaces of the lung (Fig. 18.4). Lymphocytes are present in the normal lung in nodules or aggregates in the submucosa of conducting airways. Lymphocytes are also scattered throughout the interstitium, interalveolar septae, and pleurae, as well as on epithelial surfaces. Although larger, follicularlike aggregations of lymphocytes with germinal centers and surrounding lymphocytes are found in many mammalian species, these larger aggregates, which are referred to as bronchus-associated lymphoid tissue (BALT) and are thought to be analogous to Peyer's patches in the intestinal mucosa, do not appear to be present in the normal human lung. A profuse network of lymphatic vessels drain the airway mucosa and distal lung parenchyma, with efferent lymphatic vessels coursing through the bronchovascular bundles and terminating in regional lymph nodes. The hilar and paratracheal lymph nodes that drain bronchial and alveolar tissues play a critically important role in immune responses to foreign antigenic substances that reach the lower respiratory tract. In addition to lymphocytes in the lung interstitium and bronchoalveolar spaces, a large pool of intravascular lymphocytes exists that is estimated to be much larger than that of the liver or kidney.

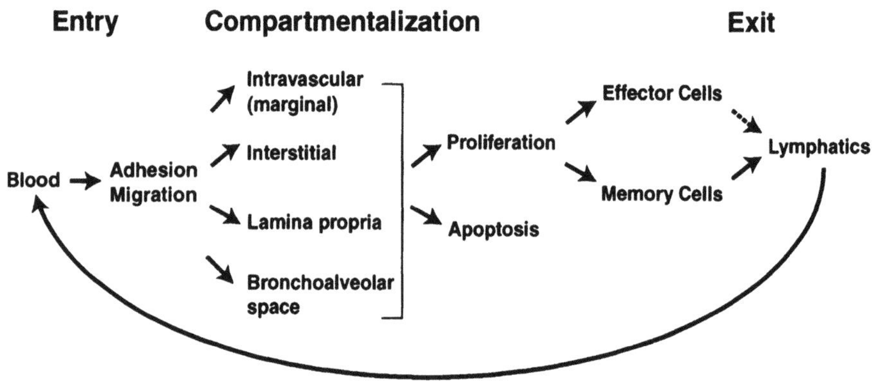

FIGURE 18.4. Lymphocyte traffic and compartmentalization in the lung. Lymphocytes can enter, exit, and reenter the lung. They can enter lung tissues via adhesion to vascular endothelial cells, migrate into extravascular spaces, and become compartmentalized. Some lymphocytes remain within microvessels adherent to endothelium (marginal intravascular pool), some enter the interstitium, some the lamina propria, and some reach the epithelium and can cross into bronchoalveolar spaces. Lymphocytes can proliferate and become effector or memory cells, although some undergo apoptosis. Lymphocytes exit the lung via lymphatics and may then reside in regional lymph nodes or reenter circulating blood.

Foreign materials absorbed from the respiratory tract potentially encounter multiple collections of lymphoid tissues while passing from alveoli or bronchioles proximally to hilar lymph nodes. Lymphocytes may migrate into the lung lymphoid aggregates via the high endothelium of postcapillary venules, reside for a variable time period in such tissues, and then reenter the extrapulmonary environment by entering lymphatic vessels and regaining access to the systemic circulation (Fig. 18.4). This "trafficking" of lymphocytes is an active, ongoing process with lymphocytes entering the lung via specific adhesion molecules on endothelial cells. Lymphocytes can rapidly and massively infiltrate lung parenchyma in response to inflammatory stimuli. Hilar lymph nodes play a particularly important role in generating antigen-specific immune responses because they contain both antigen-presenting dendritic cells and macrophages, as well as antigen-reactive B and T lymphocytes.

Monoclonal antibodies that detect specific antigens on cell surfaces are used to identify distinct types of T lymphocytes. T lymphocytes exist in the lung predominantly as two distinct phenotypes: helper/inducer T cells (CD4+ cells), and cytotoxic/suppressor T-cells (CD8+ cells) (Fig. 18.5). T lymphocytes can recognize foreign antigens, yet avoid recognition of self-antigens by expressing antigen receptors that are highly specific for single antigenic epitopes. Two types of T-cell antigen receptors (TCRs) have been identified. The more common of the two is the $\alpha\beta$ receptor, which is found on 90–95% of lung T cells. T lymphocytes bearing $\alpha\beta$ TCRs are able to recognize a much more diverse array of antigens than are cells bearing the $\gamma\delta$ TCR, which is the other major TCR expressed by T cells. TCRs, however, bind only to antigens that are presented to lymphocytes in association with MHC proteins on the surfaces of antigen-presenting cells. Helper/inducer (CD4+) T lymphocytes recognize antigen associated with MHC class II molecules, while cytotoxic/suppressor (CD8+) T cells recognize antigen in association with class I MHC molecules.

CD4+ and CD8+ T cells differ both in their responses to MHC/antigen combinations as well as in their effector functions once stimulated. The CD4+ T lymphocyte population provides signals that promote the proliferation of both CD4+ and CD8+ T cells, provide help for B cells to proliferate and differentiate, and release factors that activate eosinophils and mast cells. They also activate mononuclear phagocytes to express MHC molecules, present antigens, and kill parasites. Many of these functions are carried out via lymphokines that affect target cells. Some of the lymphokines released by stimulated CD4+ lymphocytes include interleukin-2 (IL-2), interferon-γ, lymphotoxin, granulocyte-macrophage colony-stimulating factor (GM-CSF), IL-3, IL-4, IL-5, IL-6, and IL-10.

T lymphocytes that are of the CD8+ lineage may either mediate cell lysis or dampen immune responses (hence the name, *cytotoxic/suppressor T cell*), but they do not appear to be capable of carrying out both functions. Cytotoxic CD8+ cells can kill cells that express foreign antigens in conjunction with MHC class I molecules, for example cells infected with virus or intracellular parasites. When CD8+ cytotoxic cells interact with

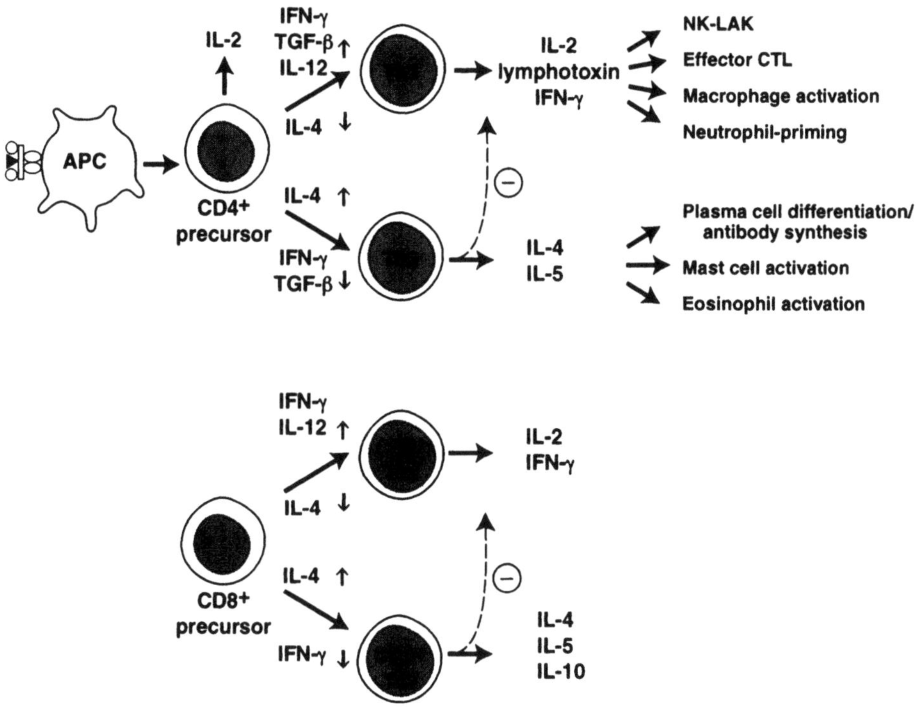

FIGURE 18.5. Lymphocyte subsets in the lung. Most T cells in the lung can be identified as either CD4+ (helper/inducer phenotypes) or CD8+ (cytotoxic suppressor phenotypes) lymphocytes. Differentiation of CD4+ lymphocytes to T helper 1 or T helper 2 phenotypes is influenced by the cytokine milieu, and some cytokines produced by a given T helper subset or other cytokine-secreting cells can suppress proliferation of other subsets. T helper 1 lymphocytes secrete characteristic lymphokines (IL-2, lymphotoxic, interferon-γ) and can give rise to NK-LAK (natural killer-lymphocyte activated killer cells), effector CTL (effector cytotoxic T lymphocytes), activate macrophages, and prime neutrophils for an enhanced respiratory burst. T helper 2 cells secrete different lymphokines (IL-4, IL-5) and play an important role in the differentiation of B-cell precursors to plasma cells that synthesize and secrete antibodies, activation of mast cells, and eosinophil recruitment and activation. Cytotoxic/suppressor (CD8+) T cells can also be influenced to differentiate into phenotypes that have characteristic patterns of lymphokine production and activities.

a target expressing MHC class I molecules in association with an antigen, adhesion molecules facilitate cell–cell contact and granule contents are released. Important mediators of cytotoxicity released from cytotoxic CD8+ T cell granules include serine proteases and complementlike molecules called *perforins*. It is less clear how CD8+ T lymphocytes carry out their suppressor functions; however, CD8+ T cells can suppress CD4+ T lymphocyte proliferation in response to allogeneic cells or soluble antigen and can also suppress CD4+ T lymphocyte help for immunoglobulin synthesis.

Other investigations have identified T-cell phenotypes that have distinctive cytokine release profiles and are implicated in specific lung disorders. T helper 1 (Th1) cells produce IL-2, IFN-γ, and lymphotoxin. T helper 2 (Th2) cells produce IL-4, IL-5, and IL-10, and both Th1 and Th2 cells express cytokines such as IL-3, tumor necrosis factor-α, and GM-CSF. T cells that express cytokines of both subtypes have been termed Th0 cells and are thought to be precursors of Th1 and Th2 lymphocytes, and cells which produce high amounts of transforming growth factor-β (TGF-β) have been termed Th3. Cytokines derived from Th1 cells (e.g., IFN-γ) inhibit generation of Th2 cells, and Th2-derived cytokines such as IL-4 inhibit generation of Th1 cells. The functions of Th1 and Th2 cells appear to correlate with their distinctive cytokine profiles. Th1 cells are involved in cell-mediated inflammatory reactions and induce delayed-type hypersensitivity reactions. Th1 cells are also likely important in pulmonary granulomatous responses found in mycobacterial infection or pulmonary sarcoidosis. The Th2 cell-derived cytokines promote antibody production, particularly IgE responses, and enhance eosinophil proliferation and function. Th2 cells are also strongly linked to parasitic infections and to asthma.

Soluble, antigen-specific immunoglobulins synthesized by cells of the B lymphocyte lineage mediate humoral immunity. The variable portion (domain) of immunoglobulin molecules binds antigen, whereas the Fc region of the constant region (domain) binds specifically to immunoglobulin receptors on macrophages, T cells and other cell types which play important roles in immunoglobulin-mediated immune responses. Although humoral immunity is important for pulmonary immune responses, relatively little is known about B cells in the human lung, and relatively few B cells are found on alveolar epithelial surfaces in the normal human lung.

Immune Responses in the Normal Lung

Integrated Immune Responses

Pulmonary host defenses should prevent a given foreign substance from penetrating deeply into the respiratory tract or should neutralize and eventually remove an invading foreign substance if deposited in the lower respiratory tract. Responses to foreign substances would ideally be adequate to protect the host and to maintain normal lung function, but they should not cause irreversible tissue injury while attempting to eradicate pathogens. Despite continual deposition of antigenic substances in the respiratory tract, mucosal and alveolar clearance mechanisms are highly efficient. Nonspecific clearance mechanisms (mechanical barriers and macrophage phagocytosis) constantly remove particulates and prevent them from evoking an immune response.

Antigen-specific immune responses may be necessary to cope with infectious agents that overwhelm nonspecific defenses including elements of innate immunity and gain access to the lower respiratory tract. Such immune responses, however, are complex and involve activation of many different cell types and the interaction of acquired and innate immune responses. Cytokines are of key importance in orderly immune responses, intercellular messengers that mediate cell–cell interactions and stimulate target cells or facilitate translocation of immune cells into the lung via chemotactic properties. Although cells of the hematopoietic lineage receive great emphasis in discussions of pulmonary immune responses, numerous cell types in addition to lymphocytes or macrophages are important participants in an orderly immune response. Lymphocytes, macrophages, or polymorphonuclear leukocytes interact directly via adhesion molecules or indirectly via cytokines with endothelial cells, epithelial cells, or fibroblasts. Epithelial cells release cytokines (e.g., interleukin-8, which is a potent neutrophil chemoattractant). In addition to releasing numerous cytokines, endothelial cells, express adhesion molecules for blood lymphocytes, neutrophils, eosinophils, and monocytes, facilitating recruitment of white cells from extrapulmonary sites and their entry into areas of inflammation. Other cell types (e.g., fibroblasts, mast cells, neuroendocrine epithelial cells, and platelets) also play important roles in immune responses and lung inflammation.

Regulation of Immune Responses in the Lung

Mechanisms must exist to suppress inflammatory reactions and maintain normal lung homeostasis in the setting of frequent deposition of inhaled antigenic material in the respiratory tract. Efficient and rapid clearance of foreign substances from mucosal and alveolar surfaces via mucociliary action and phagocytic responses helps to prevent immune stimulation. In addition, the local milieu of the lower respiratory tract tends to inhibit immune reactivity. Alveolar macrophages present antigen rather ineffectively in the normal host and tend to suppress immune responses. Normal pulmonary surfactant also suppresses immune reactivity, particularly lymphocyte responses; however, if nonspecific defenses are overwhelmed, specific antigen-driven responses must be upregulated. In addition to other cell types (e.g., endothelial and epithelial cells), lymphocytes and macrophages secrete both proinflammatory and inhibitory cytokines that modulate immune responses. The balance of proinflammatory and inhibitory cytokines likely determines the outcome of the inflammatory response to a specific antigen. Indeed, granulomatous inflammatory disorders such as pulmonary sarcoidosis or non-granulomatous disorders such as idiopathic interstitial pneumonia may represent overexuberant responses to inhaled environmental antigens. These responses may occur only in individuals who are susceptible to immune stimulation by a given antigen combined with insufficient immunosuppressive mechanisms. Genetically determined HLA alleles may predispose certain individuals to mount immune responses to a given stimulus, as has been suggested for workers exposed to beryllium compounds.

Altered Host Defenses and Lung Disease

The integrated system of host defenses can be perturbed at many levels. Pneumonia often occurs in a setting of impaired host defenses. Intubation of the trachea compromises mechanical barriers and is commonly associated with nosocomial pneumonia. Depressed mucociliary function, as seen in patients with Kartagener's syndrome or cystic fibrosis, leads to a vicious cycle of depressed clearance mechanisms predisposing to persistent bacterial colonization of the airways and recurrent infection. The destruction of airway mucosa and lung parenchyma caused by recurrent infection in turn causes further impairment of clearance mechanisms and perpetuation of this cycle. Pneumonia due to *Staphylococcus aureus* is often associated with recent influenza viral infection that can depress clearance mechanisms and lead to subsequent bacterial pneumonia. Other exogenous agents (e.g., inhalation of tobacco smoke or excessive intake of alcohol) can suppress clearance mechanisms and immune responses, predisposing individuals to develop bacterial pneumonia. Specific immune defects (e.g., complement deficiency, chronic granulomatous disease, or hypogammaglobulinemia) also predispose individuals to a variety of pulmonary infections. Malnutrition, diabetes mellitus, corticosteroid, and other immunosuppressive therapy, malignancy, cytotoxic chemotherapy, and organ transplantation can all depress immune defenses, leading to bacterial, fungal, or mycobacterial pneumonia.

Although it is important in controlling infection, immune responses to various stimuli may cause severe and irreparable lung injury. Inhaled inorganic dusts (e.g., asbestos or beryllium compounds) do not uniformly cause lung disease in all exposed individuals. The probability of developing pneumoconiosis from inorganic dust exposure is likely related to a combination of adequacy of clearance mechanisms and the ability of such dusts to activate immune responses in a given individual. Inhalation of organic substances may evoke a hypersensitivity response (e.g., farmers' lung) with lymphocytic infiltration of the lung parenchyma in susceptible individuals. Other chronic inflammatory lung diseases of unknown etiology (e.g., sarcoidosis) may be caused by exaggerated immune responses in susceptible individuals to a variety of inhaled antigens from environmental sources. Last, depletion or genetic lack of antiprotease defenses (e.g., α_1-antitrypsin deficiency leading to emphysema) or antioxidant defenses (e.g., glutathione depletion in idiopathic pulmonary fibrosis or antioxidant depletion in cystic fibrosis) may cause or accentuate lung damage in the inflamed lung.

Summary

Many of the defense mechanisms that protect the lung from infection or injury are relatively nonspecific and involve innate immune mechanisms, but others involve immune responses to specific antigens. Despite constant exposure to a variety of environmental agents, lung homeostasis is readily maintained in normal individuals. Immune responses to foreign substances that reach the lower respiratory tract involve both mobile, hemopoietic lineage cells (dendritic cells, macrophages, lymphocytes) and relatively nonmobile cells (epithelial cells, endothelial cells, neuroendocrine cells, and fibroblasts) that interact with and regulate each other via adhesion molecules and cytokines. Specific T cell phenotypes (e.g., Th1 and Th2 lymphocytes) appear to play important roles in pulmonary infection and noninfectious inflammatory lung diseases (e.g., sarcoidosis and asthma). Inhaled particulates or infectious agents may be cleared without antigen-driven activation of acquired immune responses. Infection may be prevented by intact innate immune responses, and infection may be limited and controlled by intact innate and adaptive (antigen-specific) immune mechanisms. Various conditions or disease states, however, can perturb this system of mechanical and immune defenses, leading to infectious pneumonitis or inflammatory, noninfectious disorders of the airways or lung parenchyma. A better understanding of pulmonary immune responses (e.g., mediator cascades, adhesion molecules, cytokine networks) and protective mechanisms that limit lung damage (e.g., antiproteases and antioxidants) will lead to the development of strategies that enhance immunity (e.g., reducing susceptibility to pulmonary infection in elderly or immunocompromised patients) or lessen lung inflammation in noninfectious inflammatory diseases (e.g., asthma or sarcoidosis).

Recommended Readings

Agostini, C., Chilosi, M., Zambello, R., Trentin, L., and Semenzato, G. (1993) Pulmonary immune cells in health and disease: lymphocytes. Eur. Respir. J. 6, 1378–1401.

Dar, K.J., and Crystal, R.G. (1999) Inflammatory lung disease: molecular determinants of emphysema, bronchitis, and fibrosis. In: Gallin, J.I., and Snyderman, R. (eds.) pp. 1061–1081. Inflammation: basic principles and clinical correlates. Raven Press, New York.

Delves, P.J., and Roitt, I.M. (2000) The immune system. New Engl. J. Med. 343, 37–49, 108–117.

Fraser, R.S., Colman, N., Müller, N.L., and Paré, P.D. (1999) Pulmonary defense and other nonrespiratory functions. In: Fraser and Parés' diagnosis of diseases of the chest. pp. 125–135. W.B. Saunders Co., Philadelphia.

Huttner, K.M., and Bevins, C.L. (1999) Antimicrobial peptides as mediators of epithelial host defense. Pediatr. Res. 45, 785–794.

Lipscomb, M.F., Bice, D.E., Lyons, R., Schuyler, M.R., and Wilkes, D. (1995) The regulation of pulmonary immunity. Adv. Immunol. 59, 369–455.

Medzhitov, R., and Janeway, C. Jr. (2000) Innate immunity. New Engl. J. Med. 343, 338–344.

Meyer, K.C. (2001) The role of immunity in susceptibility to respiratory infection in the aging lung. Respir. Physiol. 128, 23–31.

Reynolds, H.Y. (1991) Integrated host defense against infections. In: Crystal, R.G., and West, J.B. (eds.) pp. 1899–1911. The lung: scientific foundations. Raven Press, New York.

19 Acute Lung Injury

Barbara A. Cockrill and Homayoun Kazemi

The acute respiratory distress syndrome (ARDS) is a syndrome of diffuse lung injury characterized by high permeability pulmonary edema, diffuse alveolar infiltrates, hypoxia, and reduced lung compliance that occurs within hours to days following a variety of precipitating conditions. ARDS is now understood as a characteristic response of the lung to injury, and as part of the multiple organ dysfunction syndrome (MODS). Despite extensive research into the pathogenesis and treatment of this disorder, our understanding remains incomplete and mortality remains high.

The European–American Consensus Conference developed a working definition of acute lung injury (ALI) as distinct from the fully developed syndrome of ARDS (Bernard, 1994). These definitions are designed to allow easy classification of patients with the goal of more standardized research in this area. The definition of ALI is comprised of three parts: (1) hypoxemia, which is characterized by the ratio of the partial pressure of arterial oxygen (Pao_2) and fraction of inspired oxygen (Fio_2) of ≤ 300 regardless of the level of positive end-expiratory pressure (PEEP); (2) diffuse bilateral infiltrates on chest radiograph; and (3) the absence of left ventricular failure. If the degree of hypoxemia is more severe as defined by a Pao_2:Fio_2 ratio of less than or equal to 200, then the patient is considered to have ARDS. In recognition that this syndrome is not limited to adults, the name was changed from *adult* respiratory distress syndrome to *acute* respiratory distress syndrome.

A number of predisposing factors have been identified and include direct injury to the lung (e.g., gastric aspiration or pulmonary contusion) as well as remote injury (e.g., sepsis, pancreatitis, or multiple blood transfusion). With a single insult, the risk of developing ARDS is substantial, and risk increases in the presence of multiple risk factors (Fowler, 1983; Hudson, 1995) (Table 19.1). Approximately 150,000 cases of ARDS occur in the United States every year. Mortality remains greater than 50% in many series (Montgomery, 1985), although some improvement in the mortality rate has been noted (Milberg, 1995). Many patients require prolonged support and intensive care, and survivors need intensive rehabilitation; thus, the cost of caring for these patients can be staggering.

TABLE 19.1. Incidence of ARDS by clinical risk factor.*

Risk factor	Alone	With other risks	Total
Sepsis syndrome	56/136 (41.2)	19/40 (47.5)	75/176 (42.6)
Multiple transfusions	28/77 (36.4)	18/38 (47.4)	46/115 (40.0)
Near drowning	2/6 (33.3)	1/2 (50.0)	3/8 (37.5)
Pulmonary contusion	12/55 (21.8)	21/56 (37.5)	33/111 (29.7)
Aspiration	13/59 (22.0)	12/36 (33.3)	25/95 (26.3)
Multiple fractures	7/63 (11.1)	24/72 (33.3)	31/135 (23.0)
Drug overdose	14/164 (8.5)	12/36 (33.3)	26/200 (13.0)

* Number with ARDS/number at risk (%).
[*Source*: Modified from Hudson et al. (1995) Clinical risk for development of the acute respiratory distress syndrome. Am. J. Respir. Crit. Care Med. 151,293–301; with permission.]

ARDS is a progressive syndrome that can be divided into three pathological stages: early or exudative, middle or proliferative, and late or fibrotic. These stages are not distinct, and considerable overlap exists. These divisions, however, aid in understanding both the clinical findings as well as the histologic changes.

This chapter will review the clinical manifestations and pathology of ARDS with emphasis on the three pathological stages, as well as the current understanding of pathogenesis, and the management of this disorder.

Pathology

The histologic findings in human ARDS can be divided into three stages: exudative, proliferative, and fibrotic (Tomashefski, 1990). These distinctions are somewhat artificial in that the three processes often coexist in the same pathologic specimen.

The early or exudative phase is apparent at the onset of ARDS, and features are still apparent up to 1 week later. The histologic findings during this phase include widespread necrosis of Type-1 alveolar lining cells and adjacent endothelial cells with resultant exposure of the basement membrane and loss of alveolar surface integrity (Pratt, 1979). Interstitial edema is often out of proportion to the apparently mild ultrastructural damage, although functional abnormalities of the endothelial tight junctions are present and may not be apparent structurally (Albertine, 1985). Intraparenchymal hemorrhage is widespread. Thus, the alveolar barrier is lost, leading to the exudation of fluid and plasma proteins into alveolar spaces. In pathological preparations, this protein-rich edema fluid lining the alveoli is seen as the characteristic hyaline membrane. There is extensive neutrophil adhesion to pulmonary capillaries and prominent neutrophil infiltration of the interstitial space. Microthrombi begin to be seen in the microvasculature, although infarction in the early phase is unusual (Tomashefski, 1983). This paradox is likely due to the dual vascular supply of the lung by both the pulmonary and bronchial systems (Reid, 1986). The origin of the thrombi is not certain, but probably is the result of *in situ* thrombosis secondary to exposure of the basement membrane and non-specific activation of clotting cascades. It is important to note that lung architecture is still intact, and that the process is completely reversible without sequelae with treatment of the initial insult [Fig. 19.1(A)]. Not all patients go on to develop the later, fibroproliferative phases of ARDS.

Within 48 hours, the proliferative phase begins. This intermediate phase of ARDS is characterized by proliferation of Type-2 alveolar epithelial cells, which proliferate along

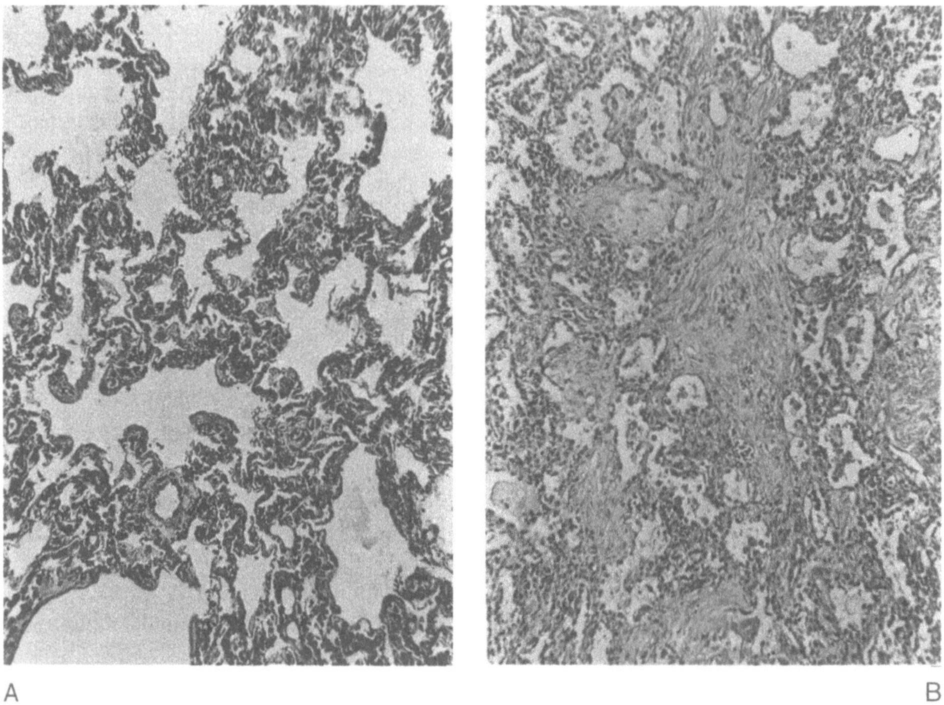

FIGURE 19.1. Pathologic changes in the early and late stages of ARDS. In the early stage (A), diffuse microatelectasis, alveolar wall damage, hyaline membranes, congestion, and increased cellularity of alveolar walls are evident. Open alveolar spaces allow for gas exchange and PEEP responsiveness. In late stage disease (B), patchy areas of organizing fibrosis are interspersed with open residual and regenerated air spaces. In this phase, the patient is much less likely to be PEEP responsive. [*Source*: From Lamy, et al. (1976) Pathologic features and mechanisms of hypoxia in adult respiratory distress syndrome. Am. Rev. Respir. Dis. 114, 267–269; with permission.]

the denuded alveolar basement membrane. The abundant neutrophillic infiltration persists, and mononuclear cells and lymphocytes are also present. This is accompanied by the appearance of an exuberant granulation tissue that is most prominent in alveolar ducts, and may obliterate the airspace (Tomashefski, 1983).

The third or fibrotic phase is characterized by intraalveolar and peribronchial fibrosis with a reduction in cellular infiltrate over a period of days to weeks [Fig. 19.1(B)]. The progression from proliferation to fibrosis may occur rapidly and collagen deposition in the lung has been documented as early as 3 days after the onset of ARDS, and the proliferative and fibrotic areas usually coexist. Epithelial cells eventually migrate over the granulation tissue which appeared during the proliferative phase, and it is thus incorporated into the interstitial space. This so-called fibrosis by accretion is a nonspecific healing pattern of the lung, and is the major cause of lung architecture distortion in ARDS (Spencer, 1985; Pratt, 1979). Two other mechanisms also account for fibrosis seen in ARDS. Fibroblasts and myofibroblasts proliferate within the alveolar wall, thus leading to widened alveolar septa. Finally, atelectatic segments may become fused by deposition of organizing fibrin or hyperplastic epithelium (Tomashefski, 1990). Distortion and obliteration of the alveolar and bronchial spaces occurs, with loss of normal lung architecture.

Collagen synthesis is increased in late ARDS. The level of procollagen III, which is a byproduct of collagen synthesis, is increased in both the serum and in the bronchoalveolar lavage fluid, and increased lavage fluid levels correlate with the degree of histologic evidence of intraalveolar fibrosis (Entzian, 1990; Waydhas, 1993). Fibrosing alveolitis is well recognized in most patients dying with ARDS (Lamy, 1976, 1987). The stimulus for progression to increased collagen synthesis and fibrosis is unknown; however, lung fibroblasts isolated from patients dying with ARDS are capable of proliferation in the absence of growth factors required by normal lung fibroblasts (Bitterman, 1992). Increased levels of procollagen III have been documented in bronchoalveolar lavage fluid by as early as day 3 after the onset of ARDS before clinical evidence of fibrosis (i.e., markedly decreased compliance) is appreciated, and procollagen III levels correlate with survival (Clark, 1994). It is likely that once a significant degree of fibrosis is present, mortality is greatly increased.

The pulmonary vasculature also shows progressive changes. As ARDS advances, there is an overall decrease in the cross-sectional luminal area due to *in situ* thrombosis and fibrous intimal proliferative lesions in preacinar and intraacinar arteries (Tomashefski, 1983) (Fig. 19.2). In addition to a significant increase in the arterial muscularity that extends to smaller vessels, these findings are likely to account for the progressive increase in pulmonary artery pressure during the course of ARDS in patients who do not survive (Zapol, 1977). Areas of infarction occur in the later stages and are most prominent adjacent to visceral pleura probably because collateral circulation is less effective in these sites (Jones, 1985). The ischemic areas are preferentially ventilated, as indicated by injection of dye into the bronchial tree, presumably because the ischemic tissue is more compliant. This predisposes the lung to barotrauma (i.e., pneumothorax and bronchopleural fistula) and contributes to the increase in dead-space ventilation.

Clinical Features

Throughout the clinical course, ARDS is characterized clinically by hypoxia requiring supplemental oxygen, decreased lung compliance, increased dead-space ventilation, and often increases in pulmonary vascular resistance. The pathophysiological mechanisms for these findings changes, however, during the course of the syndrome, and the clinical manifestations correspond with what is seen pathologically.

The first stage of ARDS begins after a latent period following an initial insult, usually within 12–24 hours. The patient is tachypneic and cyanotic, often with a paucity of other physical findings. Hypoxemia is present and the lungs demonstrate reduced compliance. Diffuse infiltrates are found on chest radiographs. Impaired pulmonary perfusion manifested by mild pulmonary hypertension may be found. The primary problem is noncardiogenic protein-rich pulmonary edema fluid due to capillary leak causing flooding of the interstitium and alveoli. The boggy fluid-filled lungs are thus prone to widespread atelectasis that results in areas where perfusion occurs in the absence of ventilation (pulmonary shunt). Hypoxia, therefore, is typically more troublesome than hypercarbia in early ARDS. At this point, hypoxia will usually respond to increased PEEP. PEEP acts by opening and maintaining atelectatic segments, making them available for gas exchange, and thereby decreasing the shunt fraction. Lung compliance is also reduced early in the course of ARDS due to interstitial edema fluid, as well as to surfactant dysfunction. The pulmonary vascular resistance may be elevated early on due to hypoxic vasoconstriction and possibly overdistention imposed by mechanical ventilation.

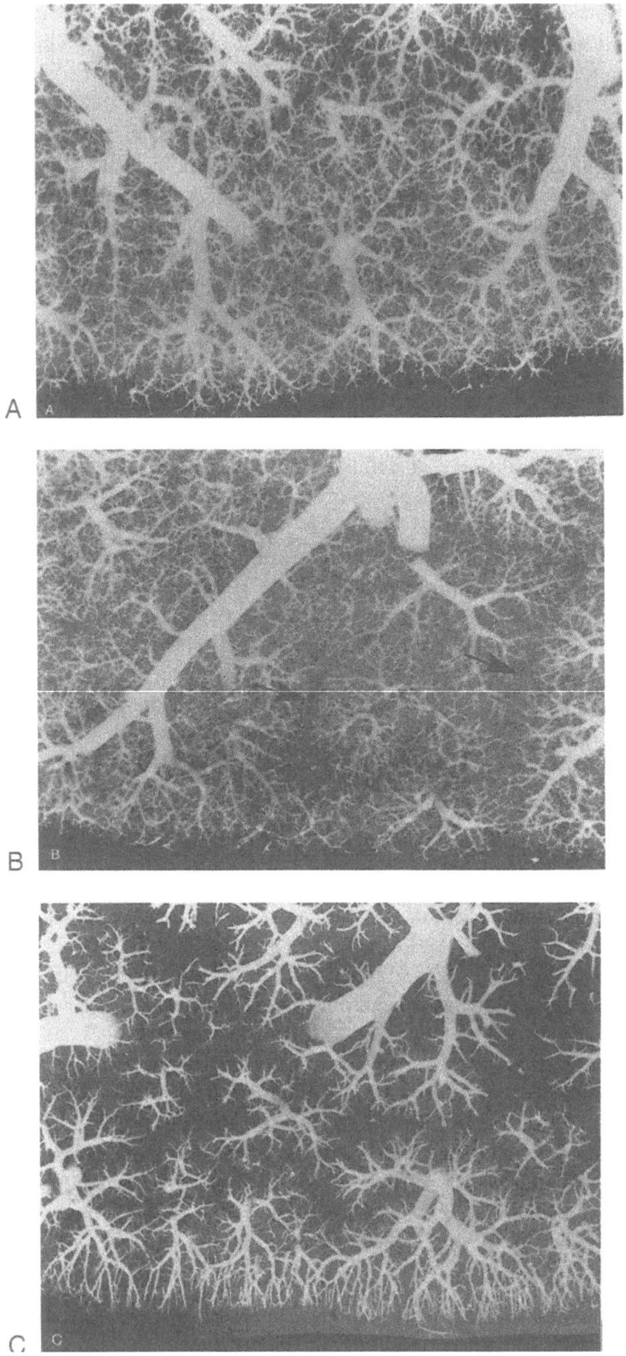

FIGURE 19.2. Postmortem angiographic patterns in ARDS. (A) Normal lung. (B) Patient with early ARDS (6 days after aspiration). There is reduced filling of small arteries and prominent edematous interlobular septa (arrows). Platelet fibrin microthrombi were present histologically. (C) Extensive reduction of filled peripheral arteries because of intimal obliteration. Subpleural branches are stretched about dilated air spaces with "picket fence" appearance (16 days after toxic inhalation). [*Source*: From Tomashefski, J.F., Jr. (1990) Pulmonary pathology of the adult respiratory distress syndrome. Clin. Chest Med. 11 (4), 593–619; with permission.]

Thrombocytopenia and disseminated intravascular coagulation are systematically often present due to activation of clotting factors by the injured pulmonary endothelium. The clinical distinction between the proliferative and fibrotic phases of ARDS is very difficult to make, and these two phases are often lumped as the fibroproliferative phase for clinical discussion. Indeed, the two phases are pathologically present simultaneously. The development of fibroproliferative stage is often accompanied by signs of systemic inflammation (e.g., fever and hypotension). Oxygenation remains a problem; however, in contrast to the early phase hypoxia is less responsive to the application of PEEP. Hypoxia often responds early on to increasing PEEP, which acts by stenting open edematous small air spaces. In the fibroproliferative phase, the influx of inflammatory cells and fibrosis makes the application of PEEP less effective in improving oxygenation. Lung compliance continues to worsen as edema is replaced by the proliferation of inflammatory tissue and eventually fibrosis.

The increasing vascular obstruction, both from intraluminal clot and from perivascular fibrosis, increases the areas that are ventilated but not perfused (i.e., dead space). Ventilation–perfusion inequalities and increased airways resistance develop due to peribronchial fibrosis. Higher-minute ventilation is therefore required to remove CO_2, and usually results in higher inspiratory pressures if a normal $Paco_2$ is maintained. These alterations make effective mechanical ventilation difficult, and lead to refractory hypercarbia or hypoxemia, prolonged ventilator dependency, and predispose patients to complications (e.g., barotrauma and nosocomial pneumonia).

It is likely that many of the pathophysiological changes that occur in the lungs during ARDS are occurring in capillary beds of other organ systems. Most patients with ARDS suffer at least transient abnormalities in the function of other systems, and mortality is directly related to the number and length of organs system dysfunction (Cockrill, 1989; Knaus, 1989).

Pathogenesis

ARDS is a manifestation of generalized inflammation. As such it is part of MODS (Goris, 1989). The appearance of ARDS and MODS in response to injury is not uniform. With a given insult, only a minority of patients will develop the syndromes (Hudson, 1995). The factors that determine which patients will respond to a particular insult with the progression to ARDS and MODS are not yet understood.

Many biological mediators have been implicated in the pathogenesis of ARDS and MODS, and the exact roles are an area of active research. In general, the process is initiated by exposure to a stimulus (e.g., endotoxin), which activates macrophages to release both tumor necrosis factor (TNF-α) and interleukin 1 (IL-1). An inflammatory cascade in then set in motion that results in widespread inflammation and tissue damage. Once the process has been initiated, however, many other factors are involved in sustaining and causing progression of ARDS. These factors include the injury caused by mechanical ventilation and supervening infection. This section will focus on the central initiating roles of endotoxin, TNF-α, and IL-1, as well as the synergistic role of ventilator-induced lung injury.

In cases of gram-negative sepsis, a role for endotoxin in the pathogenesis of systemic inflammatory response and the subsequent development of ARDS and MODS is certain. The mechanism by which endotoxin is associated with ARDS serves as a model for other factors that initiate the inflammatory cascade (Fig. 19.3). Other inciting agents include toxins from gram-positive bacteria, fungi, and parasites, as well as mediators released after tissue damage (e.g., in trauma or burn injury). Animal models confirm

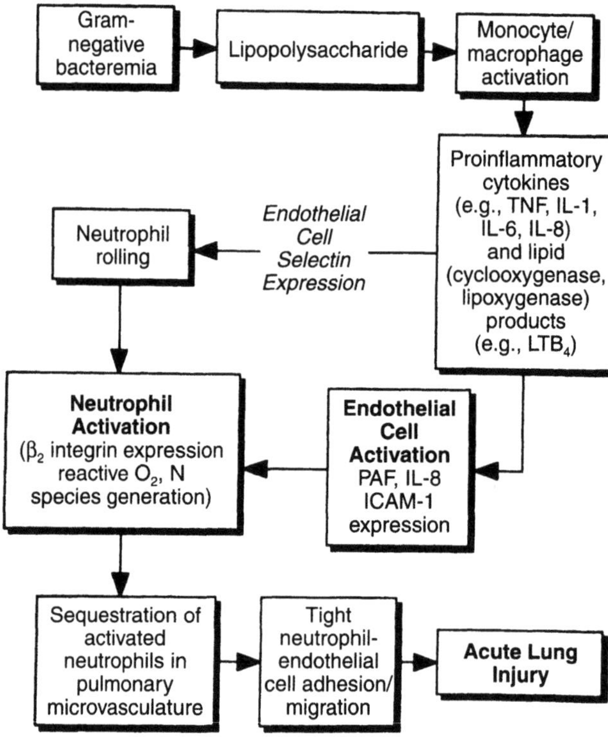

FIGURE 19.3. A complex sequence of biologic events occurs following the initial insult, which in this case is bacteremia. The steps detailed describe a likely mechanism that leads to acute lung injury. TNF = tumor necrosis factor; IL = interleukin 1; LTB4 = leukotriene B4; PAF = platelet-activating factor; ICAM-1-intercellular adhesion molecule-1; O_2 = oxygen. [*Source*: From Sessler et al. (1996) Current concepts in sepsis and acute lung injury. Clin. Chest Med. 17, 213–235; with permission.]

that injection of endotoxin at high doses is sufficient for the development of ARDS-like pathology, and the mechanisms involve both direct effects and, more importantly, the initiation of an inflammatory cascade (Brigham, 1986). Serum endotoxin levels are elevated in both at risk patients and those with established ARDS (Parsons, 1991).

Interleukin-1, which is a mediator produced by activated macrophages, is also implicated early in ARDS. Endotoxin injection increases mRNA for IL-1 transcription within 30 minutes. Injection of IL-1 in animal models reproduces the shock syndrome and subsequent development of organ system dysfunction (Okusawa, 1988). IL-1 receptor antagonists (IL-1ra) injection blocks the systemic hypotension and increase in pulmonary leakage of albumin and water in rats with gram-negative and endotoxin-induced sepsis (Rose, 1994). A number of studies in septic animal models have demonstrated decreased mortality when animals receive IL-1 receptor antagonists either as pretreatment or early treatment after sepsis is induced (Ohlsson, 1990; Alexander, 1991,1992; Fisher, 1992). In humans, persistent elevations in blood IL-1 levels over time in patients with ARDS correlate with decreased survival and likely indicate a persistent inflammatory state. In a longitudinal study of patients with ARDS, the combination of elevated serum IL-1 and IL-6 levels were consistent and efficient predictors of mortality (Meduri, 1995). Clinical trials of IL1-ra in patients with sepsis at risk for the development of ARDS have unfortunately not shown a difference in survival. Subgroup analysis, however, indicates that patients who are most severely ill with organ dysfunction may benefit (Fisher, 1994). This suggests that only patients with a persistent generalized inflammatory state will actually benefit from having the inflammatory cascade shutdown.

The most complete evidence for a role in the initiation of acute lung injury and multiple organ dysfunction syndrome, is for tumor necrosis factor α (TNF-α). Human TNF-α is a polypeptide that is produced by macrophages in response to a variety of stimuli, including endotoxin. The effects of IL-1 and TNF-α are closely linked, and the two mediators act synergistically (Waage, 1988). TNF-α, in concert with IL-1, stimulates the formation and release of a number of other inflammatory mediators, and therefore sets up a cascade of inflammation, which may lead to the development of fulminant ARDS if it proceeds unchecked. In patients at risk for the development of ARDS or MODS, sustained elevations of TNF and IL-1 can be demonstrated in those who do go on to develop the syndromes (Meduri, 1995).

The animal data supporting the role of TNF-α in early ARDS is extensive. The development of increased capillary membrane permeability following the administration of endotoxin is largely mediated by the subsequent release of TNF-α. Infusion of TNF into animals induces a shock syndrome that is identical to that seen after infusion of *E. coli* or endotoxin. Hypotension, third-spacing of fluids due to increased capillary permeability, and stress hormone responses occur early, with development of progressive pulmonary dysfunction, renal tubular necrosis and hemorrhagic bowel necrosis within hours (Tracey, 1986; 1987). The histology of the lung lesion is identical to the pathology seen in early ARDS: extensive neutrophil adhesion to pulmonary capillaries, diffuse edema, interstitial inflammation, and parenchymal hemorrhage (Stephens, 1988).

The expression of neutrophil and monocyte adhesion molecules in vascular endothelium is also increased by TNF-α which allows the inflammatory cells to be held in a "primed" state in the lung. TNF-α has been shown to increase endothelial neutrophil adhesion by causing an increase in expression of the complement component C3bi (CR3) (Gamble, 1985).

In humans, TNF-α levels are increased in the serum and bronchoalveolar lavage fluid of patients with early ARDS, and levels are higher in nonsurvivors of ARDS (Meduri, 1995). Lavage levels are higher than serum indicating local production within the lungs (Hyers, 1991). Increased expression of TNF mRNA by alveolar macrophages is confirmed by *in situ* hybridization studies: In patients with early ARDS, 66% of alveolar macrophages expressed the TNF gene without stimulation compared with 31% in non-ARDS lung disease and less than 8% in normal controls (Tran Van Nhieu, 1993).

TNF may also be involved in the initiation of the small vessel *in situ* thrombosis that is observed in the transition phase between early and the proliferative phase of ARDS. Human and bovine endothelial cells demonstrate a dose-dependent increase in the procoagulant tissue factor expression that is associated with a decrease in endothelial cell–dependent protein C activation (Nawroth, 1986). These findings indicate a shift in the balance of anticoagulant and procoagulant activity toward an activated procoagulant state.

Finally, TNF may contribute to the pathophysiology of ARDS by inhibiting the synthesis of surfactant. It is known that surfactant dysfunction occurs, and administration of exogenous surfactant results in transient improvement in gas exchange in patients with ARDS (Spragg, 1994). *In vitro* studies demonstrate decreased phosphatidylcholine synthesis by human Type-II pneumocytes when exposed to TNF, which is an effect that is partially blocked by indomethacin (Arias-Diaz, 1994) and by pentoxifylline (Balibrea-Cantero, 1994).

An early role of TNF-α and IL-1 in the initiation of the ARDS and MODS is certain. This knowledge has thus far unfortunately not translated into an effective treatment. As early mediators in the inflammatory cascade, TNF-α and IL-1 cause primary injury, but, more importantly, also serve as a *trigger* for the release of a complicated series of subsequent inflammatory mediators. The problems in translating this knowledge into clinical medicine are likely three: (1) to stop the process before the cascade has been initiated is impractical outside of the laboratory because early identification of patients who are likely to progress onto ARDS is extraordinarily difficult; (2) there are many repeated and ongoing insults for patients, and not simply a single, isolated initiating event whose effects can be neutralized by antiinflammatory therapy; and (3) oversuppression of the inflammatory process may have deleterious effects (e.g., increased infection and decreased injury healing).

The role of ventilator-induced lung injury has been recognized. In a rat model, mechanical ventilation with high peak inspiratory pressures resulted in severe pulmonary injury in less than 60 minutes, and the degree of injury was attenuated by the addition of PEEP (Webb, 1974). In studies on adult sheep, high pressure and large volume ventilation caused both the physiologic and pathologic changes of human ARDS after a short period (Kolobow, 1987). In an elegant series of experiments, Dreyfuss et al. sought to determine if the injurious effects of mechanical ventilation were due to high airway pressures or the resulting overdistention and large alveolar volumes. Dreyfuss and colleagues compared five groups of rats: a control group ventilated at normal tidal volume; a second group ventilated with high pressure and large volume; a third group ventilated with high pressure and volume, but with the addition of PEEP; a fourth group ventilated with high pressure and low volume (the animals had the thoracic cage strapped to prevent overdistention); and the last group, which was ventilated with negative external pressure sufficient to result a tidal volume similar to the high-pressure/high-volume group. All three high-volume groups showed marked interstitial edema and ultrastructural damage to Type-1 pneumocytes after only 20 minutes of mechanical ventilation. The addition of PEEP mediated the damage somewhat. In contrast, the control group and high-pressure/low-volume group did not show evidence of significant parenchymal injury. This indicates that the alveolar *overdistention*, and not the actual *pressure*, is most important in the development of the lung injury. The application of positive PEEP had a protective effect. Two mechanisms for this protective effect are postulated: (1) By preventing atelectasis, the application of PEEP avoided overdistention of nonatelectatic segments, and/or (2) PEEP prevented the shear stress of opening atelectatic alveoli with each ventilator breath.

The mechanism of lung injury induced by mechanical ventilation is likely to be related to initiation of inflammation, as well as to direct cellular injury. Different patterns of mechanical ventilation are associated with differences in lung neutrophil activation and associated lung injury. Surfactant-depleted rats ventilated in a conventional high-volume mode demonstrated marked influx of activated neutrophils into the lung and associated marked structural damage. Consistent with the protective effect of PEEP, rats with the same initial injury, but ventilated by high-frequency ventilation (which avoids atelectasis), had a similar influx of neutrophils; however, the cells were not activated and there was only minimal structural damage (Sugiura, 1994).

It is of note that more recent studies indicate that overdistention of the lung by mechanical ventilation is associated with an increase in TNF-α, IL-1, and other inflammatory cytokines in bronchoalveolar lavage fluid in both rats (Tremblay, 1997) and in humans (Ranieri, 1998). These studies support the concept that different mechanical ventilation strategies can significantly influence the inflammatory cascade in the lung, and possibly contribute to the initiation and/or propagation of lung injury.

Therapy

With advances in therapy over the last 25 years, survival has improved, but mortality remains high (Milberg, 1995). This section will discuss the approach to ventilator management, including the use of permissive hypercapnia and PEEP. Three promising newer modes of treatment will also be discussed: the use of inhaled nitric oxide, the use of corticosteroids in late ARDS, and perfluorocarbon liquid ventilation. The reader is referred to other reviews for a more complete discussion of therapy of ARDS (Koleff, 1995).

Ventilator Management

It has become increasingly recognized that mechanical ventilation itself promotes lung damage, and that ARDS may in part be a product of therapy rather than simply the progression of the underlying disease. The mechanism of ventilator-induced lung injury has been elucidated in animal models (discussed earlier) and appears to be overdistention of lung units and sheer stress imposed on atelectatic segments. It is clear that the conditions that are associated with ventilator-induced injury in practice are unavoidable in the ARDS patient if a normal alveolar ventilation is maintained. This has lead to the strategy of permissive hypercapnia: In this approach attention is given to avoiding overdistention and ventilator-induced injury at the expense of alveolar hypoventilation and an elevated $Paco_2$. Relative hypoventilation is accepted in order to ventilate the patient with lower distending volumes.

The problem that arises in the management of ARDS patients is determining what inspired tidal volume will result in overdistention. Despite the early description of ARDS as a diffuse lung disease, studies using computed tomography demonstrate that the lung involvement is heterogeneous: Both areas of dense consolidation and areas that appear relatively normal are present (Maunder, 1986; Gattinoni, 1987) (Fig. 19.4). Because much of the lung is consolidated in severe ARDS, the application of traditional tidal volumes (12–15 ml/kg) will preferentially ventilate and overdistend the more compliant, relatively spared regions, thus leading to ventilator-induced injury. In clinical practice, the nonoverdistending volume is difficult to determine and is constantly changing; therefore, lower distending *pressures* are targeted in order to avoid overdistention.

Ventilation with smaller tidal volumes inevitably results in a rising $Paco_2$. The physiologic consequences of hypercarbia include a decrease in alveolar Po_2, respiratory acidosis, and an increase in intracranial pressure. The effect on alveolar oxygen can generally be overcome by a slight increase in the inspired oxygen concentration; however, in some very severe patients ill this may be a limiting factor. A key to the successful use of permissive hypercapnia is to allow the $Paco_2$ to rise slowly, thereby allowing for intracellular and renal compensation of the acidosis. If the $Paco_2$ rises too rapidly, severe intracellular acidosis will develop. Head injury and increased intracranial pressure are contraindications to the use of permissive hypercapnia due to the concern regarding further increases in intracranial pressure. The increase in intracranial pressure caused by hypercarbia is mostly due to vasodilation of cerebral vessels in the face of maintained intravascular pressure. Thus, the intracranial blood volume is increased. When faced with a head-injured ARDS patient, some have advocated tra-

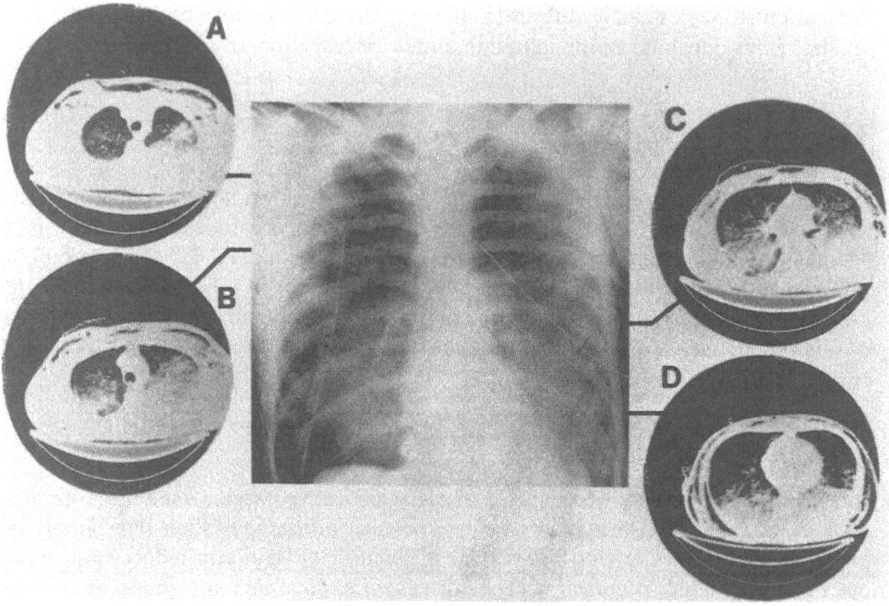

FIGURE 19.4. Standard anteroposterior chest roentgenogram in a patient with ARDS and corresponding computed tomographic (CT) images at levels indicated (A through D). Despite appearance of diffuse, symmetric involvement by standard chest x-ray film, CT images demonstrate sparing of substantial portion of lung parenchyma. Note lack of homogeneity and tendency toward posterior involvement on CT images. Cross-sectional images obtained by CT are oriented with anterior chest wall facing up and right-to-left orientation identical to that of convention chest x-ray film. [*Source*: From Maunder, R.J., et al. (1986) Preservation of normal lung regions in the adults respiratory distress syndrome: analysis by computed tomography. JAMA 255, 2463–2465.]

cheal gas insufflation as a means to lower the P_{CO_2} while still using small volume lower pressure ventilation (Levy, 1995). Tracheal gas insufflation is a technique whereby gas (oxygen) is continuously instilled through a catheter positioned near the distal end of the endotracheal tube; thus, it "flushes" out the alveolar CO_2 (Ravenscraft, 1993).

Clinical studies of the effects of lung-protective ventilator strategies are now available. When compared with mortality with retrospective controls, the use of pressure-limited ventilation in ARDS resulted in a decrease in mortality (Hickling, 1990). In a nonrandomized study of trauma patients with ARDS, the permissive hypercapnia strategy resulted in a marked increase in survival (91% vs. 48%, $p < 0.01$) (Gentilello, 1995). In a prospective randomized trial, permissive hypercapnia in conjunction with PEEP adjusted to recruit maximum lung volume without causing overdistention resulted in better survival and improved gas exchange and lung compliance when compared with conventional ventilation (Amato, 1998). Decreased mortality with lower lung volumes has not been found by all groups however (Brochard, 1998; Stewart, 1998). All of these studies had fewer than 150 patients.

In the largest study of protective ventilator strategies to date, the National Heart Lung and Blood Institute ARDSNetwork compared high versus low tidal volumes in more than 800 patients with ARDS. This study randomly assigned low tidal volume (6 ml/kg of ideal body weight) or high tidal volume (12 ml/kg) mechanical ventilator treatment to patients with ALI within 36 hours of onset. The study, originally targeted for 1000 patients, was stopped early due to a markedly lower mortality in the low tidal volume group (30% vs. 40% in the high tidal volume group, $p = 0.0054$) (ARDSNetwork, 1999). This study was not designed to investigate the effects of PEEP.

A much earlier advance in the ventilator management of ARDS was the application of PEEP (Pepe, 1984). By applying pressure at end-expiration, PEEP acts by essentially stenting open small airways and thus recruiting otherwise atelectatic segments for gas exchange. The shunt fraction (i.e., areas of perfused but nonventilated lung) is reduced and oxygenation improves. A lower inspired oxygen concentration can therefore be used to maintain adequate oxygenation. Oxygen in high concentrations is directly toxic to lung parenchyma; therefore, a reduction in the inspired concentration is likely to be beneficial. In animal models, the application of positive end expiratory pressure also provides some protection from ventilator-induced injury, presumably by recruiting a greater lung volume for ventilation and avoiding the shear stress associated with opening of atelectatic lung units with each inspiration (Dreyfuss, 1988).

Nitric Oxide

Nitric oxide (NO), which is the molecule responsible for the vasorelaxant effects of endothelial-derived relaxing factors, has shown utility in the management of ARDS patients (Rossaint, 1993). When given by inhalation in ARDS, NO causes modest decreases in pulmonary artery pressure and improvement in oxygenation presumably by improving ventilation perfusion matching. Inhaled NO is delivered only to ventilated lung units, and the resulting pulmonary vasodilation is also limited to these units (Zapol, 1994). Thus, a "steal" of blood from poorly ventilated units toward well-ventilated units occurs, and ventilation perfusion matching is improved.

The administration of NO allows a decrease in Fio_2 and PEEP, while maintaining the same level of oxygenation. Patients with more severe pulmonary vasoconstriction tend to respond with a greater improvement in Pao_2 and fall in pulmonary artery pressure, although some very severely ill patients may show no response (Bigetello, 1994). Tachyphylaxis is not observed, although rarely it may be difficult to wean NO off without rebound increases in pulmonary artery pressure and deterioration in oxygenation. Discontinuation of the NO will lead to a prompt return of Pao_2 and pulmonary artery pressure to baseline values. When used in conjunction with the strategy of permissive hypercapnia, NO can be used to reverse the vasoconstrictive effects of an elevated $Paco_2$ (Puybasset, 1994).

There may be other reasons to treat ARDS patients with inhaled NO: *In vitro* data indicate that NO has some antiinflammatory effects. NO decreases platelet aggregation in *in vivo* models (Radomski, 1990), and may decrease the tendency for the pulmonary capillary *in situ* thrombosis seen in ARDS. Furthermore, NO modulates some inflammatory responses of the endothelium. NO exposure decreases the expression of VCAM, which is a neutrophil adhesion molecule in cytokine-stimulated endothelial cells. It is possible that inhaled NO, if given early, may modulate the autodestructive inflammation responsible for ARDS. The clinical importance of these observations remains unknown.

Although the addition of inhaled NO improves oxygenation and pulmonary artery pressure in ARDS patients, the effect on mortality is not certain. Rossaint et al. (1993) treated 14 patients with severe ARDS with NO and found improved survival compared with historical controls. All of the study patients, however, were relatively young and were trauma victims, and both factors are known to be associated with better survival in ARDS (Cockrill, 1989). In addition, many of the patients were also managed with extracorporeal membrane oxygenation and carbon dioxide removal (ECMO/

$ECCO_2R$). Others have not found any improvement in survival when treating patients with severe, late-stage ARDS with NO (Bigatello, 1994). The role for inhaled NO in ARDS is evolving. It is clear that its use often allows a modest decrease in Fio_2, perhaps helping to avoid the effects of oxygen toxicity. It may be that NO, applied early, will affect survival due to both the antiinflammatory properties and better tolerance of permissive hypercapnia.

Corticosteroids in Late ARDS

The recognition that ARDS is the result of a widespread inflammatory response led to investigation of high doses of corticosteroids to try to prevent and treat the syndrome. Corticosteroids are unfortunately *not* beneficial early in the course of the syndrome. High doses of methylprednisilone do not prevent ARDS in patients at risk for the syndrome (Luce, 1988). If given within the first 24 hours after the onset of ARDS, there is no improvement in clinical course or survival. In fact, when given early there is a trend toward increased overwhelming infection (Bernard, 1987). There is clearly no role for high doses of corticosteroids in the treatment of early ARDS.

The pathology of late ARDS is very similar to that of other fibrosing lung diseases (e.g., idiopathic interstitial fibrosis) for which corticosteroids may be effective treatment. In uncontrolled, nonrandomized investigations, high doses of corticosteroids appeared to improve survival in patients with late fibroproliferative ARDS as compared with historical survival data (Ashbaugh, 1985; Hooper, 1990; Meduri, 1991). Corticosteroids in high doses is a potentially harmful treatment; therefore, a large, randomized, placebo-controlled trial is currently ongoing.

Liquid Ventilation

A new and intriguing area currently under investigation is the use of fluorinated hydrocarbons (perfluorocarbons) as a medium for gas exchange in injured lungs. The perfluorocarbons are clear, colorless, and odorless liquids in which both O_2 and CO_2 are extremely soluble. In contrast, neither water nor lipids are soluble in perfluorocarbons. Ventilation and oxygenation can therefore take place through this liquid. Two methods of gas exchange are used: complete or partial ventilation. In complete liquid ventilation, the lungs are filled with the perfluorocarbon and the entire volume is exchanged with each "breath." This method has been effective in small animal models; however, because of the technical difficulties in moving a large volume of liquid in and out of the lungs, it is difficult with patients. In partial liquid ventilation, the lungs are filled to functional residual capacity with perfluorocarbon, and a tidal volume of gas is then delivered to the remaining lung.

In animal models of ARDS, liquid ventilation is associated with improvement in gas exchange, pulmonary mechanics, and survival (Hirschl, 1995). Lung biopsy specimens reveal a reduction in pulmonary hemorrhage, lung fluid accumulation, and inflammatory infiltration in sheep with oleic acid–induced respiratory failure treated with liquid ventilation as compared with animals supported with gas ventilation (Hirschl, 1995). The mechanism by which gas exchange improves mostly appears to be due to the recruitment of atelectatic segments. The liquid enters dependent atelectatic segments and recruits them for gas exchange. Because these segments are held open by the liquid, overdistention and sheer stress from positive pressure ventilation is minimized. Animal studies have also shown a decrease in lung inflammatory infiltrate (Hirschl, 1995), the mechanism of which is unclear, but may be due to less barotrauma and/or direct effects on pulmonary macrophages. Finally, perfluorocarbon may aid in pulmonary toilet and decrease infection rates. Because perfluorocarbon is heavier than water, pulmonary secretions "float" on top of the layer of perfluorocarbon, thereby facilitating their removal by tracheal suctioning. A limitation in the use of liquid ventilation is that perfluorocarbon is radioopaque; therefore, the chest radiograph becomes useless in the evaluation of the patient.

In premature infants with infant respiratory distress syndrome, perfluorocarbon is well tolerated, rapidly improves gas exchange, and appears to improve survival when compared with historical controls (Leach, 1996). Preliminary studies in patients with ARDS indicate that liquid ventilation is well tolerated and apparently without adverse effects. A multicenter trial of partial liquid ventilation is currently underway in adults with ARDS.

Summary

This chapter reviews the mechanisms of ARDS, its pathogenesis, and currently available technical and other therapeutic options. Acute injury to the lung which can lead to development of the adult acute respiratory distress syndrome (ARDS) is a multifaceted disorder of the respiratory tract caused by a host of agents (with sepsis and pneumonia as leading causes), and ultimately manifested by significant hypoxemia, shortness of breath, diffuse infiltrates on chest x-ray and diminished lung compliance. ARDS affects some 150,000 people in the United States annually and has high mortality rates (in excess of 40%). In many instances, ARDS is also associated with multiple organ failure, involving heart, kidney, liver, central nervous system and hematopoietic system. Following injury to the lung there is extravasation of protein rich fluid across the alveolar-capacity membrane, and an intense inflammatory response is present in the lung that can lead to diffuse fibrosis. The pathogenesis of ARDS is not completely known but it clearly involves release of cytokines in the lung, particularly TNFα and IL-1. In most subjects with ARDS there is significant pulmonary hypertension, remodeling of pulmonary vascular bed, and loss of capillary bed. Pathology in the lung is divided into three stages: early or exudative, middle or proliferative, and late or fibrotic. However, many of these stages overlap. The mortality in the disease increases with involvement of other organs. Management of the disease at present is supportive with mechanical ventilation, and PEEP and increased oxygen content of inspired air. Recent studies have shown that over distention of alveoli by mechanical means in ARDS will lead to further lung injury and release of cytokines. This has led to development of new approaches to management of ARDS, the so-called "lung protective strategies", which minimizes the rise in pressure during ventilatory support, and limits the tidal volume to 6 ml/kg of ideal body weight. This approach has reduced mortality in ARDS significantly. Other approaches in patient management have included the use of nitric oxide, which reduces pulmonary hypertension and improves oxygenation, although it is not clear that the mortality is significantly reduced with nitric oxide. Liquid ventilation with fluorocarbons has also been used in some patients with ARDS and appears to make some difference in outcome, although long-term studies are not available at this time. Some studies have suggested that steroid use in the fibrotic stage of ARDS improves clinical outcome and mortality.

References

Albertine, K.H. (1985) Ultrastructural abnormalities in increased-permeability pulmonary edema. Clin. Chest Med. 6, 345–366.

Alexander, H.R., Doherty, G.M., Venzon, D.J., Merino, M.J., Fraker, D.L., and Norton, J.A. (1992) Recombinant interleukin-1 receptor antagonist (IL1ra): effective therapy against gram-negative sepsis in rats. Surgery 112, 188–194.

Alexander, J.R., Doherty, G.M., Buresh, C.M., Venzon, D.J., and Norton, J.A. (1991) A recombinant human receptor antagonist to interleukin 1 improves survival after lethal endotoxemia in mice. J. Exp. Med. 173, 1029–1032.

Amato, M.B.P., Barbas, C.S.V., Medeiros, D.M., Magaldi, R.B., Schettino, G. de P., Lorenzi-Filho, G., et al. (1998) Effect of a protective-ventilation strategy on mortality in the acute respiratory distress syndrome. N. Engl. J. Med. 338, 347–354.

Arias-Diaz, J., Vara, E., Garcia, C., and Balibrea, J.L. (1994) Tumor necrosis factor-alpha-induced inhibition of phosphatidylcholine synthesis by human type II pneumocytes is partially mediated by prostaglandins. J. Clin. Invest. 94(1), 244–250.

Ashbaugh, D.G., and Maier, R.V. (1985) Idiopathic pulmonary fibrosis in adult respiratory distress syndrome. Arch. Surg. 120, 530–535.

Balibrea-Cantero, J.L., Arias-Diaz, J., Barcia, C., Torres-Melero, J., Simon, C., Rodriguez, J.M., and Vara, E. (1994) Effect of pentoxifylline on the inhibition of surfactant synthesis induced by TNF-α in human type II pneumocytes. Am. J. Respir. Crit. Care Med. 149, 699–706.

Bernard, G., Artigas, A., Brigham, K., et al. (1994) The American European Consensus Conference on ARDS: definitions, mechanisms, relevant outcomes, and clinical trial coordination. Am. J. Respir. Crit. Care Med. 149, 818–824.

Bernard, G.R., Luce, J.M., Sprung, C.D., et al. (1987) High-dose corticosteroids in patients with the adult respiratory distress syndrome. N. Engl. J. Med. 317, 1565–1570.

Bigatello, L.M., Hurford, W.E., Kacmarek, R.M., Roberts, J.D. Jr, and Zapol, W.M. (1994) Prolonged inhalation of low concentrations of nitric oxide in patients with severe adult respiratory distress syndrome. Effects on pulmonary hemodynamics and oxygenation. Anesthesiology 80(4), 761–770.

Bitterman, P.B. (1992) Pathogenesis of fibrosis in acute lung injury. Am. J. Med. 92(Suppl. 6A), 6A39S–6A43S.

Brigham, K.L., and Meyrick, B. (1986) Endotoxin and lung injury. Am. Rev. Respir. Dis. 133, 913–927.

Brochard, L., Roudot-Thoraval, F., Roupie, E., Delclaux, C., Chastre, J., Fernandez-Mondejar, E., et al. (1998) Tidal volume reduction for prevention of ventilator-induced lung injury in acute respiratory distress sydrome. Am. J. Respir. Crit. Care Med. 158, 1831–1838.

Clark, J.G., Milberg, J.A., Steinberg, K.P., and Hudson, L.D. (1994) Elevated lavage levels of N-terminal peptide of type III procollagen are associated with increased fatality in adult respiratory distress syndrome. Chest 105 (Suppl. 3), 126S–127S.

Dreyfuss, D., Soler, P., Basset, G., and Saumon, G. (1988) High inflation pressure pulmonary edema. Respective effects of high airway pressure, high tidal volume and positive end-expiratory pressure. Am. Rev. Respir. Dis. 137, 1159–1164.

Entzian, P., Huckstadt, A., Kreipe, H., and Barth, J. (1990) Determination of serum concentrations of type III procollagen peptide in mechanically ventilated patients. Pronounced augmented concentrations in the adult respiratory distress syndrome. Am. Rev. Respir. Dis. 142(5), 1079–1082.

Fisher, C.J., Jr., Opal, S.M., Lowry, S.F., Sadoff, J.C., LaBrecque, J.F., Donovan, H.C., et al. (1994) Role of interleukin-1 and the therapeutic potential of interleukin-1 receptor antagonist in sepsis. Circ. Shock 44, 1–8.

Fisher, C.J., Dhainaut, J.F., Opal, S.M., Pribble, J.P., Balk, R.A., Slotman, G.J., et al. (1994) Recombinant human interleukin 1 receptor antagonist in the treatment of patients with sepsis syndrome. JAMA 271, 1836–1843.

Fisher, E., Marano, M.A., Van Zee, K.J., Rock, C.S., Hawes, A.S., Thompson, W.A., DeForge, L., et al. (1992) Interleukin-1 receptor blockade improves survival and hemodynamic performance in *E. coli* septic shock, but fails to alter host responses to sublethal endotoxemia. J. Clin. Invest. 89, 1551–1557.

Fowler, A.A., Hamman, R.F., Good, J.T., Benson, K.N., Baird, M., Eberle, D.J., et al. (1983) Adult respiratory distress syndrome: risk with common predispositions. Ann. Intr. Med. 98, 593–597.

Gamble, J.R., Harlan, J.M., Klebanoff, S.J., and Vadas, M.A. (1985) Stimulation of the adherence of neutrophils to umbilical vein endothelium by human recombinant tumor necrosis factor. Proc. Natl. Acad. Sci. USA 82(24), 8667–8671.

Gattinoni, L., Pesenti, A., Avalli, L., Rossi, F., and Bombino, M. (1987) Pressure-volume curve of total respiratory system in acute respiratory failure: computed tomographic scan study. Am. Rev. Respir. Dis. 136, 730–736.

Gentilello, L.M., Anardi, D., Mock, C., Arreola-Risa, C., and Maier, R.V. (1995) Permissive hypercapnia in trauma patients. J. Trauma 39, 856–853.

Goris, R.J.A. (1990) Mediators of multiple organ failure. Int. Care Med. 16, S192–S196.

Hickling, K.G., Henderson, S.J., and Jackson, R. (1990) Low mortality associated with low volume pressure limited ventilation with permissive hypercapnia in severe adult respiratory distress syndrome. Int. Care Med. 16, 372–377.

Hirschl, R.B., Parent, A., Tooley, R., McCracken, M., Johnson, K., Shaffer, T.H., et al. (1995) Liquid ventilation improves pulmonary function, gas exchange and lung injury in a model of respiratory failure. Ann. Surg. 221, 79–88.

Hooper, R.G., and Kearl, R.A. (1990) Established ARDS treated with a sustained course of adrenocortical steroids. Chest 97, 138–143.

Hudson, L.D., Milberg, J.A., Anardi, D., and Maunder, R.J. (1995) Clinical risks for development of the actue respiratory distress syndrome. Am. J. Respir. Crit. Care Med. 151, 293–301.

Hyers, T.M., Tricomi, S.M., Dettenmeier, P.A., and Fowler, A.A. (1991) Tumor necrosis factor levels in serum and bronchoalveolar lavage fluid of patients with the adult respiratory distress syndrome. Am. Rev. Respir. Dis. 144(2), 268–271.

Jones, R., Reid, L.M., Zapol, W.M., et al. (1985) Pulmonary vascular pathology: human and experimental studies. In: Zapol, W.M., Falke, K.J. (eds.) pp. 23–160. Acute respiratory failure. Marcel Dekker, New York.

Knaus, W.A., and Wagner, D.P. (1989) Multiple systems organ failure: epidemiology and prognosis. Crit. Care Clin. 5(2), 221–232.

Koleff, M.H., and Schuster, D.P. (1995) The acute respiratory distress syndrome. N. Engl. J. Med. 332, 27–37.

Kolobow, T., Moretti, M.P., Fumagalli, R., Mascheroni, D., Prato, P., Chen, V., and Joris, M. (1987) Severe impairment in lung function induced by high peak airway pressure during mechanical ventilation. Am. Rev. Respir. Dis. 312–318.

Lamy, M., Fallat, R.J., Roeniger, E., Dietrich, H.P., Ratliff, J.L., Eberhart, R.C., et al. (1976) Pathologic features and mechanisms of hypoxemia in adult respiratory distress syndrome. Am. Rev. Respir. Dis. 114(2), 267–284.

Leach, C.L., Greenspan, J.S., Rubenstein, S.D., Shaffer, T.H., Wolfson, M.R., Jackson, J.C., et al. (1996) Partial liquid ventilation with perflubron in premature infants with severe respiratory distress syndrome. The LiquiVent Study Group. N. Engl. J. Med. 335(11), 761–767.

Levy, B., Bollaert, P.E., Nace, L., and Larcan, A. (1995) Intracranial hypertension and adult respiratory distress syndrome: usefulness of tracheal gas insufflation. J. Trauma. 39, 799–801.

Luce, J.M., Montgomery, A.B., Marks, J.D., et al. (1988) Ineffectiveness of high-dose methylprednisolone in preventing parenchymal lung injury and inporving mortality in patients with septic shock. Am. Rev. Respir. Dis. 138, 62–68.

Marini, J.J., and Kelsen, S.G. (1992) Re-targeting ventilatory objectives in adult respiratory distress syndrome. Am. Rev. Respir. Dis. 148, 2–3.

Maunder, R.J., et al. (1986) Preservation of normal lung regions in the adults respiratory distress syndrome: Analysis by computed tomography. JAMA 255, 2463–2465.

Meduri, G.U., Belenchia, J.M., Estes, R.J., Wunderink, R.G., El Torky, M., and Leeper, K.V. (1991) Fibroproliferative phase of ARDS clinical findings and effects of corticosteroids. Chest 100, 943–952.

Meduri, G.U., Headley, S., Kohler, G., Stentz, F., Tolley, E., Unberger, R., and Leeper, K. (1985) Persistent elevation of inflammatory cytokines predicts a poor outcome in ARDS. Chest 108, 1303–1314.

Milberg, J.A., Davis, D.R., Steinberg, K.P., et al. (1995) Improved survivial of patients with actue respiratory distress syndrome (ARDS): 1983–1993. JAMA 273, 306–309.

Montgomery, A.B., Stager, M.A., Carrico, C.J., and Hudson, L.D. (1985) Causes of mortality in patients with the adult respiratory distress syndrome. Am. Rev. Respir. Dis. 132, 485–489.

Nawroth, P.P., and Stern, D.M. (1986) Modulation of endothelial cell hemostatic properties by tumor necrosis factor. J. Exp. Med. 163(3), 740–745.

National Institutes of Health National Heart Lung Blood Institute ARDS Network Presentation. American Thoracic Society Annual Meeting, San Diego, April 26, 1999.

Ohlsson, K., Bjork, P., Bergenfeldt, M., Hageman, R., and Thompson, R.C. (1990) Interleukin-1 receptor antagonist reduces mortality from endotoxin shock. Nature 348, 550–552.

Okusawa, S., Gelfand, J.A., Ikejima, R., Connolly, R.J., and Dinarello, C.A. (1988) Interleukin 1 induces a shock-like state in rabbits. Synergism with tumor necrosis factor and the effect of cyclooxygenase inhibition. J. Clin. Invest. 81, 1162–1172.

Parker, J.C., and Snow, R.L. (1972) Influence of external ATP on permeability and metabolism of dog red blood cells. Am. J. Physiol. 223(4), 888–893.

Parsons, P.P., Worthen, G.S., Moore, E.E., Tate, R.M., and Henson, P.M. (1989) The association of circulation endotoxin with the development of the adult respiratory distress syndrome. Am. Rev. Respir. Dis. 140, 294–301.

Pepe, P.E., Hudson, L.D., and Carrico, C.J. (1984) Early application of positive end-expiratory pressure in patients at risk for the adult respiratory distress syndrome. N. Engl. J. Med. 311, 281–286.

Pratt, P.C., Vollmer, T.R., Shelburne, J.D., and Crapo, J.D. (1979) Pulmonary morphology in a multihospital collaborative extracorporeal membrane oxygenation project: light microscopy. Am. J. Pathology 95, 191–208.

Puybasset, L., Stewart, T., Rouby, J.J., Cluzel, P., Mourgeon, E., Belin, M.F., et al. (1994) Inhaled nitric oxide reverses the increase in pulmonary vascular resistance induced by permissive hypercapnia in patients with acute respiratory distress syndrome. Anesthesiology 80, 1254–1267.

Radomski, M.W., Palmer, R.M., and Moncada, S. (1990) An L-arginine/nitric oxide pathway present in human platelets regulates aggregation. Proc. Natil. Acad. Sci. USA 87(13), 5193–5197.

Ranieri, V.M., Tortorella, D., DeTullio, R., Puntillo, F., Grasso, S., Mascia, L., et al. (1998) Limitation of mechanical lung stress decreases BAL cytokines in patients with ARDS. Am. J. Respir. Crit. Care Med. 157, A694.

Ravenscraft, S.A., Burke, W.C., Nahum, A., Adams, A.B., Nakos, G., Marcy, T.W., and Marini, J.J. (1993) Tracheal gas insufflation augments CO_2 clearance during mechanical ventilation. Am. Rev. Respir. Dis. 148, 345–351.

Reid, L.M., and Jones, R.C. (1986) Pathology of pulmonary vascular bed in adults respiratory distress syndrome (ARDS). In: Kazemi, H., Hymen, A.L., and Kadowitz, J., (eds.) Acute Lung Injury. PSG Publishing, Littleton, MA, pp. 7–24.

Rose, E.C., Juliano, C.A., Tracey, D.E., Yoshimure, T., and Fu, S.M. (1994) Role of interleukin-1 in endotoxin induced lung injury in the rat. Am. J. Respir. Cell Mol. Biol. 10, 214–221.

Rossaint, R., Falke, K.J., Lopez, F., Slama, K., Pison, U., and Zapol, W.M. (1993) Inhaled nitric oxide for the adult respiratory distress syndrome. N. Engl. J. Med. 328, 399–405.

Sloane, P.J., Gee, M.H., Gottleib, J.E., Albertine, K.H., Peters, S.P., Burns, J.R., et al. (1992) A multicenter registry of patients with acute respiratory distress syndrome: physiolgy and outcome. Am. Rev. Respir. Dis. 146, 419–426.

Snow, R.L., Daives, P., Pontoppidan, H., Zapol, W.M., and Reid, L. (1982) Pulmonary vascular remodeling in adult respiratory distress syndrome. Am. Rev. Respir. Dis. 126, 887–892.

Spencer, J. (1985) Pathology of the lung, fourth ed. pp. 261–267. Oxford Pergamon Press, Oxford.

Spragg, R.G., Gilliar, N., Richman, P., Smith, R.M., Hite, R.D., Pappert, D., et al. (1994) Acute effects of a single dose of porcine surfactant on patients with the adults respiratory distress syndrome. Chest 105, 195–202.

Stephens, K.E., Ishizaka, A., Larrick, J.W., Raffin, T.A. (1988) Tumor necrosis factor causes increased pulmonary permeability and edema. Comparison to septic acute lung injury. Am. Rev. Respir. Dis. 137(6), 1364–1370.

Stewart, T.E., Meade, M.O., Cook, D.J., Granton, J.T., Hodder, R.V., Lapinsky, S.E., et al. (1998) Evaluation of a ventilation strategy to prevent barotrauma in patients at high risk for acute respiratory distress syndrome. N. Engl. J. Med. 338, 355–361.

Sugiura, M., McCulloch, P.R., Wren, S., Dawson, R.H., and Groese, A.B. (1994) Ventilator pattern influences neutrophil influx and activation in atelectasis-prone rabbit lung. J. Appl. Physiol. 77, 1355–1365.
Tomashefski, J.F., Jr. (1990) Pulmonary pathology of the adult respiratory distress syndrome. Clin. Chest Med. 11, 593–619.
Tomashefski, J.R., Jr., Davies, P., Boggis, C., Greene, R., Zapol, W.M., and Reid, L.M. (1983) The pulmonary vascular lesions of the adult respiratory distress syndrome. Am. J. Pathol. 112, 112–126.
Tracey, K.J., Beutler, B., Lowry, S.F., Merryweather, J., Wolpe, S., Milsark, I.W., et al. (1986) Shock and tissue injury induced by recombinant human cachectin. Science 234(4774), 470–474.
Tracey, K.L.J., Lowry, S.F., Fahey, T.J., 3d, Albert, J.D., Fong, Y., Hesse, D., et al. (1987) Cachectin/tumor necrosis factor induces lethal shock and stress hormone responses in the dog. Surg. Gynecol. Obstet. 164(5), 415–422.
Tran Van Nhiu, J., Misset, B., Lebargy, F., Carlet, J., and Bernaudin, J.F. (1993) Expression of tumor necrosis factor-alpha gene in alveolar macrophages from patients with adult respiratory distress syndrome. Am. Rev. Respir. Dis. 147(6 Pt1), 1585–1589.
Tremblay, L., Valenza, F., Ribeiro, S.P., Li, J., and Slutsky, A.S. (1997) Injurious ventilatory strategies increase cytokines and c-fos m-RNA expression in an isolated rat lung model. J. Clin. Invest. 99, 944–952.
Waage, A., and Espevik, T. (1988) Interleukin-1 potentiates the lethal effects of tumor necrosis factor/cachetin in mice. J. Exp. Med. 167, 1987–1992.
Waydhas, C., Nast-Kolb, D., Trupka, A., Lenk, S., Duswald, K.H., Schweiberer, L., and Jochum, M. (1993) Increased serum concentrations of procollagen type III peptide in severely injured patients: an indicator of fibrosing activity? Crit. Care Med. 21(2), 240–247.
Webb, H.H., and Tierney, D.F. (1974) Experimental pulmonary edema due to intermittent positive pressure ventilation with high inflation pressures. Protection by positive end-expiratory pressure. Am. Rev. Respir. Dis. 110(5), 556–565.
Zapol, W.M., and Hurford, W.E. (1994) Inhaled nitric oxide in adult respiratory distress syndrome. Adv. Pharmacol. 31, 513–530.
Zapol, W.M., and Snider, M.T. (1977) Pulmonary hypertension in severe acute respiratory failure. N. Engl. J. Med. 296, 476–480.

Recommended Readings

Abraham, A., and Terada, L. (eds.) (1998) Acute lung injury. American College of Chest Physicians, Northbrook, IL.
Malalor, S., and Sznajder, J.I. (1998) Acute respiratory distress syndrome. Plenum Press, New York.
Ware, L.B., and Matthay, M.A. (2000) The acute respiratory distress syndromes. N. Engl. J. Med. 342, 1334–1349.

20
Asthma

PETER J. BARNES

Asthma is one of the commonest diseases in industrialized countries, and there is convincing evidence to suggest that its prevalence and morbidity are increasing, despite better recognition and increased prescriptions for antiasthma therapy (Burney, 1992). It is somewhat paradoxical that the morbidity and mortality of asthma should be increasing at a time when there is increased understanding of the pathophysiology of asthma and when effective therapies are available. It points to the need for an even better understanding of the underlying mechanisms involved in asthma and elucidation of the mode of action of currently available antiasthma therapies.

It is now clear that chronic inflammation underlies the clinical syndrome. In the past it was assumed that the basic defect in asthma lay in abnormal contractility of airway smooth muscle, giving rise to variable airflow obstruction and the common symptoms of intermittent wheeze and shortness of breath. Studies of airway smooth muscle from asthmatic patients, however, have shown no convincing evidence for increased contractile responses to spasmogens (e.g., histamine *in vitro*), indicating that asthmatic airway smooth muscle is not fundamentally abnormal and suggesting that it is the *control* of airway caliber *in vivo* that is abnormal (Barnes, 1996).

It had been assumed for many years that mast cells play a critical role in asthma and that mast cell mediators could largely account for the pathophysiology of asthma. It has become clear more recently that many different inflammatory cells are activated in asthmatic airways, and that these cells produce a variety of mediators that act in a complex manner on target cells of the airways to produce the abnormal pathophysiological features typical of asthma. Research has established that asthma, even in its mildest clinical forms, involves a special type of inflammation in the airways.

Genes versus Environment

The role of genes in asthma is controversial. There is clearly a strong genetic determinant to atopy, and several genes have been associated by genetic linkage and genomewide searches (Sandford et al., 1996). These include a gene coding for the high-affinity IgE receptor on chromosome 11q and the cytokine cluster on chromosome 5q (IL-4, IL-5, GM-CSF, IL-9). An even more complete understanding of the genetic influences on atopy, however, would provide no useful information about genetic influences in asthma. The atopic state is found in approximately 40% of the population and is the highest risk factor known for the development of asthma. The chances of developing asthma in atopic individuals is minimal. It is much more likely that environmental factors are important in switching atopy to asthma. These environmental factors are not yet certain, but they include viral infections (e.g., respiratory syncytial virus in infancy) and early exposure to allergens and environmental tobacco smoke. Twin studies have shown a relatively low concordance for identical twins (20–30%), emphasizing the importance of environmental switches.

It is likely, however, that genetic factors may be important in determining the severity of asthma and its response to treatment. Many genes involved in asthmatic inflammation and response to treatment are polymorphic, and it is likely that certain combination of alleles are associated with more severe disease. Thus, for the β_2-adrenergic receptor, a common polymorphism Gln27Glu in which glutamine at position 27 is substituted for by glutamic acid, is more difficult to downregulate on exposure to β_2-agonists and is associated with lesser degrees of airway hyperresponsiveness among asthmatic patients (Hall et al., 1995), whereas a Arg16Gly polymorphism, in which β_2-receptors are more easily downregulated, is associated with increased nocturnal asthma (Turki et al., 1995).

Asthma as an Inflammatory Disease

It had been recognized for many years that patients who die of asthma attacks have grossly inflamed airways. The airway lumen is occluded by a tenacious mucus plug composed of plasma proteins exuded from airway vessels and mucus glycoproteins secreted from surface epithelial cells. The airway wall is edematous and infiltrated with inflammatory cells, which are predominantly eosinophils and lymphocytes. The airway epithelium is invariably shed in a patchy manner and clumps of epithelial cells are found in the airway lumen. There have occasionally been opportunities to examine the airways of asthmatic patients who die accidentally, and similar, though less marked, inflammatory changes have been observed (Dunnill, 1960). It has been possible more recently to examine the airways of asthmatic patients by fiberoptic and rigid bronchoscopy, by bronchial biopsy, and by bronchoalveolar lavage. Direct bronchoscopy reveals that the airways of asthmatic patients are often reddened and swollen, indicating acute inflammation. Lavage has revealed an increase in the numbers of lymphocytes, mast cells and eosinophils, and evidence for activation of macrophages in comparison with nonasthmatic controls. Biopsies have revealed evidence for increased numbers and activation of mast cells, macrophages, eosinophils, and T-lymphocytes (Djukanovic et al., 1990). These changes are found even in patients with mild asthma who have few symptoms, and this suggests that asthma is an inflammatory condition of the airways.

The relationship between inflammation and clinical symptoms of asthma is not clear. There is evidence that the degree of inflammation is related to airway hyperrespon-

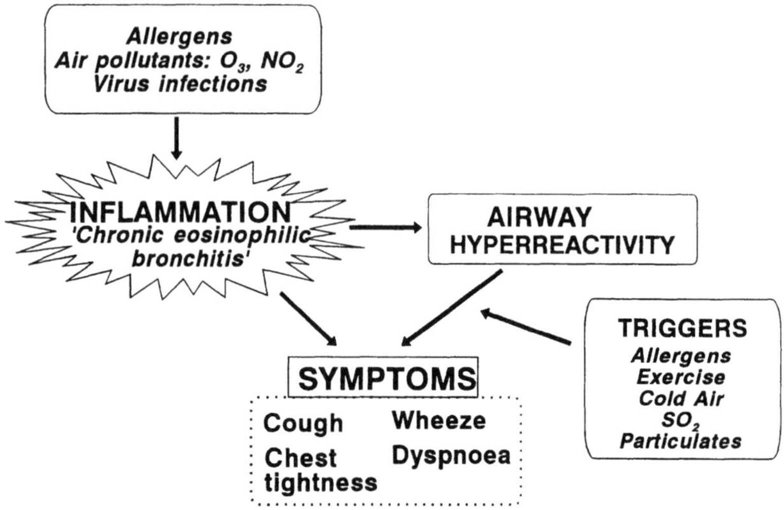

Figure 20.1. Relationship between allergic inflammation and symptoms in asthma.

siveness (AHR), as measured by histamine or methacholine challenge. Increased airway responsiveness is an exaggerated airway narrowing in response to many stimuli that is characteristic of asthma and the degree of AHR relates to asthma symptoms. Inflammation of the airways may increase airway responsiveness, which thereby allows triggers that would not narrow the airways to do so; however, inflammation may also lead directly to an increase in asthma symptoms (e.g., cough and chest tightness), by activation of airway sensory nerve endings (Fig. 20.1).

Although most attention has been focused on the acute inflammatory changes seen in asthmatic airways, asthma is a *chronic* inflammatory disease, with inflammation persisting over many years in most patients. Superimposed on this chronic inflammatory state are acute inflammatory episodes that correspond to exacerbations of asthma. It is clearly important to understand the mechanisms of acute and chronic inflammation in asthmatic airways and to investigate the long-term consequences of this chronic inflammation on airway function. It is also important to consider the effects of therapy on the inflammatory process.

Inflammatory Cells

Many different inflammatory cells are involved in asthma, although the precise role of each cell type is not yet certain (Barnes, 1992a) (Fig. 20.2). It is evident that no single inflammatory cell is able to account for the complex pathophysiology of asthma, but some cells are predominant in asthmatic inflammation.

Mast Cells

Mast cells are clearly important in initiating the acute responses to allergens and probably to other indirect stimuli (e.g., exercise and hyperventilation, via osmolality or thermal changes, and fog). There are questions, however, about the role of mast cells in more chronic inflammatory events, and it seems more probable that other cells (e.g.,

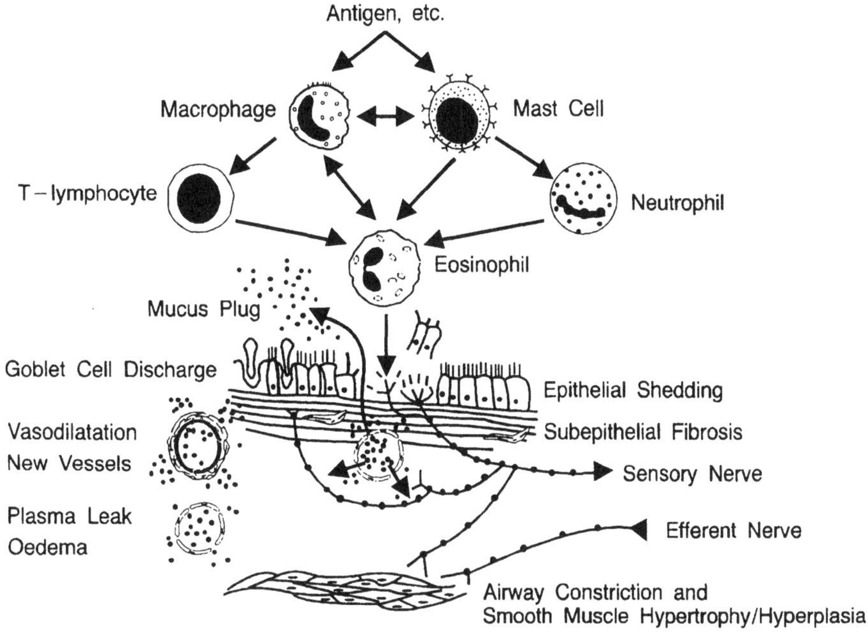

Figure 20.2. The complex pathophysiology of asthma.

macrophages, eosinophils, and T-lymphocytes) are more important in the chronic inflammatory process, including AHR.

Macrophages and Dendritic Cells

Macrophages, which are derived from blood monocytes, may traffic into the airways in asthma and may be activated by allergens via low-affinity IgE receptors (Lee and Lane, 1992). The enormous repertoire of macrophages allows these cells to produce many different products, including a large variety of cytokines that may orchestrate the inflammatory response. Macrophages therefore have the capacity to initiate a particular type of inflammatory response via the release of a certain pattern of cytokines. Macrophages may both increase and decrease inflammation, depending on the stimulus. Alveolar macrophages normally have a *suppressive* effect on lymphocyte function, but this may be impaired in asthma after allergen exposure (Spiteri et al., 1994). Macrophages may therefore play an important antiinflammatory role, preventing the development of allergic inflammation. Macrophages may also act as antigen-presenting cells that process allergens for presentation to T-lymphocytes, although alveolar macrophages are far less effective in this respect than are macrophages from other sites (e.g., the peritoneum) (Holt and McMenamin, 1989). By contrast, dendritic cells that are specialized macrophagelike cells in the airway epithelium are very effective antigen-presenting cells (Holt and McMenamin, 1989); therefore, they may play a very important role in the initiation of allergen-induced responses in asthma.

Eosinophils

Eosinophil infiltration is a characteristic feature of asthmatic airways. It differentiates asthma from other inflammatory conditions of the airway. Indeed, asthma might more accurately be termed *chronic eosinophilic bronchitis*. Allergen inhalation results in a marked increase in eosinophils in bronchoalveolar lavage fluid at the time of the late reaction, and there is a close relationship between eosinophil counts in peripheral blood or bronchial lavage and AHR. Eosinophils were originally viewed as beneficial cells in asthma because they have the capacity to inactivate histamine and leukotrienes, but it now seems more likely that they may play a damaging role, and that they may be linked to the development of airway hyperresponsiveness through the release of basic proteins and oxygen-derived free radicals (Gleich, 1990).

An important area of research is now concerned with the mechanisms involved in *recruitment* of eosinophils into asthmatic airways. Eosinophils are derived from bone marrow precursors. After allergen challenge eosinophils appear in bronchoalveolar lavage fluid during the late response, and this is associated with a decrease in peripheral eosinophil counts and with the appearance of eosinophil progenitors in the circulation. The signal for increased eosinophil production is presumably derived from the inflamed airway. Eosinophil recruitment initially involves adhesion of eosinophils to vascular endothelial cells in the airway circulation, both their migration into the submucosa and their subsequent activation. The role of individual adhesion molecules, cytokines, and mediators in orchestrating these responses has yet to be clarified. Adhesion of eosinophils involves the expression of specific glycoprotein molecules on the surface of eosinophils (integrins) and their expression of such molecules as intercellular adhesion molecule-1 (ICAM-1) on vascular endothelial cells. An antibody directed at ICAM-1 markedly inhibits eosinophil accumulation in the airways after allergen exposure. It also blocks the accompanying hyperresponsiveness (Wegner et al., 1990). ICAM-1, however, is not selective for eosinophils and cannot account for the selective recruitment of eosinophils in allergic inflammation. The adhesion molecules VLA4 expressed on eosinophils and VCAM-1 appear to be more selective for eosinophils (Pilewski and Albelda, 1995) and IL-4 increases the expression of VCAM-1 on endothelial cells. Eosinophil migration may be due to the effects of platelet activating factor (PAF), which is selectively chemoattractant to eosinophils (Barnes et al., 1988a), and to the effects of such cytokines as GM-CSF, IL-3, and IL-5 (Barnes, 1994a). These cytokines may be very important for the survival of eosinophils in the airways and may "prime" eosinophils to exhibit enhanced responsiveness. Eosinophils from asthmatic patients show greatly exaggerated responses to PAF and phorbol esters than do eosinophils from atopic nonasthmatic individuals (Chanez et al., 1990). This is further increased by allergen challenge (Evans et al., 1996), which, suggests that they may have been primed by exposure to cytokines in the circulation. There are several mediators involved in the migration of eosinophils from the circulation to the surface of the airway. The most potent and selective agents appear to be chemokines [e.g., RANTES, which is expressed in epithelial cells (Kwon et al., 1995), and eotaxin (Jose et al., 1994)].

Neutrophils

The role of neutrophils in human asthma is less clear. Neutrophils are found in the airways of some patients with chronic bronchitis and bronchiectasis that do not have the same degree of AHR found in asthma, but there is increasing evidence that neutrophils may be important in acute exacerbations of asthma. There is some evidence that neutrophils may be involved in acute exacerbations of asthma and are prominent in the airways of patients who die of sudden unexpected asthma attacks (Sur et al., 1993).

T-Lymphocytes

CD4+ T-lymphocytes (T-helper cells) play a very important role in coordinating the inflammatory response in asthma through the release of specific patterns of cytokines, resulting in the recruitment and survival of eosinophils and in the maintenance of mast cells in the airways. T-lymphocytes are coded to express a distinctive pattern of cytokines that may be similar to that described in the murine Th2 type of T-lymphocytes, which characteristically expresses IL-3, IL-4 and IL-5 (Kay, 1991; Mosman and Sad, 1996). This programming of T-lymphocytes is presumably due to such antigen-presenting cells as dendritic cells, which may migrate from the epithelium to regional lymph nodes or which interact with lymphocytes resident in the airway mucosa. The reason why antigens (e.g., house dust mite) preferentially lead to Th2 cell proliferation is not yet known, but the role of co-stimulatory molecules (e.g., B7–2 on antigen presenting cells that interact with CD28 on T cells) may be critical (Kuchroo et al., 1995). CD8+ lymphocytes (T suppressor cells) may also have a Th2 type profile (Mosman and Sad, 1996).

Structural Cells

There is increasing recognition that several structural cells in the airway may play an important role in the inflammatory response as a source of mediators. Indeed, these cells may be particularly important in maintaining the inflammatory response in asthma. These cells include airway epithelial and endothelial cells, fibroblasts and myofibroblasts, and airway smooth muscle cells. Epithelial cells also produce inflammatory mediators (e.g., endothelins, proinflammatory cytokines, chemokines and growth factors) (Devalia and Davies, 1993) (Fig. 20.3). Epithelial cells may play a key role in translating inhaled environmental signals into an airway inflammatory response and are probably the main target cell for inhaled glucocorticoids.

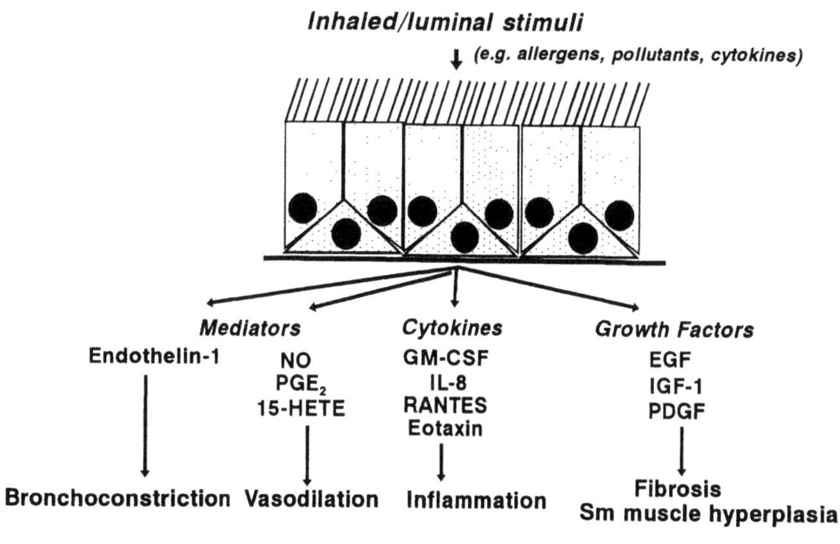

Figure 20.3. Epithelial cells as a source of inflammatory mediators.

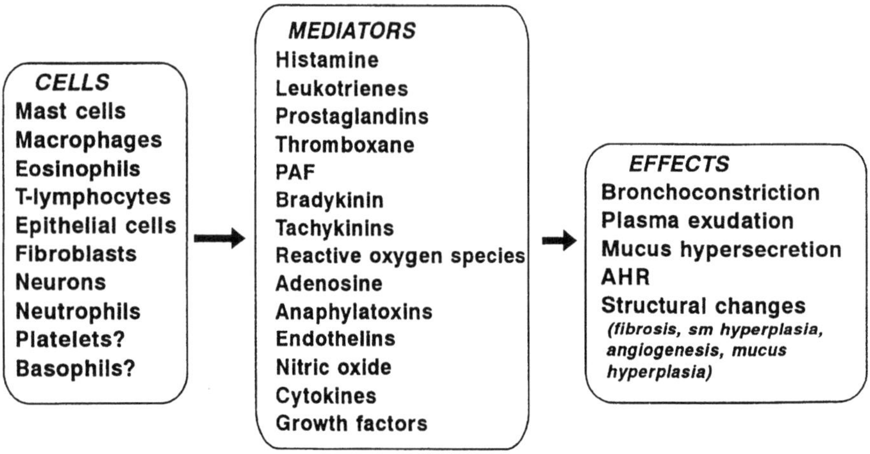

Figure 20.4. Multiple inflammatory mediators are produced in asthma from many cell types, resulting in complex inflammatory effects.

Inflammatory Mediators

Many different mediators have been implicated in asthma and they may have a variety of effects on the airways that could account for the pathological features of asthma (Barnes et al., 1988b; Barnes et al., 1996) (Fig. 20.4). Such mediators as histamine, prostaglandins, and leukotrienes contract airway smooth muscle, increase microvascular leakage, increase airway mucus secretion, and attract other inflammatory cells. Because each mediator has many effects, the role of individual mediators in the pathophysiology of asthma is not yet clear. Indeed, the multiplicity of mediators makes it unlikely that antagonizing a single mediator will have a major impact in clinical asthma.

The cysteinyl-leukotrienes LTC_4, LTD_4, and LTE_4 are potent constrictors of human airways. They have been reported to increase AHR, and they may play an important role in asthma (Henderson, 1994). The development of potent specific leukotriene antagonists has made it possible to evaluate the role of these mediators in asthma. Potent LTD_4 antagonists protect (by about 50%) against exercise- and allergen-induced bronchoconstriction, which suggests that leukotrienes contribute to bronchoconstrictor responses (Chung, 1995). Chronic treatment with leukotriene antagonists improve lung function and symptoms in asthmatic patients, although the degree of improvement is modest compared with what would be expected of inhaled glucocorticoids. The role of leukotrienes in chronic asthma remains to be defined, but leukotriene antagonists have been introduced for asthma therapy in some countries.

A mediator that attracted considerable attention is platelet-activating factor (PAF) because it mimics many of the features of asthma, including AHR (Barnes et al., 1988a). Although PAF appears to be produced by the inflammatory cells involved in asthmatic inflammation and mimics many of the pathophysiological features of asthma, its role in asthma will only become apparent with the use of potent and specific antagonists. However, initial results with such potent PAF antagonists as apafant (WEB 2086) and modipafant in chronic asthma have been disappointing (Spence et al., 1994; Kuitert et al., 1995).

Cytokines

Cytokines are increasingly recognized to be important in chronic inflammation and to play a critical role in orchestrating and perpetuating the inflammatory response (Fig. 20.5). Many inflammatory cells (e.g., macrophages, mast cells, eosinophils, and lymphocytes) are capable of synthesizing and releasing these proteins, and such structural cells as epithelial cells, endothelial cells, and even airway smooth muscle cells may also release a variety of cytokines and may therefore participate in the chronic inflammatory response (Barnes, 1994a). Each cytokine may release several other cytokines, and these act together in a complex network. Although inflammatory mediators like histamine and leukotrienes may be important in the acute and subacute inflammatory responses and in exacerbations of asthma, it is likely that cytokines play a dominant role in chronic inflammation. Almost every cell is capable of producing cytokines under certain conditions. Research in this area is hampered by a lack of specific antagonists, although important observations have been made using specific neutralizing antibodies. The cytokines that appear to be of particular importance in asthma include the lymphokines secreted by T-lymphocytes: IL-3, which is important for the survival of mast cells in tissues; IL-4, which is critical in switching B-lymphocytes to produce IgE and for expression of VCAM-1 on endothelial cells; and IL-5, which is of critical importance in the differentiation, survival, and priming of eosinophils. There is evidence for increased gene expression of IL-5 in lymphocytes in bronchial biopsies of patients with symptomatic asthma (Hamid et al., 1991). Other cytokines (e.g., IL-1, IL-6, TNF-α, and GM-CSF) are released from a variety of cells, including macrophages and epithelial cells, and may be important in amplifying the inflammatory response. TNF-α may be an important amplifying mediator in asthma and is produced in increased amounts in asthmatic airways. Inhalation of TNF-α increased airway responsiveness in normal individuals (Thomas et al., 1995).

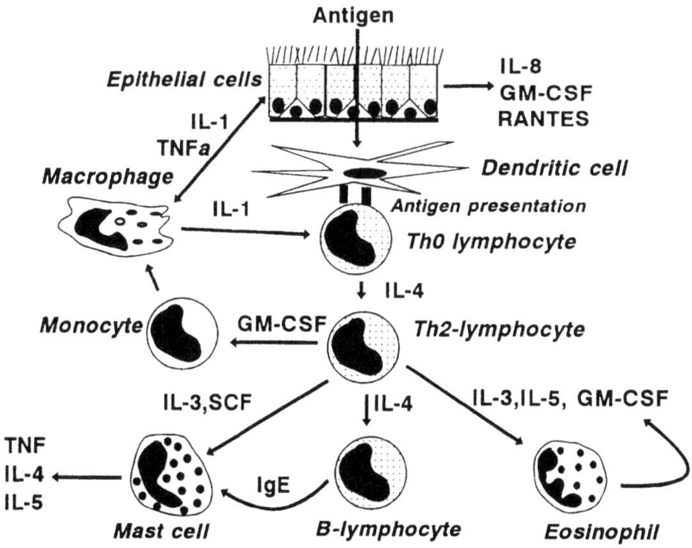

Figure 20.5. The cytokine network in asthma.

Endothelins

Endothelins are potent peptide mediators that are potent vasoconstrictors and bronchoconstrictors (Barnes, 1994b). They also induce airway smooth muscle cell proliferation and fibrosis and may therefore play a role in the chronic inflammation of asthma. There is evidence for increased expression of endothelin-1 in asthma, particularly in airway epithelial cells (Springall et al., 1991).

Nitric Oxide

Nitric oxide (NO) is a fee radical gas produced by several cells in the airway by NO synthases (Barnes 1995b). An inducible form of the enzyme (iNOS) is expressed in epithelial cells of asthmatic patients (Hamid et al., 1993) and can be induced by cytokines in airway epithelial cells (Robbins et al., 1994). This may account for the increased concentration of NO in the exhaled air of asthmatic patients (Kharitonov et al., 1994). NO itself is a potent vasodilator and this may increase plasma exudation in the airways; it may also amplify the Th2-lymphocyte mediated response (Barnes and Liew, 1995). More recent evidence supports the view that NO plays a key role in regulating airway responsiveness and bronchial relaxation. It also regulates apoptosis (Laskir and Laokir, 2000).

Effects of Inflammation

The chronic inflammatory response has several effects on the target cells of the airways, resulting in the characteristic pathophysiological changes associated with asthma. Important advances have been made in understanding these changes, although their role in asthma symptoms is often not clear.

Airway Epithelium

Airway epithelial shedding may be important in contributing to airway hyperresponsiveness and may explain how several different mechanisms (e.g., ozone exposure, certain virus infections, chemical sensitizers and allergen exposure) can lead to its development because all of these stimuli may lead to epithelial disruption. Epithelium may be shed as a consequence of inflammatory mediators (e.g., eosinophil basic proteins and oxygen-derived free radicals) together with various proteases released from inflammatory cells. Epithelial cells are commonly found in clumps in the BAL or sputum (Creola bodies) of asthmatics, which suggests that there has been a loss of attachment to the basal layer or basement membrane. Epithelial damage may contribute to AHR in a number of ways, including loss of its barrier function to allow penetration of allergens, loss of enzymes (e.g., neutral endopeptidase) that normally degrade inflammatory mediators, loss of a relaxant factor (so-called epithelial-derived relaxant factor), and exposure of sensory nerves that may lead to reflex neural effects on the airway.

Fibrosis

An apparent increase in the basement membrane has been described in fatal asthma, although similar changes have been described in the airways in other conditions (Dunnill, 1960). Electron microscopy of bronchial biopsies in asthmatic patients demonstrates that this thickening is due to subepithelial fibrosis (Djukanovic et al., 1990). Type-III and -V collagen appear to be laid down and may be produced by myofibroblasts that are situated under the epithelium. The mechanism of fibrosis is not yet clear, but several cytokines, including TGFβ and platelet-derived growth factor (PDGF), which activate fibroblasts, may be produced by epithelial cells or macrophages in the inflamed airway (Barnes, 1994a).

Airway Smooth Muscle

There is still debate about the role of abnormalities in airway smooth muscle in asthmatic airways. *In vitro* airway smooth muscle from asthmatic patients usually shows no increased responsiveness to spasmogens. Reduced responsiveness to β-agonists has also been reported in postmortem or surgically removed bronchi from asthmatics, although the number of β-receptors is not reduced, which suggests that β-receptors have been uncoupled (Bai et al., 1992). These abnormalities of airway smooth muscle may be a reflection of the chronic inflammatory process. For example, the reduced β-adrenergic responses in airway smooth muscle could be due to phosphorylation of the stimulatory G-protein coupling β-receptors to adenylyl cyclase, resulting from the activation of protein kinase C by the stimulation of airway smooth muscle cells by inflammatory mediators (Grandordy et al., 1994).

There is also a characteristic *hypertrophy* and *hyperplasia* of airway smooth muscle in asthmatic airways (Ebina et al., 1990) that is presumably the result of stimulation of airway smooth muscle cells by various growth factors (e.g., PDGF, or endothelin-1 released from inflammatory cells) (Hirst et al., 1996).

Vascular Responses

Vasodilatation occurs in inflammation, yet little is known about the role of the airway circulation in asthma, partly because of the difficulties involved in measuring airway blood flow. The bronchial circulation may play an important role in regulating airway caliber because an increase in the vascular volume may contribute to airway narrowing. Increased airway blood flow may be important in removing inflammatory mediators from the airway, and may play a role in the development of exercise-induced asthma.

Microvascular leakage is an essential component of the inflammatory response, and many of the inflammatory mediators implicated in asthma produce this leakage (Chung et al., 1990). There is good evidence for microvascular leakage in asthma and it may have several consequences on airway function, including increased airway secretions, impaired mucociliary clearance, formation of new mediators from plasma precursors (e.g., kinins), and mucosal edema, which may contribute to airway narrowing and increased airway hyperresponsiveness.

Mucus Secretion

Mucus hypersecretion is a common inflammatory response in secretory tissues. Increased mucus secretion contributes to the viscid mucus plugs that occlude asthmatic airways, particularly in fatal asthma. There is evidence for hyperplasia of submucosal glands that are confined to large airways and of increased numbers of epithelial goblet cells. This increased secretory response may be due to inflammatory mediators acting on submucosal glands and due to stimulation of neural elements. Little is understood about the control of goblet cells, which are the main source of mucus in peripheral airways, although studies investigating the control of goblet cells in guinea pig airways suggest that cholinergic, adrenergic, and sensory neuropeptides are important in stimulating secretion (Kuo et al., 1990).

Neural Mechanisms

There has been a revival of interest in neural mechanisms in asthma (Barnes, 1995c). Autonomic nervous control of the airways is complex because along with classical cholinergic and adrenergic mechanisms, nonadrenergic noncholinergic (NANC) nerves and several neuropeptides have been identified in the respiratory tract (Barnes et al., 1991). Several studies have investigated the possibility that defects in autonomic control may contribute to airway hyperresponsiveness and asthma, and abnormalities of autonomic function (e.g., enhanced cholinergic and α-adrenergic responses or reduced β-adrenergic responses) have been proposed. Current thinking suggests that these abnormalities are likely to be secondary to the disease, rather than primary defects (Barnes, 1995c). It is possible that airway inflammation may interact with autonomic control by several mechanisms.

Inflammatory mediators may act on various prejunctional receptors on airway nerves to modulate the release of neurotransmitters (Barnes, 1992b). Thus, thromboxane and PGD_2 facilitate the release of acetylcholine from cholinergic nerves in canine airways, whereas histamine inhibits cholinergic neurotransmission at both parasympathetic ganglia and postganglionic nerves via H_3-receptors. Inflammatory mediators may also activate sensory nerves, which results in reflex cholinergic bronchoconstriction or release of inflammatory neuropeptides. Inflammatory products may also sensitize sensory nerve endings in the airway epithelium, so that the nerves become hyperalgesic. Hyperalgesia and pain (*dolor*) are cardinal signs of inflammation. In the asthmatic airway both may mediate cough and dyspnea, which are characteristic symptoms of asthma. The precise mechanisms of hyperalgesia are not yet certain, but such mediators as prostaglandins and certain cytokines may be important.

Bronchodilator nerves that are nonadrenergic are prominent in human airways. It has been suggested that these nerves may be defective in asthma (Lammers et al., 1992). In animal airways, vasoactive intestinal peptide (VIP) has been shown to be a neurotransmitter of these nerves, and a striking absence of VIP-immunoreactive nerves has been reported in the lungs from patients with severe fatal asthma (Ollerenshaw et al., 1989). It is likely, however, that this loss of VIP-like immunoreactivity is due to degradation by tryptase released from degranulating mast cells in the airways of asthmatics (Barnes, 1989b). In human airways the bronchodilator neurotransmitter appears to be the fee radical gas NO (Belvisi et al., 1992).

Airway nerves may also release neurotransmitters that have inflammatory effects. Thus, such neuropeptides as substance P (SP), neurokinin A, and calcitonin-gene–related peptide may be released from sensitized inflammatory nerves in the airways, which increase and extend the ongoing inflammatory response (Barnes, 1991) (Fig. 20.6). There is evidence for an increase in SP-immunoreactive nerves in airways of patients with severe asthma (Ollerenshaw et al., 1991), which may be due to proliferation of sensory nerves and increased synthesis of sensory neuropeptides as a result of nerve growth factors released during chronic inflammation. There may also be a reduction in the activity of enzymes (e.g., neutral endopeptidase), which degrade such neuropeptides as SP (Nadel, 1991). There is also evidence for increased gene expression of the receptor that mediates the inflammatory effects of SP (Adcock et al., 1993). Thus, chronic asthma may be associated with increased neurogenic inflammation, which

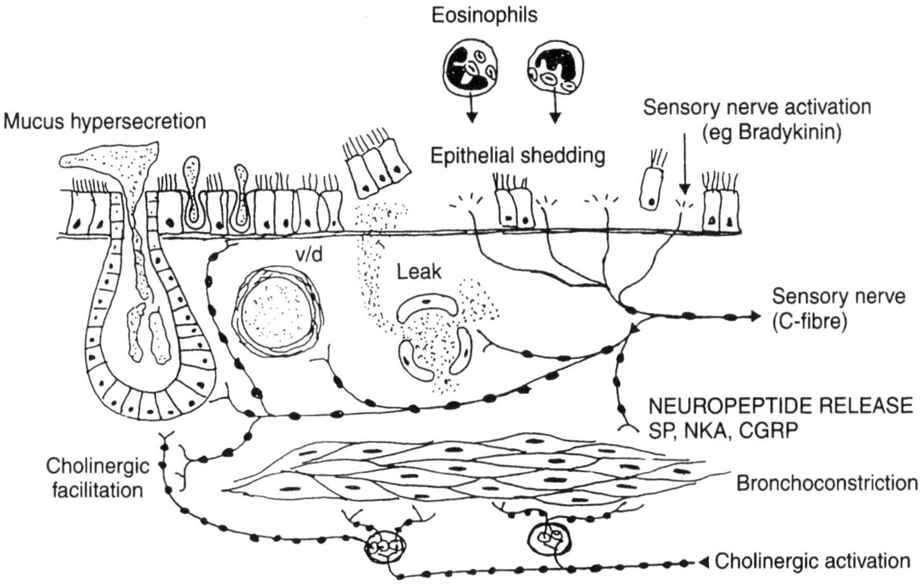

Figure 20.6. Neurogenic inflammation is due to the inflammatory effects of neuropeptides released from unmyelinated sensory nerves (C-fibers) at the allergic inflammatory site.

may provide a mechanism for perpetuating the inflammatory response even in the absence of initiating inflammatory stimuli.

Transcription Factors

The chronic inflammation of asthma is due to increased expression of multiple inflammatory proteins (e.g., cytokines, enzymes, receptors, adhesion molecules). In many cases these inflammatory proteins are induced by transcription factors, DNA binding factors that increase the transcription of selected target genes (Barnes and Adcock, 1995). One transcription factor that may play a critical role in asthma is nuclear factor-kappa B (NF-κB), which can be activated by multiple stimuli, including protein kinase C activators, oxidants, and proinflammatory cytokines (e.g., IL-1β and TNF-α). NF-κB is the predominant transcription regulating the expression of iNOS, the inducible form of cyclooxygenase (COX-2), chemokines (IL-8, RANTES, MIP-1α), proinflammatory cytokines (TNF-α, GM-CSF), and adhesion molecules (ICAM-1, VCAM-1) (Siebenlist et al., 1994). NF-κB in epithelial cells may play a pivotal role in amplifying inflammation in such diseases as asthma (Barnes and Karin, 1996) (Fig. 20.7). This is further supported by studies showing that NF-κB potently inhibited by glucocorticoids (Barnes and Adcock 1993).

Implications for Therapy

The advances in our understanding of asthma have important implications for the way in which asthma therapy should be used. In the past, asthma was treated primarily with bronchodilators, which act predominantly by relaxing airway smooth muscle; however

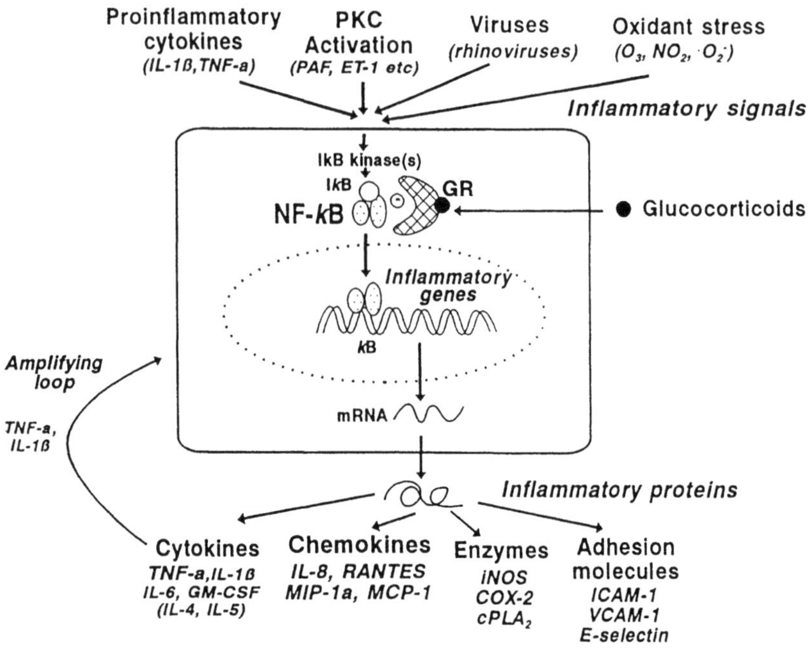

Figure 20.7. Nuclear factor-kappa B (NF-κB) may play a pivotal role in inflammation of chronic allergic disease because it is activated by many factors that increase inflammation, and results in the coordinate expression of multiple inflammatory genes and proteins.

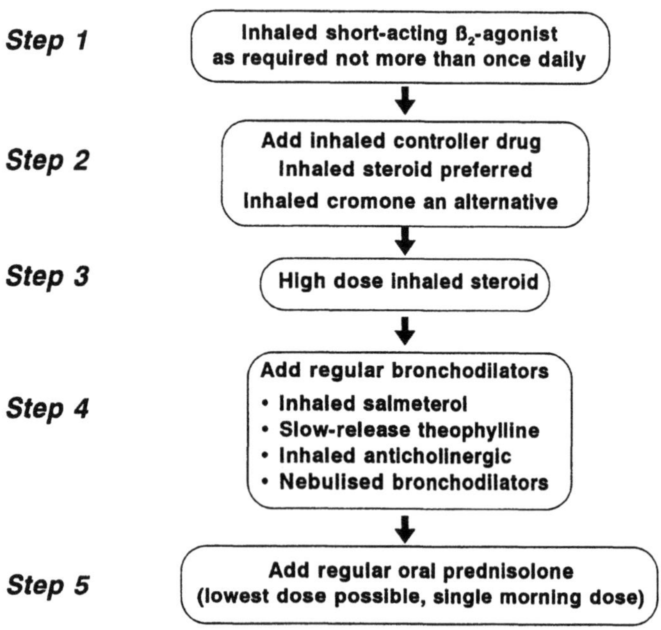

Figure. 20.8. Step-wise approach to asthma treatment.

because it is now apparent that even the mildest of asthmatic patients have airway inflammation that leads to airway narrowing, it would seem more appropriate to use antiinflammatory treatment as first-line therapy (Barnes, 1989a). For patients with mild asthma cromolyn sodium or nedocromil sodium may be sufficient, but inhaled glucocorticoids will be needed for many patients with asthma. Inhaled glucocorticoids are highly effective in controlling asthma and may work on many aspects of the chronic inflammatory response (Barnes, 1995a). By contrast, bronchodilators (e.g., β_2-agonists) may relieve the symptoms of asthma acutely, but fail to control the underlying chronic inflammatory response. Indeed, there is evidence to suggest that regular use of inhaled β-agonists may even *increase* AHR and chronic asthma symptoms (Barnes and Chung, 1992). It is now recommended that they should be used only for acute symptom control and that asthma should be controlled primarily with inhaled antiinflammatory agents. It is hoped that the earlier use of inhaled steroids may reduce the morbidity and mortality of asthma, and that it may also prevent the irreversible structural changes that may underlie the gradual decline in lung function seen in patients with poorly controlled asthma. There is some evidence to support this (Haahtela et al., 1994; Selroos et al., 1995).

Based on our current understanding of asthma, treatment guidelines have now been introduced in many countries, with the early introduction of inhaled glucocorticoids and a step-wise increase in therapy with increasing disease severity (British Thoracic Society, 1993; Sheffer, 1992; NHLBL/WHO 1995) (Fig. 20.8). This has led to better control of asthma and reduced hospital admissions.

References

Adcock, I.M., Peters, M., Gelder, C., Shirasaki, H., Brown, C.R., and Barnes, P.J. (1993) Increased tachykinin receptor gene expression in asthmatic lung and its modulation by steroids. J. Mol. Endocrinol. 11, 1–7.

Bai, T.R., Mak, J.C.W., and Barnes, P.J. (1992) A comparison of beta-adrenergic receptors and in vitro relaxant responses to isoproterenol in asthmatic airway smooth muscle. Am. J. Respir. Cell Mol. Biol. 6, 647–651.

Barnes, P.J., Chung, K.F., and Page, C.P. (1988a) Platelet-activating factor as. a mediator of allergic disease. J. Allergy Clin. Immunol. 81, 919–934.

Barnes, P.J., Chung, K.F., and Page, C.P. (1988b) Inflammatory mediators and asthma. Pharmacol. Rev. 40, 49–84.

Barnes, P.J. (1989a) A new approach to asthma therapy. N. Engl. J. Med. 321, 1517–1527.

Barnes, P.J. (1989b) Vasoactive intestinal peptide and asthma. New Engl. J. Med. 321, 1128–1129.

Barnes, P.J. (1991) Sensory nerves, neuropeptides and asthma. Ann. NY Acad. Sci. 629, 359–370.

Barnes, P.J., Baraniuk, J., and Belvisi, M.G. (1991) Neuropeptides in the respiratory tract. Am. Rev. Respir. Dis. 144, 1187–1198, and 1391–1399.

Barnes, P.J. (1992a) New aspects of asthma. J. Int. Med. 231, 453–461.

Barnes, P.J. (1992b) Modulation of neurotransmission in airways. Physiol. Rev. 72, 699–729.

Barnes, P.J. (1994a) Cytokines as mediators of chronic asthma. Am. J. Resp. Crit. Care Med. 150, S42–S49.

Barnes, P.J. (1994b) Endothelins and pulmonary diseases. J. Appl. Physiol. 77, 1051–1059.

Barnes, P.J. (1995a) Inhaled glucocorticoids for asthma. New Engl. J. Med. 332, 868–875.

Barnes, P.J. (1995b) Nitric oxide and airway disease. Ann. Med. 27, 389–393.

Barnes, P.J. (1995c) Is asthma a nervous disease? Chest 107, 119S–124S.

Barnes, P.J. (1996) Pathophysiology of asthma. Br. J. Clin. Pharmacol. 42, 3–10.

Barnes, P.J., Chung, K.F., and Page, C.P. (1996) Inflammatory mediators of asthma—2. Pharm. Rev.

Barnes, P.J., and Adcock, I.M. (1993) Anti-inflammatory actions of steroids: molecular mechanisms. Trends Pharmacol. Sci. 14, 436–441.

Barnes, P.J., and Adcock, I.M. (1995) Transcription factors in asthma. Clin. Exp. Allergy 27 (Suppl. 2), 46–49.
Barnes, P.J., and Chung, K.F. (1992) Questions about inhaled b2-agonists in asthma. Trends Pharmacol. Sci. 13, 20–23.
Barnes, P.J., and Karin, M. (1996) NF-kappa B:. a pivotal role in chronic inflammation. New Engl. J. Med.
Barnes, P.J., and Liew, F.Y. (1995) Nitric oxide and asthmatic inflammation. Immunol. Today 16, 128–130.
Belvisi, M.G., Stretton, C.D., and Barnes, P.J. (1992) Nitric oxide is the endogenous neurotransmitter of bronchodilator nerves in human airways. Eur. J. Pharmacol. 210, 221–222.
British Thoracic Society (1993) Guidelines on the management of asthma. Thorax. 48S (Suppl.), S1–S24.
Burney, P.G.J. (1992) Epidemiology. Br. Med. Bull. 48, 10–22.
Chanez, P., Dent, G., Yukawa, T., Barnes, P.J., and Chung, K.F. (1990) Generation of oxygen free radicals from blood eosinophils from asthma patients after stimulation with PAF or phorbol ester. Eur. Respir. J. 3, 1002–1007.
Chung, K.F., Rogers, D.F., Barnes, P.J., and Evans, T.W. (1990) The role of increased airway microvascular permeability and plasma exudation in asthma. Eur. Respir. J. 3, 329–337.
Chung, K.F. (1995) Leukotriene receptor antagonists and biosynthesis inhibitors: potential breakthrough in asthma therapy. Eur. Respir. J. 8, 1203–1213.
Devalia, J.L., and Davies, R.J. (1993) Airway epithelial cells and mediators of inflammation. Respir. Med. 6, 405–408.
Djukanovic, R., Roche, U.R., Wilson, J.W., Ceasley, C.R.W., Twentyman, O.P., Howarth, P.H., and Holgate, S.T. (1990) Mucosal inflammation in asthma. Am. Rev. Respir. Dis. 142, 434–457.
Dunnill, M.S. (1960) The pathology of asthma, with special reference to the changes in the bronchial mucosa. J. Clin. Pathol. 13, 27–33.
Ebina, M., Yaegashi, H., Chiba, R., Takahashi, T., Motomiya, M., and Tanemura, M. (1990) Hyperreactive site in the airway tree of asthmatic patients recoded by thickening of bronchial muscles: a morphometric study. Am. Rev. Respir. Dis. 141, 1327–1332.
Evans, D.J., Lindsay, M.A., O'Connor, B.J., and Barnes, P.J. (1996) Priming of circulating human eosinophils following exposure to allergen challenge. Eur. Respir. J. 9, 703–708.
Gleich, G.J. (1990) The eosinophil and bronchial asthma: current understanding. J. Allergy Clin. Immunol. 85, 422–436.
Global strategy for asthma management and prevention (1995) NHLBI/WHO Workshop Report 3659.
Grandordy, B.M., Mak, J.C.W., and Barnes, P.J. (1994) Modulation of airway smooth muscle b-adrenoceptor function by a muscarinic agonist. Life Sci. 54, 185–191.
Haahtela, T., Järvinsen, M., Kava, T., et al. (1994) Effects of reducing or discontinuing inhaled budesonide in patients with mild asthma. New Engl. J. Med. 331, 700–705.
Hall, I.P., Wheatley, A., Wilding, P., and Liggett, S.B. (1995) Association of Glu 27 b2-adrenoceptor polymorphism with lower airway reactivity in asthmatic subjects. Lancet 345, 1213–1214.
Hamid, Q., Azzawi, M., Sun Ying, et al. (1991) Expression of mRNA for interleukin-5 in mucosal bronchial biopsies from asthma. J. Clin. Invest. 87, 1541–1549.
Hamid, Q., Springall, D.R., Riveros-Moreno, V., et al. (1993) Induction of nitric oxide synthase in asthma. Lancet 342, 1510–1513.
Henderson, W.R. (1994) The role of leukotrienes in inflammation. Ann. Intern. Med. 121, 684–697.
Hirst, S.J., Barnes, P.J., and Twort, C.H.C. (1996) PDGF isoform-induced proliferation and receptor expression in human cultured airway smooth muscle cells. Am. J. Physiol. 14, L415–L428.
Holt, P.G., and McMenamin, C. (1989) Defence against allergic sensitization in the healthy lung: the role of inhalation tolerance. Clin. Exp. Allergy 19, 255–262.
Jose, P.J., Griffiths-Johnson, D.A., Collins, P.D., et al. (1994) Endotoxin: a potent eosinophil chemoattractant cytokine detected in a guinea pig model of allergic airways inflammation. J. Exp. Med. 179, 881–887.
Kay, A.B. (1991) Asthma and inflammation. J. Allergy Clin. Immunol. 87, 893–913.

Kharitonov, S.A., Yates, D., Robbins, R.A., Logan-Sinclair, R., Shinebourne, E., and Barnes, P.J. (1994) Increased nitric oxide in exhaled air of asthmatic patients. Lancet 343, 133–135.

Kuchroo, V.K., Das, M.P., Brown, J.A., et al. (1995) B7-1 and B7-2 costimulatory molecules activate differentially the Th1/Th2 developmental pathways: application to autoimmune disease therapy. Cell 80, 707–718.

Kuitert, L.M., Angus, R.M., Barnes, N.C., et al. (1995) The effect of a novel potent PAF antagonist, modipafant, in chronic asthma. Am. J. Respir. Crit. Care Med. 151, 1331–1335.

Kuo, H., Rhode, J.A.L., Tokuyama, K., Barnes, P.J., and Rogers, D.F. (1990) Capsaicin and sensory neuropeptide stimulation of goblet cell secretion in guinea pig trachea. J. Physiol. 431, 629–641.

Kwon, O.J., Jose, P.J., Robbins, R.A., Schall, T.J., Williams, T.J., and Barnes, P.J. (1995) Glucocorticoid inhibition of RANTES expression in human lung epithelial cells. Am. J. Respir. Cell Mol. Biol. 12, 488–496.

Lammers, J.W.J., Barnes, P.J., and Chung, K.F. (1992) Non-adrenergic, non-cholinergic airway inhibitory nerves. Eur. Respir. J. 5, 239–246.

Laskir, J.D., and Laskir, D.L. (eds.) (1999) Cellular and molecular biology of nitric oxide Marcel Dekker, New York.

Lee, T.M., and Lane, S.J. (1992) The role of macrophages in the mechanisms of airway inflammation in asthma. Am. Rev. Respir. Dis. 145, S27–30.

Mosman, T.R., and Sad, S. (1996) The expanding universe of T-cell subsets: Th1, Th2 and more. Immunol. Today 17, 138–146.

Nadel, J.A. (1991) Neutral endopeptidase modulates neurogenic inflammation. Eur. Resp. J. 4, 745–754.

Ollerenshaw, S., Jarvis, D., Woolcock, A., Sullivan, C., and Scheibner, T. (1989) Absence of immunoreactive vasoactive intestinal polypeptide in tissue from the lungs of patients with asthma. N. Engl. J. Med. 320, 1244–1248.

Ollerenshaw, S.L., Jarvis, D., Sullivan, C.E., and Woolcock, A.J. (1991) Substance P immunoreactive nerves in airways from asthmatics and non-asthmatics. Eur. Resp. J. 4, 673–682.

Pilewski, J.M., and Albelda, S.M. (1995) Cell adhesion molecules in asthma: homing activation and airway remodelling. Am. J. Respir. Cell Mol. Biol. 12, 1–3.

Robbins, R.A., Barnes, P.J., Springall, D.R., et al. (1994) Expression of inducible nitric oxide synthase in human bronchial epithelial cells. Biochem. Biophys. Res. Commun. 203, 209–218.

Sandford, A., Weir, T., and Pare, P. (1996) The genetics of asthma. Am. J. Respir. Crit. Care Med. 153, 1749–1765.

Selroos, O., Pietinalcho, A., Lofroos, A.-B., and Riska, A. (1995) Effect of early and late intervention with inhaled corticosteroids in asthma. Chest 108, 1228–1234.

Sheffer, A.L. (1992) International consensus report on diagnosis and management of asthma. Clin. Exp. Allergy 22 (Suppl. 1), 1–72.

Siebenlist, U., Franzuso, G., and Brown, R. (1994) Structure, regulation and function of NF-κB. Ann. Rev. Cell Biol. 10, 405–455.

Spence, D.P.S., Johnston, S.L., Calverley, P.M.A. et al. (1994) The effect of the orally active platelet-activating factor antagonist WEB 2086 in the treatment of asthma. Am. J. Resp. Crit. Care Med. 149, 1142–1148.

Spiteri, M.A., Knight, R.A., Jeremy, J.Y., Barnes, P.J., and Chung, K.F. (1994) Alveolar macrophage-induced suppression of peripheral blood mononuclear cell responsiveness is reversed by in vitro allergen exposure in bronchial asthma. Eur. Resp. J. 7, 1431–1438.

Springall, D.R., Howarth, P.H., Counihan, H., Djukanovic, R., Holgate, S.T., and Polak, J.M. (1991) Endothelin immunoreactivity of airway epithelium in asthmatic patients. Lancet 337, 697–701.

Sur, S., Crotty, T.B., Kephart, et al. (1993) Sudden onset fatal asthma: a distinct entity with few eosinophils and relatively more neutrophils in the airway submucosa. Am. Rev. Respir. Dis. 148, 713–719.

Thomas, P.S., Yates, D.H., and Barnes, P.J. (1995) Tumor necrosis factor-α increases airway responsiveness and sputum neutrophils in normal human subjects. Am. J. Respir. Crit. Care Med. 152, 76–80.

Turki, J., Pak, J., Green, S., Martin, R., and Liggett, S.B. (1995) Genetic polymorphism of the b_2-adrenergic receptor in nocturnal and nonnocturnal asthma: evidence that Gly 16 correlates with the nocturnal phenotype. J. Clin. Invest. 95, 1635–1641.

Wegner, C.D., Gundel, L., Reilly, P., Haynes, N., Letts, L.G., and Rothlein, R. (1990) Intracellular adhesion molecule-1 (ICAM-1) in the pathogenesis of asthma. Science 247, 456–459.

Racommended Readings

Asthma (1992) Lancet Supplement 2, Vol. 350, London.
Barnes, P.J., Leff, A.R., Grunstein, M., and Woolcock, A.J. (1997) Asthma. Lippincott-Raven, Philadelphia.
Brenner, B.E. (ed.) (1999) Emergency asthma. Marcel Dekker, New York.
Fahy, J.V., Corry, D.B., and Boushey, H.A. (2000) Airway inflammation and remodeling in Asthma. Curr. Opin. Pulm. Med. 6, 15–20.
Harrison, S., Page, C.P., and Spina, D. (1999) Airway nerves and protein phosphates. Gen. Pharmacol. 32, 287–298.
Linden, A., Cardell, L.O., Yoshihara, S., and Nadel, J.A. (1999) Bronchodilation by pituitary adenlate cyclase-activating peptide and related peptides. Eur. Respir. J. 14, 443–451.
Weiss, K.B., Buist, A.S., and Sullivan, S.D. (2000) Asthma's impact on society: the social and economic burden. Marcel Dekker, New York.

21
Pathobiology of Emphysema

Gordon L. Snider

Emphysema is a condition of the lung characterized by permanent and abnormal enlargement of the airspaces distal to the terminal bronchioles accompanied by destruction of their walls and without obvious fibrosis. *Destruction* is defined as nonuniformity in the pattern of respiratory air space enlargement; the orderly appearance of the acinus and its components is disturbed and may be lost.

Emphysema cannot readily be diagnosed accurately in life unless it is severe. As a result, the term *chronic obstructive pulmonary disease* (COPD) has come into wide use to characterize patients who do not have asthma, but who have a syndrome of chronic cough, wheezing, and chronic airflow obstruction. COPD is defined as a disease state characterized by the presence of airflow obstruction due to chronic bronchitis or emphysema. The airflow obstruction is generally progressive, may be accompanied by airways hyperreactivity, and may be partially reversible. *Chronic bronchitis* is defined as the presence of chronic productive cough that does not result from a medically discernible cause (e.g., tuberculosis or lung cancer) and that has been present for more than 3 months in two successive years. Emphysema is almost always associated with chronic bronchitis.

This chapter will focus on the pathobiology of emphysema; however, before proceeding to do so, the epidemiology of COPD will be briefly reviewed in order to make clear the information relating to emphysema that has come from epidemiologic studies.

Epidemiology of COPD

Prevalence and Mortality

In 1994 an estimated 16 million persons in the United States suffered from COPD, representing an increase of 60% since 1982. In 1993, 95,910 deaths were attributed to COPD in the United States, representing the fourth most frequent cause of death and double the 47,335 deaths from this cause in 1979. Prevalence, incidence, and mortality rates for COPD increase with age. Overall, the age-adjusted death rate increased from 14.0 per 100,000 population in 1979 to 26.5 in 1991; during this time the death rate for cardiovascular disease declined. These data reflect that COPD mortality rates, in contrast to cardiovascular mortality rates, are relatively insensitive to smoking cessation. The mortality rate for whites was almost 1.5 times that for nonwhites (20.7 vs. 14.3). Mortality is higher for men than for women, but the gap is narrowing. Between 1979 and 1993, age-adjusted mortality increased for white males by 10.1%; for white females by 126%; for black males by 35%; and black females by 130%. The dramatic increase in mortality for females likely reflects a lag in initiation of tobacco smoking.

Risk Factors

Tobacco Smoking

Tobacco smoking and age account for more than 85% of the risk of developing COPD in the United States. Only homozygous α1-antitrypsin (AAT) deficiency presents a comparable risk, but this factor accounts for less than 1% of patients with COPD. Data from longitudinal, cross-sectional, and case control studies show that compared with nonsmokers, cigarette smokers have higher COPD mortality. They also have higher prevalence and incidence of productive cough, other respiratory symptoms, and spirometrically shown airways obstruction. A dose–response relationship exists for tobacco smoking; differences between smokers and nonsmokers increase as daily cigarette consumption and number of years smoked increase. Pipe and cigar smokers have higher COPD mortality and morbidity than do nonsmokers, although their rates are lower than those of cigarette smokers. For reasons not yet known, only about 15–20% of cigarette smokers develop clinically significant COPD. Familial aggregation of COPD has been established, and genetic factors may account for increased susceptibility of a segment of the population.

Passive Smoking

Passive, or involuntary, smoking, which is also known as "second-hand smoke," is the exposure of nonsmokers to cigarette smoke indoors. Cigarette smoke in indoor air can produce eye irritation and may incite wheezing in asthmatic persons. An increased prevalence of respiratory symptoms and disease, and small but measurable decreases in lung function, have been shown in the children of smoking compared with nonsmoking parents; however, the significance of these findings for the future development of COPD is unknown. Despite these uncertainties, children should be protected from environmental tobacco smoke.

Air Pollution

It is established that high levels of environmental air pollution are harmful to persons with chronic heart or lung disease; peaks of urban air pollution are strongly associated with increases in morbidity and mortality due to COPD. Although the exact role of air pollution in producing COPD is unclear, its role is small compared with that of cigarette smoking.

Sex, Race, and Socioeconomic Status

Prevalence and mortality rates are higher in males than females, and they are higher in whites than nonwhites. Incidence and mortality are generally higher in blue-collar workers than they are in white-collar workers and in those with fewer years of formal education.

Occupation

Working in an occupation in which the air is polluted with chemical fumes or a biologically inactive dust leads to increased prevalence of chronic airflow obstruction, increased rates of decline in forced expiratory volume for 1 second (FEV_1), which is a measure of ventilatory capacity, and increased mortality from COPD. Interaction between cigarette smoking and exposure to hazardous dust (e.g., silica or coal dust) results in increased rates of COPD. In all these studies, however, smoking effects are much greater than occupational effects. Except for coal and gold dust exposure, which alone can cause COPD, the effects of occupational pollution and smoking appear to be additive.

Hyperresponsive Airways

It has been proposed, but not proved, that the atopic state or nonspecific airways hyperresponsiveness (usually measured as responsiveness to methacholine inhalation) predisposes smokers to the development of airways obstruction. In the absence of asthma, studies have failed to show a relation of manifestations of COPD in smokers to standardized levels of IgE, eosinophilia, or skin-test reactivity to allergens. Airways hyperreactivity in COPD is inversely related to FEV_1 and is predictive of an increasing rate of decline of FEV_1 in smokers. It is not clear, however, whether airways hyperreactivity is a cause of development of airflow obstruction or results from the airway inflammation that occurs in smoking-related airflow obstruction. Nonspecific airway hyperreactivity occurs in a significantly higher proportion of women than men.

Infection

Childhood respiratory infection may predispose to COPD. There is no compelling evidence to support a role for adult respiratory infections in accelerating the rate of pulmonary function decline in persons with COPD.

The Structural Basis of COPD

Patients with COPD whose lungs are examined pathologically after surgical excision or after autopsy show three main types of changes: enlargement of submucosal glands in the trachea and large airways, emphysema, and bronchiolitis. The enlarged submucosal bronchial glands are likely to be responsible for the mucus hypersecretion that characterizes chronic bronchitis, but they do not contribute to airflow obstruction. Emphysema causes airflow obstruction by giving rise to decreased elastic recoil with diminished lung emptying during expiration due to bronchiolar collapse at large lung volumes because of decreased numbers of alveolar attachments to distal, poorly supported airways. Bronchiolitis causes narrowing of alveoli as a result of luminal secretion, goblet cell metaplasia, inflammation of the bronchiolar wall and fibrosis. The airflow obstruction of COPD is due to a variable mixture of emphysema and bronchiolar disease.

Emphysema

Elastic Fiber Damage and Emphysema

It has been known for many years that loss of elastic recoil and anatomically demonstrated disruption of the elastic fiber network of the lungs are a key part of the pathobiology of emphysema. Other connective tissue elements (e.g., collagen and the proteoglycans) are also involved, but, as will be discussed in detail later, elastin damage appears to be a key element in the causation of emphysema.

Anatomic Types of Emphysema

Localization of the lesions of mild emphysema in the acinus, which is the unit of lung structure distal to the terminal bronchiole, serves as the basis for classification of the disease (see Table 21.1). The acinus is composed of three to five orders of respiratory bronchioles, the alveolar ducts, and their alveoli. All of the structures distal to the terminal bronchiole participate in gas exchange and constitute the respiratory tissues of the lungs.

Centriacinar Emphysema

Centriacinar emphysema begins in the respiratory bronchioles. Scarring and focal dilation of the bronchioles and of the adjacent alveoli result in the development of an enlarged air space or microbulla in the center of the secondary lung lobule. Airspace enlargement spreads peripherally from the centriacinar region.

TABLE 21.1. Classification of respiratory airspace enlargement.

Simple airspace enlargement (no destruction)
Developmental
Down's syndrome
Congenital
Congenital lobar inflation (due to bronchial atresia)
Acquired
Secondary to loss of lung volume (atelectasis, lung resection)
Emphysema (Airspace enlargement with destruction)
Proximal acinar emphysema
Centrilobular emphysema
Focal emphysema
Panacinar emphysema
Distal (subpleural, paraseptal) emphysema
Airspace enlargement with fibrosis

Focal Emphysema

A widespread form of centriacinar emphysema, occurring in individuals who have had heavy exposure to a biologically inactive dust (e.g., coal dust) is called *focal emphysema.* Large numbers of pigment-laden macrophages are noted in association with focal emphysema, which is uniformly distributed through the lungs.

Centrilobular Emphysema

This type of emphysema is the form of centriacinar emphysema most frequently associated with prolonged cigarette smoking. This lesion involves the upper and posterior portions of the lungs more than the lower portions.

Panacinar Emphysema

The term *panacinar emphysema* refers to dilation of all of the respiratory air spaces of the secondary lung lobule. This form of emphysema may be either focal or diffuse. The focal form is more common at the lung bases and commonly occurs in association with centribobular emphysema (CLE) in smokers. Diffuse panacinar emphysema is the lesion most often associated with AAT deficiency; the emphysema is usually worse at the bases than at the apices.

Distal Acinar Emphysema

Distal acinar emphysema, which is also known as paraseptal or subpleural emphysema, is localized along fibrous interlobular septa or beneath the pleura. The remainder of the lung is often spared; thus, pulmonary function may be normal or nearly so despite the presence of many superficial areas of locally severe emphysema. This is the type of emphysema that produces the apical bullae that gives rise to simple spontaneous pneumothorax in young persons.

Air Space Enlargement with Pulmonary Fibrosis

Air space enlargement with fibrosis, formerly termed *irregular* or *paracicatricial emphysema,* occurs with obvious fibrosis in the lungs. The process is usually an inconsequential lesion adjacent to scars, but it may be quite extensive and clinically important, occurring as a complication of such fibrosing diseases as tuberculosis, silicosis, and sarcoidosis. The underlying fibrosing lung disease is usually evident radiographically, with extensive linear or nodular shadows interspersed among enlarged airspaces.

Bullae

Bullae are areas of marked focal dilation of respiratory airspaces that may result from the coalescence of adjacent areas of emphysema, from locally severe panacinar emphysema, or from a ball-valve effect in the bronchi supplying an emphysematous area. The bullae may be simple airspaces, or they may retain the trabeculae of the emphysema that led to them. Although locally severe emphysema of any type can give rise to bullae, giant bullae are particularly likely to complicate distal acinar emphysema.

Pathogenetic Implications of the Heterogeneity of Emphysema

The different anatomic patterns, as well as age and sex distributions of the various types of emphysema, suggest that each may have a different etiology and pathogenesis. Another way of thinking about the problem is to characterize emphysema as one of the stereotyped ways that the lungs heal after certain injuries. After all, the lungs are organs in which about 300 million air cells are formed by a complexly organized, delicate, membranous tissue. We should not be surprised if a lung that has healed after some injuries retains a relation between air spaces (albeit enlarged and distorted) and the architecturally altered surrounding tissues.

Lung Function in Emphysema

Airflow Obstruction

As already noted, emphysema may occur in the absence of airflow obstruction, but its clinical importance lies in its frequent association with obstruction of airflow. Airflow obstruction may be diagnosed clinically by delay in forced expiration or the auscultation of wheezes. Forced expiratory spirometry, however, is the usual tool used for diagnosing and quantifying airflow obstruction clinically. A decreased FEV_1 as compared with the normal value for age, height, and sex, coupled with a decreased ratio of FEV_1 to forced vital capacity (FVC) is diagnostic of airflow obstruction. The criteria for diagnosis of COPD are FEV_1 less than 70% predicted, FEV_1/FVC less than 90% predicted, and postbronchodilator increases of FEV_1 less than 15% baseline value or less than 300 ml.

Compliance and Lung Volumes

The compliance of the lungs is consistently decreased in emphysema. There is increased lung distensibility with an accompanying increase in total lung capacity, functional residual capacity and residual volume, and a decrease in vital capacity. The volume pressure curve of the lungs is usefully described by an exponential function,

$$V = A - Be^{-KP}$$

where V = lung volume
P = static recoil pressure
A, B, and K = constants

The constant K is an index of lung distensibility that describes the shape of the volume–pressure curve over the upper half of the lung volume. It is inversely related to the bulk elastic modulus of the lungs. The constant K is a better predictor of emphysema than are other measures of elastic recoil or lung volumes.

The correlations of all measures of the volume–pressure relationship with pathologic severity of emphysema are relatively weak. The reason for this weak relation may be that the volume pressure relations of the whole lung is a net measurement that reflects the contributions of regions with varying degrees of alteration in lung compliance. The airspaces of CLE have been shown to be less compliant than normal and the overall compliance of the lungs that contain them. Compliance may also be increased in some areas of destruction with little airspace enlargement.

CO Diffusing Capacity

The carbon monoxide (CO) diffusing capacity, whether measured by the single breath or steady-state method, and whether values are expressed in absolute terms, as a percentage predicted, as a function of alveolar volume, or as a fractional CO uptake, have consistently been the best predictor of the severity of emphysema. This relation likely reflects the loss of internal surface area and of capillary bed of the lung that occurs in emphysema. This test, however, is neither specific nor sensitive as a test of mild emphysema; the diffusing capacity is decreased in young smokers within a few years of starting smoking.

Arterial Blood Gases

The loss of lung homogeneity with emphysema gives rise to ventilation–perfusion mismatch. Increased dead-space ventilation and hypoxemia occur. In the early stages of the disease net alveolar hyperventilation is able to maintain normal carbon dioxide levels; however, with increasing airflow obstruction and ventilation–perfusion mismatch, hypercapnia with worsening hypoxemia supervenes.

α-1 Antitrypsin Inhibitor (AAT) Deficiency

The only known genetic abnormality that leads to the development of COPD is homozygous α_1-antitrypsin inhibitor (AAT); phenotypic expression of the genetic abnormality is rare, accounting for less than 1% of COPD. AAT is a serum protein that is normally found in the lungs and whose main role is the inhibition of neutrophil elastase. It is a glycoprotein, coded for by a single gene on chromosome 14. The serum protease inhibitor phenotype (PI* type) is determined by the independent expression of the two parental alleles. The AAT gene is highly pleomorphic. Some 75 alleles are known, and they have been classified into normal (associated with normal serum levels of normally functioning AAT), deficient (associated with serum AAT levels lower than normal), null (associated with undetectable AAT in the serum), and dysfunctional (AAT is present in normal amount but does not function normally). The variants of AAT occur because of point mutations that result in a single amino acid substitution. The normal M alleles occur in about 90% of persons of European descent with normal serum AAT levels; their phenotype is designated Pi MM. Normal values of serum AAT are 150–350 mg/dL (commercial standard) or 20–48 μM (true laboratory standard).

More than 95% of persons in the severely deficient category are homozygous for the Z allele, designated PI* ZZ, and have serum AAT levels of 2.5–7 μM (mean, 16% normal). Most of these persons are Caucasians of northern European descent because the Z allele is rare in oriental and blacks. Phenotypes that are rarely observed are associated with low levels of serum AAT, including PI* SZ and persons with nonexpressing alleles, PI* null. The latter occur in homozygous form, PI* null–null; or, in heterozygous form with a deficient allele, PI* Znull. Persons with phenotype PI* SS have AAT values ranging from 15 to 33 μM (mean, 52% of normal). The threshold protective level of 11 μM or 80 mg/dL (35% of normal) is based on the knowledge that PI* SZ heterozygotes, with serum AAT values of 8–19 μM (mean, 37% of normal), rarely develop emphysema.

Severe AAT Deficiency and Lung Disease

Severe AAT deficiency leads to premature development of emphysema, often with chronic bronchitis and occasionally with bronchiectasis. The onset of pulmonary disease is accelerated by smoking. Dyspnea begins at a median age of 40 years in smokers as compared with a median age of 53 years in nonsmokers. Panacinar emphysema usually begins radiographically at the bases. The severity of lung disease varies markedly. Nonindex cases (i.e., those discovered in population surveys) tend to have better lung function, whether they smoke or not, than index cases (i.e., those discovered because they have lung disease). Nonindex cases may live into their eighth or ninth decade. Airflow obstruction occurs more frequently in men; asthma, recurrent respiratory infections, and familial factors are also risk factors for airflow obstruction.

Heterozygous AAT Deficiency and Lung Disease

PI* MZ heterozygotes have serum AAT levels that are intermediate between PI* MM normals and PI* ZZ homozygotes (12–35 μM; mean, 57% of normal). There is controversy as to whether they are at increased risk for emphysema. In population studies there appears to be no increased risk for COPD; however, there has been an increased frequency of heterozygotes in family studies and in surveys of some populations of COPD patients.

Pathogenesis of Emphysema

The Elastase–Antielastase Hypothesis

The hypothesis that emphysema was caused by an imbalance between elastases and antielastase in the lungs evolved soon after the discovery of homozygous AAT deficiency. Support for this theory quickly came from the simultaneous discovery that papain, which is a plant enzyme with elastolytic properties, could induce emphysema in animals.

The elastase–antielastase hypothesis of emphysema may be briefly summarized as follows. The lung is protected from elastolytic damage by AAT, a protein that is present in the lungs of persons with normal serum proteins but is sharply reduced in persons with homozygous AAT deficiency. α-2-macroglobulin enters the lungs only if capillary permeability is increased. Secretory leucoprotease inhibitor is secreted by bronchial glands and goblet cells; its main role is to protect the airway epithelium from proteolytic injury, but it may also play a role in preventing alveolar elastic fiber injury. Elafin, which is a more recently discovered antielastase, also plays a protective role.

Neutrophils elaborate elastase, which is a serine protease. Macrophages elaborate an elastaselike metalloprotease and may ingest, and later release, neutrophil elastase; macrophage-derived cysteine proteinases, cathepsin B and cathepsin L may also play a role. Oxidants derived from neutrophils and macrophages or from cigarette smoke may inactivate AAT inhibitor and may interfere with lung matrix repair. Endogenous antioxidants (e.g., superoxide dismutase, glutathione, and catalase) protect the lung against oxidant injury. The elastase–antielastase hypothesis is still in a state of evolution; supporting evidence comes from a variety of sources.

Animal Models of Emphysema with Elastin Damage

Instillation into the lungs of elastolytic enzymes induces experimental emphysema; proteolytic but nonelastolytic enzymes do not do so. Studies of lung function in animals with elastase-induced emphysema have shown increases in lung compliance, functional residual capacity, residual volume, and total lung capacity. Airflow limitation, which occurs during forced expiration, is due to loss of elastic recoil and alveolar attachments to airways resulting from the absence of bronchiolitis. The diffusing capacity for CO is lowered in proportion to the severity of the emphysema. Arterial blood gas studies show hypoxemia without carbon dioxide retention. Right ventricular hypertrophy occurs in hamsters with elastase-induced emphysema.

Treatment with elastase leads to a diminution of lung elastin content at 24 hours, followed by rapid restoration of total lung elastin to normal levels; however, the anatomic arrangement of the elastic fibers remains grossly disordered. There is biochemical evidence of damage and subsequent repair of collagen in experimental, elastase-induced emphysema.

Intravenous and intratracheal administration of endotoxin causes pulmonary neutrophilia. Mild emphysema has been induced by multiple intravenous injections of endotoxin in Rhesus monkeys and in dogs and by multiple intratracheal instillations of endotoxin in hamsters.

It has been reported that rats with pancreatic elastase-induced emphysema treated with all-*trans*-retinoic acid underwent reversal of the emphysematous changes, as indicated by reversal of the lung volume and morphometric changes in their lungs. If this observation is confirmed in mammalian species that do not grow throughout their lifetimes, then a new approach for the treatment of human emphysema may emerge.

Emphysema and Impaired Cross-Linking of Elastin

In the blotchy mouse (a genetic model of emphysema) and young animals fed lathyrogens, emphysema results from impaired cross-linking of elastin. Elastase-induced emphysema is made worse by treatment of the animals with a lathyrogen. These data support the importance of elastic fiber damage in the pathogenesis of emphysema.

Evidence from Humans of Elastin Degradation in Emphysema

There seems little question that the emphysema in severe AAT deficiency is caused by insufficient protease inhibitor to balance the effects of elastases. Some investigators have found an association between blood leukocyte elastase concentration and airflow obstruction in this disease.

Desmosine, which is a cross-link amino acid that gives elastin its great stability, is not metabolized in the body. It comes only from elastin, thus enabling it to serve as a biologic marker of elastin degradation. Studies using an HPLC technique and isotope dilution, have found ratios of urinary desmosine to creatinine that are higher than normal in persons with COPD and are intermediate in smokers without airflow obstruction. Studies using ELISA have shown that patients with COPD have elevated plasma and urinary elastin peptide levels as compared with levels in normal nonsmokers. Normal smokers have intermediate values.

Immunocytochemical Localization of Elastase in the Lung

Direct evidence supporting the protease–antiprotease hypothesis of emphysema in smokers is rather sparse. One group has shown immunoultrastructurally that neutrophil elastase is bound to elastic fibers in the lungs of smokers, and that the amount of elastase is proportional to the amount of emphysema present. Another group failed to substantiate these findings, whereas a third provided confirmation.

Neutrophils versus Macrophages as the Source of Elastase

Neutrophil elastase at first appears to be the most plentiful intracellular elastase. All circulating neutrophils pass through the pulmonary capillaries many times each hour, and neutrophils may enter the lungs in large numbers in response to chemotactic gradients. The elastase in the neutrophils, however, is packaged in the azurophilic granules, and cannot readily reach the external milieu. Macrophages are concentrated in the centriacinar zones of the lungs, which is the region where the centrilobular emphysema of smokers occurs. When the potential capability of the macrophage to solubilize elastin is considered, along with its manyfold greater frequency, its greater longevity, and its concentration in the centriacinar zone in smokers, the primacy of neutrophil elastase is no longer so certain.

Bronchoalveolar lavage provides an *in vivo* technique for readily sampling the contents of the respiratory airspaces of humans. The numbers of lavageable lung cells in smokers is increased by four- to fivefold. Neutrophils are rich in elastolytic proteinases, including neutrophil elastase, cathepsin-G and proteinase-3. The percentage of neutrophils in BAL fluid of smokers is unchanged, but their absolute number is increased because of a four-to-fivefold increase in the total number of cells in the BAL fluid of smokers.

Macrophages represent 90% of lavageable cells. They secrete chemotactic factors that are likely responsible for the increased number of neutrophils. Macrophages also secrete a neutrophil-degranulating agent and can ingest and later release neutrophil elastase.

Small amounts of elastaselike esterase activity occur in BAL fluid of smokers compared with nonsmokers. Much of this activity is attributable to metalloprotease. Human macrophages do not elaborate a measurable amount of metalloelastase into culture medium, but they are capable of degrading elastin when cultured on elastin-rich ^{3}H-lysine-labeled extracellular matrixes deposited by rat smooth muscle cells or when cultured in direct contact with radiolabeled elastin. The elastin degradation is inhibited by the tissue inhibitor of metalloproteases, but it is inhibited minimally or not at all by inhibitors of cysteine proteases, or by the serine protease inhibitor, eglin-c.

Evidence has been obtained that a 92-kDa gelatinase is a major product of human neutrophils that has elastolytic properties. Human macrophage metalloelastase is also expressed by macrophages from cigarette smokers. It is 64% identical to mouse metalloelastase (MME). In contrast to wild-type mice, knock-out mice that are deficient in MME ($MME^{-/-}$) failed to develop increased numbers of macrophages in their lungs, as well as emphysema in response to long-term exposure to cigarette smoke. $MME^{-/-}$ mice that received monthly instillations of monocyte chemoattractant protein-1 accumulated macrophages in their lungs, but they did not develop emphysema. These experiments provide convincing evidence that macrophage elastase can account for the emphysema that results from chronic cigarette smoke inhalation in wild-type mice.

Macrophages synthesize and express cathepsin B, whose levels in healthy cigarette smokers are approximately thrice those in the macrophages of nonsmokers, and approximately 10 times higher in the bronchoalveolar lavage fluid of smokers. Cathepsin B induces emphysema on instillation into hamster lungs. Cathepsins L and S are elastolytic; they may also play a role in elastin degradation by macrophages.

Animal Models of Emphysema Without Elastin Damage

Several animal models of emphysema have been developed that do not depend on elastin damage.

Cadmium

Experimental airspace enlargement with fibrosis is produced by aerosolized or intratracheally instilled cadmium chloride ($CdCl_2$). The addition of the lathyritic agent, β-amino-propionitrile (BAPN), to the intratracheal $CdCl_2$ treatment resulted in marked enhancement of the airspace enlargement. In a study in which lung elastin was radiolabeled during the neonatal period, it was demonstrated that neonatally formed lung elastin is not destroyed in the model of $CdCl_2$ airspace enlargement with fibrosis in hamsters. Rendering hamsters neutropenic with antineutrophil globulin did not inhibit the airspace enlargement produced by combined $CdCl_2$-BAPN treatment. Thus, it seems likely that $CdCl_2$-induced airspace enlargement with fibrosis is not caused by neutrophil elastase.

Oxidants

Exposure of rats to hyperoxia has been found to give rise to airspace enlargement and lung volumes without altering compliance or alveolar number. There was also an almost 50% increase in collagen content of the lungs. A number of laboratories have observed airspace enlargement with exposure of animals to NO_2 or ozone. The mechanism of development of oxidant-induced emphysema is not clear, but a response of collagen seems more important than that of elastin. For example, treatment of the oxygen-exposed rats with an analog of proline that prevents synthesis of cross-linked collagen prevented emphysema development.

Starvation

Severe starvation induces airspace enlargement in rats and hamsters. The mechanism underlying this form of emphysema is obscure, but it might be related to disordered growth of the rodent lung, with an imbalance between lung and chest wall growth. It does not appear to be related to neutrophil elastase.

Tobacco Smoke

Emphysema with typical structural and physiologic characteristics has been produced in guinea pigs exposed to the smoke of 10 cigarettes daily for up to 12 months. The structural changes appear to be related to collagen breakdown and repair with an apparent increase in elastin concentration in the lung tissue. Exposure of adult mice to the smoke of two unfiltered cigarettes per day, 6 days per week, for up to 6 months, also produces emphysema.

Synthesis of Current Information on Pathogenesis of Emphysema

Significance of Models of Airspace Enlargement with Fibrosis

The different anatomic patterns of the various types of emphysema suggest that each may have a different etiology and pathogenesis. For example, a number of observations suggest that CLE in smokers may be a variant of airspace enlargement with fibrosis: Fibrosis is a part of the microbullae of CLE. Cigarette smoke is an important source of cadmium accumulation in humans; cadmium occurs in emphysematous lungs in direct proportion to the severity of emphysema. The compliance of the microbullae of CLE is less than that of normal lungs, and much less than that of the lungs that contain them. Elastin concentration is decreased in panacinar emphysema, but collagen concentration is normal. Collagen concentration is increased in mild and severe CLE, whereas elastin concentration is decreased only in severe CLE, which may be due to an increase in nonelastin connective tissue. Elastase–antielastase imbalance may account for panlobular emphysema and the widespread discontinuities in alveolar walls, whereas the centrilobular lesions may represent focal airspace enlargement with fibrosis.

Emphysema—A Stereotyped Response of the Lungs to Injury

Another way of thinking about the problem of variable pathogenesis of different types of emphysema is to characterize emphysema as one of the stereotyped ways that the lungs heal after multiple injuries that results in inflammation (Table 21.2). After all, the lungs are organs in which about 300 million air cells are formed by a complexly organized, delicate, membranous tissue. We should therefore not be surprised if a lung that has healed after some injuries retains a relation between air spaces (albeit enlarged and distorted) and the architecturally altered surrounding tissues.

The production of experimental emphysema with diverse agents reviewed in this chapter supports the preceding concept. The airspace enlargement may be due to a repair process that attempts to preserve the basic structure of the gas-exchanging portion of the lung parenchyma. That is, airspaces communicating with the external environment, created by walls consisting of a capillary *rete* supported by a complex connective tissue framework and invested by a highly specialized thin epithelium. It seems quite clear from the experimental data that not all airspace enlargements occur on a background of elastin destruction, although elastin damage results in emphysema. Injury and repair of connective tissue (i.e., collagen, elastin, and proteoglycans) play a central role in all disease processes, thus ending in airspace enlargement.

In some of these models, connective tissue proliferation in response to severe injury stops after levels of lung connective tissue are restored to initial values, and the alveolar wall remains thin and delicate despite its partial destruction. In other models, grossly excessive amounts of connective tissue are produced, with marked fibrous thickening of the alveolar walls. Local distending forces give rise to airspace enlargement. Less growth of alveoli than of the chest wall is the likely explanation for airspace enlargement in severe starvation. Altered growth is also likely to play a role in airspace enlargement due to hyperoxia, although just how collagen metabolism relates to this process is obscure. In both starvation and hyperoxia models, an imbalance of connective tissue elements rather than loss of a particular element may be important.

The concept of emphysema as a stereotyped response of the lung to injury should not be considered a denial of the elastase–antielastase hypothesis of the pathogenesis of human emphysema. The premature development of emphysema in individuals with homozygous API deficiency, especially if they smoke tobacco, is strong evidence in support of the elastase–antielastase hypothesis of the pathogenesis of human emphysema. Physiologic findings of increased lung compliance and pathologic evidence of fraying and rupture of elastic fibers in severe emphysema are indicative of elastic fiber network damage. A great deal of information from animal models of emphysema provides strong but indirect support of this hypothesis; however, there is virtually no direct evidence to support the hypothesis of elastase–antielastase imbalance as a cause of elastic fiber destruction in the pathogenesis of common emphysema of cigarette smokers. It thus seems likely that an imbalance of elastases and antielastases is not the only mechanism for production of emphysema in humans, and, indeed, more than one mechanism may well be operative in a single individual.

TABLE 21.2. Current concepts of pathogenesis of emphysema.

Type of emphysema	Concept
Centrilobular emphysema	Inflammation plus fibrosis
Panacinar emphysema	Inflammation plus elastase-antielastase imbalance
Focal emphysema	Dust Inhalation plus inflammation
Developmental emphysema	Impaired alveologenesis

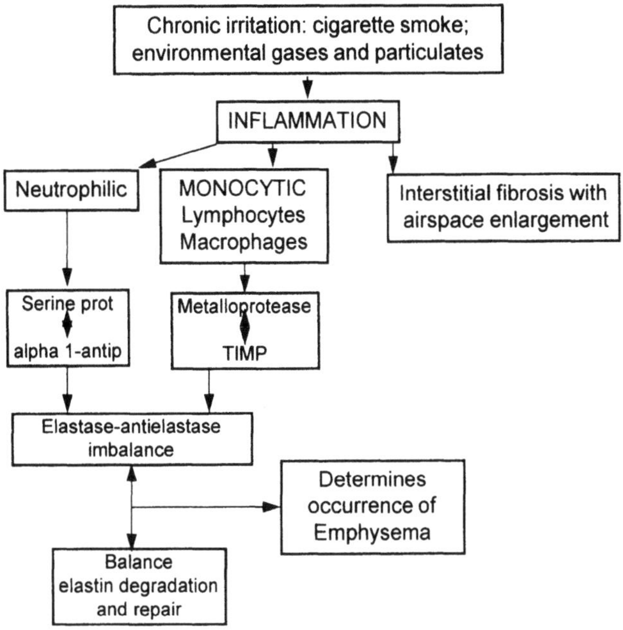

FIGURE 21.1. The pathogenesis of emphysema.

Summary

Emphysema is a condition of the lungs characterized by enlargement of respiratory airspaces due to destructive changes, that along with chronic bronchitis and airflow obstruction constitute the defining characteristics of COPD. In 1994 about 16 million Americans suffered from COPD; almost 100,000 died from the disease making it the fourth most frequent cause of death. The major risk factor for COPD is tobacco smoking, with urban and occupational air pollution playing secondary roles. Unknown genetic influences probably determine that only 15–20% of smokers develop the disease. AAT deficiency is the most important genetic defect associated with emphysema; gene frequency is estimated to be high in Caucasian, but less than 10% of the exacted number of persons with AAT-COPD have been identified, and the genetic defect accounts for only 1% of COPD. The airflow obstruction of COPD is due to a variable mix of bronchiolitis and of emphysema that causes closure of distal airways at larger than normal lung volumes.

Elastic fiber damage is present in emphysema. A number of different types of emphysema, each with its own anatomic pattern and associations, suggests that the pathogenesis of emphysema differs among these types and that emphysema may be one of the stereotyped responses of the lungs to different lung injuries. Airflow obstruction may or may not be present with emphysema, but it is the cause of impairment in this disease. Lung compliance and diffusing capacity are decreased; hypoxemia and, in severe forms of the disease, hypercapnia, are two important consequences of COPD.

There are two main hypotheses of the pathogenesis of emphysema: the inflammation fibrosis hypothesis and the elastase–antielastase hypothesis. Inflammation in response to cigarette smoke is the staring point for both hypotheses. The inflammation-fibrosis hypothesis posits that inflammation is followed by fibrosis and collapse of some airspaces with enhanced recoil forces, causing dilation of adjacent airspaces. The elastase–antielastase hypothesis posits that an excess of elastases overcomes the antielastase defenses of the lungs. Neutrophils and macrophages are the source of serine and metalloproteases, respectively. AAT and secretory leukocyte protease inhibitor are the major serine antiproteases; tissue inhibitors of metalloproteases are the major protections against metalloproteases. If elastase activity predominates, elastin degradation occurs and is followed by emphysema (see Fig. 21.1).

Recommended Readings

Hogg, J.C. (2001) Chronic obstructive pulmonary disease: an overview of pathology and pathogenesis. Novartis Found. Symp. 234, 4–19, discussion 19–26.

Piquette, C.A., Rennard, S.I., and Snider, G.L. (2000) Chronic bronchitis and emphysema. In: Murray, J.F., Nadel, N.J., Mason, R.J., and Boushey, H.A. Jr. (eds.), (2000) Textbook of Respiratory Medicine. 3rd edn. pp. 1187–1246. W.B. Saunders Co., Philadelphia.

Saetta, M., Turato, G., Maestrelli, P., Mapp, C.E., and Fabbri, L.M. (2001) Cellular and structural bases of chronic obstructive pulmonary disease. Am. J. Respir. Crit. Care. Med. 163, 1304–1309.

Silverman, E.K. (2001) Genetics of chronic obstructive pulmonary disease. Novartis Found. Symp. 234, 45–58, discussion 58–64.

Stockley, R.A. (2001) Proteases and antiproteases. Novartis Found. Symp. 234, 189–199, discussion 199–204.

Turato, G., Zuin, R., and Saetta, M. (2001) Pathogenesis and pathology of COPD. Respiration 68, 117–128.

22
Inhaled Noninfectious Toxicants and Their Effects on the Lung

Dean E. Schraufnagel, Suman Goel, and Nai-San Wang

The primary function of the respiratory system is to bring in air and allow gas diffusion to the blood through a thin barrier. Our air unfortunately contains more than the desired diffusible gases and nitrogen. Particles and inorganic and organic molecules come in contact with various parts of the respiratory system. Most of these substances are benign and exhaled or readily cleared but some are toxic. The damage done by inhaled toxicants appear in several forms. It may be as predictable as a chemical reaction, or as variable as an idiosyncratic allergic reaction, the injury may be immediate or delayed, and the damage may require large or small amounts of toxicants. An overview of the common inhaled toxicants and their effects on the lungs will be given. Texts of environmental medicine and reviews of individual agents will provide more detailed information.

Particle Deposition, Clearance, and Lung Defense Mechanisms

A human's lung filters about 500 L of air per hour, 10,000 L/day, 4 million L/year, and 250 million L/lifetime. This air contains particles, gases, and infectious agents that maybe inert and noxious. To damage the lung the inhaled substances must be deposited on its mucosal surface or dissolve in its epithelial fluid. The deposition depends on the particles' *inertia*, *sedimentation*, and *diffusion*. Inhaled particles change their direction often and their inertia causes impaction onto the epithelium when a turn is required. Inertial forces increase with air velocity and breathing frequency. Large particles

(>10 μm, aerodynamic equivalent diameter) usually contact nasal tissues; small particles and fibers (<3 μm aerodynamic equivalent diameter) enter deeply into the lung. Particles of 10–2 μm settle out gradually as the airways branch and taper. Particles less than 2 μm may precipitate in the peripheral airways and alveoli; those less than 0.2 μm are usually exhaled while still suspended. Sedimentation is the main force in the distal lung units where air is relatively still. It is augmented by slow, deep breathing and acts primarily on particles between 5 μm and 0.2 μm in diameter. Particles less than 0.5 μm move by diffusion in the small airways and undergo Brownian movement. Gases are distributed in proportion to their solubility and concentration and the volume of tissue to which they are exposed. In addition, the size, shape, concentration, duration of exposure, and chemical nature of the particles influence retention in and damage to the lung.

The lung is structured to defend itself efficiently. Mechanical factors affect clearance of particles at every point in the respiratory system—from nasal hair and turbinate geometry to the extensive ciliary system, bronchiolar mucus, and alveolar surfactant. Inhaled particles deposit widely but not evenly. Particles deposited in the nasal cavity are swept backward over the mucus-lend ciliated epithelium to the nasopharynx, where they are swallowed. Particles deposited in the trachea and bronchi are escalated to the pharynx and esophagus. The nose clears particles at about 6 mm/minute; the trachea clears them at 5–20 mm/minute, and small bronchioles clear at 0.5–1 mm/minute (McClellan et al., 1995). The mucous layer is made up of a thick upper (gel) layer that traps particles and a thin lower (sol) layer. Cilia beat the thick gel forward and their back stroke is in the thin sol layer. Respiratory tract mucus arises from several cells and its physical properties change in different bronchopulmonary and cardiac diseases. If inhaled particles pierce the epithelium, they are cleared through lymphatic drainage. Particles deposited in the alveoli stick to the surfactant-rich liquid and are engulfed by macrophages that carry them to the terminal bronchiole to ascend the mucociliary escalator. When particles are abundant in the airspace, bronchial epithelial cells and both Type-I and Type-II alveolar lining cells phagocytize particles. The phagocytized particles are later released either into the interstitium or back into the air space. Particles arriving in the interstitial space may enter the capillaries to be engulfed by intravascular monocytic cells (macrophages) or, if close to the terminal airways, by interstitial macrophages. Particles in the alveolar space may also reach juxta-alveolar lymphatics through the cytoplasm of the apposed Type-I alveolar cells and lymphatics, inducing inflammatory and fibrotic reactions in the interstitium, pleura, and bronchial wall. Particles that reach the pleura or lobular septum are cleared through the lymphatics. Clearance may be less efficient if the epithelial barrier and mesenchymal tissue are damaged. In addition to the mucociliary system, bronchial branching, smooth muscle activity, and the integrated neural system prevent entry and forcefully expel undesirable particles by bronchoconstriction and cough.

Following long or heavy inhalation, collections of particles are found around the small bronchioles and in the hilar lymph nodes. These foci vary in size and shape and can be found in the bronchial wall, lobular septum, pleura, and even in such distant organs as spleen, bone marrow, and liver.

Macrophages that scavenge infectious agents, particles, and tissue debris are the most important nonspecific cellular component of defense in the lung. Lymphocytes, eosinophils, mast cells, plasma cells, neutrophils, and other mononuclear cells, as well as their products of immunoglobulins, lymphokines, and mediators, also participate in the nonspecific defenses. They are more important, however, in defending against specific antagonists. The specific or immune defense is an extensive topic that is not discussed here.

The defense of the epithelium against bacteria and viruses is enhanced by secretions containing such materials as lactoferrin, interferon, lysozyme, and other enzymes. The membrane zinc metallopeptidases, neutral endopeptidase, and carboxypeptidase M detoxify inhaled substances (Jackman et al., 1995). Macrophages also contain proteolytic enzymes that act in concert with other mechanisms. Deamidase, which is a serine peptidase, has high activity in alveolar macrophages, although dipeptidyl peptidase IV and prolylcarboxypeptidase are also important (Dragovic et al., 1995). Surfactant opsonizes and enhances intracellular killing of microorganisms by alveolar macrophages (O'Neill et al., 1984).

Both circulating and tissue phagocytes generate toxic oxygen byproducts that kill organisms after phagocytosis. If these metabolites escape, they may damage neighboring tissues. To protect against this injury, lungs have a rich antiproteolytic and antioxidant system. Protease inhibitors block several enzymes, including plasmin, thromboplastin antecedent, and kallikrein, which are integral parts of the enzyme systems that affect coagulation, vascular permeability, kinin generation, fibrinolysis, and complement formation. Intracellular and extracellular protective mechanisms include the antioxidant enzymes superoxide dismutase, catalase, glutathione peroxidase, NADPH, and cytochrome C reductase, which interact with many cytokine pathways.

Because each terminal bronchiole collects and concentrates the content of about 10,000 alveoli, the smallest bronchiole is often overwhelmed first after most inhalational assaults. It is the weakest link in the lung defense and is often the site to show the earliest changes. Examples of such changes include "alveolar ductitis" of young cigarette or marijuana smokers, and "small-airway disease" in the early state of asbestosis, many other types of pneumoconiosis, and cystic fibrosis. Alveolar ductitis is characterized by an accumulation of macrophages in alveolar ducts and alveolar spaces immediately distal to the bronchiole.

The bronchi and bronchioles may collect particles at duct openings and other areas of ciliary defects. The particles elicit inflammation that results in further expansion of this space leading to cystic enlargement. Further accumulation of inhaled particles, macrophages, and bacteria may lead to bronchial diverticula, often at the sites of bronchial bifurcation (Yao et al., 1984). Bronchial diverticula may perforate to form a new bronchoalveolar communication (e.g., the canal of Lambert's). They more frequently perpetuate and propagate focal chronic inflammation and fibrosis in the bronchial wall. At times the bronchial lumen can be obliterated completely by the propagating fibrosis, which may result in an obstructive pneumonia distally.

Collections of inhaled particles with chronic inflammation and fibrosis can be found anywhere in the lung, but especially in the lobular septum and pleura. Fibrosis in the lung parenchyma interferes with the elastic recoil of the lung and impairs the normal cleansing processes. The lesion, therefore, may propagate even without continuous particle exposure.

Common histologic changes of a toxicant inhalation include loss of cilia and ciliated cells, damage and necrosis of epithelial cells with repair, mucous cell proliferation, mucous plugging, and collections of macrophages and other inflammatory cells in the airway. Inflammation and macrophage retention can be mild and persisting or severe with disruption of the basal lamina, necrosis, massive collection of inhaled particles, and fibrosis. The changes can be so extensive that they destroy or obliterate normal structure. Smooth muscle cell proliferation is common in scarred lung tissues.

The pathological changes of inhaled toxicants depend on their physical and chemical properties. Particles can be classified into organic and inorganic. Inorganic particles can be further divided into silica-rich, silica-poor, and fibrous particles. Both silica-rich and silica-poor particles often accumulate in upper lung fields, whereas fibrous particles tend to accumulate in lower lung fields.

Particle Analysis

Identification and quantitative analysis of the particulate burden may require an interdisciplinary approach involving patients, clinicians, epidemiologists, toxicologists, and microscopists for recognition, diagnosis, and prevention of disease. Most foreign material that enters the lung is removed, so analyzing it for the particles and fibers that remain only gives an estimate of exposure. Correlating the deposited particles with focal tissue changes may suggest a relationship. Fibers can be analyzed morphologically and chemically. Most studies have used autopsied or surgically removed lung tissue, avoiding areas of consolidation, congestion, and tumor.

Digestion techniques for mineral analysis involve sodium or potassium hydroxide, hydrogen peroxide, 5.25% sodium hypochlorite (bleach), formamide, or proteolytic enzymes to dissolve the lung. Tissue ashing is also effective, but not in a high temperature (400–500°C) furnace, which causes fragmentation of the fibers, artifactually increasing their number and decreasing their length (Roggli et al., 1992). After digestion, the inorganic residue is collected on an acetate or polycarbonate filter and analyzed. Conventional bright-field light microscopy can identify many fibers, but it cannot easily distinguish different fiber types of the same mineral. Phase-contrast light microscopy can improve resolution to detect fibers with a diameter of only 0.2 μm. Transmitted polarized light microscopy is excellent for identifying silica, talc, and other minerals. It can differentiate such deposits of inert fibers as rutile and titanium dioxide pigment from asbestos (Rode et al., 1981).

The electron microscope can precisely locate and determine the chemical composition of particles. Transmission electron microscopy gives the highest resolution and can be coupled with electron diffraction or x-ray analysis to determine chemical composition, but it is less practical than scanning electron microscopy. The scanning electron microscope's usual imaging results from secondary electrons, which have low energy and give great depth of focus and a range of magnifications from 10 to 500,000 times. With scanning electron microscopy fibers as small as 0.3 μm long and 0.05 μm in diameter can be detected (Roggli, 1989) (Fig. 22.1). Sample preparation is not difficult and analysis can be automated with image analyzers and software programs that discriminate between mineral fibers and organic particles.

In the scanning electron microscope, electrons are either generated by heating a tungsten or lanthanum thermionic filament or by a field emission tip. With field emission systems, electrons are drawn off a sharp-tipped tungsten crystal. Resolution from a field emission scanning electron microscope is better than 1 mm for secondary images and about 3 nm for backscattered images. With thermionic filaments the best resolutions are about 2 nm for secondary images and 5 nm for backscattered images. As with transmission electron microscopy, the electrons are drawn toward the anode by an accelerating voltage and focused by electromagnetic lenses. The electron stream goes through coils that scan the electrons across the surface of the specimen. When the beam electrons hit the specimen, secondary electrons result from inelastic collisions of the primary electrons with electrons backscattered from the atomic nuclei of the sample. The energies of secondary electrons are less than 50 eV and they travel only a short distance. Those within a few nanometers of the sample surface may escape and be detected. The number of secondary electrons collected is dependent on the topography of the specimen (Schraufnagel et al., 1990).

FIGURE 22.1. Scanning electron microscopy and x-ray energy dispersive analysis show silica particles in different forms. At the left is a particle with sharp edges primarily of silicon dioxide. The angled particle in the center is a magnesium silicate and the feathered particles on the right are potassium feldspar. The minerals may often be identified by the combination of their elements. [*Source:* Yao et al. (1984).]

Backscattered electrons result from the primary beam's interaction with the atomic nuclei in the sample. The electrons are scattered back with energies close to that of the incident beam (≥50 eV) in proportion to the atomic number of the sample—only about 5% of the incident electrons is backscattered from carbon, but about 50% is scattered back from gold (Becker et al., 1981). These high-energy electrons can originate from deeper in the specimen than secondary electrons. An early biological application of backscattered electron imaging was to detect foreign materials within the lung in pneumoconiosis. Backscattered electron imaging can instantly show heavy elements and their location, but it does not directly distinguish between specific elements as x-ray analysis does. It is also sensitive to surface topography.

X-rays are produced when accelerated electrons in the electron microscope collide with and displace orbiting electrons of the atoms in the specimen. Inner shell vacancies are replaced by higher-energy electrons accompanied by the emission of x-rays with wavelengths characteristic for the elements bombarded. In addition to characteristic x-rays, background radiation, called *Bremsstrahlung*, is produced. The most common type of x-ray spectroscopy, *energy-dispersive x-ray analysis*, can detect elements as light as sodium without difficulty. Windowless detectors can analyze elements as light as boron or beryllium. X-rays are collected by a silicon crystal diffused with lithium known as an Si(Li) detector, which is placed near the sample. The electron hole-pairs formed in the Si(Li) crystal by the incoming x-rays are proportional to the x-ray energies, and their charges can be amplified and sorted with a multichannel analyzer. The information is displayed as a series of peaks that indicate the elements present and their composition. Elemental amounts as small as 10^{-10} g can be detected with spatial resolutions of 10 nm (Hall, 1988). All elements can be collected simultaneously. When attached to a scanning electron microscope, particles may be identified, counted, sized, and analyzed for chemical composition. Limitations of x-ray analysis include its insensitivity to light elements and long collection times. X-ray analysis only detects elements; compounds must be inferred by the ratio of elements, which is often difficult.

Additional limitations are imposed by the specimens and their preparation. Cryotechniques should be used for diffusible elements because soluble elements may be displaced during preparation. Particle detection alone may not uncover the mechanisms of the lung injury. A particle may not be harmful, but it may transport noxious agents. For example, certain silica particles may carry nitric acid. The toxicity of nitric acid then would not be measured by simple particle counting. Among the general population, nonasbestos mineral fibers outnumber asbestos fibers by at least four to one (Churg, 1983). Churg reported that calcium phosphate (apatite), talc, silica, rutile, kaolinite, micas, feldspar, and other silicates (in decreasing order of frequency) accounted for most nonasbestos mineral fibers recovered from the human lung. The biologic significance of most of the nonasbestos mineral fibers is unknown, although they have been found more commonly in patients with lung cancer (Yao et al., 1984).

Air Pollution

The most important inhaled toxicant is cigarette smoke, which accounts for more than three quarters of the respiratory disability seen by physicians. More than 50,000 reports on smoking have established its harmful effects as incontrovertible, although other inhaled agents also cause cancer (Table 22.1) and other respiratory illness. Urban air contains mixtures of carbon, sulfur, nitrogen oxides, organic residues that contain tars and aldehydes, vaporized metals and acids, and various organic and inorganic dust particles. Health effects of inhaling these pollutants have been extensively reported and emissions and concentration standards have been established to protect public health. In the United States, the Clean Air Act defined the permissible levels of particulates, sulfur dioxide, and nitrogen dioxide. The literature is increasingly emphasizing the importance of respirable particulates as a cause of health problems (Xu et al., 1993). The clean air standards for particulate matter less than or equal to 10 μm in aerodynamic diameter (PM_{10}) for 24 hours as the annual average can be found on the webpage of the U.S. Environmental Protection Agency: http://www.epa.gov/oar/

Particulate air pollution may contribute to respiratory disease even below government standards (Utell, 1993). A high level of these fine particulates (PM_{10}) is correlated with an increased incidence of respiratory symptoms, increased hospitalizations and health care visits for respiratory and cardiovascular disease, increased respiratory morbidity as measured by absenteeism from work or school, and increased respiratory and cardiovascular mortality. Particulate air pollution decreases lung function and causes decreased activity of patients with lung and heart disease. Nitrogen dioxide, ozone, and sulfur dioxide are also associated with respiratory symptoms. The association between the health problem and pollution is often not recognized and offending pollution is not removed from the person's environment.

The effect of air pollution depends on the affected person's health, location, and activity. Outdoor pollution is not closely linked to respiratory health because most people spend little time outside and there is variation in the penetration of the particles into indoor environments (Spengler et al., 1985). The variation depends on physical and chemical characteristics of the pollutant, such as particle size and solubility. Certain individuals and disease states are more susceptible to air pollution. Exercising asthmatics exposed to acidic particles at concentrations as low as $75\,\mu g/m^3$ develop bronchoconstriction, but controls with chronic obstructive lung disease do not (Morrow et al., 1993).

TABLE 22.1. Lung carcinogens.

Compounds	Where found
Arsenic compounds	Metal smelting Electrical devices Medications Fungicides Herbicides
Asbestos	Insulation Heat protection Construction
Beryllium compounds	Missile fuels Metal alloys Aerospace and nuclear
Bis-chlormethyl ether	Laboratories Alkylating agents Solvents Ion-exchange resins Polymers Water repellents
Chromium compounds	Metal alloys Paints, pigments Preservatives
Melphalan (Nitrogen mustard)	Antineoplastic and alkylating agents
Mustard gas (sulphur mustard)	Poison war gas
Nickel compounds	Plating Iron alloys Ceramics Batteries Stainless steel arc welding
Radon	Quarries Underground mines Homes
Tobacco smoke	Active and passive smoking

Asbestos-Related Disease

Asbestos is a class of fibrous, magnesium-silicate minerals that induce several lung diseases. The lung fibrosis caused by asbestos is similar to that found with usual interstitial pneumonitis, except that asbestos bodies are present (Fig. 22.2). Asbestos bodies are complexes of organic material and iron that form around asbestos particles. Asbestos particles are usually defined as mineral pieces greater than 1 μm in length with roughly parallel long sides and an aspect ratio of at least 5:1. Asbestos can be classified into two forms; serpentine (e.g., chrysotile) and amphibole (e.g., crocidolite and amosite). Most asbestos bodies isolated from human lung have an amphibole core, which at first may be puzzling because most asbestos used in industry and construction

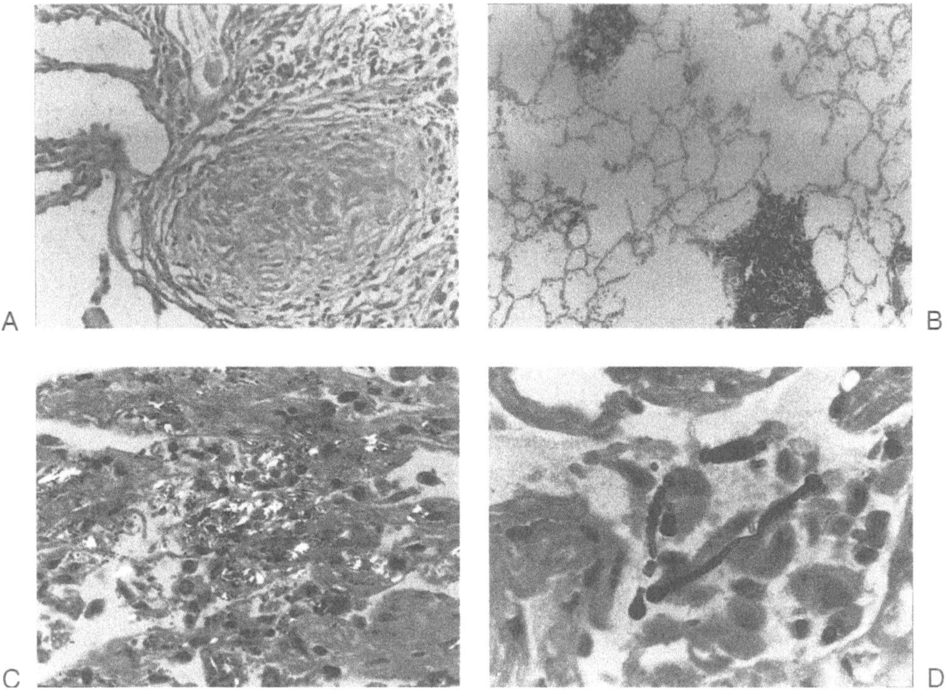

FIGURE 22.2. (A) This micrograph shows a silicotic nodule of concentric fibrosis. (B) Coal miners may have stellate collections of carbon pigments in distal airways. With silica-poor dust exposure, the neighboring alveoli are usually normal as in this micrograph. (C) Polarized light shows bright birefringent talc particles in the lung. These magnesium silicates particles are about 0.5–10 μm in length. They can be inhaled or injected intravenously and produce characteristic lung damage. (D) Asbestos bodies are golden brown beaded, often dumb-bell shaped structures with a central core of asbestos and iron-rich attachments.

is the nonamphibole chrysotile. Chrysotile asbestos bodies account for only about 2% of asbestos bodies analyzed (Roggli et al., 1988). Their rarity may result from the fragmentation and clearance of chrysotile particles because asbestos bodies generally only form on fibers 20 μm or more in length.

The extent of fibrosis is positively correlated with fiber concentration and negatively correlated with fiber size, with most having a concentration of 10^6 particles/g of dry lung. Amosite is more fibrogenic than tremolite, which is more fibrogenic than chrysotile (Churg et al., 1990). Commercial amosite and crocidolite asbestos form the cores of most asbestos bodies in manual laborers and those occupationally exposed to asbestos (Churg et al., 1979), whereas amphiboles form the core of typical asbestos bodies from the lungs of the general population (Churg et al., 1979) (Fig. 22.2). Nonasbestos fibers that are commonly present in lungs of the general population often have a low aspect ratio that makes them invisible by light microscopy.

Mesothelioma is a tumor that has been highly associated with asbestos exposure. Crocidolite was thought to be the main asbestos type responsible for mesotheliomas in the United States until Roggli and colleagues found that 81% of the mesothelioma cases had amosite. Twenty-one percent had chrysotile, 16% had crocidolite, 55% had other forms, and 71% had nonasbestos mineral fibers (Roggli et al., 1993).

Silica-Related Lung Disease

Silica induces lung fibrosis that often manifests itself as disability 10–20 years after exposure. Inhalation of the crystalline forms of silica is more toxic than the amorphous forms, and impurities of iron or aluminum in the silica crystals reduce their toxicity. Sandblasting is particularly dangerous because the quartz crystals are cleaved and their freshly cut edges interact with epithelial surfaces. The SiOH groups of the crystal surface form hydrogen bonds with cellular macromolecules and denature lipid membranes. Free radicals are generated by the crystal–membrane interaction. Alveolar macrophages phagocytize the silica particles, but are damaged or killed by this action. Injured macrophages stimulate fibroblast proliferation and collagen production. The permissible exposure limit for respirable silica set by the Occupational Safety and Health Administration is available at their website:
http://www.osha-slc.gov/OshStd_data/1910_1000_TABLE_Z-3.html

The clinical diagnosis of silicosis relies on a history of silica exposure (Table 22.2), consistent with chest radiographic abnormalities, and the absence of other causes for the illness. An open lung biopsy is usually not needed, although when clinical presentations are atypical, a biopsy, bronchoalveolar lavage, and scanning electron microscopy with energy-dispersive x-ray analysis may be important. Silicosis usually results from chronic exposure for 10–20 years or from a remote, but more intense, exposure followed by a long exposure-free interval. Accelerated silicosis results from exposure to high concentrations of silica over short period (5–10 years). Progression after exposure continues even if the worker is removed from the workplace. Antinuclear antibodies and clinical autoimmune connective tissue diseases are frequently associated with this form of silicosis. Acute silicosis, which is the most infrequent yet most devastating form of this disease, results from exposure to overwhelming concentrations of free crystalline silica for a short time. The invariable downhill clinical course of acute silicosis includes progressive dyspnea, cor pulmonale, and cachexia. Death usually results from respiratory failure.

Acute exposure to high concentrations of silica can result in granulomatous inflammation or silicoproteinosis. These changes may be transient. The more commonly encountered histologic reactions are diffuse fibrosis, anthracosis, and the silicotic nodule. The silicotic nodule is characterized by dense, often concentric lamellae of collagen with central necrosis and surrounding scar-induced irregular emphysema (Fig. 22.2). These nodules result in progressive massive fibrosis when conglomerated. Silicoproteinosis is characterized by alveolar filling with lipoproteinaceous material similar

TABLE 22.2. Workers at risk for silicosis.

Occupation	Exposure hazard
Sandblaster	Shipbuilding and iron-working
Miner or tunneler	Underground miners; drillers
Miller	Finely milled silica for fillers and abrasives; "silica flour workers"
Pottery worker	Crushing flint and fettling
Glassmaker	Sand for polishing and enameling
Foundry worker	Silica in mold making, fettling
Quarry worker	Slate, sandstone, granite
Abrasives worker	Finely ground aerosolized particles

Source: Adapted from Balaan et al. (1998).

to that seen in idiopathic alveolar proteinosis. Detecting birefringent silica-containing crystals is helpful in making the diagnosis of silicosis.

Patients with silicosis are more susceptible to pulmonary tuberculosis, and a goal of therapy is to prevent or promptly treat tuberculosis. This may be difficult at times because of the underlying chest radiographic abnormalities and subtleness of tuberculosis. Because it is important not to treat active tuberculosis with a single medication, patients with abnormal chest radiographs and a suspicion of tuberculosis receive multiple drug therapy. Systemic sclerosis, rheumatoid arthritis, and glomerulonephritis with nephrotic syndrome are also occasionally seen with silicosis.

Other Pneumonoconioses

Many other inorganic materials may enter the lung and can cause fibrosis. Silica-poor pneumoconiosis (e.g., coal miner's lung) often involves focal stellate-shaped fibrosis surrounded by centrilobular emphysema (Fig. 22.2). Inert dusts, oxides of titanium, tin, antimony, and barium sulfate or carbonate are other common types of inhaled silica-poor particles. Workers engaged in production of Al_2O_3 abrasives, aluminum alloys, and ores may develop pulmonary fibrosis (Jederlinic et al., 1990). When large amounts of particles are inhaled (e.g., in processes such as arc-welding), the mucociliary clearance system may be overwhelmed. Uncleared particles remain in peribronchial aggregates and macrophage clusters. With more massive exposure, nodular peribronchial, subpleural, perivascular residues, and alveolar parenchymal retention of particles are found.

Hypersensitivity Pneumonitis

Hypersensitivity pneumonitis is an immune-mediated pulmonary disease that presents in several ways, depending on the nature and duration of exposure to the offending agents. The clinical presentation may be acute respiratory and systemic symptoms starting about 6 hours after exposure. The most common symptoms are chills, fever, malaise, myalgia, cough, and dyspnea. They last 12–18 hours, but recur with reexposure. Recurrent and prolonged exposure leads to irreversible pulmonary damage. Progressive dyspnea may develop without acute episodes and is associated with cough, malaise, weakness, and weight loss. The characteristic laboratory feature is serum precipitating antibodies against the offending antigen, but a variety of cytokines and cell-mediated immune responses take place that lead to lymphocytic and granulomatous inflammation often with inflammatory bronchiolitis obliterans (Fig. 22.3). Once the antigen or environment is identified, avoidance usually results in no further episodes.

Inhalation Fevers

Inhalation exposures from environmental or occupational settings that result in flulike illnesses with respiratory and constitutional symptoms are referred to as *inhalation fevers*. The syndromes follow an inhalant exposure and result in a pulmonary inflammatory response. The dose, duration of exposure, particle size, site of deposition, and host factors help determine the clinical outcome. Similarities of the various inhalation fevers suggest that the mechanisms underlying these conditions may be similar. In

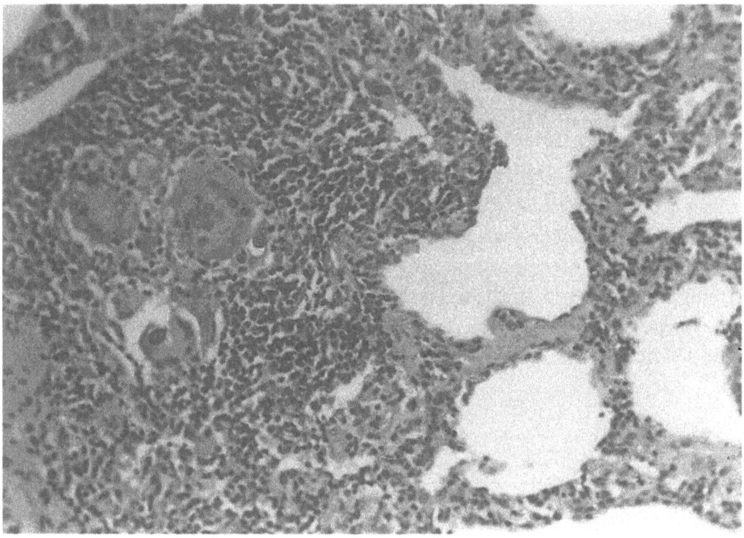

FIGURE 22.3. Extrinsic allergic alveolitis or hypersensitivity pneumonitis results from exposure to an organic antigen. There are collections histologically of lymphocytes and plasma cells, often with Langhans giant cells, and poorly formed nonnecrotizing granulomata. Alveolar ductitis and acute inflammatory cells are also frequently found. The micrograph is from a patient with pigeon breeder's disease.

addition, fumes of the heavy metals lead, mercury, and cadmium, and arsine gas are toxic but may produce slightly different reactions.

Metal Fume Fever

Finely dispersed metal oxide particles less than 1 μm in diameter can be produced when aluminum, antimony, arsenic, cadmium, chromium, cobalt, copper, iron, lead, magnesium, manganese, nickel, selenium, silver, tin, or zinc are heated in such processes as welding, galvanizing, brazing, and smelting (Blount, 1990). Individuals inhaling these fumes in poorly ventilated areas may develop a metallic or sweet taste associated with throat irritation in 3–10 hours. Several hours later they develop a nonproductive cough and dyspnea, and occasionally chills, fever, nausea, vomiting, malaise, headache, and myalgias. Symptoms peak in 18 hours and resolve in 1–2 days, but they recur with return to the contaminated environment. Symptoms are less prominent as more time is spent in the area. Zinc oxide, which is a common causative agent, has been associated with IgE-mediated immediate and late-phase reactions, recurrent urticaria, and angioedema (Farrell, 1987). Zinc fume exposure can result in diffuse inflammation of the respiratory bronchioles and alveolar ducts. Cadmium can also cause delayed-onset pulmonary edema. Occupational asthma can occur in workers with recurrent metal fume fever.

Organic Fevers

Humidifiers and Air Conditioning

Exposure to aerosols generated from air conditioning and humidification systems may lead to fever, chills, headache, malaise, anorexia, cough, chest tightness, dyspnea, and occasionally polyuria, 4–12 hours later. The symptoms are often worse with reexposure, but tolerance develops. Lung function tests are usually normal, although vital capacity may decrease after an inhalation challenge. Chest radiographs are normal in early phase and help to separate the fever syndrome from the hypersensitivity pneumonitis.

A high level of microorganisms within humidification or air conditioning systems is the common element of these cases. It is not always clear which organisms are responsible, although *Thermoactinomyces vulgaris* and *Saccharopolyspora rectivirgula* (formerly *Micropolyspora faenii*) are commonly isolated (Table 22.3). Serum precipitating antibodies against bacteria and fungi isolated from contaminated humidifiers are often negative. The period of time between exposure and symptoms suggests an immune sensitization. The treatment is avoidance or frequent cleaning of the humidifier or air

TABLE 22.3. Representative causes of hypersensitivity pneumonitis by inhaled antigens.

Disease	Source of antigens	Probable antigen
Plant products		
Farmer's lung	Moldy hay	Thermophilic actinomycetes, *Saccharopolyspora rectivirgula*, *Thermoactinomyces vulgaris*, Aspergillus species
Bagassosis	Moldy pressed sugarcane (bagasse)	Thermophilic actinomycetes, *S. rectivirgula*, *T. vulgaris*
Mushroom-worker's disease	Moldy compost	Thermophilic actinomycetes, *S. rectivirgula*, *T. vulgaris*, mushroom
Malt-worker's lung	Contaminated barley	*Aspergillus clavatus*
Maple bark disease	Contaminated maple logs	*Cryptostroma corticale*
Sequoiosis	Contaminated wood dust	Graphium species, pullularia species
Pulp-worker's disease	Contaminated wood pulp	Alternaria species
Humidifier lung	Contaminated humidifiers, dehumidifiers, air conditioners	Thermophilic actinomycetes, *Thermoactinomyces candidus*, *T. vulgaris*, penicillium species, cephalosporium species, amoeba
Familial hypersensitivity pneumonitis	Contaminated wood dust in walls	*Bacillus subtilis*
Compost lung	Compost	Aspergillus species
Cheese-washer's disease	Cheese casings	Penicillium species
Wood-trimmer's disease	Contaminated wood trimmings	Rhizopus species, mucor species
Tea-grower's disease	Tea plants	Unknown
Coffee-workers lung	Green coffee beans	Unknown
Cephalosporium hypersensitivity pneumonitis	Contaminated basement (sewage)	Cephalosporium
Sauna-taker's disease	Sauna water	Pullularia species
Detergent-worker's disease	Detergent	*B. subtilis* enzymes
Paprika-splitter's lung	Paprika dust	*Mucor stolonifer*
Dry rot lung	Infected wood	*Merulius lacrymans*

TABLE 22.3. *Continued*

Disease	Source of antigens	Probable antigen
Potato-riddler's lung	Moldy straw around potatoes	Thermophilic actinomycetes, *S. rectivirgula*, *T. vulgaris*, aspergillus species
Tobacco-worker's disease	Mold on tobacco	Aspergillus species
Hot tub lung	Mold on ceiling	Cladosponum species
Tap water lung	Contaminated water	Bacteria or fungi
Wine-grower's lung	Mold on grapes	*Botrytis cinerea*
Suberosis	Cork dust	Penicillium
Saxophone lung	Saxophone mouthpiece	*Candida albicans*
Grain-worker's lung	Grain dust	*Erwinia herbicola*
Fish meal-worker's lung	Fish meal dust	Unknown
Animal products		
Pigeon-breeder's disease	Pigeon droppings	Altered pigeon serum
Duck fever	Duck feathers	Duck proteins
Turkey-handler's lung	Turkey products	Turkey proteins
Bird-fancier's lung	Bird products	Bird proteins
Dove-pillow's lung	Bird feathers	Bird proteins
Laboratory-worker's pneumonitis	Rat fur	Male rat urine
Pituitary snuff-taker's disease	Pituitary powder	Bovine and porcine proteins
Mollusk shell pneumonitis	Mollusk shells	Animal proteins
Insect products		
Miller's lung	Wheat weevils	*Sitophilus granarius*
Reactive simple chemicals		
TDI pneumonitis	Toluene diisocyanate	Altered proteins (albumin and others)
TMA pneumonitis	Trimellitic anhydride	Altered proteins
MDI pneumonitis	Diphenylmethane diisocyanate	Altered proteins
Epoxy resin lung	Heated epoxy resin	Phthalic anhydride
Pauli's HP	Pauli's reagent	Sodium diazobenzene-sulfonate
Drugs		
Drug-induced hypersensitivity pneumonitis	Gold, thiazides, amiodarone	Altered proteins

Source: Adapted from Cormier (1998).

conditioner. Endotoxin-containing aerosols may cause a granulomatous lung reaction in lifeguards (Rose et al., 1998).

Aerosols of water contaminated with different legionella species can lead to a self-limited illness called *Pontiac fever*. Fever, chills, myalgia, headache, malaise, nonproductive cough, diarrhea, nausea, vomiting, chest pain, dizziness, and sore throat are self-limiting and may arise from inhaling organisms (Kaufmann et al., 1981). Since the original outbreak many legionella species from different heated water sources have been shown to cause similar clinical symptoms. It is not known what determines whether a legionella species will cause Legionnaires' disease or Pontiac fever.

Bacterial strain differences have been postulated because outbreaks generally cause one or the other illness, but no important difference in toxins has been found. The number of viable organisms and host factors, including alcohol use, chronic lung disease, and immunosuppressive states, increase the risk of Legionnaires' disease.

Grain Fevers

Workers newly exposed to the dusts of cotton, flax, soft hemp, or kapok may develop fever, chills, malaise, nausea, cough, and rhinitis within 1–6 hours. Symptoms resolve within hours to days. Tolerance develops, and symptoms cease with continued exposure. Cotton workers who develop this mill fever may later develop byssinosis. Gram-negative bacterial endotoxins contaminating the vegetable dusts may be responsible (Rylander et al., 1985).

Organic dust toxic syndrome (also mycotoxicosis, silo unloader's syndrome) is a noninfectious illness caused by inhalation of organic dust from moldy silage, hay, or other agricultural dusts. A variant grain fever occurs after exposure to massive concentrations of grain dust. These organic dust syndromes are common and often unrecognized because the fever, chills, cough, dyspnea, chest tightness, myalgia, malaise, nausea, headache, and mucous membrane irritation that occur after exposure are nonspecific symptoms.

Polymer Fume Fever

Polymers heated to more than 300°C may give off such complex fluorinated chemicals as octafluoroisobutylene, tetrafluoroethylene, hexafluoropropylene, oxygen difluoride, carbonyl fluoride, carbon tetrachloride, acids, and olefins that are hazardous to health. Similar symptoms occur several hours after inhalation. The symptoms usually resolve within 12–48 hours, but long-term sequelae have been reported (Shusterman et al., 1986). Recurrent attacks may be debilitating. Because of the association between cigarette smoking and exposure to polytetrafluoroethylene pyrolysis products, hand washing before breaks and elimination of smoking in the work area have been successful in eliminating occupational episodes of polymer fume fever.

Noxious Gases

Inhalation of noxious gases (e.g., chlorine or ammonia) may cause corrosive necrosis to the upper airway and large bronchi followed by lung edema and a fatal outcome. In nonfatal cases symptoms usually resolve within weeks, although asthma symptoms and histological changes may persist. Inflammation and bronchial epithelial denudation, infiltration of eosinophils, lymphocytes, and an increase in mast cells are probably responsible for airway hyperresponsiveness. The gases of air pollution (i.e., sulfur dioxide, nitrogen dioxide, and ozone) cause more subtle damage to the airways and lung parenchyma (Fig. 22.4).

Sulfur Dioxide

Sulfur dioxide, which is a highly soluble, irritating gas that is absorbed quickly in the nose and upper airways, is generated primarily by the burning of fossil fuels that contain sulfur. Inhalation of sulfur dioxide causes mild bronchial constriction that is dependent on intact parasympathetic innervation. When exposed to 5 ppm of sulfur dioxide for

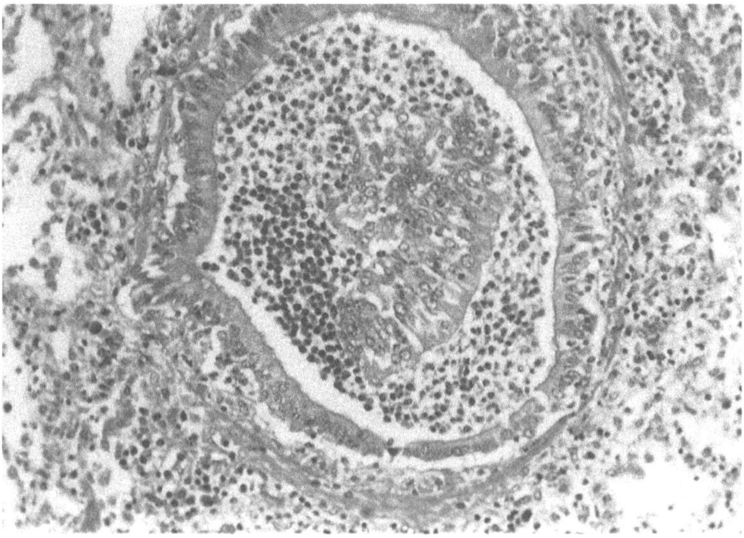

FIGURE 22.4. A variety of noxious agents damage bronchial and bronchiolar epithelium. This micrograph came from a victim of smoke inhalation. There was extensive airway damage and alveolar inflammation centered in the proximal acinus. This micrograph shows desquamated bronchial epithelium in a large bronchiole.

10 minutes, most human subjects show increased resistance to the flow of air. Asthmatics have an increased sensitivity to sulfur dioxide; bronchoconstriction may occur at concentrations as low as 0.25 ppm (Sheppard et al., 1981). Increase in atmospheric sulfur dioxide significantly affects morbidity and mortality, as seen with the large-scale pollution disasters in Donora, Pennsylvania, in 1948 (Shrenk et al., 1949) and in London in 1952 (Logan, 1953).

Nitrogen Dioxide

Nitrogen dioxide is a red–orange–brown highly corrosive and oxidizing gas with a pungent odor, produced by oxidization of nitric oxide. This pollutant is a particular risk to farmers because it can be liberated from silage to produce silo-filler's disease. Like ozone, nitrogen dioxide is an irritant that is capable of causing pulmonary edema. Like ozone, acute exposure also causes changes in pulmonary function testing and damages alveolar Type-I cells. The nitrogen oxides that are formed endogenously from NO_2 react in many cellular reactions. In low doses, the beneficial effects of nitric oxide may be useful in treating such lung conditions as pulmonary hypertension and the respiratory distress syndrome (Gaston et al., 1994).

Ozone

Ozone is a protector in the upper atmosphere and a pollutant in the lower atmosphere. It irritates the airways and lung and can cause death from pulmonary edema. Ozone causes desquamation of the epithelium throughout the ciliated airways, degenerative changes in Type-I cells, and swelling and rupture of the alveolar capillary endothelium.

Carbon Monoxide

Carbon monoxide is a colorless, odorless, tasteless, and nonirritating gas that results from incomplete combustion of organic matter. Its principal source outdoors is motor vehicle emissions. When inhaled, carbon monoxide diffuses quickly into the blood and reversibly binds to hemoglobin, taking over available binding sites for oxygen. Sustained exposure to high concentrations of carbon monoxide leads to interruption of oxygen transport.

Other Inhaled Toxicants

Lead

In addition to its association with fume fever, inhaling airborne lead even at low levels can result in developmental and central nervous system problems in children and hypertension in adults. Children are more susceptible because of their smaller body size and higher rates of absorption and growth. The concentrations of airborne lead and blood levels in children have dropped since the prohibition of lead in gasoline.

Aldehydes

Aldehydes are formed by oxidation of hydrocarbons by sunlight and incomplete combustion and are released from formaldehyde-containing resins. About 50% of the total aldehyde in polluted air is formaldehyde and 5% is acrolein. Formaldehyde irritates mucous membranes of the nose, upper respiratory tract, and eyes. Occupational exposure to formaldehyde may infrequently cause asthma (Nordman et al., 1985). Acrolein is more irritating than formaldehyde and is often a component of house fires. It increases airway resistance and tidal volume, and decreases respiratory frequency.

Radon and Radioactive Elements

The widespread production and use of radioactive materials for electricity, nuclear weapons, laboratory research, manufacturing, and medical diagnosis have generated problems in dealing with poisoning by such metals. Radon is the greatest source of radiation exposure from natural radiation in the United States. Radon is an inert gas, but its progeny are solid, charged particles. Two of these decay products, ^{218}Po and ^{214}Po, emit α particles that damage lung cells' DNA. Underground miners have more lung cancer than cohorts, and the greater the exposure, the greater the risk. The U.S. Environmental Protection Agency estimated that 4–8 million U.S. homes have increased levels. Radon is the second-leading cause of environmental lung cancer, accounting for up to 30% of lung cancer in nonsmokers. The risk of cancer from radon is multiplicative with the risk from smoking (Chaffey et al., 1994).

Other agents can be inhaled and produce chronic health effects. For example, pulmonary arterial hypertension and cor pulmonale have been associated with chronic domestic wood smoke inhalation (Sandoval et al., 1993).

Occupational Asthma

Asthma affects about 5% of the world's population, and about 15% of asthma is related to occupational exposure. It is the most prevalent occupational lung disease in developed countries. Occupational asthma is asthma (reversible air-flow reduction and

TABLE 22.4. Occupational asthma causes.

Agent	Workers at risk
High molecular weight agents	
Cereals	Bakers, millers
Animal-derived allergens	Animal handlers
Enzymes	Detergent users, pharmaceutical workers
Gums	Carpet makers, pharmaceutical workers
Latex	Health professionals
Seafoods	Seafood processors
Low molecular weight agents	
Isocyanates	Spray painters; insulation installers; manufacturers of plastics, rubbers, and foam
Wood dusts	Forest workers, carpenters, cabinetmakers
Anhydrides	Users of plastics, epoxy resins
Amines	Shellac and lacquer handlers, solderers
Fluxes	Electronics workers
Chloramine-T	Janitors, cleaners
Dyes	Textile workers
Persulfate	Hairdressers
Formaldehyde, glutaraldehyde	Hospital staff
Acrylate	Adhesive handlers
Drugs	Pharmaceutical workers, health professionals
Metals	Solderers, refiners

Source: Adapted from Chan-Yeung et al. (1995).

bronchial hyperresponsiveness) that results from conditions in the work environment. Work-aggravated asthma is pre-existing asthma that is worse in the workplace environment. Occupational asthma may occur immediately after an acute exposure to irritant gases or chemicals (e.g., chlorine or ammonia), but more commonly occurs after a latency period. The latency may be from weeks to years (Chan-Yeung et al., 1995).

The characteristic feature of occupational asthma is bronchial hyperresponsiveness, which can be measured by an inhalation challenge to methacholine. Hyperresponsiveness may decrease with time away from the inciting agent and may increase with reexposure. The response to the agent may be acute, delayed, or both. The early asthmatic response typically occurs in a few minutes, reaches maximal intensity in about 30 minutes and lasts 60–90 minutes. A delayed response usually occurs 4–6 hours after the stimulus, reaches maximal intensity in 8–10 hours, and lasts 24–28 hours.

About 250 natural and synthetic chemicals can cause occupational asthma (Table 22.4) with isocyanates being the most common. Occupational asthma may or may not be IgE dependent. Many high molecular weight compounds cause asthma by producing specific IgE antibodies, whereas low molecular weight chemicals (e.g., anhydrides, platinum salts, dyes, and isocyanates) act as haptens that combine with a body protein to produce specific IgE antibodies. IgE-associated asthma usually has an acute response. Complex biological molecules that induce asthma include animal excretions, microbial agents, wheat and rye flour, enzymes from Bacillus subtilis, and grain and wood dusts. Very low concentrations of such agents as toluene and diisocyanate cause asthma by idiosyncratic reactions.

Exposure is the most important determinant of whether occupational asthma develops. In occupational asthma with latency, the greater the exposure, the greater the prevalence of asthma (Chan-Yeung et al., 1995). About 40% of patients with occupational asthma have exposure for less than 2 years, and about 20% have exposure for more than 10 years before they develop asthma. Smoking and previous atopy are also important. For example, smoking may predispose sensitization to platinum salts. The pathology of asthma includes airway wall thickening with eosinophils and other inflammatory infiltrates, edema, hypertrophy of smooth muscle cells and mucous glands, fibrosis and thickening of the basal lamina, surface mucus cell hyperplasia, and obstruction of the airway lumen by mucus.

Most patients with occupational asthma with latency do not recover even after years away from exposure. They may have persistent hyperresponsive airways and airway inflammation even after removal of the inciting agent. The duration of symptoms after removal from the environment is an indicator of prognosis. Early removal from the environment enhances the chance and completeness of recovery.

The diagnosis of occupational asthma rests on the presence of asthma or airway hyperresponsiveness, although this may disappear if the patient is removed from the workplace. There should be a history of exposure, usually with a known inciting agent, and symptoms usually improve with removal from the work environment. A challenge with the offending agent usually results in recurrence of symptoms.

Summary

The lungs filter an enormous amount of air every day, which makes them vulnerable to toxic inhaled organic and inorganic particles and gases. Whether a toxicant results in injury depends primarily on the agent, but the dose, mechanism of contact, and body defense are also important. Injuries may be immediate or delayed, transient or permanent. The agent may damage directly or stimulate the host to harm itself. Different agents predictably injure different parts of the respiratory tree but the smallest bronchioles are vulnerable to the greatest number of agents.

Smoking and air pollution are the most common causes of lung injury. Asbestos and silica are the most culpable inorganic agents. Organic agents can produce hypersensitivity pneumonitis by complicated immunologic processes. Inhalational fevers may be variants of these reactions even when they do not involve inhaling organic material because they trigger a similar cascade of immune responses. Noxious gases often cause direct damage on the respiratory tissues by a chemical reaction, but inhaled gases such as carbon monoxide and radon cause systemic damage.

Inhaled toxicants can cause asthma in previously healthy persons or exacerbate it in persons with preexisting asthma. As with hypersensitivity pneumonitis, the injurious agent can cause a delayed response.

Diagnosing, treating, and preventing disease from inhaled toxicants can be a formidable challenge that brings together primary clinicians, pulmonary and occupational specialists, pathologists, radiologists, basic science researchers, epidemiologists, public health officials, legislators and public-minded citizens. Awareness, avoidance, education, and preparedness are important tools to manage these public health hazards.

References

Balaan, M.R., and Banks, D.E. (1998) Silicosis. In: Rom, W.N. (ed.) pp. 435–448. Environmental and Occupational Medicine. Third ed. Lippincott-Raven, Philadelphia.

Becker, R.P., and Geoffroy, J.S. (1981) Backscattered electron imaging for the life sciences: introduction and index to applications. Scan. Electron Microscop. 4, 195–206.

Blount, B.W. (1990) Two types of metal fume fever: mild vs. serious. Mil. Med. 155, 372–377.

Chaffey, C.M., and Bowie, C. (1994) Radon and health—an update. J. Publ. Health Med. 16, 465–470.

Chan-Yeung, M., and Malo, J.L. (1995) Occupational asthma. N. Engl. J. Med. 333, 107–112.

Churg, A. (1983) Nonasbestos pulmonary mineral fibers in the general population. Environ. Res. 31, 189–200.

Churg, A., and Warnock, M.L. (1979) Analysis of the cores of asbestos bodies from members of the general population: patients with probable low-degree exposure to asbestos. Am. Rev. Respir. Dis. 120, 781–786.

Churg, A., Wright, J., Wiggs, B., and Depaoli, L. (1990) Mineralogic parameters related to amosite asbestos-induced fibrosis in humans. Am. Rev. Respir. Dis. 142(6 Pt. 1), 1331–1336.

Churg, A.M., and Warnock, M.L. (1979) Analysis of the cores of ferruginous (asbestos) bodies from the general population. III. Patients with environmental exposure. Lab. Inv. 40, 622–626.

Cormier, Y. (1998) Hypersensitivity pneumonitis. In: Rom, W.N. (ed.) pp. 457–465. Environmental and Occupational Medicine. Lippincott-Raven, Philadelphia.

Dragovic, T., Schraufnagel, D.E., Becker, R.P., Sekosan, M., Votta-Velis, E.G., and Erdös, E.G. (1995) Carboxypeptidase M activity is increased in bronchoalveolar lavage in human lung disease. Am. J. Respir. Cell Mol. Biol. 152, 760–764.

Farrell, F.J. (1987) Angioedema and urticaria as acute- and late-phase reactions to zinc fume exposure, with associated metal fume fever-like symptoms. Am. J. Ind. Med. 12, 331–337.

Gaston, B., Drazen, J.M., Loscalzo, J., and Stamler, J.S. (1994) The biology of nitrogen oxides in the airways. Am. J. Respir. Crit. Care Med. 149(2 Pt. 1), 538–551.

Hall, T.A. (1988) Capabilities and limitations of probe methods for the microanalysis of chemical elements in biology: a brief introduction. Ultramicroscopy 24, 181–184.

Jackman, H.L., Tan, F., Schraufnagel, D., Dragovic, T., Dezsö, B., Becker, R.P., and Erdös, E.G. (1995) Plasma membrane-bound and lysosomal peptidases in human alveolar macrophages. Am. J. Respir. Cell Mol. Biol. 13, 196–204.

Jederlinic, P.J., Abraham, J.L., Churg, A., Himmelstein, J.S., Epler, G.R., and Gaensler, E.A. (1990) Pulmonary fibrosis in aluminum oxide workers. Investigation of nine workers, with pathologic examination and microanalysis in three of them. Am. Rev. Respir. Dis. 142, 1179–1184.

Kaufmann, A.F., McDade, J.E., Patton, C.M., et al. (1981) Pontiac fever: isolation of the etiologic agent (Legionella pneumophila) and demonstration of its mode of transmission. Am. J. Epidemiol. 114, 337–347.

Logan, W.P.D. (1953) Mortality in London fog incident. Lancet 1, 336–338.

McClellan, R.O., and Henderson, R.F. (1995) Concepts in inhalation toxicology. McClellan, R.O., and Henderson, R.F. (eds.) Taylor and Francis, Philadelphia.

Morrow, P.E., Utell, M.J., Pauer, M.A., Speers, D.M., and Gibb, F.R. (1993) Effects of near ambient levels of sulfuric acid aerosol on lung function in exercising subjects with asthma and COPD. Ann. Occup. Hyg. 38, 933.

Nordman, H., Keskinen, H., and Tuppurainen, M. (1985) Formaldehyde asthma-rare or overlooked? J. Allergy Clin. Immunol. 75, 91–99.

O'Neill, S., Lesperance, E., and Klass, D.J. (1984) Rat lung lavage surfactant enhances bacterial phagocytosis and intracellular killing by alveolar macrophages. Am. Rev. Respir. Dis. 130, 225–230.

Rode, L.E., Ophus, E.M., and Gylseth, B. (1981) Massive pulmonary deposition of rutile after titanium dioxide exposure: Light-microscopical and physico-analytical methods in pigment identification. Acta Pathol. Microbiol. Scand. [A] 89, 455–461.

Roggli, V.L. (1989) Scanning electron microscopic analysis of mineral fibers in human lungs. In: Ingram, P., Shelburne, J.D., and Roggli, V.L. (eds.) pp. 97–110. Microprobe Analysis in Medicine. Hemisphere, New York.

Roggli, V.L., and Brody, A.R. (1988) Imaging techniques for application to lung toxicology. In: Gardener, D.E., Crapo, J.D., and Massaro, E.J. (eds.) pp. 117–145. Toxicology of the Lung. Raven Press, New York.

Roggli, V.L., Pratt, P.C., and Brody, A.R. (1992) Analysis of tissue mineral fiber content. In: Roggli, V.L., Greenberg, S.D., and Pratt, P.C. (eds.) pp. 299–339. Pathology of Asbestos-associated Diseases. Little, Brown, Boston.

Roggli, V.L., Pratt, P.C., and Brody, A.R. (1993) Asbestos fiber type in malignant mesothelioma: an analytical scanning electron microscopic study of 94 cases. Am. J. Ind. Med. 23(4), 605–614.
Rose, C.S., Martyny, J.W., Newman, L.S., Milton, D.K., King, T.E., Beebe, J.L., et al. (1999) "Lifeguard lung": endemic granulomatosis pneumonitis in an indoor swimming pool. Am. J. Publ. Health 88, 1785–1800.
Rylander, R., Haglind, P., and Lundholdm, M. (1985) Endotoxin in cotton dust and respiratory function decrement among cotton workers in an experimental cardroom. Am. Rev. Respir. Dis. 131, 209–213.
Sandoval, J., Salas, J., Martinez-Guerra, M.L., Gomez, A., Martinez, C., Portales, A., et al. (1993) Pulmonary arterial hypertension and cor pulmonale associated with chronic domestic woodsmoke inhalation. Chest 103, 12–20.
Schraufnagel, D.E., Roggli, V.L., Ingram, P., and Shelburne, J. (1990) An introduction to analytical electron microscopy. In: Schraufnagel, D.E. (ed.) pp. 1–46. Electron microscopy of the lung. Marcel Dekker, New York.
Sheppard, D.A., Saisho, A., Nadel, J.A., and Boushey, H.A. (1981) Exercise increases sulfur dioxide induced bronchocontriction in asthmatic subjects. Am. Rev. Respir. Dis. 123, 486–491.
Shrenk, H.H., Heimann, H., Clayton, G.D., Gafafen, W., and Wexler, H. (1949) Air pollution in Donora, Pennsylvania: epidemiology of the unusual smog episodes of October 1948. Publ. Health Ser. Bull. 306. (Complete Vol.)
Shusterman, D., and Neal, E. (1986) Prolonged fever associated with inhalation of multiple pyrolysis products. Ann. Emerg. Med. 15, 831–833.
Spengler, J.D., Treitman, R.D., Tosteson, T., Mage, D.T., and Soczek, M.L. (1985) Personal exposure to respirable particulates and implications for air pollution epidemiology. Environ. Sci. Technol. 19, 700–707.
Utell, M.J. (1993) Particulate air pollution and health. New evidence on an old problem. Am. Rev. Respir. Dis. 147, 1334–1335.
Xu, X., and Wang, L. (1993) Association of indoor and outdoor particulate level with chronic respiratory illness. Am. Rev. Respir. Dis. 148, 1516–1522.
Yao, T.T., Wang, N.S., Michel, R.P., and Poulsen, R.S. (1984) Mineral dusts in lungs with scar or scar cancer. Cancer 54, 1814–1823.

Recommended Readings

Adamson, I.Y., Prieditis, H., and Vincent, T.R. (1999) Pulmonary toxicology of an atmospheric particulate sample is due to the soluble fraction. Toxic. Appl. Pharmacol. 157, 43–50.
Crapo, J.D., McClennan, R.D., and Gardner, D.E. (1999) Toxicology of the lung. Third ed. Taylor and Francis, Philadelphia.
Swift, D., and Foster, W.M. (1999) Air pollutants and the respiratory tract. Marcel Dekker, New York.

Index

C

MIX
Papier aus verantwortungsvollen Quellen
Paper from responsible sources
FSC® C105338

If you have any concerns about our products,
you can contact us on
ProductSafety@springernature.com

In case Publisher is established outside the EU,
the EU authorized representative is:
Springer Nature Customer Service Center GmbH
Europaplatz 3, 69115 Heidelberg, Germany

Printed by Libri Plureos GmbH
in Hamburg, Germany